AF675540

Market Design for a High-Renewables Electricity System

Market Design for a High-Renewables Electricity System

Special Issue Editor

Wadim Strielkowski

MDPI • Basel • Beijing • Wuhan • Barcelona • Belgrade • Manchester • Tokyo • Cluj • Tianjin

Special Issue Editor
Wadim Strielkowski
Prague Business School
Czech Republic

Editorial Office
MDPI
St. Alban-Anlage 66
4052 Basel, Switzerland

This is a reprint of articles from the Special Issue published online in the open access journal *Energies* (ISSN 1996-1073) (available at: https://www.mdpi.com/journal/energies/special_issues/MD_HRES).

For citation purposes, cite each article independently as indicated on the article page online and as indicated below:

LastName, A.A.; LastName, B.B.; LastName, C.C. Article Title. *Journal Name* **Year**, *Article Number*, Page Range.

ISBN 978-3-03936-256-1 (Hbk)
ISBN 978-3-03936-257-8 (PDF)

© 2020 by the authors. Articles in this book are Open Access and distributed under the Creative Commons Attribution (CC BY) license, which allows users to download, copy and build upon published articles, as long as the author and publisher are properly credited, which ensures maximum dissemination and a wider impact of our publications.
The book as a whole is distributed by MDPI under the terms and conditions of the Creative Commons license CC BY-NC-ND.

Contents

About the Special Issue Editor

Wadim Strielkowski is a Professor with the Prague Business School and an Assistant Director of the Centre for Energy Studies there. Previously, he was a Visiting Scholar with the Department of Agricultural and Resource Economics at UC Berkeley; a Research Associate at the Energy Policy Research Group, University of Cambridge; an Assistant Professor, Charles University; a Senior Research Fellow, the Global Change Research Institute of the Czech Academy of Sciences; a Deputy Director for Development at CERGE-EI Prague; a Vice-Chancellor of the College of Economics and Management in Prague; and a Research Fellow at the University of Nottingham. He has authored and co-authored more than 200 academic publications in international peer-reviewed journals and is one of the most highly-cited Czech energy economists.

Preface to "Market Design for a High-Renewables Electricity System"

Recent targets for tackling climate change proposed by most of the world's governments envisage dramatic cuts to greenhouse gas emissions as well as increases in the share of renewable energy in total gross energy production. However, despite these developments, the electricity sector will continue to bear the most significant burden stemming from economy-wide decarbonization which will in turn require high shares of renewable energy sources (RES) in the electricity system. The good news is that technological progress in wind and solar energy coupled with the increased use of interconnection, hydro resources, and new battery technologies, and the growing importance of smart meters and smart grids might make the high-proportion renewables electricity system a realistic future scenario. Increasing the share of RES will be challenging without substantial modifications to the current market design. This book provides assessments and evaluations of the emerging trends in electricity markets, with a focus on high-renewables electricity systems. Various issues are analyzed, such as wind and solar energy, interconnection, smart meters, smart grids of the future (including their social implications), or the peer-to-peer (P2P) electricity trading, which is closely connected to the principle of sharing economy. One of the main issues this volume attempts to address is how the market design for a high-renewables electricity system would be different from the classical post-liberalization market design. The studies published in this book contemplate the problem of how to encourage penetration of RES in electricity markets with the help of policies targeted at promoting renewables on the supply and demand sides to address the external benefits of renewables. Support for RES should integrate public preferences and these can be addressed by assessing willingness-to-pay (WTP) for renewable energy sources. Policies to promote renewables closely interact with other climate change mitigation efforts in energy sector such as energy efficiency improvements on demand and supply and therefore should be developed considering energy and climate targets.

Wadim Strielkowski

Special Issue Editor

energies

MDPI

Article

Internet of Energy (IoE) and High-Renewables Electricity System Market Design

Wadim Strielkowski [1,*], Dalia Streimikiene [2], Alena Fomina [3] and Elena Semenova [4]

1 Department of Trade and Finance, Faculty of Economics and Management, Czech University of Life Sciences Prague, 16500 Praha-Suchdol, Prague, Czech Republic
2 Lithuanian Energy Institute, Breslaujos 3, LT-44403 Kaunas, Lithuania; dalia@mail.lei.lt
3 JSC—Central Research Institute of Economy Management and Information Systems "Electronics", Kosmonavta Volkova str. 12, 127299 Moscow, Russia; fomina@ymservices.ru
4 Department of Innovation and Integrated Quality Systems, Saint-Petersburg State University of Aerospace Instrumentation, Bolshaya Morskaia str. 67, 190000 Saint Petersburg, Russia; eg.semenova@mail.ru
* Correspondence: strielkowski@pef.czu.cz

Received: 13 November 2019; Accepted: 15 December 2019; Published: 16 December 2019

Abstract: The growing importance of the Internet of Energy (IoE) brands the high-renewables electricity system a realistic scenario for the future electricity system market design. In general, the whole gist behind the IoE is developed upon a somewhat broader idea encompassing the so-called "Internet of Things" (IoT), which envisioned a plethora of household appliances, utensils, clothing, smart trackers, smart meters, and vehicles furnished with tiny devices. These devices would record all possible data from all those objects in real time and allow for a two-way exchange of information that makes it possible to optimize their use. IoT employs the Internet Protocol (IP) and the worldwide web (WWW) network for transferring information and data through various types of networks and gateways as well as sensor technologies. This paper presents an outline stemming from the implications of the high-renewables electric system that would employ the Internet of Energy (IoE). In doing so, it focuses on the implications that IoE brings into the high-renewables electricity market inhabited by smart homes, smart meters, electric vehicles, solar panels, and wind turbines, such as the peer-to-peer (P2P) energy exchange between prosumers, optimization of location of charging stations for electric vehicles (EVs), or the information and energy exchange in the smart grids. We show that such issues as compatibility, connection speed, and most notoriously, trust in IoE applications among households and consumers would play a decisive role in the transition to the high-renewables electricity systems of the 21st century. Our findings demonstrate that the decentralized approach to energy system effective control and operation that is offered by IoE is highly likely to become ubiquitous as early as 2030. Since it may be optimal that large-scale rollouts start in the early 2020s, some form of government incentives and funding (e.g. subsidies for installing wind turbines or solar panels or special feed-in-tariffs for buying renewable energy) may be needed for the energy market to make early progress in embracing more renewables and in reducing the costs of later investments. In addition, there might be some other alternative approaches aimed at facilitating this development. We show that the objective is to minimize the overall system cost, which consists of the system investment cost and the system operating cost, subject to CO2 emissions constraints and the operating constraints of generation units, network assets, and novel carbon-free technologies, which is quite cumbersome given the trend in consumption and the planned obsolescence. This can be done through increasing energy efficiency, developing demand side management strategies, and improving matching between supply and demand side, just to name a few possibilities.

Keywords: renewable energy sources; sustainability; Internet of Energy; smart meters; smart grid

1. Introduction

Smart grid technologies would make it possible to be more efficient in terms of using energy sources and optimizing them, whenever necessary, with regard to the environmental or power system limitations. With all that, various energy strategies can be implemented for creating benefits for all system users and for providing them with clean and cheap energy at all times. As renewable energies mature, prices fall, education improves, and competitiveness improves, the likelihood of technology spreading across national borders increases.

Given the current state of technological development and the energy market, it is likely that the high-renewables electricity system market design of the future would be based on smart grids powered by the Internet of Energy (IoE) [1]. The term "smart grid" is characteristically used for describing an electricity system that supports four basic operations encompassing electricity generation, electricity transmission, electricity distribution, and electricity control [2–4]. A smart grid is based on the bidirectional exchange of information and energy within the electricity networks. Using its unique qualities, it is capable of optimizing, saving, and delivering energy precisely where it is needed [5].

Smart grids of the future would involve large shares of renewable energy sources (RES). Generating electricity from renewable energy sources would provide direct and indirect economic benefits beyond cost, as well as environmental benefits from reducing CO_2 emissions. Moreover, generation of electricity from renewable energy sources integrated into the smart grid system can be one of the best options for future energy security. The smart grid system addresses the deterioration of the power source and the modern information technology for communication and improves the efficiency of power distribution.

However, this renewable energy is likely to be generated not only at the industry level (e.g., by large state or private companies) but also at the household or individual level. In the future, every energy consumer would become a "prosumer" (an agent at the electric energy market that is simultaneously buying, producing, and often selling electric energy) as described, for example, by Mengelkamp et al. [6]. Thence, it would be very important to link all pro-active prosumers as well as large energy producers and users into efficient networks that would allow a two-way flow of information and energy [7]. To meet future energy needs, the smart grid system can be used as an efficient energy security system. Nevertheless, this cannot be achieved without the profound use of information and communication technologies (ICTs).

Here is where the Internet of Energy (IoE) comes in being the fastest in all current energy transfers because the actual speed as well as the efficiency of the energy transfer [8]. Even though IoE might seem like a very novel idea, it is largely based on the advancements, rules and the general architecture of the "old-fashioned" information and communication technologies (ICTs) and Internet. Putting things very simply, IoE consists of millions of energy-generating installations, as well as devices and household appliances that report back to the power grid using peer-to-peer or server-based network for receiving information, running an analysis, and sending commands [9–11]. Thence, in the nearest future, high-renewables smart grids would enable the two-way flow of information and energy with a purpose of providing power for all system users [12,13].

With regard to the above, for achieving the high-renewables electricity system market design it would be crucial to move to the rapid energy transfer and planning in the future. Renewable energy sources are being introduced in an unequal environment where their energy prices do not fully reflect the externalities. The global subsidies for traditional fuels and nuclear energy remain high despite the benefits of renewable energy and concerns about environmental quality. Much of the expansion of renewable capacity occurs in countries with large subsidy systems that can compensate investors for the relatively high cost of renewable energy technologies.

Many aspects, such as electromagnetics, materials science, information science, automation, and the like are involved in the generation, conversion, transmission, distribution and power consumption of the smart grids. Therefore, it requires a lot of talent to work together or to accompany [14,15]. With the breakthrough of materials science and power electronics, the advantage of some advanced technologies that would further boost the smart grids and high-renewables systems such as direct current (DC) transmission is obvious. It is highly likely that DC transmission would become the most important type of energy transfer in the future. In the last ten years or so, several product categories more than doubled, including home-based energy management systems, smart lighting controls, residential demand response, and building information modelling, as pointed out by Luca de Tena et al. [16]. Led by solar, wind and gas turbines, this segment represents more than a quarter of the advanced and RES-focused energy market.

High-renewables electric smart grids of the future powered by the IoE would largely benefit from the optimal solutions applied to smart homes, electric vehicles (EVs), solar panels, wind turbines, as well as peer-to-peer (P2P) flow of electricity and information between prosumers. Nevertheless, the transition to the high-renewables electricity systems of the 21st century would have to tackle many technical issues such as compatibility, connection speed, as well as social acceptance.

This paper is structured as follows: Section 2 provides a thorough literature review focusing on the innovative policies for promoting high-renewables smart electricity systems and smart grids. Section 3 describes smart network technologies. Section 4 provides scenarios for the high-renewables electricity system market design that would employ IoE. Finally, Section 5 concludes with outlaying some final outcomes and policy implications.

2. Literature Review

IoE represents a global interconnected network that is comprised of various household and industrial appliances, electric devices, large and small, as well as smart grids that interconnect them all together. Another important element of IoE is, of course, the presence of smart meters, or sensors that constantly monitor all processes within this network and send signals across the grid, helping the IoE to understand the appliances schedule and the consumers to adjust their energy consumption patterns and usage. [17–19]. All in all, it becomes quite apparent that IoE is capable of helping in achieving and further increasing the sustainability for the smart grids through making the production, transmission, and consumption of electric energy more efficient and economically feasible. With all that benefits smart grids and IoE are offering, it is quite surprising that many agents at the contemporary energy market (e.g., car producers, energy suppliers, or utility companies), still have certain troubles in adjusting to the new technological advancements [20,21] (even though it is apparent that their deployment requires substantial capital investments). However, it quite clear that new types of energy demand and supply that would include more renewable energy sources and prosumers would inevitably lead to the profound changes in world's electrical networks. With regard to that, smart grids offer a whole scale of opportunities how to tackle these changes depending on the concrete situation, business models, regulation, and power infrastructure [22,23].

Figure 1 that follows, shows how the flexibility and innovations trends in IoE framework and its applications to high renewables penetrations of electricity market are driving energy transition. There are three main aspects to be considered that mark the transformation of the sector (digitalisation, decentralisation, and electrification).

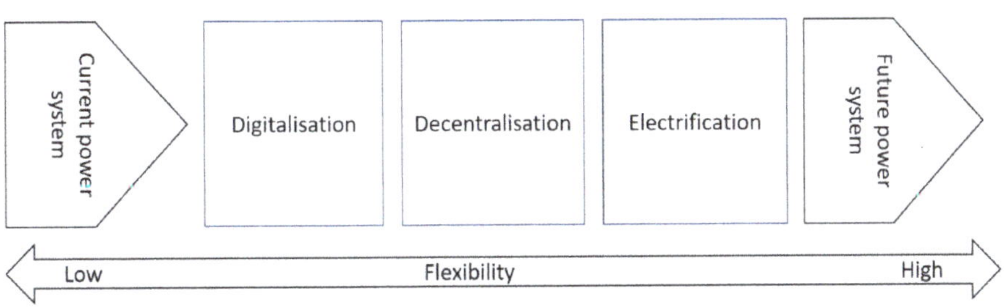

Figure 1. Flexibility and innovations in the Internet of Energy (IoE) and high renewables framework. Source: Own results.

With regard to the above, one can see how the aspects outlined above impact the transition to the high renewables penetrations of electricity market: digitalisation includes ICT solutions to exchanging data and energy in the framework of IoE, decentralisation embeds the distributed energy resources (DERs) that would help to further decentralise the future power system, while electrification means the increasing share of electric transport, industry, and buildings.

Furthermore, high penetration of renewables also makes the IoE an indispensable tool for the dynamic demand response, when the utilities need to reschedule or delay the operation of the appliances located in households and businesses during the periods when demand levels spike. This is particularly useful due to the intermittent nature of RES.

In the future, high-renewables electricity system market design would encompass the new concepts of energy consumers' and producers' relationships, market operations, as well as electric energy trading [24–27]. IoE and the smart grid would allow for the two-way flow of information and energy in the real time as we know it from the world wide web (WWW) and the Internet [28]. There would be many interesting and useful applications that might be helpful in mitigating power system operation hurdles and natural challenges. For example, Pina et al. [29] analyze the impact of demand side management strategies in the evolution of the electricity mix of Flores Island in the Azores and conclude that IT solutions might improve the operation of the existing installed capacity. Moreover, Strielkowski et al. [30] explain in greater detail how photovoltaic system owners can control their use of electricity using the "power manager" gateways and battery storage for achieving the economically effective outcome. Another interesting case study involving the IoE and its implications for the smart grids is the autonomic power system (APS), that presents a novel concept of "self*" (self-configuring, self-healing, self-optimizing, and self-protecting) system [31]. APS constituted a system-wide approach with the decentralized intelligence making autonomous decisions required for meeting the priorities of the system's stakeholders, employing the integration of a vast number of flexible, diverse, and independently controlled technologies in system operation and planning [4,32,33].

When it comes to the technical details and characteristics for the IoE applied to the electricity market in terms of high variable renewable resources penetration, various layers of IoE in smart grids covering management services (security control, data monitoring, customer or market data) and applications (smart homes, electric vehicles, demand response, and demand side management) can be considered: e.g., transport layer, physical layer, network layer, or application layer [34,35].

IoE and smart grids would allow system operators to promptly react to peaks or failures in electric energy demand and also to forecast these issues well in advance and to adjust to these situations by optimizing energy generation from, say, RES, accordingly. All these would increase energy market efficiency and profitability [36–39]. In a way, it is quite similar to the so-called "cashless economy", when the Internet is used to optimize payments and money transfers, and banks, financial institutions, as well as government regulators have an instant snapshot of all of those activities and transactions.

Energy companies operating on the future high-renewables and smart solutions-driven electricity market would be involved into the generation of energy from renewable sources and natural gas, energy trading and tailor-made energy services and developments for companies [40,41]. They would offer their customers a reliable and environmentally friendly energy supply based on the sustainable use of renewable energies.

Another important aspect is smart meters. Their numbers are increasing, and their usage is becoming notorious in all aspects of energy generation and saving. However, in some cases, as for example, Rausser et al. [42] demonstrate using a case study from the Republic of Ireland, their deployment has little effect of the energy consumption behaviour of the households and individuals. Nevertheless, smart meters would also be a very important pillar of the high-renewables sustainable energy system [43,44]. In the future market design, the peak load shave would be achieved by shifting the usage time of the energy without changing the total energy consumption [45]. It works on the application of smart meters to collect data and optimize energy production [46]. Although this strategy improves the conventional grid, the IoE framework is not considered a mathematical model and the inclusion of renewable energy sources is lacking. The introduction of smart meters and IoE connected power supplies has allowed consumers to track and monitor their energy consumption and save energy costs. This has many practical applications and technical solutions, in particular in Smart Cities that represent a communication infrastructure offering a concise, unified and affordable access to municipal services including energy supply [47].

Similar technology is expected to make aggressive progress in the areas of energy production and transmission [48,49]. This massive and rapid growth is aimed at efficient use of resources in power generation and higher operational efficiency to meet growing energy needs [50–52].

With limited participation of demand resources, the wholesale market functioned mainly with network operators selling large central station equipment to meet the steady demand. Renewable energy today is cheaper than coal and nuclear power in most parts of such advanced economies as, for example, the United States and more cost-competitive with natural gas [53,54].

Furthermore, it becomes obvious that the targeted rapid increase in power supply from intermittent renewable sources in many countries is a fundamental challenge to the smooth functioning of many power systems. Wind and solar power are the fastest growing forms of renewable energy [55,56]. The supply of wind and solar energy is largely determined by wind speed and solar radiation which can be correlated only slightly with the times of electricity demand [57–59]. It is this feature of renewable energy intermittent power supply that adds cost to the entire generation system that is implicitly paid for by either other producers, consumers or taxpayers [60,61]. With the constantly increasing size and quantity, today's generation and energy costs are often competitive with coal and nuclear without taking into account the reserve capacity and complexity of grid connectivity that affect their value in a system [62]. If it burns out and displaces electricity from other sources, it can reduce the profitability of these sources and increase supply prices. It has been proposed to use all electricity from wind and solar power which greatly simplifies the management of the electricity grid [63,64].

Effective wholesale electricity markets are crucial for rapid and affordable decarbonization, as they demonstrably invest efficiently and rapidly in new technologies. However, the electricity markets will only support massive investments in clean energy if they are able to send efficient price signals as decarbonization increases [65,66]. The scale and pace of investment required to halt the climate crisis means that wind and solar energy will almost certainly play an important role in future power systems because of their low cost and speed of deployment. However, variable power and the marginal cost of wind, solar and other forms of variable renewable energy (VRE) jeopardize the ability of current market designs to send the required price signals [67,68]. New variable resources such as sun and wind are one of the biggest drivers for more flexibility. There are many ways to unlock flexibility: new and more flexible gas systems, storage of all sizes, power electronics to regulate wind and solar power, and a constellation of connected devices ready to consume electricity smarter. Restructuring wholesale electricity markets, which work best by avoiding specific technology revenues, must find new and improved ways to assess flexibility and allow current and future market participants to provide them at minimal cost. The uneven geographic distribution of wind and solar potential is likely to burden the grid at some sites, resulting in transmission and distribution restrictions. Some electricity markets, such as the California Independent System Operator (CAISO) in Germany and the United Kingdom, have begun to recognize variable and resilient electrical resources to varying degrees [69,70]. In addition, the Federal Energy Regulatory Commission (FERC) and Pennsylvania-New Jersey-Maryland (PJM) Interconnection policy makers in the United States are also shifting their focus to the role that battery energy storage and flexible resources such as distributed resource aggregators (DRAs) play in the development of electricity markets [71]. However, regulatory, economic and technological barriers have largely prevented the participation of demand resources such as batteries and smart thermostats in the wholesale electricity market. With limited participation of demand resources, the wholesale energy market functioned mainly with network operators selling large central station equipment to meet the steady demand.

3. Promoting High-Renewables Smart Grids

High-renewables smart grids would bring in many useful solutions to the existing energy market problems. Flexible demand, micro-generation and energy storage technologies can reduce the emerging demand peaks, while smart network technologies increase the utilization of existing network assets. The concept represents a shift from asset redundancy to more intelligent operation through real-time coordination of new flexible technologies. Figure 2 that follows offers the comparison of the costs of electricity obtained from the renewable energy sources.

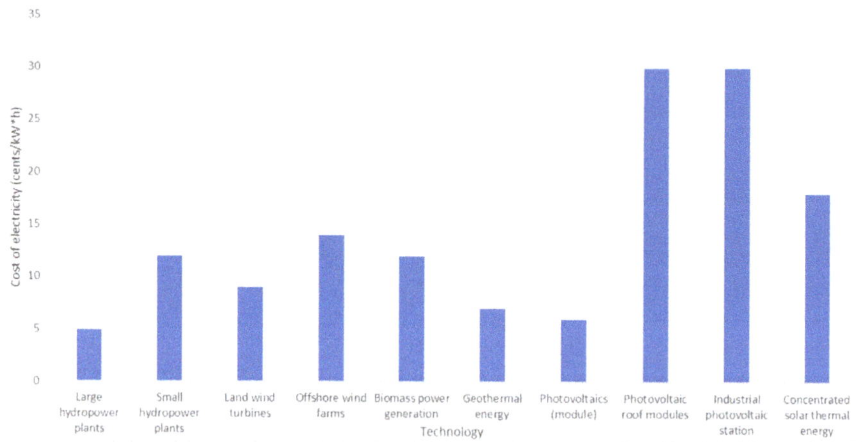

Figure 2. Comparison of the costs of electricity obtained from the renewable energy sources. Source: [72].

Overall, it becomes obvious that in spite of the popularity of RES-based technologies, the costs of electricity are the lowest when it comes from the large hydropower plants, which is followed by the geothermal energy and biomass. Photovoltaics and wind turbines are just starting to catch up in terms of economic efficiency.

In general, it would be quite straightforward to simultaneously and holistically assess the impact of high-renewables smart grid technologies across all timescales and system levels via capturing the overall economic value of these technologies for three milestone dates (2020, 2030, and 2050). In the current situation (2020) the value of flexibility is not that high, the deployment of energy storage is not justified for a cost higher than €1100/kW with the optimal storage capacity of 2 GW, and its overall economic value for the power system for the European country a size of the United Kingdom or Germany being around €0.5 billion per year.

There have been recent developments—in the UK with electricity market reform, in Germany with the Energiewende, and in the State of New York with its Reforming the Energy Vision—for ideas from the new round of natural experiments in electricity market organization, currently underway in jurisdictions with 80% or more low carbon electricity targets.

A fundamental feature of the high-renewables electricity system market design's vision is the integration of a vast number of flexible, diverse, and independently controlled technologies in system operation and planning. Figure 3 that follows depicts the investments into smart grids by areas. It is apparent that over the past few years the share of investments into the rest of the network is declining, while the investments into the power equipment as well as to the smart grid infrastructure and smart meters are slowly but steadily increasing.

Table 1 that follows shows the contrasting description of two market designs—namely the Internet of Energy and high-renewables market design versus the current (state-of-the-art) design that employs current technological advancements.

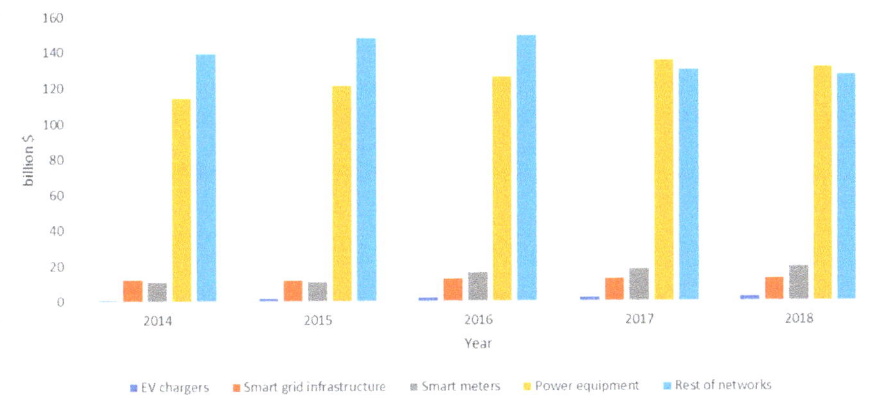

Figure 3. Investments into smart grids by areas. Source: [73].

Table 1. IoE and high-renewables market design versus current state-of-the-art.

Major Issues	Positive Attributes	Negative Attributes
Flexible technologies	Flexible generation	Inflexible generation
	Interconnection and flexible network technologies	Isolated and conventional generation technologies
	Demand side response	Traditional distribution and consumption
	Energy storage	Obsolete energy storage
Market operation	Improving system control	Inefficient and obsolete system control
	Big Data	"Old-style" approach to Big Data
	Integration of transmission and distribution	Isolated transmission and distribution
	Decentralized control	Centralized control
Market design	Revisiting the current design principles	Existing design principle
	Moving beyond "like-for-like" replacements	Like-for-like replacements
	Strategic investments	No strategic investment or planning
	Managing uncertainty	Old-fashioned traditionalist approach
	Improving resilience of high impact events	Vulnerability of high impact events
Coordination and implementation	Transmission coordination	No harmonization of regimes
	Whole systems approach	Fragmented system approach
	Coordination across energy vectors	Single-energy system planning
Novel commercial and regulatory outline	Altering the regulator's role	Traditional regulator's role
	Providing incentives for the smart grid	Asset-based regulatory philosophy
	Setting level playing field markets	No level playing field markets
	Merging the wholesale and retail markets and cost-reflective charging	No integration of wholesale and retail markets
	Altering the role of system operators	Traditional roles of independent system operators (ISO) and distribution system operators (DSO)
	Acknowledging the higher risk of smart technologies	No recognition of risk from new technologies
	Deepening the EU integration	Split-up of the EU

Source: Own results based on [74].

With regard to the above, it appears important to consider how one can better balance supply and demand, aiming towards an electricity market where prices are reflective of costs to the overall system. For example, in the United Kingdom alone, smart power could save consumers up to £8 billion a year by 2030 [75]. Moreover, it can also ensure that the UK meets its 2050 carbon targets and secures its energy supply for generations to come. As a result, three major innovations can be identified as the key drivers of success: interconnection, storage, and demand flexibility [76].

Another issue that has to be mentioned in connection with promoting and sustaining high-renewables smart grids is the Big Data which can be used for many activities and innovations within the IoE framework design. In the energy markets, Big Data represents a valuable asset and a source of all possible information about the consumption and behaviour of users, households, and companies [77]. Thanks to the smart meters, Internet and fast mobile technologies, all that information can be collected and transferred quickly to be analysed. However, one has to realize that due to the extensive sizes of such datafiles, traditional methods (e.g., econometric or statistical analysis) are no longer applicable. Instead, computer algorithms and artificial intelligence (AI) are employed which is similar to the recent developments in the "regular" Internet nowadays.

The applications of the Big Data might range from building more accurate predictions and extrapolations that would forecast demand and output more precisely to minimizing asset failures of the system. There is a need for system operators (SO) to become data led organizations (as distinct from asset availability managers) which constitutes an important argument in the debate of the value of independent system operators.

IoE and Big Data are also crucial for energy prosumers that are more entangled into peer-to-peer (P2P) energy market arrangements through buying, selling, and producing their own energy from the renewables. The sharing economy-type of high-renewables energy market that is being created and would be likely to dominate in many countries in the nearest future calls for optimized and fast collection and processing of all available data.

4. Smart Network Technologies

With all of the above explained, one also has to point out some noteworthy differences between the smart grids and the Internet, to which they are often compared to. Similar to the Internet (although with a wider diversity of resources), smart grids constitute an interconnected system of devices and controls, nevertheless, all these systems are dynamically evolving and changing as the demand of electricity of the different types of consumers is shifting due to the changing market conditions. The blockchain technology that is used in Bitcoin, the world's most popular cryptocurrency, might provide help in dealing with this highly sophisticated environment for achieving cost-effective energy solutions.

One good example of this is the optimization systems that involve office and residential buildings and electric vehicles (EVs). Managers of these buildings face the tasks of providing a logistically balanced system of charging stations for the EVs parked or stationed on their premises. Another technical solution associated with EVs is the possibility of some of them not only to absorb but also to inject power back to the grid (a so-called "vehicle-to-grid" concept, or V2G). Without smart grids and IoE, it would be quite cumbersome to come up with the optimal mix of accurate solutions due to supply and demand uncertainties: e.g., inherent uncertainty of RES or load shifting for reducing peak power consumption [78,79].

Thanks to IoE smart grids can achieve better efficiency and learn how to optimize the energy needs for both prosumers operating at peer-to-peer (P2P) markets and industrial companies in the production chain. All of that gives a boost for the new sectors such as the solar photovoltaics, intelligent distribution networks, and electric vehicles charging, just to give a few examples. In addition, the rising competition and technological advancements are slowly but gradually changing the current system of distribution networks and old-fashioned grids.

At the planning timescale, flexible demand, micro-generation, and energy storage technologies can reduce the emerging demand peaks, while smart network technologies increase the utilization of existing network assets; both these effects are translated in reduced requirements for capital-intensive generation and network capacity investments in the future [80]. Moreover, dynamic demand response shifts the amount of consumer electricity it to when it more optimal from the grid operation point of view [81].

The shift towards carbon-free transport which aims at reducing car emissions and electrify both personal and communal transportation introduced the concept of electric vehicles (EV). Currently, EVs are becoming quite a mundane thing even though their costs and operation might be somewhat higher than in the case of traditional vehicles. Comparing the number of EVs and internal combustion engine (ICE) vehicles, one can find out that in 2019 there were about 5.6 million EVs all around the world (including passenger cars and light commercial vehicles with battery-electric drive, range extender, as well as plug-in hybrids) [82]. This is a steady increase from the 3.2 million EVs in 2018, 1.9 million in 2017, and 1.1 million in 2016 [83]. However, EVs still constitute only a margin of all registered internal ICE vehicles in the world which yielded around 1.4 billion in 2019 [84]. Nevertheless, IoE is becoming one of the key concepts in the electrification of transport, since it promises to coordinate, manage, and match the growing fleet of electric vehicles. For example, one of the pressing issues with EVs today is the occurrence of the charging stations. Nowadays, there are about 160 thousand charging stations worldwide with the expected number to be slightly above 200 thousand in the early 2020 [85,86] Not too frequent in comparison with their fuel counterparts, they often make it difficult to plan a trip over large distances. One of the solutions might be charging EVs from one another in an interconnected network of drivers who are also energy prosumers (the V2G concept described above). Figure 4 that follows shows a graphic representation of V2G within IoE framework marked by the high renewables penetrations of electricity market. It outlines the key role of the IoE that helps the transition of the power sector through the monitoring the performance of its assets, maintaining more precise operations and control in real time, setting up novel market designs, as well as defining new business models and solutions.

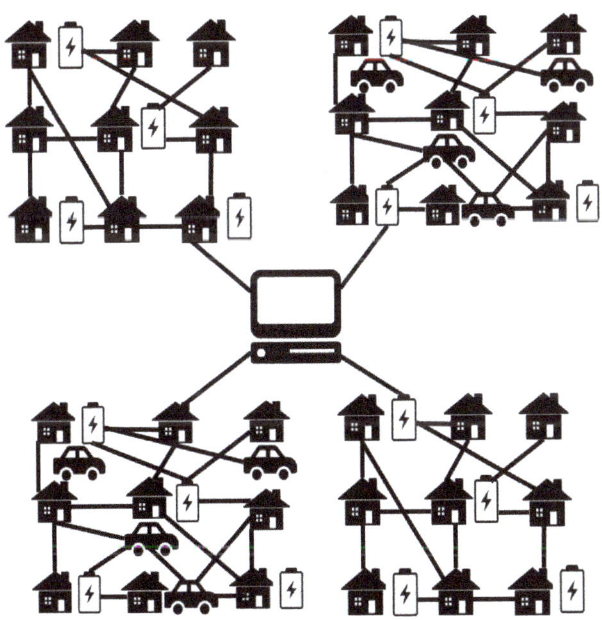

Figure 4. Vehicle-to-grid (V2G) within IoE framework and high renewables penetrations of electricity market. Source: Own results.

One of the pioneers in introducing the electric transportation to the daily lives of the people is the European Union that committed to having over 250 million EVs by 2025 all across its Member States, which, although an impressive figure in itself, might not even be enough for the effective transition to the climate targets of 2030 [87].

One can see that due to the existing economic and social constraints, electric vehicles might also constitute a problem and distort the power system. They are still too expensive for the average citizen to buy and operate, the charging options are rather limited, and the battery storage is quite modest with the driving range being in a couple of hundreds of kilometres. Thence, technical operation and market optimization need to be used in order for them to become a ubiquitous part of the future smart grids [88].

With the increasing sales of EVs in United States, European Union, China, and India, the vehicle charging issues would accumulate and would need to be solved [89]. For example, in the absence of proper regulations, the majority EV owners would charge their vehicles during the day's peaks, thus placing a burden on the electricity system [90]. IoE might provide solutions for that offering demand-side response and optimizing the charging times scheduling and planning. In addition, the rising EV fleet might become a kind of a giant energy storage system that would be able not only accumulate, but also to move large amounts of energy over distance. This might be an interesting solution for the distributed renewable energy generation, even though strict and precise on-time planning is needed, which should involve Big Data, smart meters of the latest generations, as well as faster mobile networks.

Let us show this on a practical example using the data from the United Kingdom (UK). Table 2 that follows shows an example of tariffs and metered import for the three selected UK regions represented by the Northeast, Central Scotland, and East Midlands (chosen for the best geographical spread and representativeness).

Table 2. Tariffs, distribution charges, and metered import in three selected UK regions.

Region	F, £	v, £	Number of Customers	Metered Import, kWh
Northeast	0.0483	0.027	1,500,000	3400
Central Scotland	0.0373	0.04114	2,200,000	3800
East Midlands	0.0147	0.02266	2,400,000	3900

Source: Own calculations based on [30].

Furthermore, one might model the hypothetical household demand profile (expressed in GW) with various levels of deployment of electric vehicles (EV)—from 0% to 100%. This is a rather simplified model since it neglects some relevant benefits of the restructuring market, since more factors might be needed, as well as the relationship between EV and charging stations. Some of the influential factors in this case might be represented by the location of the charging facility, local subsidiary policies, as well as the battery level of the EV. However, despite this limitation, we can draw the model framework that would constitute an example of reorganizing the energy market.

The model in question is based on our methodology described in our earlier paper (for more details see [87]).

$$(1 + x)\,F + (1 + x)\,v * (EC \times n) = TR \tag{1}$$

where:

x—energy tariffs and charges component;

F—household daily unit power fee (kW);

V—energy unit fee (kWh) for a given household connected to the network;

EC—energy consumption for a given household;

N—number of households in the system;
TR—total electricity revenues;

Table 3 that follows shows the differences in tariffs (per household per year) for residential EV and non-EV households in the three UK regions presented above. It becomes apparent that the inclusion of RES (represented here by solar panels) and most importantly EVs shift the balance and make profound changes to the economic situation.

Table 3. Differences for residential solar photovoltaic (PV), non-PV, electric vehicle (EV) and non-EV households in UK regions.

Region	**Northeast**			
(PV, EV), %	Tariff, £ (per household per year)			
	(no EV, no PV)	(EV, no PV)	(no EV, PV)	(EV, PV)
(1, 1)	179.19	282.02	103.16	205.99
(1, 50)	140.82	219.24	82.85	161.26
(50, 1)	227.50	361.08	128.73	262.32
(50, 50)	165.84	260.18	96.09	190.43
Region	**Central Scotland**			
(PV, EV), %	Tariff, £ (per household per year)			
	(no EV, no PV)	(EV, no PV)	(no EV, PV)	(EV, PV)
(1, 1)	122.42	172.93	65.15	115.67
(1, 50)	102.25	143.39	55.60	96.74
(50, 1)	160.27	228.35	83.08	151.16
(50, 50)	125.09	176.84	66.41	118.16
Region	**East Midlands**			
(PV, EV), %	Tariff, £ (per household per year)			
	(no EV, no PV)	(EV, no PV)	(no EV, PV)	(EV, PV)
(1, 1)	112.28	157.90	54.71	100.33
(1, 50)	93.77	131.49	46.17	83.89
(50, 1)	160.27	228.35	83.08	15116
(50, 50)	117.80	165.78	57.26	105.24

Source: Own results.

In addition, let us show the changes in the household demand profile with various levels of EV deployment stemming from the changes. Figure 5 that follows depicts some basic (rather "back-of-the-envelope") results of our calculations. Generally, it becomes apparent that with the rising share of EV in day-to-day transportation decisions, the demand profile can increase by 40%–50% during the peaks. All of these might cause serious issues for the energy systems that are not prepared for such scale of EV penetration.

Overall, there are two cases when EVs do not exhibit smart charging capability and when the EVs exhibit smart charging capability and are scheduled through dynamic pricing. In the former case, EVs are assumed to start charging immediately after returning home; given that most users return home during late afternoon and/or evening hours [17–21] when the non-EVs demand peak occurs, the system peak demand at these hours is significantly increased. In the latter case, the demand response of smart-charging EVs is concentrated and creates a new demand peak at the late-night hours of the day since the latter exhibit the lowest prices due to their low inflexible demand levels.

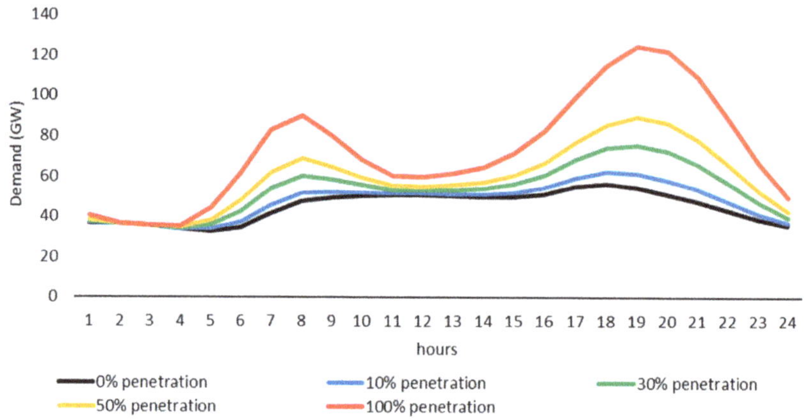

Figure 5. Household demand profile with various levels of EV deployment. Source: Own results.

5. Conclusions and Implications

The commitment to tackle global warming and climate change, with profound decarbonisation of the world's economies, is going to have many economic and social impacts and we all should be ready for them. Thence, political commitments that have an impact on the energy markets should also reflect the existing state of technological development and social acceptance of this development. This social acceptance is a serious issue because not all "traditional" energy providers and customers would be willing to change the old ways. Rapid flow of information and AI-like solutions that allow to optimize the flow of energy from producers to consumers and the flow of information the other way around (and vice versa) might be an issue for some individuals and companies. Some might even bring up an analogy to the "Big Brother" that is constantly monitoring each and every transaction and watching over each and every agent in the grid.

However, the trend might be also quite different as it happened with the ascent of Internet (which also served as an inspiration for the smart grids). Neglected and underestimated at first, the Internet quickly marched into our daily lives and fundamentally changed many principles of the world economy, quickly becoming a mundane issue for the majority of world's population. Its social acceptance is almost absolute, even though some people are starting to comprehend what threats it might cause. Similar things might happen to IoE and smart grids—once their potential is apparent and popularized, there will be no way back in reforming the current energy market design.

One would probably agree with the fact that the transition to the high-renewables electricity markets is going to be accompanied by the increasing share of RES in electricity generation. This will fundamentally alter the power markets within several decades and revert the existing rules that have been obeyed by energy market players for years. Renewable energy sources are gaining in importance with increasing power generation, and could easily cover half of the total energy demand by 2050. The hope of accelerating the energy transition depends on the electrification of a larger part of the total energy consumption. Renewable energy advocates have received much good news over the past few years about falling solar and wind power prices and rapid growth rates in these industries.

With regard to the above, it is important for policymakers and stakeholders to understand the potential and the power of the ongoing "RES revolution" (which, one can argue, still represents an "evolution") and to prepare for the new market reality with all its outcomes and consequences. The profile of renewable energy development is becoming ever more diverse and in some regions of the world more and more challenges will have to be overcome. While global new investments in renewable energies are still relatively high, there has been some decline in recent years.

In addition, it also appears that the deployment of different technologies and policy measures in the high-renewables energy system that would involve IoE would lead to the decentralized approach to system control and operation in the next few decades. Since it may be optimal that large-scale roll-out starts in the early 2020s if one is to reduce the cost of later investments, some government incentives may be needed for industry to make early moves.

Moreover, the concept of the "Internet of Energy" is very crucial for the demand flexibility (i.e., the case when demand for the electricity is steered to the grid off-peak periods using, for example, price incentives or reduced tariffs). Flexible energy revolution should be supported and implemented due to the growing demand and wide usage of electric vehicles (EV) and electric heating in the future. These technologies would help us to save carbon emissions, but they would be impossible to sustain without the major changes in the smart grid.

All in all, IoE and high-renewables electricity system market design represent a shift from asset redundancy to more intelligent operation through real-time coordination of new flexible technologies. With flexible demand, micro-generation, and energy storage technologies being able to reduce the emerging demand peaks, smart network technologies will increase the utilization of existing network assets, all representing a shift from asset redundancy to more intelligent operation through real-time coordination of new flexible technologies. Future demand and distributed generation connections are increasingly hard to predict due to uncertainty in timing, volume, and location of new demand, and connections of renewable generation due to their dependence on future technology developments, government policies, and market arrangements. Flexible planning frameworks are required to account for uncertainty in future development and identify investment strategies that are cost-efficient under all possible future realisations. Given all that above, we nevertheless have to stress that the outcomes of this paper mostly emerge from the energy policy aspects of high-renewables smart grids, and therefore does not attribute too much attention to the technical issues of Internet of Energy technique and the related framework applied to high renewables penetrations of electricity market.

Last but not least, high-renewables smart grids-focused electricity markets can increase economic efficiency and profits throughout the whole energy sector. While grid operators today hold natural monopolies, the shift to the high-renewables smart grids populated by millions and billions of prosumers might change the rules of the game and create the new revenue stream that would be beneficial for both energy suppliers (large and small) and energy consumers. Therefore, some forms of government funding are appropriate nowadays in order to boost this positive and highly desirable development towards the fundamental change of the existing energy system. This might include, for example, subsidies to the households and businesses for shifting towards electric transport, special preferential tariffs for renewable energy (feed-in-tariffs to buy out the "green" electricity), or special programs to increase environmental awareness, and the learning of how to increase energy efficiency by using advanced smart metering that would allow to constantly monitor the energy intake for each individual appliance and to assess it in a concise and comprehensive way (similar to the health data many of us assess using various fitness trackers and smart watches).

Supporting renewable energy demonstration projects to disseminate information in remote areas, training microfinance leaders, and decentralizing the implementation of renewable energy projects can promote the dissemination of renewable energy projects. Most of the support for renewable energy strategies and technologies in developing countries comes from local governments or from international donors, undermining their sustainability as funds vary with changing priorities and crises. Finally, the introduction of innovative strategies and the sustainability of renewable energy markets and technologies can benefit from the adoption of a comprehensive energy governance framework.

All in all, it appears that high-renewables electricity system market design would provide solutions to many issues including the CO_2 reduction of mitigating environmental solution and climate change, however it needs more support from relevant stakeholders and policymakers aimed at securing its economic viability and helping to support its social acceptance and awareness.

Author Contributions: Conceptualization, W.S., A.F., and E.S.; methodology, W.S. and D.S.; formal analysis, W.S., A.F., D.S., and E.S.; investigation, A.F., D.S., and E.S.; resources, A.F. and E.S.; data curation, W.S., A.F., and E.S.; writing—original draft preparation, W.S., D.S., A.F., and E.S.; writing—review and editing, W.S. and D.S.; visualization, W.S. and A.F.; supervision, W.S.; funding acquisition, A.F. and E.S. All authors have read and agreed to the published version of the manuscript.

Funding: This research received no external funding.

Conflicts of Interest: The authors declare no conflict of interest.

References

1. Newbery, D.; Pollitt, M.G.; Ritz, R.A.; Strielkowski, W. Market design for a high-renewables European electricity system. *Renew. Sustain. Energy Rev.* **2018**, *91*, 695–707. [CrossRef]
2. ILO, A. Design of the Smart Grid Architecture According to Fractal Principles and the Basics of Corresponding Market Structure. *Energies* **2019**, *12*, 4153. [CrossRef]
3. Ahmad, T.; Chen, H.; Wang, J.; Guo, Y. Review of various modeling techniques for the detection of electricity theft in smart grid environment. *Renew. Sustain. Energy Rev.* **2018**, *82*, 2916–2933. [CrossRef]
4. Strielkowski, W. Social and economic implications for the smart grids of the future. *Econ. Sociol.* **2017**, *10*, 310–318. [CrossRef]
5. Alahakoon, D.; Yu, X. Smart electricity meter data intelligence for future energy systems: A survey. *IEEE Trans. Ind. Inform.* **2015**, *12*, 425–436. [CrossRef]
6. Mengelkamp, E.; Gärttner, J.; Rock, K.; Kessler, S.; Orsini, L.; Weinhardt, C. Designing microgrid energy markets: A case study: The Brooklyn Microgrid. *Appl. Energy* **2018**, *210*, 870–880. [CrossRef]
7. Parag, Y.; Sovacool, B.K. Electricity market design for the prosumer era. *Nat. Energy* **2016**, *1*, 16032. [CrossRef]
8. Rana, M. Architecture of the Internet of energy network: An application to smart grid communications. *IEEE Access* **2017**, *5*, 4704–4710. [CrossRef]
9. Vrba, P.; Mařík, V.; Siano, P.; Leitão, P.; Zhabelova, G.; Vyatkin, V.; Strasser, T. A review of agent and service-oriented concepts applied to intelligent energy systems. *IEEE Trans. Ind. Inform.* **2014**, *10*, 1890–1903. [CrossRef]
10. Bui, N.; Castellani, A.P.; Casari, P.; Zorzi, M. The internet of energy: A web-enabled smart grid system. *IEEE Netw.* **2012**, *26*, 39–45. [CrossRef]
11. Wissner, M. The Smart Grid-a saucerful of secrets? *Appl. Energy* **2011**, *88*, 2509–2518. [CrossRef]
12. Dudin, M.N.; Frolova, E.E.; Protopopova, O.V.; Mamedov, A.A.; Odintsov, S.V. Study of innovative technologies in the energy industry: Nontraditional and renewable energy sources. *Entrep. Sustain. Issues* **2019**, *6*, 1704–1713. [CrossRef]
13. Pavić, I.; Capuder, T.; Kuzle, I. Low carbon technologies as providers of operational flexibility in future power systems. *Appl. Energy* **2016**, *168*, 724–738. [CrossRef]
14. Kabalci, Y. A survey on smart metering and smart grid communication. *Renew. Sustain. Energy Rev.* **2016**, *57*, 302–318. [CrossRef]
15. Palensky, P.; Kupzog, F. Smart grids. *Annu. Rev. Environ. Resour.* **2013**, *38*, 201–226. [CrossRef]
16. de Tena, D.L.; Pregger, T. Impact of electric vehicles on a future renewable energy-based power system in Europe with a focus on Germany. *Int. J. Energy Res.* **2018**, *42*, 2670–2685. [CrossRef]
17. Nyberg, R.A. Using 'smartness' to reorganise sectors: Energy infrastructure and information engagement. *Int. J. Inf. Manag.* **2018**, *39*, 60–68. [CrossRef]
18. Vlasov, A.I.; Shakhnov, V.A.; Filin, S.S.; Krivoshein, A.I. Sustainable energy systems in the digital economy: Concept of smart machines. *Entrep. Sustain. Issues* **2019**, *6*, 1975–1986. [CrossRef]
19. Akhavan-Hejazi, H.; Mohsenian-Rad, H. Power systems big data analytics: An assessment of paradigm shift barriers and prospects. *Energy Rep.* **2018**, *4*, 91–100. [CrossRef]
20. Koirala, B.P.; Koliou, E.; Friege, J.; Hakvoort, R.A.; Herder, P.M. Energetic communities for community energy: A review of key issues and trends shaping integrated community energy systems. *Renew. Sustain. Energy Rev.* **2016**, *56*, 722–744. [CrossRef]
21. Osório, G.J.; Shafie-khah, M.; Carvalho, G.C.; Catalão, J.P. Analysis Application of Controllable Load Appliances Management in a Smart Home. *Energies* **2019**, *12*, 3710. [CrossRef]

22. Kakran, S.; Chanana, S. Smart operations of smart grids integrated with distributed generation: A review. *Renew. Sustain. Energy Rev.* **2018**, *81*, 524–535. [CrossRef]
23. Salmerón-Manzano, E.; Manzano-Agugliaro, F. The Role of Smart Contracts in Sustainability: Worldwide Research Trends. *Sustainability* **2019**, *11*, 3049. [CrossRef]
24. Li, Y.; Cheng, X.; Cao, Y.; Wang, D.; Yang, L. Smart Choice for the Smart Grid: Narrowband Internet of Things (NB-IoT). *IEEE Internet Things J.* **2018**, *5*, 1505–1515. [CrossRef]
25. Valtanen, K.; Backman, J.; Yrjölä, S. Blockchain-Powered Value Creation in the 5G and Smart Grid Use Cases. *IEEE Access* **2019**, *7*, 25690–25707. [CrossRef]
26. Kuzmin, E.A.; Volkova, E.E.; Fomina, A.V. Research on the concentration of companies in the electric power market of Russia. *Int. J. Energy Econ. Policy* **2019**, *9*, 130–136. [CrossRef]
27. Ma, R.; Chen, H.H.; Huang, Y.R.; Meng, W. Smart grid communication: Its challenges and opportunities. *IEEE Trans. Smart Grid* **2013**, *4*, 36–46. [CrossRef]
28. Huang, B.; Li, Y.; Zhang, H.; Sun, Q. Distributed optimal co-multi-microgrids energy management for energy internet. *IEEE CAA J. Autom. Sin.* **2016**, *3*, 357–364. [CrossRef]
29. Pina, A.; Silva, C.; Ferrão, P. The impact of demand side management strategies in the penetration of renewable electricity. *Energy* **2012**, *41*, 128–137. [CrossRef]
30. Strielkowski, W.; Štreimikienė, D.; Bilan, Y. Network charging and residential tariffs: A case of household photovoltaics in the United Kingdom. *Renew. Sustain. Energy Rev.* **2017**, *77*, 461–473. [CrossRef]
31. McArthur, S.; Taylor, P.; Ault, G.W.; King, J.; Athanasiadis, D.; Alimisis, V.; Czaplewski, M. The autonomic power system-network operation and control beyond smart grids. In Proceedings of the 3rd IEEE PES Innovative Smart Grid Technologies (ISGT Europe), Berlin, Germany, 14–17 October 2012; pp. 1–7. [CrossRef]
32. Papadaskalopoulos, D.; Strbac, G. Nonlinear and Randomized Pricing for Distributed Management of Flexible Loads. *IEEE Trans. Smart Grid* **2016**, *7*, 1137–1146. [CrossRef]
33. Xu, Y.; Milanovic, J.V. Artificial-Intelligence-Based Methodology for Load Disaggregation at Bulk Supply Point. *IEEE Trans. Power Syst.* **2015**, *30*, 795–803. [CrossRef]
34. Reka, S.S.; Dragicevic, T. Future effectual role of energy delivery: A comprehensive review of Internet of Things and smart grid. *Renew. Sustain. Energy Rev.* **2018**, *91*, 90–108. [CrossRef]
35. Hwang, L.C.; Chen, C.S.; Ku, T.T.; Shyu, W.C. A bridge between the smart grid and the Internet of Things: Theoretical and practical roles of LoRa. *Int. J. Electr. Power Energy Syst.* **2019**, *113*, 971–981. [CrossRef]
36. Holttinen, H.; Meibom, P.; Orths, A.; Lange, B.; O'Malley, M.; Tande, J.O.; Smith, J.C. Impacts of large amounts of wind power on design and operation of power systems, results of IEA collaboration. *Wind Energy* **2011**, *14*, 179–192. [CrossRef]
37. Connor, P.M.; Baker, P.E.; Xenias, D.; Balta-Ozkan, N.; Axon, C.J.; Cipcigan, L. Policy and regulation for smart grids in the United Kingdom. *Renew. Sustain. Energy Rev.* **2014**, *40*, 269–286. [CrossRef]
38. Jenkins, J.D.; Pérez Arriaga, I.J. Improved regulatory approaches for the remuneration of electricity distribution utilities with high penetrations of distributed energy resources. *Energy J.* **2017**, *38*, 63–91. [CrossRef]
39. Lo, C.H.; Ansari, N. The progressive smart grid system from both power and communications aspects. *IEEE Commun. Surv. Tutor.* **2011**, *14*, 799–821. [CrossRef]
40. Defeuilley, C. Energy transition and the future (s) of the electricity sector. *Util. Policy* **2019**, *57*, 97–105. [CrossRef]
41. Mitchell, C. Momentum is increasing towards a flexible electricity system based on renewables. *Nat. Energy* **2016**, *1*, 15030. [CrossRef]
42. Rausser, G.; Strielkowski, W.; Štreimikienė, D. Smart meters and household electricity consumption: A case study in Ireland. *Energy Environ.* **2018**, *29*, 131–146. [CrossRef]
43. Reddy, K.S.; Kumar, M.; Mallick, T.K.; Sharon, H.; Lokeswaran, S. A review of Integration, Control, Communication and Metering (ICCM) of renewable energy based smart grid. *Renew. Sustain. Energy Rev.* **2014**, *38*, 180–192. [CrossRef]
44. Byun, J.; Hong, I.; Park, S. Intelligent cloud home energy management system using household appliance priority based scheduling based on prediction of renewable energy capability. *IEEE Trans. Consum. Electron.* **2012**, *58*, 1194–1201. [CrossRef]
45. Gelazanskas, L.; Gamage, K.A. Demand side management in smart grid: A review and proposals for future direction. *Sustain. Cities Soc.* **2014**, *11*, 22–30. [CrossRef]
46. Daki, H.; El Hannani, A.; Aqqal, A.; Haidine, A.; Dahbi, A. Big Data management in smart grid: Concepts, requirements and implementation. *J. Big Data* **2017**, *4*, 1–19. [CrossRef]

47. Zanella, A.; Bui, N.; Castellani, A.; Vangelista, L.; Zorzi, M. Internet of Things for Smart Cities. *IEEE Internet Things J.* **2014**, *1*, 22–32. [CrossRef]
48. Cerruela García, G.; Luque Ruiz, I.; Gómez-Nieto, M.Á. State of the art, trends and future of bluetooth low energy, near field communication and visible light communication in the development of smart cities. *Sensors* **2016**, *16*, 1968. [CrossRef]
49. Jaradat, M.; Jarrah, M.; Bousselham, A.; Jararweh, Y.; Al-Ayyoub, M. The internet of energy: Smart sensor networks and big data management for smart grid. *Procedia Comput. Sci.* **2015**, *56*, 592–597. [CrossRef]
50. Kaygusuz, K. Energy situation, future developments, energy saving, and energy efficiency in Turkey. *Energy Sources* **1999**, *21*, 405–416. [CrossRef]
51. Uddin, M.; Rahman, A.A. Energy efficiency and low carbon enabler green IT framework for data centers considering green metrics. *Renew. Sustain. Energy Rev.* **2012**, *16*, 4078–4094. [CrossRef]
52. Çengel, Y.A. Energy efficiency as an inexhaustible energy resource with perspectives from the US and Turkey. *Int. J. Energy Res.* **2011**, *35*, 153–161. [CrossRef]
53. Brook, B.; Blees, T.; Wigley, T.; Hong, S. Silver Buckshot or Bullet: Is a Future "Energy Mix" Necessary? *Sustainability* **2018**, *10*, 302. [CrossRef]
54. Davis, S.J.; Lewis, N.S.; Shaner, M.; Aggarwal, S.; Arent, D.; Azevedo, I.L.; Benson, S.M.; Bradley, T.; Brouwer, J.; Chiang, Y.M.; et al. Net-zero emissions energy systems. *Science* **2018**, *360*. [CrossRef] [PubMed]
55. Deng, X.; Lv, T. Power system planning with increasing variable renewable energy: A review of optimization models. *J. Clean. Prod.* **2019**. [CrossRef]
56. Zöphel, C.; Schreiber, S.; Müller, T.; Möst, D. Which Flexibility Options Facilitate the Integration of Intermittent Renewable Energy Sources in Electricity Systems? *Curr. Sustain. Renew. Energy Rep.* **2018**, *5*, 37–44. [CrossRef]
57. Yixian, L.I.U.; Roberts, M.C.; Sioshansi, R. A vector autoregression weather model for electricity supply and demand modeling. *J. Modern Power Syst. Clean Energy* **2018**, *6*, 763–776. [CrossRef]
58. Horst, D.; Jentsch, M.; Pfennig, M.; Mitra, I.; Bofinger, S. Impact of renewable energies on the indian power system: Energy meteorological influences and case study of eefects on existing power fleet for rajasthan state. *Energy Policy* **2018**, *122*, 486–498. [CrossRef]
59. Jerez, S.; Tobin, I.; Turco, M.; Jiménez-Guerrero, P.; Vautard, R.; Montávez, J.P. Future changes, or lack thereof, in the temporal variability of the combined wind-plus-solar power production in Europe. *Renew. Energy* **2019**, *139*, 251–260. [CrossRef]
60. Johansson, P.O.; Kriström, B. Welfare evaluation of subsidies to renewable energy in general equilibrium: Theory and application. *Energy Econ.* **2019**, *83*, 144–155. [CrossRef]
61. Simshauser, P. On the Stability of Energy-Only Markets with Government-Initiated Contracts-for-Differences. *Energies* **2019**, *12*, 2566. [CrossRef]
62. Ringkjøb, H.K.; Haugan, P.M.; Solbrekke, I.M. A review of modelling tools for energy and electricity systems with large shares of variable renewables. *Renew. Sustain. Energy Rev.* **2018**, *96*, 440–459. [CrossRef]
63. Ma, S.; Geng, H.; Yang, G.; Pal, B.C. Clustering-Based Coordinated Control of Large-Scale Wind Farm for Power System Frequency Support. *IEEE Trans. Sustain. Energy* **2018**, *9*, 1555–1564. [CrossRef]
64. Zeng, Y.; Zhang, R.; Wang, D.; Mu, Y.; Jia, H. A regional power grid operation and planning method considering renewable energy generation and load control. *Appl. Energy* **2019**, *237*, 304–313. [CrossRef]
65. Szabó, L.; Kelemen, Á.; Mezősi, A.; Pató, Z.; Kácsor, E.; Resch, G.; Liebmann, L. South East Europe electricity roadmap-modelling energy transition in the electricity sectors. *Clim. Policy* **2019**, *19*, 495–510. [CrossRef]
66. Kraan, O.; Kramer, G.J.; Nikolic, I.; Chappin, E.; Koning, V. Why fully liberalised electricity markets will fail to meet deep decarbonisation targets even with strong carbon pricing. *Energy Policy* **2019**, *131*, 99–110. [CrossRef]
67. Godoy-González, D.; Gil, E.; Gutiérrez-Alcaraz, G. Ramping ancillary service for cost-based electricity markets with high penetration of variable renewable energy. *Energy Econ.* **2019**, *85*, 104556. [CrossRef]
68. Sinsel, S.R.; Yan, X.; Stephan, A. Building resilient renewable power generation portfolios: The impact of diversification on investors' risk and return. *Appl. Energy* **2019**, *254*, 113348. [CrossRef]
69. Miller, L.; Carriveau, R. Energy demand curve variables-An overview of individual and systemic effects. *Sustain. Energy Technol. Assess.* **2019**, *35*, 172–179. [CrossRef]
70. Brown, M.A.; Zhou, S.; Ahmadi, M. Smart grid governance: An international review of evolving policy issues and innovations. *Wiley Interdiscip. Rev. Energy Environ.* **2018**, *7*, e290. [CrossRef]

71. Sakti, A.; Botterud, A.; O'Sullivan, F. Review of wholesale markets and regulations for advanced energy storage services in the United States: Current status and path forward. *Energy Policy* **2018**, *120*, 569–579. [CrossRef]
72. EIA. Annual Energy Outlook 2018. 2019. Available online: https://www.eia.gov/outlooks/aeo/pdf/electricity_generation.pdf (accessed on 18 October 2019).
73. IEA. Smart Grids. 2019. Available online: https://www.iea.org/tcep/energyintegration/smartgrids (accessed on 18 October 2019).
74. Strbac, G.; Konstantelos, I.; Aunedi, M.; Pollitt, M.; Green, R. Delivering Future-Proof Energy Infrastructure. Report for National Infrastructure Commission. 2016. Available online: https://www.gov.uk/government/uploads/system/uploads/attachment_data/file/507256/Future-proof_energy_infrastructure_Imp_Cam_Feb_2016.pdf (accessed on 7 October 2019).
75. Shivakumar, A.; Pye, S.; Anjo, J.; Miller, M.; Rouelle, P.B.; Densing, M.; Kober, T. Smart energy solutions in the EU: State of play and measuring progress. *Energy Strategy Rev.* **2018**, *20*, 133–149. [CrossRef]
76. Strbac, G.; Pollitt, M.; Konstantinidis, C.V.; Konstantelos, I.; Moreno, R.; Newbery, D.; Green, R. Electricity transmission arrangements in Great Britain: Time for change? *Energy Policy* **2014**, *73*, 298–311. [CrossRef]
77. Popovič, A.; Hackney, R.; Tassabehji, R.; Castelli, M. The impact of big data analytics on firms' high value business performance. *Inf. Syst. Front.* **2018**, *20*, 209–222. [CrossRef]
78. Minchala-Avila, L.I.; Garza-Castañón, L.E.; Vargas-Martínez, A.; Zhang, Y. A Review of Optimal Control Techniques Applied to the Energy Management and Control of Microgrids. *Procedia Comput. Sci.* **2015**, *52*, 780–787. [CrossRef]
79. Hakimi, S.M.; Hasankhani, A.; Shafie-khah, M.; Catalão, J.P.S. Optimal sizing and siting of smart microgrid components under high renewables penetration considering demand response. *IET Renew. Power Gener.* **2019**, *13*, 1809–1822. [CrossRef]
80. Yang, C.; Yao, J.; Lou, W.; Xie, S. On Demand Response Management Performance Optimization for Microgrids Under Imperfect Communication Constraints. *IEEE Internet Things J.* **2017**, *4*, 881–893. [CrossRef]
81. Strbac, G. Demand side management: Benefits and challenges. *Energy Policy* **2008**, *36*, 4419–4426. [CrossRef]
82. Electrive. Number of Plug-In Cars Climbs to 5.6M Worldwide. 2019. Available online: https://www.electrive.com/2019/02/11/the-number-of-evs-climbs-to-5-6-million-worldwide (accessed on 29 November 2019).
83. Statista. Worldwide Number of Battery Electric Vehicles in Use From 2012 to 2018. 2019. Available online: https://www.statista.com/statistics/270603/worldwide-number-of-hybrid-and-electric-vehicles-since-2009 (accessed on 28 November 2019).
84. Carsguide. How Many Cars Are There in the World? 2019. Available online: https://www.carsguide.com.au/car-advice/how-many-cars-are-there-in-the-world-70629 (accessed on 28 November 2019).
85. HIS Markit. Number of Fast-Charging Stations for Electric Vehicles set to Rise to Nearly 200,000 in 2020. 2019. Available online: https://news.ihsmarkit.com/press-release/design-supply-chain-media/number-fast-charging-stations-electric-vehicles-set-rise-nea?page=1 (accessed on 29 November 2019).
86. Statista. Number of Fast-Charging Stations for Electric Vehicles Worldwide in 2013 and 2020. 2019. Available online: https://www.statista.com/statistics/283531/electric-vehicles--global-number-of-fast-charging-stations/ (accessed on 29 November 2019).
87. Strielkowski, W.; Volkova, E.; Pushkareva, L.; Streimikiene, D. Innovative Policies for Energy Efficiency and the Use of Renewables in Households. *Energies* **2019**, *12*, 1392. [CrossRef]
88. Taiebat, M.; Xu, M. Synergies of four emerging technologies for accelerated adoption of electric vehicles: Shared mobility, wireless charging, vehicle-to-grid, and vehicle automation. *J. Clean. Prod.* **2019**, *230*, 794–797. [CrossRef]
89. Newbery, D.; Strbac, G. What is needed for battery electric vehicles to become socially cost competitive? *Econ. Transp.* **2016**, *5*, 1–11. [CrossRef]
90. Parsons, G.R.; Hidrue, M.K.; Kempton, W.; Gardner, M.P. Willingness to pay for vehicle-to-grid (V2G) electric vehicles and their contract terms. *Energy Econ.* **2014**, *42*, 313–324. [CrossRef]

© 2019 by the authors. Licensee MDPI, Basel, Switzerland. This article is an open access article distributed under the terms and conditions of the Creative Commons Attribution (CC BY) license (http://creativecommons.org/licenses/by/4.0/).

energies

MDPI

Article

On the Stability of Energy-Only Markets with Government-Initiated Contracts-for-Differences

Paul Simshauser

Department of Accounting, Finance & Economics, Griffith University, Brisbane 4111, Australia; p.simshauser@griffith.edu.au

Received: 19 May 2019; Accepted: 28 June 2019; Published: 3 July 2019

Abstract: Rising levels of variable renewable energy (VRE) in Australia's National Electricity Market have been driven by a 20% renewable energy target by 2020. This certificated renewable portfolio standard has successfully driven new investment, allocated risk amongst buy- and sell-side market participants and met overall policy objectives. But a policy vacuum for achieving long-term CO_2 emission targets post 2020 has led to sub-national and, potentially, national governments initiating contract-for-differences (CfDs) to drive further investment activity in new plant—with virtually no coordination between the jurisdictions. In a gross pool energy-only market setting, replacing *on-market* transactions between retailers and generators with *off-market* transactions between governments and generators may have unintended side-effects vis-à-vis market stability. In this article, an energy-only gross pool is modeled with rising levels of *off-market* government-initiated CfDs, with a specific focus on spot and forward contract market outcomes. Model results show that as VRE plant enters, coal plant exit, and *on-market* firm hedge contracts historically supplied by coal plant are progressively replaced by *off-market* CfDs. In the event, while a tractable equilibrium can be maintained in the spot market, shortages of "primary issuance" hedge contracts emerge in the forward market. Any shortage of hedge contract capacity is likely to raise forward contract price premiums above efficient levels, force price-elastic customers into accepting unwanted spot market exposures and may unintentionally foreclose non-integrated (2nd tier) energy retailers, all of which harms consumer welfare. A wide-ranging program of government CfDs may therefore not be compatible with an energy-only market design.

Keywords: renewable energy; energy policy; wholesale market design

---

1. Introduction

1.1. Background to Australia's National Electricity Market

Australia's National Electricity Market (NEM) is an energy-only gross pool in which all generators bid into a central platform and are dispatched under a uniform first-price auction clearing mechanism. Being a mandatory gross pool, all generators must sell their output in the spot market, and energy retailers must buy all load from the spot market.

The volatility that accompanies organized energy-only spot markets, particularly those with a high VoLL (NEM value of lost load at AUD $14,500/MWh is amongst the highest in the world) creates the conditions necessary for active trade in forward contracts. While there is an almost endless array of derivative instruments, the three most commonly traded contracts are swaps, $300 caps and increasingly, run-of-plant power purchase agreements (PPAs). Swaps and caps are traded both on-exchange and over-the-counter, generally over a 1–3 year tenor at quarterly resolution, with market liquidity running at ~300% of physical trade and considerable variation in liquidity by season and region. PPAs on the other hand tend to be long-dated (10–15 year), structured as run-of-plant instruments and designed specifically to underwrite the entry of variable renewable energy (VRE), that is, wind or utility-scale solar PV, with the market characterized by a small number of buyers and a large number of (potential) sellers/developers.

Australia's 20% renewable energy target (RET) is a renewable portfolio standard that requires energy retailers achieve a 20% renewable market share by 2020. The broad-based market scheme is mobilized through the trade of renewable energy certificates, and has been highly successful in facilitating VRE plant entry. The 20% RET is expected to be comfortably met by 2020.

Once the 20% RET has been met, the NEM faces a virtual policy vacuum in terms of future CO_2 emissions reductions. The most recent attempt (in 2018) at designing a bipartisan approach to reduce power sector CO_2 emissions in a manner compatible with Australia's international obligations and the NEM's energy-only market design involved a scheme known as the *National Energy Guarantee*. This market-based scheme was comprised of a reliability obligation (plant capacity) and an emissions obligation (CO_2 emissions), with energy retailers being the liable parties. But landing a united energy and climate change policy architecture to guide the NEM's ongoing transition has proven to be a complex political task [1,2]. Indeed, the politics of a two-decades-long climate change policy war has underpinned the downfall of at least three sitting Australian Prime Ministers since 2010, including Prime Minister Malcolm Turnbull (by his own party) whilst pursuing the *National Energy Guarantee*.

Policy discontinuity has led to cyclical swings in VRE plant investment commitments. The Commonwealth Government's decision to review the 20% RET in early-2014 produced sufficient policy uncertainty to induce a transient cessation in buy-side PPA activity—such instruments typically being an essential ingredient to VRE project commitment in an energy-only market [2].

One response to the Commonwealth Government's climate change policy discontinuity has been the emergence of government-initiated contracts-for-differences (CfDs) undertaken unilaterally by sub-national governments—first by the Australian Capital Territory (wind, 2015), then Queensland (solar PV, 2016), South Australia (semi-CfD for battery storage, 2017) and Victoria (wind and solar PV, 2018). These targeted, centrally planned CfD auctions have been successful at meeting their policy objectives.

The Commonwealth election in mid-2019 gave rise to two distinct policy choices at the national level. The incumbent conservative Liberal government sought to underwrite new "dispatchable" (i.e., coal, gas, pumped-hydro) generation in order to reduce energy prices by way of government-initiated CfDs. The democratic Labor opposition on the other hand sought to revive the Liberal government's *National Energy Guarantee* in order to reduce CO_2 emissions and provide the policy stability sought by market participants to guide ongoing investment in *on-market* transactions. However, if bipartisan support for the *National Energy Guarantee* could be achieved, Labor would default to a wide-ranging program of centrally planned CfDs to ensure ongoing VRE investment continuity to meet a 50% renewable target by 2030.

Government-initiated and taxpayer-funded CfDs are an interesting development [3,4]. CfDs can play a legitimate role in dealing with energy market failures relating to missing policy, missing markets and incomplete markets. As a policy mechanism, they represent a means by which to deliver generation plant capacity that, for whatever reason, the market is failing to deliver. Government-initiated CfDs have the effect of diversifying buy-side forward market liquidity and in doing so bring about certain short run benefits. Holding all else constant, apart from reducing CO_2 emissions, VRE CfDs facilitate state/regional economic development and by adding new renewable supply can reduce spot prices, at least on a transient basis.

Used modestly, the impact of government-initiated CfDs on an energy-only market is likely to be benign at worst. However, a wide-ranging program of CfDs whereby material levels of *on-market* transactions (i.e., between generators and retailers) are crowded-out by *off-market* CfD transactions (i.e., between governments and generators) is likely to give rise to unintended side effects, and this forms the primary motivation of this article. To be clear, this article does not question the welfare implications of introducing VRE plant, and forcing coal plant exit. The addition of VRE plant is taken as an exogenous policy constraint to meet a legitimate and binding CO_2 emissions reduction objective. The focus is whether as a policy instrument, government-initiated CfDs are compatible with the efficient operation of an energy-only gross pool market.

In this article, a partial equilibrium model of an energy-only market is used to analyze the effects of rising levels of VRE plant, facilitated by government-initiated CfDs rather than via *on-market* transactions typically associated with a conventional renewable portfolio standard. The power system model commences with a thermal plant stock (i.e., coal and gas plant), and then VRE plant enters continuously until a 40% market share is reached.

Key results arising from the modeling are that (1) as VRE plant enters, coal plant exits, and spot market equilibrium is maintained; however, (2) because VRE plant enters via government-initiated CfDs and displaces *on-market* transactions, when coal plant exits, the forward market experiences progressively rising shortages of "primary issuance" hedge contract capacity. This latter finding has important implications for policymakers.

Structural shortages of hedge contract capacity may not matter in an energy+capacity market in which overall price volatility remains within tight limits, or in highly meshed multi-regional power grids where inter-regional trade can "prop-up" faltering market liquidity in an adjacent region. But for an imperfectly interconnected energy-only market with an extremely high VoLL, policy-induced hedge contract shortages present certain problems because falling liquidity is unlikely to be arrested by proprietary traders. On the contrary, evidence from financial markets reveals that proprietary traders and caught with unwanted inventory and irreversible forward positions [5]. Liquidity providers can be expected to exit a market with sharply falling liquidity for fear of being.

A poorly functioning forward market in an energy-only market setting may increase operational risks of incumbent market participants, produce excess contract price premiums, force the most price-elastic (industrial) customers into unwanted spot market exposures, unintentionally *foreclose* non-integrated 2nd tier retailers, and ultimately drive investment activity above efficient levels to address hedge shortages—all of which must ultimately harm consumer welfare. In short, extensive use of centralized CfDs appears inherently incompatible with the underlying design of an energy-only gross pool market with a high VoLL. Whether the market design requires change to accommodate CfDs, or renewable portfolio standards can be extended to avoid the need for centralized transactional intervention, is an open question.

1.2. On the Importance of the Financial (Hedge Contract) Market and CfDs

Energy markets are never complete or free of market failures [6]. One of the more prominent failures inherent in energy-only markets is their seeming inability to deliver the requisite mix of derivative instruments required to facilitate efficient and timely plant entry [6–15]

Long-dated contracts are typically a pre-condition for the timely entry of project financed plant, and while Australia's NEM is noted for favorable forward market liquidity, the majority of activity spans only 1–3 years—well short of contracts that deliver optimal financing and facilitate timely and efficient ex-ante investment commitment. On reason why liquid forward markets have failed to calibrate beyond 3 years is because competitive retailers cannot afford to hold hedge portfolios dominated by inflexible long-dated contracts when large components of their customer book switch supplier every 2–3 years [8,16–22].

Government-initiated CfDs have arisen due to a combination of missing or incomplete markets, and form one of a number of policy mechanisms used by governments to meet a decarbonization objective or reliability constraint [23,24]. Government-initiated CfD's have been progressively gaining prominence amongst policymakers [25–29] and amongst academics [3,23,30,31].

Typically, a government-initiated CfD will attempt to minimize the levelized cost of electricity (LCoE) as a surrogate for maximizing value to taxpayers. At one level the use of LCoE as a prime metric is understandable because forecasting market outcomes 10–15 years in advance is notoriously difficult. But as a stand-alone metric, LCoE is flawed because it treats technology output as homogeneous products as if governed by the law of one price [24,32–35]. That is, while the *physical* properties of electricity are largely homogeneous over space and time, from a *market* perspective there is rich price variation over time, space, and lead time-to-delivery, making the traded commodity a heterogeneous good (i.e., due to an inability to arbitrage, the absence of a single dominant technology, and variations in marginal costs) [6]. The economic value of plant output is not identical and assuming otherwise introduces two biases; base plant is favored over peak, and stochastic plant is favored over dispatchable plant [6].

In real-time, the law of one price applies; stochastic output from wind and solar PV are good substitutes for thermal generation. However, each year there are 105,120 NEM dispatch intervals and associated spot prices (i.e., every 5 min) and when demand is higher than forecast, all else equal, dispatchable generators increase output and receive a higher average price. Conversely, stochastic generators rarely reduce output in periods of oversupply, and hence sell disproportionately at lower prices [6,24,34,35].

Furthermore, as VRE technologies move from niche to material market shares, deployment success becomes a significant driver of market value which is amplified when thermal plant fails to exit [24,32–38]. High levels of VRE shielded by CfDs and priority-dispatched will initially place downward pressure on price [14,24,39,40]. Given negligible marginal running costs, these so-called *merit-order effects* arising from policy-induced VRE plant entry became apparent in markets such as Germany as early as 2008 [41] and had been prominent in the SA region of the NEM [42–45]. Consequently, market values of incumbent VRE (and future) plant will be adversely affected from a stream of continual entry through a combination of production "correlation effects", "merit-order effects" and "price-impression effects" [24,35,37]. However, merit-order effects eventually unwind when thermal plant is forced to exit [39,46]. This set of market dynamics has implications for a wide-ranging program of government-initiated CfDs.

1.3. Government-Initiated CfDs: Motivation and Application

The policy objective of government-initiated CfDs is to introduce generation plant that energy markets are failing to deliver. In this sense, CfDs have the effect of bringing forward future power projects to today, with the benefits, costs, and risks of doing so allocated to electricity consumers, taxpayers, and incumbent rivals.

There are many reasons why government intervention is legitimately required in energy markets. As is well understood in economics, organized spot markets and their associated forward contract markets fail to internalize known externalities. For example, energy-only spot markets may undervalue reserve capacity until it is actually required. In the absence of an explicit price on carbon, energy markets also undervalue CO_2 emissions and will therefore only be produced at the efficient level by chance. And as with many markets, research and development (R&D) is not valued; but this is compounded in energy markets because participants are unable to capture the benefits of a first-of-a-kind plant investment—in fact, the contrary is usually the case in that the market avoids costly mistakes of the first iteration of a new technology. Absent some form of government intervention, R&D will be under-supplied by the market.

Government-initiated CfDs can have the effect of "priming" a market by helping emerging technologies overcome certain entry barriers. The Queensland government's Solar150 program in 2016 awarded CfDs to four solar PV projects totalling 150 MW at a time when solar PV struggled to compete with wind. The policy had the effect of kick-starting a wave of solar developments; by late-2018 a total of 1945 MW of solar PV had been committed *on-market* including some projects on a purely merchant basis (i.e., without a PPA). In South Australia, a policy to introduce a 100 MW utility-scale battery for system stability similarly primed the market for storage—there are now 215 MW of commissioned batteries, a further 155 MW have reached financial close, and 1897 MW are under active development across Australia. In short, while there are many policy mechanisms available to remedy energy market failures, CfD's are indeed a viable policy option.

1.4. CfDs: How They Work and Why They Work

In the classic case, a government-initiated CfD auction will specify a particular technology (e.g., solar PV) or technology set, output or rated capacity (e.g., up to 100 MW) and timing for delivery (e.g., able to reach financial close within 6 months of being awarded a long-dated CfD). The CfD is in turn a form of long-dated fixed price contract, usually expressed in $/MWh. In application, a CfD is a derivative instrument because payouts are referenced against spot prices. In a two-way CfD with a strike price of say $65/MWh, the contracting government (i.e., taxpayers) will pay the difference to the renewable project proponent whenever spot prices fall below $65, and the renewable project proponent pays the government whenever spot prices are above $65. CfDs are typically run-of-plant instruments such that difference payments only apply when the renewable plant is producing—and the plant's variable output is bid into the market and dispatched whenever it is profitable to do so (i.e., spot price > $0/MWh if the CfD has a settlement floor price of $0/MWh). Absent material plant failure, in which case some form of liquidated damages may apply, the weather-related volume risks and forward price risks associated with a VRE plant are effectively transferred to the contacting government (taxpayers). After writing a CfD, plant output cannot be sold twice, and so the plant's capacity is extracted from the forward market and taxpayers in turn hold a speculative instrument (unless the government chooses to on-sell the CfD in secondary markets).

A government-initiated CfD overcomes missing and incomplete markets and crucially in the context of Australia, can successfully navigate carbon policy uncertainty because CfDs provide revenue certainty (i.e., virtual market immunity) to the power project proponent.

Finally, because power projects are capital-intensive, the cost of debt and equity capital is an important driver of overall plant unit costs ($/MWh). The direct involvement of a government through long-dated CfDs enhances the credit quality of power projects, and this generally enables higher levels of debt, a lower cost of debt capital, and makes the task of equity capital raising easier. Consequently, holding all other variables constant, by transferring the price, volume, policy, and credit default risks of power projects to taxpayers, government-initiated CfDs are capable of producing a lower LCoE for entering projects.

1.5. The Impact of CfDs vs. Carbon Pricing and Renewable Certificate Markets

CO_2 emissions reduction policies ultimately seek to alter the plant stock in a way that reduces output from coal plant and increase output from renewable and cleaner (e.g., gas-fired) resources. Regardless of the policy mechanism used (e.g., cap and trade emissions trading scheme, emissions intensity scheme, carbon tax, renewable portfolio standard, clean energy target or government-initiated CfD), wealth transfers amongst producers occur. Carbon-intensive forms of generation are adversely affected, while low and zero emissions plant benefit from any explicit or implicit price on CO_2 emissions.

Government-initiated CfDs differ from broad-based market schemes (e.g., carbon prices or renewable portfolio standards) because of the direct involvement of government in the transaction and the reallocation of market, credit, and policy risks to taxpayers. Project bankers and the credit committees of Banks, which allocate scarce debt capital, have a strong preference for long-dated government-initiated CfDs because from a credit perspective there is negligible risk of counterparty default. By contrast to conventional NEM-based over-the-counter *on-market* transactions, a government-initiated CfD re-orientates policy and credit risk away from buy-side energy market participants, and vests this with taxpayers.

When deployed judiciously, the implications of CfDs are generally benign. In the case of the Queensland Solar150 program for example, any distortionary impacts arising from 150 MW (0.4 GWh) of solar-based CfDs in a 10,500 MW (54,000 GWh) regional market would be hard to detect. Taxpayers have a collective financial exposure to CfDs that will ultimately prove to be out-of-the-money; but this needs to be balanced with other policy objectives (e.g., state development, subsequent economic and environmental benefits of the 1945 MW of *on-market* solar PV projects that immediately followed). But what happens when CfDs are not used to "prime" a market, but rather, are used to "replace" the market; that is, replace broad-based market mechanisms and *on-market* transactions like those used in renewable portfolio standards to drive VRE entry?

Holding all else constant, so-called *merit-order effects* can be expected. That is, adding more supply, renewable or non-renewable, will reduce wholesale prices. But it will do this in the short- to medium run. Because the purpose, and effect, of the entry of VRE plant at-scale is designed to replace coal plant output, it will inevitably do so. Ultimately, the marginal coal plant will find it unprofitable, and will therefore exit. At this point, prices can be expected to rebound—and in the context of the NEM this is more than a theoretical observation. There is nothing inherently wrong with this policy objective, or the course of events that follows per se. But government-initiated CfDs undertaken at-scale may adversely impact the efficiency of an energy market (as distinct from "priming" a market) by comparison to an *on-market* renewable porfolio standard for three reasons.

First, governments are remote from power system operations and power system contract and risk management requirements. Government-initiated CfD auctions are therefore typically based on simplified metrics such as minimizing LCoE by way of open-auction, or at discriminatory price benchmark set by government bureaucrats to accommodate technological variations in production or cost. But as noted in Section 1.2, LCoE is a flawed metric and an overreliance on it in CfD auctions risks introducing an inefficient pattern of plant entry in a way that on-market transactions may have avoided. In contrast, broad-based market schemes like the *National Energy Guarantee* or a well-designed renewable portfolio standard require market participants to focus not on the LCoE, but on the timing, location, and market value of new plant output [4,24,34,35]. And to the extent that market participants introduce an inefficient pattern of plant entry vis-à-vis timing, location and market value, the risk and consequence of such errors vest with shareholders, not taxpayers.

Second, government-initiated CfDs introduce quasi-market participants that, through the design of the CfD, are almost completely sheltered from the NEM's energy-only short and medium-run locational, spot and forward price signals—the primary signals relied upon by market institutions and policymakers to regulate system performance, system reliability, investment patterns, and long run consumer prices. In contrast, on-market transactions undertaken by profit-maximizing firms under a renewable portfolio standard requires that market participants assess the relative pattern of entry, locational considerations, and absorb the risks of inadequate or excess entry relative to policy objectives. Broad-based market schemes can therefore be expected to outperform a central buyer on a risk-adjusted basis; and I must emphasize on a *risk-adjusted* basis—as one reviewer noted government-initiated CfDs should produce better financing terms and a lower overall cost of capital by comparison to on-market transactions. But this observation ignores risks transferred to taxpayers and the opportunity cost of using scarce government balance sheet resources. In comparison to a central buyer, market schemes are likely to accumulate a more optimal composition of assets and allocation of investment risks, reflecting the combination of physical power system requirements, policy-related constraints, and the risk appetite of participants to transactions.

Third, and by far the most adverse implication of a wide-ranging government-initiated CfD program is the potential distortion to forward markets and therefore market efficiency more generally. Unlike renewable portfolio standards, a wide-ranging policy of government-initiated CfD instruments that form a progressively larger share of a forward market will ultimately damage the *primary-issuance* of hedge contracts.

Following an initial or "primary loss" of on-market contract liquidity, the exit of proprietary traders that invariably follows (i.e., through fear of being caught with an illiquid position and unwanted inventory) will drive a "secondary loss" of market depth and liquidity. Combined, this is capable of culminating in a structural shortage of hedge contracts (i.e., forward market liquidity dropping below 100%).

There is, of course, nothing preventing market participants from writing their own CfDs with VRE plant and other plant capacity in order to mitigate looming (or actual) hedge shortages. Indeed, in the long run this may be the only way by which hedge shortages can be remedied in an energy-only market setting with government-initiated CfDs, noting that the end result is excess capacity. But a wide-ranging program of government-initiated CfDs is likely to *crowd-out* on-market transactions —neither of these alternatives would appear to produce efficient results.

In short, a well-intentioned wide-ranging program of government-initiated CfDs can be expected to create hedge contract shortages and at best raise forward prices above the efficient level, and at worst unintentionally foreclose 2nd tier retailers while replacing well-functioning forward markets with quasi-market participants who are indifferent to the outcomes facing final electricity consumers—all of which must ultimately harm welfare. Any response by market participants that might otherwise logically follow risks being *crowded-out* by ongoing government CfDs, or may induce excess capacity. How these shortages emerge can be demonstrated quantitatively, and this forms the focus of Sections 2 and 3.

2. Materials and Methods

In order to analyze the impact of government-initiated CfDs on forward markets, a power system simulation model (NEMESYS) has been used [24]. NEMESYS is a dynamic, security-constrained, unit-commitment model with 30-min resolution and price formation based on a uniform, first-price auction clearing mechanism consistent with the NEM design. As with [47], this partial equilibrium model assumes perfect competition, transmission and ramp-rates, free entry and exit to install any combination of (indivisible) capacity that satisfies differentiable equilibrium conditions. The modeled power system commences with a conventional thermal plant stock, with scenarios derived by exogenously determining progressively higher levels of VRE output (i.e., by way of policy). And as with [35], the focus of simulations is half-hour resolution over a single year. Model logic, drawn from [24], is as follows.

2.1. NEMESYS Model Logic

Generation plant technologies and associated plant costs are essential inputs to the security-constrained unit commitment model. Two key variables for each generation technology are (unit) marginal running costs v^i and plant fixed and sunk costs, φ^i. These inputs have been derived from a power plant cost model (PF model—the logic of which appears in Appendix A). The PF model derives generation technology (generalized) long run marginal costs $p^{i\varepsilon}$ and total revenues including normal profit R^i for a given level of output o^i.

$$\left(v^i \cdot o^i\right) + \varphi^i \equiv R^i \,\middle|\, R^i = p^{i\varepsilon} \cdot o^i \tag{1}$$

NEMESYS orders plant capacity and dispatches the fleet of power generating units to satisfy security constraints and differential equilibrium conditions given specified plant options available.

Let H be the ordered set of all half-hourly periods.

$$n \in \{1 \ldots |H|\} \wedge h_n \in H \tag{2}$$

Let E be the set of all electricity consumers in the model.

$$k \in \{1 \ldots |E|\} \wedge e_k \in E \tag{3}$$

Let $C_k(q)$ be the valuation that consumer segments are willing to pay for quantity q MWh of power. The model assumes that demand in each period n is independent of other demand periods. Let q_{nk} be the metered quantity consumed by customer e_n in each period h_k expressed in MWh.

Let Ψ be the set of existing installed power plants and available augmentation options for each relevant scenario.

$$i \in \{1 \ldots |\Psi|\} \wedge \psi^i \in \Psi \tag{4}$$

As outlined in Equation (1), let φ^i be the fixed operating and sunk capacity costs and v^i be the (unit) marginal running cost of plant ψ^i respectively. Let $\overline{o^i}$ be the maximum continuous rating of power plant ψ^i. Power plants are subject to scheduled and forced outages. $F(n,i)$ is the availability of plant ψ^i in each period h_n. Annual plant availability is therefore:

$$\sum_{j=0}^{|P|} F(n,i)\ \forall \psi^i \tag{5}$$

Let o_n^i be the quantity of power produced by plant ψ^i in each period h_n.

Objective Function

Optimal welfare will be reached by maximizing the sum of producer and consumer surplus, given by the integral of the aggregate demand curve less power production costs. The objective function is therefore expressed as:

$$Obj = \sum_{n=1}^{|H|}\sum_{i=k}^{|E|}\int_{q=0}^{e_k} C_k(q)dq - \sum_{n=1}^{|H|}\sum_{\psi=1}^{|\Psi|}\left(o_{\psi^i}\cdot v^i\right) - \sum_{\psi=1}^{|\Psi|}\left(\varphi^i\right) \tag{6}$$

Subject to

$$\sum_{i=1}^{|E|} q_{kn} \leq \sum_{\psi=1}^{|\Psi|} o_{\psi^i}\hat{}0 \leq o_n^i \leq F(n,i)\hat{}0 \leq o_n^i \leq \overline{o^i}$$

2.2. Model Inputs

Salient features of the present modeling exercise are as follows. There are five generation plant technologies available for deployment in the power system, including incumbent coal plant, combined cycle gas turbines (CCGT), open cycle gas turbines (OCGT), and VRE plant, specifically wind and Solar PV. Coal, CCGT and OCGT plant are all modeled as balance sheet-financings (gearing ca. 30%–36% to maintain BBB credit metrics). In contrast, VRE plant are assumed to be project financed (ca. 65%–70% debt), all of which are assumed to be underpinned by government-initiated CfDs (i.e., there are no on-market PPAs).

Table 1 sets out generation plant technology cost input assumptions and Table A1 in Appendix A outlines all relevant corporate and project financing assumptions. When combined, these inputs provide the data necessary to produce generalized estimates of average total cost (for incumbent coal plant) and generalized long run marginal costs (for new entrant plant). Crucially, with VRE plant—a strict annualized cost/price is used in all modeling; i.e., there is no "two-step pricing" assumed. Recent VRE transactions in the NEM have been struck in the low-$50s/MWh, and appear to reflect either of (i) unique sites with excellent resource and network connection characteristics; or more commonely, (ii) what [48] have labelled two-step pricing. Two-step pricing involves a low cost 15-year PPA followed by *assumed* elevated market prices in years 16–30 based on externally-provided market forecasts. The combination of the low contracted PPA price (years 1–15) and high expected future spot prices (years 16–30) appear to collectively meet threshold equity returns, but the implication of two-step pricing is that average total cost of such projects is higher than recent PPA pricing suggests.

Table 1. Plant cost assumptions.

Technology	Installed Capacity	Generating Units	Unit Heat Rate	Capacity Factor	Auxillary Load	Capex	Unit Fuel Cost	Fixed O&M Cost	Variable O&M	Capital Works
	(MW)	(MW)	(kJ/kWh)	(%)	(%)	($/kW)	($/GJ)	($/MW/a)	($/MWh)	(%)
Incumbent Coal	1000	2	10,000	50–90	7.1	1486	3.00	50,500	4.00	0.25
CCGT	400	1	6930	30–70	3.0	1500	8.50	10,000	7.00	0.05
OCGT	500	2	11,300	1–10	1.0	1050	10.00	7000	10.00	0.05
Wind	450	118	-	39	0.0	1975	-	45,000	3.00	0.05
Solar PV	100	-	-	26	0.0	1550	-	30,000	-	0.05

2.3. Plant Costs

Plant cost results are presented in Figure 1. These results are a high-resolution LCoE, refined through co-optimized debt-finance and taxation variables. Note in Figure 1 that Incumbent Coal plant has a generalized average total cost of $64/MWh comprising fuel ($30/MWh), Operations & Maintenance (O&M) of ($8.71/MWh), debt ($4.17/MWh), taxation ($5.44/MWh) and equity ($15.74/MWh). The OCGT cost structure focuses on the *"carrying cost"* of capacity (at $14/MWh), and has a marginal running cost of $123/MWh (including variable O&M). Finally, results in Figure 1 are based on static capacity factors, but in the NEMESYS model, plant costs arise on a dynamic basis with capacity factors determined by the dispatch necessary to meet final demand.

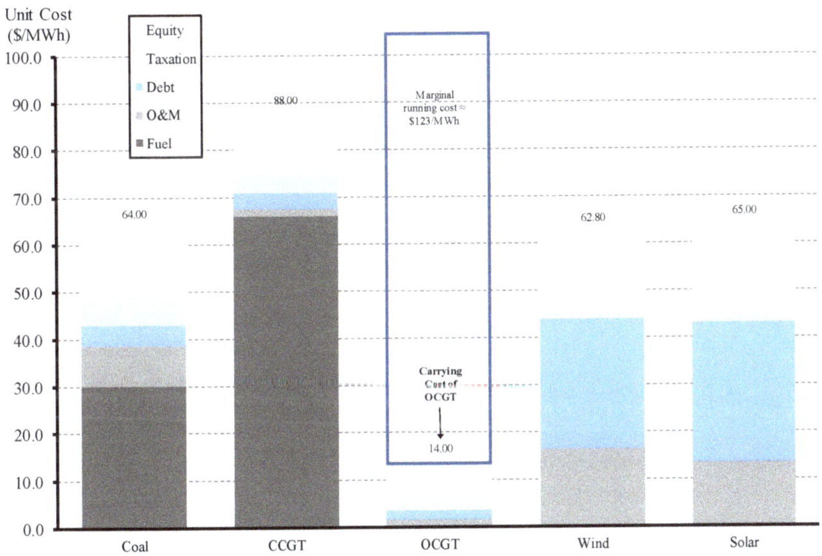

Figure 1. Generalized plant costs.

3. Model Results and Discussion

3.1. Overview of Power System

The NEMESYS Model has been populated with the plant cost results from Section 2.3, and half-hour load data (using Queensland power system final demand from 2016). From this, multiple scenarios are simulated. A long run (own-price) demand elasticity of −0.30 is applied to all variation cases [49–51]. To keep modeling results tractable, the power system is modeled as a single, non-interconnected gross-pool energy-only market. Recall also that the level of government-initiated CfDs are exogenously determined and designed to achieve a certain VRE market share. The base scenario is calibrated with

0% VRE plant (i.e., the power system commences as a thermal system with zero renewable plant), and variation scenarios span up to 40% VRE market share.

Consistent with Equation (6), the objective of the power system model is to minimize resource costs and maximize consumer welfare whilst meeting a reliability constraint of no more than 0.002% unserved energy (i.e., the NEM's long-standing reliability criteria). To assist interpretation of subsequent results, critical outputs for the two bookend scenarios (i.e., 0% and 40% VRE market share) are presented in Table 2.

Table 2. Overview of key model results.

VRE Market Share	0%	40%	Change
Energy Demand (GWh)	54,717	56,386	1669
Maximum Demand (MW)	9118	9393	275
Demand Elasticity		−0.30	
Plant Capacity			
Coal (MW)	6720	4200	−2520
CCGT (MW)	400	1200	800
OCGT (MW)	3000	3750	750
Wind (MW)	0	3797	3797
Solar (MW)	0	2711	2711
Supply of Primary Hedges (MW)	9100	7900	−1200
Power System Cost ($/MWh)	$78.74	$82.37	$3.63
Underlying System Price * ($/MWh)	$82.53	$65.84	−$16.68
CO_2 Emissions (Mt)	53.4	32.0	−21.4
Imputed Carbon Price ($/t)	n/a	$25–$35/t	
Unserved Energy (%)	0.001	0.001	

* Dispatch-weighted price, excludes any implicit or explicit value of CO_2.

Note from Table 2 that the single-region power system has an initial final Energy Demand of 54,717 GWh per annum with peak or Maximum Demand of 9118 MW. Given demand elasticity of −0.30 and the variation in wholesale prices with 40% VRE market share, energy demand and maximum demand rise to 56,386 GWh and 9393 MW, respectively. The opening plant stock is dominated by 6720 MW of coal plant, and in order to meet the reliability constraint (given plant outages) a reserve plant margin of ~11% is necessary. In order to meet a 40% VRE market share, about 3800 MW of wind and 2700 MW of solar PV capacity is added to the plant stock, and given optimal conditions, 2520 MW of coal plant retires. To meet reliability constraints, 800 MW of CCGT plant and 750 MW of OCGT plant is added—albeit operating at relatively low annual capacity factors.

Note also from Table 2 that the power system commences with a dispatch-weighted spot market price of $82.53/MWh and a system average cost of $78.74/MWh. With VRE market share of 40%, total system cost increases to $82.37/MWh whereas the underlying power system price falls to $65.84/MWh—components of this gap being underwritten by government-initiated CfDs with an imputed CO_2 value in the range of $25–$35/t. Note that each technology earns a different dispatch-weighted price according to their production profile. In any scenario, OCGT plant earns the highest average spot price (i.e., they increase output at times of high spot prices, and turn off in low spot price periods). Wind plant on the other hand earns a lower average spot price than thermal plant (i.e., coal, CCGT, and OCGT plant) since the stochastic production profile of Australian wind generators has a slight off-peak bias, and in addition, as more wind plant is added to the power system it has an impressing effect on wind's *earned price* [6,24,35]. It is worth noting that the dispatch-weighted price of Solar PV plant falls below wind once the technology reaches 7% market share due to the relatively tight correlation amongst all solar PV plant output.

Modeled power system CO_2 emissions fall from 53.4 Mt pa to 32.0 Mt pa between the 0% and 40% VRE market share scenarios. Unserved energy in both scenarios is ~0.001% of total load, and thus the plant stock in both scenarios meet the reliability criteria.

3.2. Spot Market Results

Recall that the objective of the current exercise is to analyze the implications of a wide-ranging CfD program on the functioning of an energy-only market, and the hedge market in particular. To model a wide-ranging CfD policy, the installed capacity of wind and solar PV is exogenously increased such that the market share of VRE plant progressively rises to 40% (with model results simulating in 5 percentage point increments). CfD's are assumed to be initiated by way of auction in order to minimize LCoE outcomes. Based on modeled results, this means that wind and solar PV dominate entry.

Given perfect entry, exit, and exogenously determined levels of government-initiated CfD's to drive VRE market share, all scenarios are implicitly "long-run dynamic" as measured by the time taken for the capital stock to adjust, rather than specifying a notional time period per se as in [24,35]. The thermal plant stock is therefore assumed to adjust perfectly in that VRE plant entry is accommodated by coal plant retirements ("to make room", and in line with coal plant financial distress arising from policy-induced VRE plant entry), while moderate levels of CCGT and OCGT plant enter to ensure reliability constraints are met given the intermittent nature of wind and solar PV.

Figure 2 presents the dynamic supply-side adjustment of plant capacity given a policy objective of a 40% VRE market share by way of government-initiated CfDs. Notice in Figure 2 that coal plant capacity reduces from 6720 MW to 4200 MW. Gas-fired capacity increases; CCGT plant commences at 400 MW and rises to 1200 MW while OCGT capacity commences at 3000 MW and rises by a further 750 MW. Consistent with the CfD policy objective, VRE plant increases from 0 MW to 6500 MW, comprising 2700 MW of solar (~15% market share) and 3800 MW of wind (~25% market share).

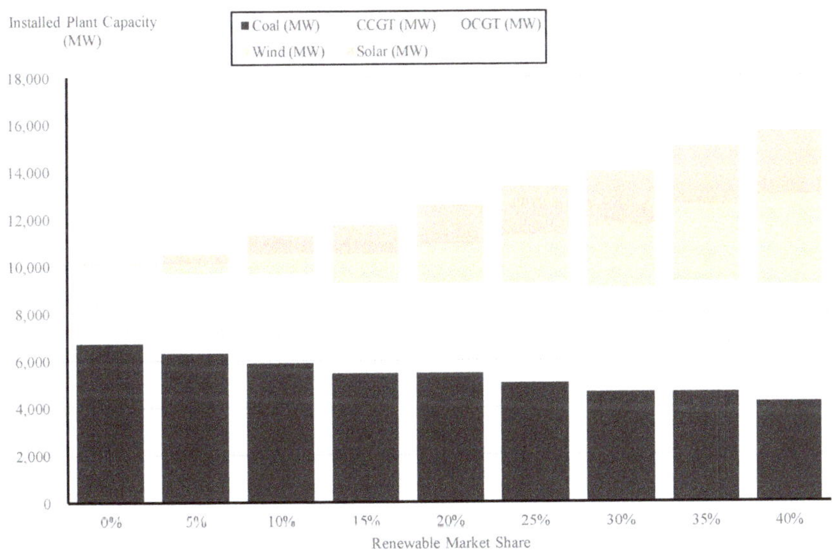

Figure 2. Installed plant capacity (MW).

Figure 3 shows how plant output adjusts with the addition of wind and solar PV plant gradually reaching a 40% market share. Given limited variations in final demand, supply-side adjustment from VRE plant entry primarily comes from the exit of coal plant. Gas-fired plant capacity continuously rises (see Figure 2) but output responds to both the entry of VRE plant and the exit of coal plant. That is, gas generation (presented as a bar series on the LHS y-axis, and a line series on the RHS y-axis with higher axis resolution to emphasize the variation) provides system flexibility, rising as coal plant exits, and falling as VRE plant increases in output.

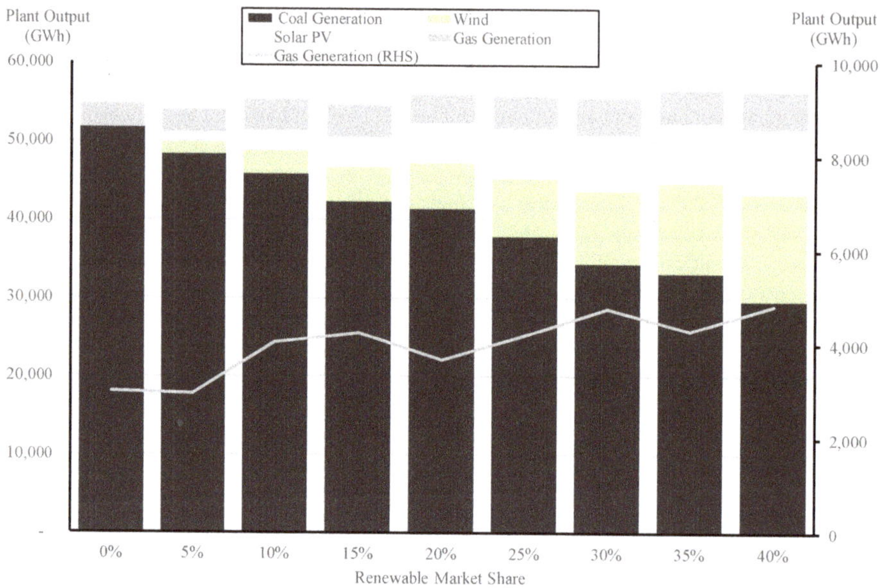

Figure 3. Plant output (GWh).

With perfect plant entry and exit, NEMESYS model results confirm the CfD policy objective can be met with the power system's spot market producing tractable results. However, what such modeling fails to reveal is an emerging structural shortage in the power system's financial market, viz. the market for forward hedge contracts used by participants to manage financial risk.

3.3. *Financial (Contract) Market Results*

Identifying the aggregate supply of hedge contracts within a NEM region is inherently difficult because in a well-functioning power system financial market, there are cross-border trades, and, more than just asset-backed portfolio managers on the sell-side. Proprietary (i.e., non-asset-backed) traders can add substantially to market depth and liquidity. The anonymity of trade makes this notoriously difficult to model (i.e., risk management of market exposures also arises from activity in tangential markets such as the market for weather derivatives, which is similarly excluded from the present analysis).

However, modeling a structural shortage of forward hedge contracts in a single (i.e., non-interconnected) region by is an easier task. The reason for this is that proprietary traders, who add to forward market liquidity, *"do not appear out of thin air"*. A necessary condition for proprietary trading is an inherent level of forward market liquidity to begin with. Consequently, hedge market analysis need only focus on contract supply from asset-backed traders, which is bounded.

To be perfectly clear on this, if a forward market is illiquid, non-asset-backed traders cannot be relied upon to enter and make-up any shortfall arising from asset-backed traders. Indeed, if forward liquidity begins to contract, proprietary traders will close out their positions and exit the market, thus accelerating any decline in liquidity. The reason for this is axiomatic; as [5] explain, holding-times of various securities is *strongly correlated* to market liquidity. Again to emphasize, in a forward market characterized by falling liquidity, proprietary traders can be expected to *close-out* positions, not open new positions, to avoid being caught with unwanted inventory.

Consequently, understanding the total *primary supply* (i.e., "primary issuance") of asset-backed forward contracts provides a basis for identifying inherent market liquidity. If the underlying supply or primary issuance of swaps and caps (nominally represented by coal and gas plant respectively) are sufficient relative to maximum demand, then the conditions necessary for trade at "multiples of physical" would appear to exist.

Conversely, if an absolute shortage of *primary issuance* hedge contracts progressively emerges, then total market liquidity will decelerate, first through the loss of primary supply (i.e., exiting dispatchable coal plant which no longer offers hedges), and then as this initial loss of liquidity becomes obvious, through the progressive exit of proprietary traders as they seek to avoid being caught with unwanted inventory. The combination of this may lead to total financial markets turnover *less than 100%*, implying some positions are virtually un-hedgeable from within the energy market.

In the analysis that follows, the plant stock outlined in Figure 2 is notionally separated into three rival oligopoly generator portfolios (two at 3520 MW, one at 3080 MW). In the NEMESYS model, individual generation plant availability is determined according to a stochastic binomial distribution with half-hour resolution given plant forced outage rates of ~5%–6%. These generating unit-level data were collated and assembled into joint probability duration curves for each of the three generator portfolios, and from there a 90th percentile confidence limit was identified as the maximum credible supply of asset-backed hedges in a manner consistent with the methodology in [24]. In essence, some plant capacity is withheld from the hedge market for self-insurance against forced plant outages, and to retain some nominal exposure to spot price outcomes. The modeled results that emerge are in turn consistent with the applied hedge market research findings in [20].

Results for Generation Portfolio #1 and Generation Portfolio #3 are presented in Figure 4. Notice that for the 3520 MW Generation Portfolio #1 (and by implication, Generation Portfolio #2 which has an identical plant portfolio) the total potential supply of hedges at the 90th percentile is about 3150 MW, and for the 3080 MW Generation Portfolio #3, total potential supply of hedges is about 2700 MW.

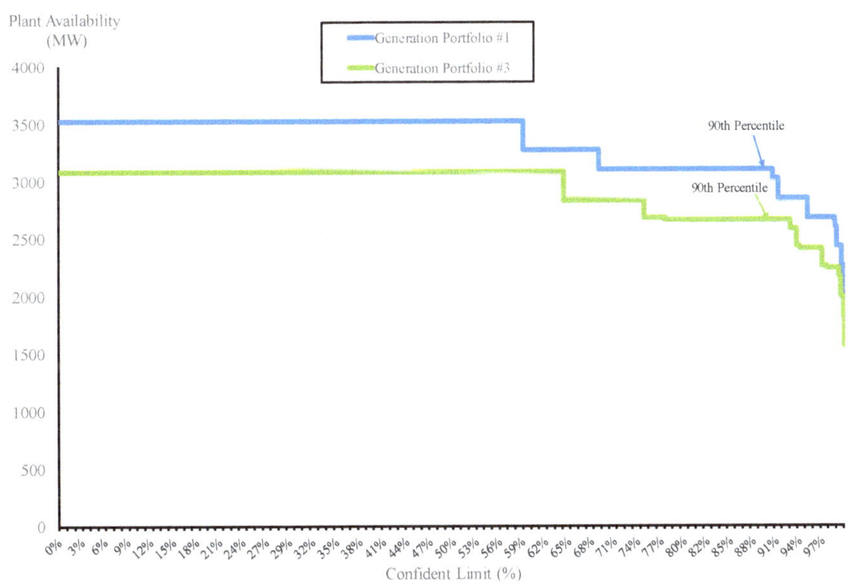

Figure 4. Primary supply of hedge contracts at 0% renewable market share.

In the model, as VRE plant enters via government-initiated CfDs, various coal generating units exit due to merit-order effects and financial distress. As coal plant exit, some level of gas-fired generation plant enters but as Figures 2 and 3 indicate, the aggregate coal and gas-fired fleet form a shrinking resource vis-à-vis aggregate final demand. Consequently, when the modeling process is undertaken for each of the three Generation Portfolios on a dynamic basis (i.e., as outlined in Figure 2 for VRE = 0%–40%), *primary issuance* hedge supply begins to decline, and this accelerates as VRE plant entry (by way of CfDs) approaches 40%. The dynamic analysis is presented in Figure 5, which reveals a growing structural shortage of primary issuance hedge contract capacity:

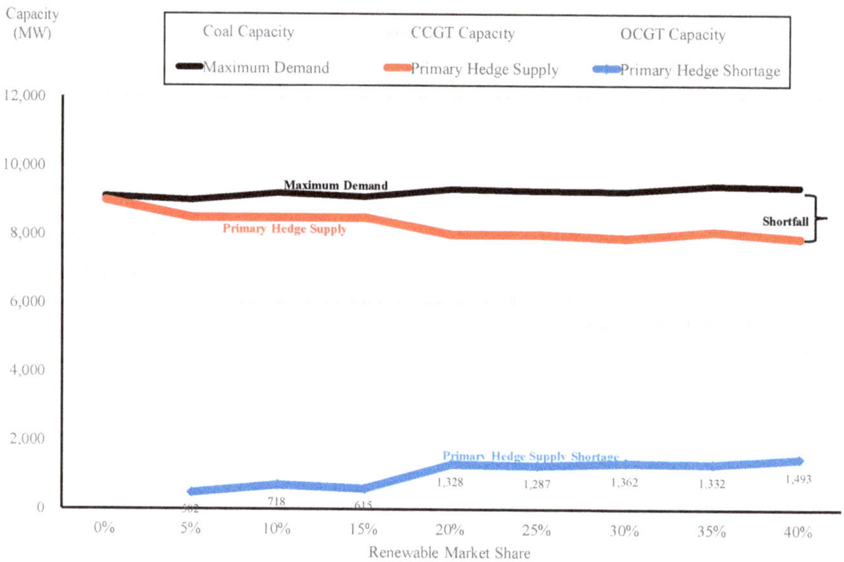

Figure 5. Primary supply of hedge contracts vs maximum demand (0%–40% VRE).

In Figure 5, the x-axis measures VRE plant market share noting that all plant has been facilitated by government-initiated CfDs (i.e., there are no *on-market* PPAs). The solid black line series depicts maximum demand, and the solid red line series presents aggregate *primary issuance* hedge contracts. The gap between the black and red lines highlights the magnitude of any hedge shortfall, also illustrated by solid blue line series—culminating in a hedge market shortfall of ~1500 MW or 16% of final market demand at 40% VRE market share. Note that even with a 5% VRE market share, the impact of government-initiated CfDs produces a non-trivial hedge shortfall *if* the thermal plant stock adjusts perfectly.

In Figure 5, the dynamic change in plant capacity is also captured by the area chart (grey for coal, dark blue for CCGT plant, and light blue for OCGT plant—essentially a reproduction of Figure 2). In response to the wide-ranging policy of government-initiated CfDs, coal plant contracts from 6700 MW to 4200 MW, while CCGT and OCGT plant capacity expands by 800 MW and 750 MW, respectively. Note that overall there is a net loss of dispatchable plant, and when combined with the extraction of hedge contract capacity from government-initiated CfDs, combines and drives the shortage of *primary issuance* hedge contracts.

3.4. Hedge Contract Shortages in an Energy-Only Market

The quantitative analysis in Figure 5 in particular revealed that pursuing a wide-ranging program of government-initiated CfDs is likely to produce an "unstable zone" in the forward market for hedge contracts. That is, while the spot market is consistently able to reach equilibrium for any level of

VRE output up to 40% market share (given certain dispatch constraints), with government-initiated CfDs the forward hedge market becomes increasingly intractable as thermal plant exits and adjusts. Remaining coal and gas plant are unable to originate sufficient hedge contracts in their own right—as Figure 5 illustrates (recall that coal and gas generators do not hedge 100% of installed capacity due to outage rates and their own desire to maintain some nominal exposure to spot price).

Once a government initiates a wide-ranging program of CfDs, it will have the effect of adding capacity to the spot market which in the short run will lower prices and force coal plant out (consistent with underlying policy objectives), but in the long run will extract 100% of the CfD plant output from the power system's financial market. Crucially, VRE projects cannot sell their output twice in the forward market.

Consequently, results suggest that a wide-ranging program of government-initiated CfDs is not compatible with the NEM's energy-only market design. On the contrary, such a policy is likely to collide with the systemic stability of the market. NEM market participants operate in one of the world's most volatile commodity markets, and access to forward hedge contracts is essential for managing operating risk exposures to sustained critical event price spikes.

Hedge shortage events in energy-only markets with a high VoLL are far more than theory. The South Australian (SA) region of the NEM was known to enter an episode of hedge contract shortages (i.e., hedge contracts <100% of physical) in 2016 and 2017 soon after the final SA coal plant exited (i.e., Northern Power Station during 2016). The surprising sophistication, and level of energy market literacy now displayed by large industrial customers in South Australia explains how the SA market temporarily adjusted. In the short run when hedge contract prices and premiums rose sharply, contract volumes and premiums appear to have been allocated and rationed across the SA power system given segment-level elasticities of demand. That is, prices in the residential consumer segment rose in line with elevated contract premiums. Through discussions with senior NEM policymakers and various industrial customers in SA, it would appear that hedge market shortages were largely absorbed by industrial customers in the short run, with a typical strategy being to secure some minimum level of hedging and run the balance of manufacturing load to the spot market (while keeping a close eye on exposed load to pre-dispatch prices).

The SA hedge market shortage was ultimately caused by the sudden and uncoordinated exit of coal plant—replacement hedge capacity is slowly being rebuilt through various new entrant VRE plant, battery storage, and gas turbines. But if the new VRE entrants were underwritten by way of government-initiated CfDs, it is not immediately obvious how such a shortfall could be rebuilt.

A wide-ranging program of government-initiated CfDs may adversely impact the residential and SME business market. The effect of extracting hedge contract capacity from the forward market may, in time, weigh heavily on retail competition. Large vertical retailers can be expected to manage positions using a combination of physical plant and forward markets—and these large utility firms have the financial capacity to allocate resources seamlessly between the two. But 2nd tier non-integrated retailers do not have the same financial resources and may in the event be inadvertently foreclosed by a wide-ranging government policy of CfDs as financial market liquidity deteriorates and hedge shortages bind. At this point, retail-level consumer pricing can also be expected to be adversely impacted through less competition.

And, un-hedgeable positions may introduce risks to the systemic stability of power market more generally. If a sufficiently large utility experienced financial distress due to excessive exposure to VoLL events because they were not able to allocate resources between physical plant and forward markets quickly enough, it could lead to cascading failures across the power market economy; unlike Australian financial institutions which can access lender of last resort facilities with the Commonwealth Government, there is no centralized financial backstop for Australia's organized energy markets.

3.5. Are Hedge Shortages Inevitable with Rising VRE Plant?

Modeling results in Section 3.3 explicitly assume VRE plant enters exclusively by way of government-initiated CfDs, and that these CfDs crowd-out on-market (bilateral) private sector PPAs. The result was a shortage of hedges. A logical question that follows is whether hedge contract shortages are inevitable in a world of rising VRE market share and coal plant exit regardless of how entry occurs—whether by government-initiated CfDs or by on-market PPAs amongst NEM market participants? The short answer is *no*. Results in Figure 5 would look different if VRE plant was able to provide its output, in spite of intermittency, into the hedge books of market participants (as is the case under a certificate-based renewable portfolio standard). That is, participants and portfolio traders can synthetically (or physically) reconstruct firm hedge contract positions by combining run-of-plant PPAs with dispatchable plant (e.g., incumbent coal, and existing or new, CCGT, OCGT, pumped-hydro, battery storage) and rely on gains from exchange in the spot market to balance positions on a risk-adjusted basis. Whether forward hedge market demand ultimately clears without materially higher reserve plant margins in the long run is an open question.

3.6. Not A Short-Run Problem

The use of government-initiated CfDs will *not* create hedge contract shortages in the short run. It is a long run problem [24,35]. If thermal plant fails to exit or thermal plant capacity remains above efficient levels, shortages in the hedge market are unlikely to appear. Indeed, in the short run government-initiated CfDs may well result in consumers benefiting from a surplus of hedge contract capacity (i.e., if thermal plant does not exit they are still available to supply hedge contracts), and short-run prices will be lower reflecting merit-order effects of adding VRE plant to the power system.

However, and to be clear, as coal plant exits the opposite occurs as evidenced by the South Australian NEM region in 2016–2017. Thermal plant must exit due to inevitable financial distress caused by VRE plant entry at-scale. And the exit of coal plant causes spot prices to rebound. Furthermore, in such a CfD scenario spot prices will rebound just as hedge contract shortages appear; thus, consumers would be unable to hedge against the very reason for hedging in the first place—viz. the risk of sharply rising wholesale market prices. And in the modeling results in Section 3.3, the reason consumers cannot fully hedge in forward markets is because hedge capacity has been extracted through a wide-ranging program of government-initiated CfDs.

3.7. Validity of Government-Initiated CfD Reallocation Mechanisms

One reviewer queried whether, to remain consistent with the energy-only market design, government-initiated CfD "receipts and payments" could be allocated to energy retailers on a pro-rata basis as is done in Great Britain's net pool market. The line of reasoning would be that energy retailers could impute a hedge position based on the output of the VRE plant portfolio contracted by the government.

This would represent a poor outcome. When energy retailers enter into bilateral PPAs under a renewable portfolio standard or emissions obligation, the terms and conditions are invariably non-vanilla and designed to suit the allocation of risk between VRE producer and energy retailer, and, with a known project, project location and real-time project-specific output and associated constraints. Under a government reallocation, the shareholders of energy retailers would be forced to digest a *"blind hedge book"* and simultaneously lose control over the timing, cost, location, and magnitude of VRE plant commitments. How this could add to market efficiency is not immediately obvious.

Another reviewer queried whether governments could initiate CfDs, and then on-sell the CfD on a project-by-project basis in the open market at market rates (i.e., effectively creating a secondary market to recycle CfD's as a PPA, and in the process crystalize taxpayer profits or losses on each transaction). This is a plausible solution to the modeling results outlined in Section 3.3. But apart from adding a layer of transaction costs into the NEM and using scarce government balance sheet resources for needless intermediation, such a policy strategy implicitly presumes a central government agency will purchase more efficiently than an entire energy market comprising no less than 50 highly sophisticated organizations. A key reason energy markets exist is government failures in the central planning of power systems. The better view is for government to set the policy objective and policy targets (i.e., renewable portfolio standard or emissions obligation), and let the market deliver the policy objective and allocate the various power system financial and operating risks to those best able to manage them.

4. Conclusions

Australia has an international obligation to reduce its CO_2 emissions under the Paris Agreement, and Australia's NEM is a CO_2-intensive power system by OECD standards [52]. In this article, a requirement to lower the CO_2 intensity of the NEM has been taken as an exogenous policy constraint in the form of a 40% VRE market share. The purpose of this modeled constraint was not to test optimal levels or the composition of VRE plant, nor the structural adjustment task and welfare implications of exiting coal plant at-scale. The specific line of inquiry was to determine whether the wide-spread use of *off-market* government-initiated CfDs to achieve higher VRE and lower CO_2 emissions is compatible with the NEM's energy-only market design.

Australia's renewable energy target (i.e., renewable portfolio standard) required that energy retailers meet a 20% renewable market share by 2020. The certificated scheme will be comfortably met—by 2018 the NEM had 8000 MW of new VRE plant (5000 MW wind, 2800 MW solar) and a further 5500 MW was under construction as at the start of 2019 (2500 MW wind and 3000 MW solar). On-market transactions had delivered more than $20 billion of investment without any need for government involvement or intermediation beyond setting the policy framework. The market was given a target, and market participants (i.e., utilities, investors, and customers) delivered the capacity.

Used carefully, CfDs present policymakers with a reliable tool which can be used to overcome an array of market failures, including those associated with missing or incomplete markets (including emergency plant for security of supply reasons, certain positive, or negative externalities including CO_2 emissions, R&D and externalities arising from first-of-a-kind commercialization investments). In the NEM, CfDs have been used selectively and effectively by state governments to "prime" emerging markets, navigate Commonwealth Government policy discontinuity, with material on-market transactions following. The Australian Capital Territory government CfDs pioneered nominal-price transactions, the Queensland government's CfDs led to more than 1900 MW of follow-on solar PV projects, and the SA government's semi-CfD for battery storage in has since resulted in more than a dozen battery projects either under active development or commitment. From a project execution perspective, the effectiveness of government-initiated CfDs are unquestionable.

But in an energy-only market setting, government-initiated CfDs must be used judiciously because they introduce "quasi-market participants" who do not respond to spot market signals, and do not participate in forward markets at all. Quasi-market participants are indifferent or substantially immune from future outcomes in spot and forward markets. This can result in plant entry that is poorly timed, poorly sized, poorly located, and above all, poorly motivated to respond to the electricity and frequency control ancillary service spot price signals which keep the power system operating in a stable manner.

CfD plant also benefits from otherwise unachievable credit metrics owing to a taxpayer-financed and credit-wrapped CfD instrument—with the risks transferred to taxpayers. Prima facie this may deliver lower project specific LCoE's, but this is only because risk has been transferred to taxpayers (which is not costless), and whether this is an efficient allocation and use of government balance sheet capacity is, in my opinion, questionable. A wide ranging program of government-initiated CfDs can be expected to crowd-out on-market rival merchant/bilateral investments.

If there is an upside to the present analysis, it is that the number of alternate policy instruments available to government to achieve policy objectives has expanded very rapidly [53]. A wide array of policy instruments exist to deal with the market failures which CfDs are intended to remedy; renewable energy policy objectives can be achieved by an emissions obligation or well-designed renewables portfolio standard [48]; the need for emergency capacity can be (and in the NEM recently has been) dealt with by establishing minimum exit notification periods for plant intending to exit the system; resource adequacy (i.e., adequate plant capacity including an appropriate reserve plant margin) can be maintained by ensuring the level of VoLL remains appropriate or by pursuing reliability obligations if this becomes necessary. All of these options work with an energy-only market design, including the forward market for contracts.

Throughout most of 2018, Australian policymakers developed a policy known as the National Energy Guarantee which had two embedded policy mechanisms for energy retailers to comply with; (i) an emissions obligation which was consistent with Australia's international CO_2 commitments under the Paris Agreement, and (ii) a reliability obligation which was consistent with the NEM's reliability criteria and was designed to ensure resource adequacy. The former was designed to encourage hedge contract activity with new renewable projects, and the latter was designed to be acquitted via ensuring adequate forward contracts were committed—both mechanisms were thus designed to add to liquidity rather than detract from it.

In contrast, as quantitative results and analysis in this article explain, a wide-ranging program of government-initiated CfDs can be expected to impair market efficiency in an energy-only market setting. While adding to buy-side liquidity, government-initiated CfDs replace on-market transactions and thus subtract from sell-side liquidity, and this matters in loosely-interconnected energy-only markets because as coal plant exits, *primary issuance* hedge contract shortages become predictable. Shortages in the forward markets may harm consumer welfare by raising contract premiums—the primary input into consumer tariffs—and by forcing the most price-elastic industrial customers into accepting spot market exposures, which at best disrupts manufacturing efficiency. Further, CfD-driven hedge market shortages may unintentionally foreclose non-integrated 2nd tier retailers—deeming such a program of government-initiated CfDs to be (unintentionally) anti-competitive. Consequently, the *National Energy Guarantee* or an equivalent suite of policies seems a better place for Australian policymakers to focus on.

Funding: This research received no external funding.

Conflicts of Interest: The author declares no conflicts of interest.

Abbreviations

The following abbreviations are used in this manuscript:

AUD	Australian Dollars
CCGT	Combined Cycle Gas Turbine
CfD	Contract for Differences
CO_2	Carbon Dioxide
LCoE	Levelized Cost of Electricity
NEM	National Electricity Market
OCGT	Open Cycle Gas Turbine
O&M	Operations & Maintenance
PV	Photovoltaic
RET	Renewable Energy Target
VoLL	Value of Lost Load
VRE	Variable Renewable Energy (i.e., wind and solar PV)

Appendix A

The purpose of the PF model is to produce plant cost estimates for various generating technologies. The PF model is essentially a dynamic, multi-period post-tax discounted cash flow optimization model which solves for multiple generating technologies, business combinations, and revenue possibilities through simultaneous convergent price, debt-sizing, taxation and equity return variables. These outputs are similar in nature to levelized cost estimates but with a level of detail beyond the typical LCoE model because corporate or project financing, credit metrics and taxation constraints are co-optimized. Model logic, derived in [48,54], is as follows:

Costs increase annually by a forecast general inflation rate (CPI). Prices escalate at a discount to CPI. Inflation rates for revenue streams π_j^R and cost streams π_j^C in period (year) j are calculated as follows:

$$\pi_j^R = \left[1 + \left(\frac{CPI \times \alpha_R}{100}\right)\right]^j, \text{ and } \pi_j^C = \left[1 + \left(\frac{CPI \times \alpha_C}{100}\right)\right]^j \tag{A1}$$

Energy output o_j^i from each plant (i) in each period (j) is a key variable in driving revenue streams, unit fuel costs and variable operations and maintenance costs. Energy output is calculated by reference to installed capacity k^i, capacity utilisation rate CF_j^i for each period j. Plant auxiliary losses Aux^i arising from on-site electrical loads are deducted.

$$o_j^i = CF_j^i \cdot k^i \cdot \left(1 - Aux^i\right) \tag{A2}$$

Electricity price for the ith plant $\left(p^{i\varepsilon}\right)$ is calculated in year one and escalated per Eq. (A1). Thus revenue for the ith plant in each period j is defined as follows:

$$R_j^i = \left(o_j^i \cdot p^{i\varepsilon} \cdot \pi_j^R\right) \tag{A3}$$

As outlined above, plant marginal running costs are a key variable and used extensively in *NEMESYS-PF*. In order to define marginal running costs, the thermal efficiency for each generation technology ζ^i needs to be defined. The constant term "3600" is divided by ζ^i to convert the efficiency result from % to kJ/kWh (i.e., the derivation of the constant term 3600 is: 1 Watt = 1 Joule per second and hence 1 Watt Hour = 3600 Joules). This is then multiplied by raw fuel commodity cost f^i. Variable operations and maintenance costs v^i, where relevant, are added which produces a pre-carbon short run marginal cost. Under conditions of externality pricing CP_j, the CO_2 intensity of output needs to be defined. Plant carbon intensity g^i is derived by multiplying the plant heat rate by combustion emissions $\dot{g}^i$ and fugitive CO_2 emissions $\hat{g}^i$. Marginal running costs in the jth period is then calculated by the product of short run marginal production costs by generation output ρ_j^i and escalated at the rate of π_j^C.

$$\vartheta_j^i = \left\{\left[\left(\frac{\left(\frac{3600}{\zeta^i}\right)}{1000} \cdot f^i + v^i\right) + \left(g^i \cdot CP_j\right)\right] \cdot o_j^i \cdot \pi_j^C \middle| g^i = \left(\dot{g}^i + \hat{g}^i\right) \cdot \frac{\left(\frac{3600}{\zeta^i}\right)}{1000}\right\} \tag{A4}$$

Fixed operations and maintenance costs FOM_j^i of the plant are measured in \$/MW/year of installed capacity FC^i and are multiplied by plant capacity k^i and escalated.

$$FOM_j^i = FC^i \cdot k^i \cdot \pi_j^C \tag{A5}$$

Earnings before interest tax depreciation and amortization (EBITDA) in the jth period can therefore be defined as follows:

$$EBITDA_j^i = \left[R_j^i - \vartheta_j^i - FOM_j^i\right] \tag{A6}$$

Capital costs $\left(X^i\right)$ for each plant i are overnight capital costs and incurred in year 0. Ongoing capital spending for each period j is determined as the inflated annual assumed capital works program.

$$x_j^i = c_j^i \cdot \pi_j^C \tag{A7}$$

Plant capital costs X_0^i give rise to tax depreciation (d_j^i) such that if the current period was greater than the plant life under taxation law (L), then the value is 0. In addition, x_j^i also gives rise to tax depreciation such that:

$$d_j^i = \left(\frac{X^i}{L}\right) + \left(\frac{x_j^i}{L+1-j}\right) \tag{A8}$$

From here, taxation payable $\left(\tau_j^i\right)$ at the corporate taxation rate (τ_c) is applied to $EBITDA_j^i$ less interest on loans $\left(I_j^i\right)$ later defined in (16), less d_j^i. To the extent $\left(\tau_j^i\right)$ results in non-positive outcome, tax losses $\left(L_j^i\right)$ are carried forward and offset against future periods.

$$\tau_j^i = Max\left(0, \left(EBITDA_j^i - I_j^i - d_j^i - L_{j-1}^i\right) \cdot \tau_c\right) \tag{A9}$$

$$L_j^i = Min\left(0, \left(EBITDA_j^i - I_j^i - d_j^i - L_{j-1}^i\right) \cdot \tau_c\right) \tag{A10}$$

The debt financing model computes interest and principal repayments on different debt facilities depending on the type, structure, and tenor of tranches. There are two types of debt facilities—(a) corporate facilities (i.e., balance-sheet financings) and (2) project financings. Debt structures include semi-permanent amortizing facilities and bullet facilities.

Corporate facilities involve 3- and 7-year money raised with an implied "BBB" credit rating. With project financings, two facilities are modeled. The first facility is nominally a 3-year bullet requiring interest-only payments after which it is refinanced with consecutive amortizing facilities and fully amortized over a 25-year period. The second facility commences with a tenor of 7 years as an amortizing facility, again set within a semi-permanent structure with a nominal repayment term of 25 years. The decision tree for the two tranches of debt is the same, so for the debt tranche where $T = 1$ or 2, the calculation is as follows:

$$if\ j\begin{cases} > 1, DT_j^i = DT_{j-1}^i - P_{j-1}^i \\ = 1,\ DT_1^i = D^i \cdot S \end{cases} \tag{A11}$$

D^i refers to the total amount of debt raised for the project. The split (S) of the debt between each facility refers to the manner in which debt is apportioned to each tranche. In the model, 35% of debt is assigned to Tranche 1 and the remainder to Tranche 2. Principal P_{j-1}^i refers to the amount of principal repayment for tranche T in period j and is calculated as an annuity:

$$P_j^i = \left(DT_j^i / \left[\frac{1-\left(1+\left(R_T^z + C_T^z\right)\right)^{-n}}{R_T^z + C_T^z}\right] \middle| z\begin{cases} = VI \\ = PF \end{cases}\right) \tag{A12}$$

In (A12), R_T is the relevant interest rate swap (3 yrs or 7 yrs) and C_T is the credit spread or margin relevant to the issued debt tranche. The relevant interest payment in the jth period $\left(I_j^i\right)$ is calculated as the product of the (fixed) interest rate on the loan by the amount of loan outstanding:

$$I_j^i = DT_j^i \times \left(R_T^z + C_T^z\right) \tag{A13}$$

Total debt outstanding D_j^i, total Interest I_j^i and total principle P_j^i for the ith plant is calculated as the sum of the above components for the two debt tranches in time j. For clarity, loan drawings are equal to D^i in year 1 as part of the initial financing and are otherwise 0.

One of the key calculations is the initial derivation of D^i. This is determined by the product of the gearing level and the overnight capital cost $\left(X^i\right)$. Gearing levels are formed by applying a cash flow constraint based on credit metrics applied by project banks and capital markets. The variable γ in the PF model relates specifically to the legal structure of the business and the credible capital structure achievable. The two relevant legal structures are vertically integrated (VI) merchant utilities (using "BBB" rated corporate facilities) and independent power producers using project finance (PF).

$$if\ \gamma \begin{cases} = VI,\ Min\left(\frac{FFO_j^i}{I_j^i}\right) \geq \delta_j^{VI} \wedge Min\left(\frac{FFO_j^i}{D_j^i}\right) \geq \omega_j^{VI} \forall\ j \left| FFO_j^i = \left(EBITDA_j^i - x_j^i\right) \right. \\ = PF, Min\left(DSCR_j^i,\ LLCR_j^i\right) \geq \delta_j^{PF},\ \forall\ j \left|\ DSCR_j = \frac{\left(EBITDA_j^i - x_j^i - \tau_j^i\right)}{P_j^i + I_j^i} \right| LLCR_j = \frac{\sum_{j=1}^{N}\left[\left(EBITDA_j^i - x_j^i - \tau_j^i\right)\cdot(1+K_d)^{-j}\right]}{D_j^i} \end{cases} \tag{A14}$$

The variables δ_j^{VI} and ω_j^{VI} are exogenously determined by credit rating agencies. Values for δ_j^{PF} are exogenously determined by project banks and depend on technology (i.e., thermal vs. renewable) and the extent of energy market exposure, that is whether a power purchase agreement exists or not. For clarity, FFO_j^i is "funds from operations" while $DSCR_j^i$ and $LLCR_j^i$ are the debt service cover ratio and loan life cover ratios. Debt drawn is:

$$D^i = X^i - \sum_{j=1}^{N}\left[EBITDA_j^i - I_j^i - P_j^i - \tau_j^i\right]\cdot(1+K_e)^{-(j)} - \sum_{j=1}^{N} x_j^i \cdot (1+K_e)^{-(j)} \tag{A15}$$

At this point, all of the necessary conditions exist to produce estimates of generalized long run marginal costs of the various power generation technologies. The relevant equation to solve for the price $\left(p^{i\varepsilon}\right)$ given expected equity returns (K_e) whilst simultaneously meeting the binding constraints of δ_j^{VI} and ω_j^{VI} or δ_j^{PF} given the relevant business combinations. The primary objective is to expand every term which contains $p^{i\varepsilon}$. Expansion of the EBITDA and tax terms is as follows:

$$\begin{aligned} -X^i + \sum_{j=1}^{N}\Big[\left(p^{i\varepsilon}\cdot o_j^i \cdot \pi_j^R\right) - \vartheta_j^i - FOM_j^i - I_j^i - P_j^i \\ -\left(\left(p^{i\varepsilon}\cdot o_j^i \cdot \pi_j^R\right) - \vartheta_j^i - FOM_j^i - I_j^i - d_j^i - L_{j-1}^i\right)\cdot \tau_c\Big]\cdot(1+K_e)^{-(j)} \\ - \sum_{j=1}^{N} x_j^i \cdot (1+K_e)^{-(j)} - D^i \end{aligned} \tag{A16}$$

The terms are then rearranged such that only the $p^{i\varepsilon}$ term is on the left-hand side of the equation: Let $IRR \equiv K_e$

$$\begin{aligned} &\sum_{j=1}^{N}(1-\tau_c)\cdot p^{i\varepsilon}\cdot o_j^i \cdot \pi_j^R \cdot (1+K_e)^{-(j)} \\ &= X^i \\ &- \sum_{j=1}^{N}\Big[-(1-\tau_c)\cdot\vartheta_j^i - (1-\tau_c)\cdot FOM_j^i - (1-\tau_c)\cdot\left(I_j^i\right) - P_j^i + \tau_c\cdot d_j^i \\ &+\tau_c L_{j-1}^i)\cdot(1+K_e)^{-(j)}\Big] + \sum_{j=1}^{N} x_j^i \cdot (1+K_e)^{-(j)} + D^i \end{aligned} \tag{A17}$$

The model then solves for $p^{i\varepsilon}$ such that:

$$\begin{aligned} &p^{i\varepsilon} \\ &= \frac{X^i}{\sum_{j=1}^{N}(1-\tau_c)\cdot P^{\varepsilon}\cdot o_j^i\cdot\pi_j^R\cdot(1+K_e)^{-(j)}} \\ &+\frac{\sum_{j=1}^{N}\left((1-\tau_c)\cdot\vartheta_j^i + (1-\tau_c)\cdot FOM_j^i + (1-\tau_c)\cdot\left(I_j^i\right) + P_j^i - \tau_c\cdot d_j^i - \tau_c L_{j-1}^i)\cdot(1+K_e)^{-(j)}\right)}{\sum_{j=1}^{N}(1-\tau_c)\cdot p^{i\varepsilon}\cdot o_j^i\cdot\pi_j^R\cdot(1+K_e)^{-(j)}} \\ &+\frac{\sum_{j=1}^{N} x_j^i\cdot(1+K_e)^{-(j)} + D^i}{\sum_{j=1}^{N}(1-\tau_c)\cdot p^{i\varepsilon}\cdot o_j^i\cdot\pi_j^R\cdot(1+K_e)^{-(j)}} \end{aligned} \tag{A18}$$

Table A1. Corporate Finance Assumptions.

Coal & Gas		Wind & Solar	
Debt Sizing Constraints		**Debt Sizing Constraints**	
FFO/I (times)	5	DSCR (times)	1.35
FFO/D (times)	3	LLCR (times)	1.35
Gearing Limit (%)	40.0	Gearing Limit (%)	70.0
		Default (times)	1.10
Corporate "BBB" Bonds		**Project Finance Facilities**	
Tranche 1 (Bullet) (Yrs)	5	Tranche 1 (Bullet) (Yrs)	5
Tranche 1 Refi (Yrs)	13–20	Tranche 1 Refi (Yrs)	13–20
Tranche 2 (Amort.) (Yrs)	7	Tranche 2 (Amort.) (Yrs)	7
Notional amortization (Yrs)	18–25	Notional amortization (Yrs)	18–25
BBB Bond Pricing		**Project Finance Facilities-Pricing**	
Tranche 1 (%)	3.60	Tranche 1 Swap (%)	2.55
Tranche 1 Margin (bps)	105	Tranche 1 Margin (bps)	200
Tranche 2 (%)	3.97	Tranche 2 Swap (%)	2.68
Tranche 2 Margin (bps)	129	Tranche 2 Margin (bps)	220
Tranche 1 (%)	3.60	Tranche 1 (%)	4.55
Tranche 2 (%)	3.97	Tranche 2 (%)	4.88
Tranche 1&2 Refi (%)	3.97	Tranche 1&2 Refi (%)	4.88
Post Tax Equity Coal (%)	12.0	Post Tax Equity (%)	10.0
Post Tax Equity Gas (%)	12.0		

References

1. Nong, D. A general equilibrium impact study of the Emissions Reduction Fund in Australia by using a national environmental and economic model. *J. Clean. Prod.* **2019**, *216*, 422–434. [CrossRef]
2. Simshauser, P.; Tiernan, A. Climate change policy discontinuity and its effects on Australia's National Electricity Market. *Aust. J. Publ. Admin.* **2019**, *78*, 17–36. [CrossRef]
3. Wild, P. Determining commercially viable two-way and one-way 'Contract-for-Difference' strike prices and revenue receipts. *Energy Policy* **2017**, *110*, 191–201. [CrossRef]
4. Bunn, D.; Yusupov, T. The progressive inefficiency of replacing renewable obligation certificates with Contracts-for-Differences in the UK electricity market. *Energy Policy* **2015**, *82*, 298–309. [CrossRef]
5. Goldstein, M.; Hotchkiss, E. Providing Liquidity in an Illiquid Market: Dealer Behavior in U.S. Corporate Bonds. July 2018. Available online: https://papers.ssrn.com/sol3/papers.cfm?abstract_id=2977635 (accessed on 5 May 2018).
6. Hirth, L.; Ueckerdt, F.; Edenhofer, O. Why wind is not coal: On the economics of electricity generation. *Energy J.* **2016**, *37*, 1–27. [CrossRef]
7. Hansen, C. Improving hedge market arrangements in New Zealand. In Proceedings of the 6th Annual National Power New Zealand Conference, Auckland, New Zealand, 30 March 2004.
8. Chao, H.; Oren, S.; Wilson, R. Re-evaluation of Vertical Integration and Unbundling in Restructured Electricity Markets. In *Competitive Electricity Markets: Design, Implementation, Performance*; Sioshansi, F.P., Ed.; Elsevier: Oxford, UK, 2008.
9. Finon, D. Investment risk allocation in decentralised markets: The need of long-term contracts and vertical integration. *OPEC Energy Rev.* **2008**, *32*, 150–183. [CrossRef]

10. Meade, R.; O'Connor, S. Comparison of long-term contracts and vertical integration in decentralised electricity markets. In *Competition, Contracts and Electricity Markets: A new perspective*; Glachant, J.M., Finon, D., De Hauteclocque, A., Eds.; Edward Elgar Publishing: Cheltenham, UK, 2011.
11. Finon, D. Investment and Competition in Decentralized Electricity Markets: How to Overcome Market Failure by Market Imperfections? In *Competition, Contracts and Electricity Markets: A new perspective*; Glachant, J.M., Finon, D., De Hauteclocque, A., Eds.; Edward Elgar Publishing: Cheltenham, UK, 2011.
12. Meyer, R. Vertical economies and the costs of separating electricity supply—A review of theoretical and empirical literature. *Energy J.* **2012**, *33*, 161–185. [CrossRef]
13. Nelson, J.; Simshauser, P. Is the merchant power producer a broken model? *Energy Policy* **2013**, *53*, 298–310. [CrossRef]
14. Newbery, D. Missing money and missing markets: Reliability, Capacity Auctions and Interconnectors. *Energy Policy* **2015**, *94*, 401–410. [CrossRef]
15. Newbery, D. Tales of Two Islands—Lessons for EU Energy Policy from Electricity Market Reforms in Britain and Ireland. *Energy Policy* **2016**, *105*, 597–607. [CrossRef]
16. Grubb, M.; Newbery, D. UK Electricity Market Reform and the Energy Transition. *Energy J.* **2018**, *39*, 1–25. [CrossRef]
17. Newbery, D. *Market Design*; EPRG Working Paper 515, University of Cambridge: Cambridge, UK, 2006.
18. Green, R. Market power mitigation in the UK power market. *Util. Policy* **2006**, *14*, 76–89. [CrossRef]
19. Chester, L. The conundrums facing Australia's National Electricity Market. *Econ. Pap.* **2006**, *25*, 262–277. [CrossRef]
20. Anderson, E.; Hu, X.; Winchester, D. Forward contracts in electricity markets: The Australian experience. *Energy Policy* **2007**, *35*, 3089–3103. [CrossRef]
21. Howell, B.; Meade, R.; O'Connor, S. Structural separation versus vertical integration: Lessons for telecommunications from electricity reforms. *Telecommun. Policy* **2010**, *34*, 392–403. [CrossRef]
22. Simshauser, P. Vertical integration, credit ratings and retail price settings in energy-only markets: Navigating the Resource Adequacy problem. *Energy Policy* **2010**, *38*, 7427–7441. [CrossRef]
23. Pollitt, M.; Anaya, K. Can current electricity markets cope with high shares of renewables? A comparison of approaches in Germany, the UK and the State of New York. *Energy J.* **2016**, *37*, 69–88. [CrossRef]
24. Simshauser, P. On intermittent renewable generation & the stability of Australia's National Electricity Market. *Energy Econ.* **2018**, *72*, 1–19.
25. Department for Business Energy & Industrial Strate. *Electricity Market Reform: Contracts for Difference*; UK Government: London, UK, 2015.
26. Department of Environment Land Water and Planning. *Victorian Renewable Energy Auction Scheme: Consultation Paper*; Victorian State Government: Melbourne, Australia, 2015.
27. Environment, Planning and Sustainable Development. *How Do the ACT's Renewable Energy Reverse Auctions Work?* ACT Government: Canberra, Australia, 2016.
28. Queensland Renewable Energy Expert Panel. *Credible Pathways to A 50% Renewable Energy Target for Queensland*; Queensland Government: Brisbane, Australia, 2016.
29. Department of the Environment and Energy. *Underwriting New Generation Investments—Call for Registrations of Interest*; Commonwealth of Australia: Canberra, Australia, 2018.
30. Kozlov, N. Contract for difference: Risks faced by generators under the new renewables support scheme in the UK. *J. World Energy Bus. Law* **2014**, *7*, 282–286. [CrossRef]
31. Onifade, T. Hybrid renewable energy support policy in the power sector: The contracts for difference and capacity market case study. *Energy Policy* **2016**, *95*, 390–401. [CrossRef]
32. Joskow, P. Comparing the costs of intermittent and dispatchable electricity generating technologies. *American Econ. Rev.* **2011**, *100*, 238–241. [CrossRef]
33. Mills, A.; Wiser, R. *Changes in the Economic Value of Variable Generation At High Penetration Levels: A Pilot Case Study of California*; LBNL-5445E; Lawrence Berkeley National Laboratory: Berkeley, CA, USA, 2012.
34. Edenhofer, O.; Hirth, L.; Knopf, B.; Pahle, M.; Schlomer, S.; Schmid, E.; Ueckerdt, F. On the economics of renewable energy sources. *Energy Econ.* **2013**, *40*, S12–S23. [CrossRef]
35. Hirth, L. The market value of variable renewables: The effect of solar & wind power variability on their relative price. *Energy Econ.* **2013**, *38*, 218–236.

36. MacGill, I. Electricity market design for facilitating the integration of wind energy: Experience and prospects with the Australian National Electricity Market. *Energy Policy* **2010**, *38*, 3180–3191. [CrossRef]
37. Nicolosi, M. The economics of renewable electricity market integration. An empirical and model-based analysis of regulatory frameworks and their impacts on the power market. Ph.D. Thesis, University of Cologne, Cologne, Germany, 2012.
38. Green, R.; Staffell, I. Electricity in Europe: Exiting fossil fuels? *Oxford Rev. Econ. Policy* **2016**, *32*, 282–303. [CrossRef]
39. Nelson, T.; Simshauser, P.; Nelson, J. Queensland solar feed-in tariffs and the merit-order effect: Economic benefit, or regressive taxation and wealth transfers? *Econ. Anal. Policy* **2012**, *42*, 277–301. [CrossRef]
40. Joskow, P. Symposium on Capacity Markets. *Econ. Energy Environ. Policy* **2013**, *2*, v–vi.
41. Sensfuβ, F.; Ragwitz, M.; Genoese, M. The merit-order effect: A detailed analysis of the price effect of renewable electricity generation on spot market prices in Germany. *Energy Policy* **2008**, *36*, 3086–3094. [CrossRef]
42. Forrest, S.; MacGill, I. Assessing the impact of wind generation on wholesale prices and generator dispatch in the Australian National Electricity Market. *Energy Policy* **2013**, *59*, 120–132. [CrossRef]
43. Cludius, J.; Forrest, S.; MacGill, I. Distributional effects of the Australian Renewable Energy Target (RET) through wholesale and retail electricity price impacts. *Energy Policy* **2014**, *71*, 40–51. [CrossRef]
44. Bell, W.; Wild, P.; Foster, J.; Hewson, M. Wind speed and electricity demand correlation analysis in the Australian National Electricity Market: Determining wind turbine generators' ability to meet electricity demand without energy storage. *Econ. Anal. Policy* **2015**, *48*, 182–191. [CrossRef]
45. Bell, W.; Wild, P.; Foster, J.; Hewson, M. Revitalising the wind power induced merit-order effect to reduce wholesale and retail electricity prices in Australia. *Energy Econ.* **2017**, *67*, 224–241. [CrossRef]
46. Felder, F. Examining electricity price suppression due to renewable resources and other grid investments. *Elec. J.* **2011**, *24*, 34–46. [CrossRef]
47. Bushnell, J. Building Blocks: Investment in Renewable and Non-Renewable Technologies. In *Harnessing Renewable Energy in Electric Power Systems: Theory, Practice, policy*; Moselle, B., Padilla, J., Schmalensee, R., Eds.; Routledge: Berkley, CA, USA, 2010.
48. Simshauser, P.; Gilmore, J. *On Entry Cost Dynamics in Australia's National Electricity Market*; EPRG Working Paper No.1841; University of Cambridge: Cambridge, UK, 2018.
49. Australian Energy Market Operator. *Electricity Demand Forecasting Methodology Information Paper*; AEMO Publication: Melbourne, Australia, 2019.
50. Narayan, K.; Smyth, R. The residential demand for electricity in Australia: An application of the bounds testing approach to cointegration. *Energy Policy* **2005**, *33*, 467–474. [CrossRef]
51. Fan, S.; Hyndman, R. The price elasticity of electricity demand in South Australia. *Energy Policy* **2011**, *39*, 709–3719. [CrossRef]
52. Grattan Institute. *Commonwealth Orange Book 2019*; Grattan Institute Publication: Melbourne, Australia, 2019.
53. Peters, B. *Governance: A Garbage Can Perspective*; Institute for Advanced Studies: Vienna, Austria, 2002.
54. Simshauser, P.; Ariyaratnam, J. What is normal profit for power generation? *J. Financ. Econ. Policy* **2014**, *6*, 152–178. [CrossRef]

© 2019 by the author. Licensee MDPI, Basel, Switzerland. This article is an open access article distributed under the terms and conditions of the Creative Commons Attribution (CC BY) license (http://creativecommons.org/licenses/by/4.0/).

energies

MDPI

Article

Emissions Trading System of the European Union: Emission Allowances and EPEX Electricity Prices in Phase III

Georg Wolff [1,*,†] and Stefan Feuerriegel [2,†]

1 Institute for Economics, University of Freiburg, Platz der Alten Synagoge, 79098 Freiburg, Germany
2 Swiss Federal Institute of Technology (ETH Zurich), Weinbergstr. 56/58, 8092 Zurich, Switzerland
* Correspondence: georg.wolff@vwl.uni-freiburg.de
† These authors contributed equally to this work.

Received: 30 June 2019; Accepted: 24 July 2019; Published: 27 July 2019

Abstract: The Emissions Trading System in the European Union was introduced to achieve the climate goal of reducing emissions by around 43% between 1990 and 2030. Accordingly, the costs of emission allowances are part of power generation and, by extension, the price of electricity. Theoretical works thus suggest a positive relationship between the price of emission allowances and electricity. However, this has not been validated empirically for phase III of the Emissions Trading System in the short run as part of the price setting mechanism of electricity producers. Our evidence suggests an opposite effect: According to our empricial results, both European Power Exchange (EPEX) day-ahead and intraday markets are negatively affected during phase III. We further test for a potentially asymmetric influence with the help of quantile regressions. Altogether, the outcome has implications for policy-makers and calls for further attention by academics and policy-makers in the future design of the Emissions Trading System, especially under larger amount of renewables in the electricity system.

Keywords: emissions trading; electricity price; econometric modeling; time series analysis; emission allowance

---

1. Introduction

The European Union introduced the first and largest trading system for greenhouse gases in the world, the Emissions Trading System (EU ETS). Its launch in 2005 came as a direct consequence of the Kyoto Protocol as a means of achieving the climate objective of reducing emissions by 43 % between 1990 and 2030 [1]. The design is based on a *cap and trade* system, which allows only a fixed amount of emissions for various greenhouse gases such as carbon dioxide (CO_2), nitrous oxide (N_2O), and perfluorocarbons (PFCs). In the past, the annual limit of greenhouse gases has been decreased in order to reduce greenhouse gases and to thus achieve climate goals.

The actual design of the cap and trade system represents a challenging task for policy-makers. On the one hand, the price for emission allowances needs to be fairly high in order to provide an incentive to reduce emissions [2]. On the other hand, the price should not threaten economic development. Therefore, policy-makers need to combine social, economic, and environmental considerations. For example, the European Commission has broadened the number of industries that are subject to the trade mechanism as part of the modifications when transitioning from phase II to III of the Emissions Trading System. Nowadays, about 11,000 heavy energy-using installations are part of the EU ETS, accounting for approximately 45 % of all emissions in Europe. In particular, the entire power production sector in Germany is obligated to participate in the EU ETS. Accordingly, every power company must hold a sufficient number of European Emission Allowances (EUA) by the end of the year, whereby each EUA entitles the energy producer to emit one ton of carbon dioxide or its equivalents.

Since 2013, phase III of the EU ETS represents the status quo. It differs from phase II (years 2008–2012) in the following key dimensions [1]:

1. **Reduction of free allocation.** Most notably, policy-makers reduced the free allocation of emission allowances once again. This has led to 40 % of allowances being sold at auction. For instance, the entire power production sector is forced to buy allowances at auction. As such, electricity producers must buy emission allowances on the energy exchange when they have exceeded their allowance, e.g., due to higher-than-expected emissions. Additionally, in the period of 2014–2016, they reduced the number of certificates by 900 million. These certificates will be used as a market stability reverse mechanism to match demand and supply.
2. **Expansion to more industries and further greenhouse gases.** The EU ETS now accounts for additional greenhouse gases that were not part of phase II, such as nitrous oxide (N_2O) and perfluorocarbons (PFCs). In addition, it adds a wider array of industry sectors, such as manufacturing industries and aircraft operators.
3. **EU registry.** While national registries collected the names of companies qualified for emissions trading during phase II, these were replaced in phase III by a registry encompassing the full European Union, which now includes 31 countries participating in the EU ETS. The European registry was introduced in order to establish a better control mechanism throughout the member states.

Based on the changes between phase II and III of the ETS, one would expect a stronger influence of the carbon price on electricity prices. In theory, the necessary expenditures on emission allowances should increase costs for operators of power plants and it is thus likely to be linked to the price of electricity. For example, lignite-fired power plants emit about 0.984 tons of carbon dioxide per 1 MWh energy, while power plants fired by natural gas produce about 0.548 tons of carbon dioxide per 1 MWh energy. (Retrieved from http://www.eia.gov/tools/faqs/faq.cfm?id=74&t=11 on 11 June 2019.) In other words, one can assume an influence of emission allowances on electricity prices if the underlying pass-through rate is nonzero [3,4]. Consequently, the price of EUA is incorporated in the price setting algorithms of electricity producers, thus resulting in electricity price fluctuations in the short run [5]. Policy-makers demand such evaluations, which allow them to assess whether emissions trading encourages high amounts of sustainable sources of electricity generation on a national or international scale.

Our literature survey later reveals that the link between carbon and European Power Exchange (EPEX) electricity prices has been studied for phase II of the EU ETS but not for phase III. Here, we follow earlier research which has focused on the short-run relationship between EUA prices and electricity prices as a means to study the price setting mechanisms of electricity firms (e.g., References [5,6]). It is thus the key contribution of this paper to quantify the impact of carbon prices on EPEX electricity prices during phase III of the EU ETS. For this purpose, we specifically compare the effect across day-ahead and intraday markets. We additionally calculate the distributional properties of the pass-through rate by means of quantile regression to test whether a price premium for EUA is only added for a certain threshold price.

Our results suggest a negative relationship between emission allowances and electricity prices. The corresponding implications are discussed by considering the intentions of policy-makers when designing phase III of the EU ETS. Especially, the findings are later set in relation to the context of various proposed changes to the EU ETS that are currently under consideration by policy-makers. Altogether, these insights can help to improve the future market design for emissions trading in order to achieve the desired climate objectives. Our findings should be seen as a starting point for future research as they call for further attention to better understand the consequences for price setting that result from the introduction of the EU ETS.

2. Related Work

This section reviews related works concerning the influence of carbon prices on the price of electricity.

2.1. Long Run vs. Short Run Relationship

Previous works have analyzed the relationship between carbon prices and the price of electricity as part of an empirical study. These reveal statistical evidence of such a relationship across several markets in both the long and short run yet not for phase III of the EU ETS.

In regard to the long run, Bunn and Fezzi [7] find a positive relationship between the carbon price and the price of electricity in the U. K. during phase I of the EU ETS. Their analysis reveals that a 1 % change in the carbon price results in an increase in electricity prices by 0.42 % in the equilibrium case. In the short run, this study observes a visible shock in electricity prices after a few days. A similar, positive relationship in the long run is observed in the Spanish market [8,9]. However, the effect here is of a lesser magnitude; i.e., a 1 % change in the carbon price yields a 0.24 % increase in the long run during phase II and in the first year of the third phase [9]. In addition, the findings suggest a smaller impact for a lower carbon price than for a higher one.

On the other hand, there is also strong evidence in favor of a short-run relationship. For example, in the Nordic electricity market, the price of EUA tends to considerably influence the electricity price in the short run [10,11]. Cotton and Mello [5] examined the Australian approach to reducing greenhouse gas emissions. The authors found no impact in the long run but did in the short run. Analogously, our analysis is concerned foremost with the short-run relationship, since we not only investigate the day-ahead market but also the price setting mechanism in the intraday market, where empirical evidence is especially limited.

According to Bannoer et al. [12], the relatively small impact may result from the almost constant carbon price, at least during the period from 2010 until 2013, and conclude, therefore, that the carbon price has only a very small effect. However, this work only studies phase II and cannot assess whether the additional adjustments from phase III have been effective.

2.2. Asymmetric Pass-Through Rate

The ratio of costs passed on to customers has been extensively discussed in the literature. From a theoretical point of view, this pass-through rate should account for 100 % in perfect markets [3]. However, the condition of perfect markets seldom holds in reality, thus resulting in a lower pass-through rate [4]. The pass-through rate depends on additional factors, such as the energy mix, the demand, as well as the supply of energy [13]. In addition, recent works [14,15] find a less than 100% but asymmetric pass-through of costs in the German market. In particular, a pass-through rate of at least 84%, ranging from 98% to 104%, is found for different load periods [14]. As a consequence, increasing the carbon price is supposed to influence the electricity price more strongly than decreasing carbon prices. As a result, the price setting mechanism might be only affected by carbon prices exceeding a certain price threshold. In Section 4.3, we control for this effect by utilizing quantile regressions in order to compute the distributional properties of the carbon effect on electricity prices.

2.3. Directional Influence

Previous research reports contradictory evidence regarding the direction of the influence. Jouvet and Solier [4] revealed significant positive as well as negative effects of EUA prices in each year from 2005 through 2011 (except 2009) for all European markets, with the negative impact mostly occurring in the Italian market. For the German and Austrian electricity markets, the effects are not consistent. Furthermore, Aatola et al. [16] found a positive but asymmetric influence of the carbon price on European electricity prices—such as U. K., France, Netherlands, Germany, Spain, and the Nordic countries. However, in the short run, they did not find a clear pattern. A work by Woo et al. [17] also points to a strong regional dependency of the emission price effect. The author's found a short-run price effect between 0.15 dollar and 0.59 dollar in Western USA with respect to a 1 dollar carbon price change.

The unclear patterns regarding the direction of the influence remain in different phases of the EU ETS, revealing mixed effects. Phase II evinces a significant positive effect of the carbon price on the hourly electricity price in the short run [6]. Paraschiv et al. [18] estimate a state space model with time-varying coefficients between 2010 and 2012 to correct the variance, finding both a positive and negative influence of the EUA price on the day-ahead electricity price. However, none of the above papers has focused on the EPEX electricity prices of both the intraday and day-ahead auction market or, on top of that, studied the changes in the relationship between EU ETS phases II and III. One of the reasons is that the introduction of phase III occurred only recently and, therefore, empirical evidence relating to it is still scarce. As a remedy, this paper examines the impact of carbon prices in phases II and III on German and Austrian day-ahead and intraday electricity prices.

3. Methods and Materials

In the following, we present our autoregressive model, as well as the underlying dataset. Additional materials and robustness checks can be found in the Supplementary Materials.

3.1. Modeling of Electricity Prices

Electricity entails several unique characteristics that differentiate it from other commodities, such as the need to instantaneously match the electricity demand and supply, the diverse array of electricity sources, as well as the lack of available storage capability. As a result, electricity prices are highly volatile and driven by seasonality [19–22], which makes the modeling of electricity prices challenging. In this respect, recent literature has yielded two dominant approaches to modeling electricity prices. First, co-integration models such as vector autoregressive (VAR) and vector error correction (VEC) models (e.g., References [6,8,9,23]). These are especially suited to research questions that seek to investigate interdependencies between multiple variables. For example, one can use them to quantify the impact of load on electricity prices and vice versa. Second, autoregressive models and their variants (e.g., with a moving-average term) are commonly used when studying the impact of several covariates on a single price variable (e.g., References [21,24–26]). For this reason, we follow the latter approach by modeling electricity prices as an autoregressive process. Specifically, we define one model for each hour, since we therefore can better capture the daily fluctuation of electricity prices [27]. This is beneficial as it allows us to focus on the short-run relationship and thus the price setting of electricity producers, i.e., how EUA is actually incorporated by decision-makers. This is also motivated by earlier research that concentrated on the short-run relationship when studying the pricing power of EUA [5,6].

Besides the choice of the model, we need to carefully consider the inclusion of covariates since a variety of exogenous factors might influence electricity prices. Common examples are fuel prices (e.g., References [10,18,28,29]), power generation (e.g., References [8,30]), feed-ins from renewables (e.g., References [8,9,31–34]), and economic factors (e.g., Reference [8]). Given the broad spectrum of considered covariates, there seems to be no consistent recommendation as to which variables to insert, especially as the significance of the impact varies from study to study. We thus follow a two-pronged approach, which first chooses the control factors that are most common in the previous literature, namely, load (as a proxy for power demand) and solar/wind power generation [29]. Moreover, we additionally use a set of additional covariates (coal price, gas price, oil price, and foreign exchange rate) as part of a robustness check.

3.2. Autoregressive Time Series Model

We now present the method by which we empirically measure the impact of EUA on electricity prices. For this purpose, we follow previous research and use autoregressive models with exogenous variables (ARX). We specifically estimate separate models for each hour of the day [18,27]. This introduces additional degrees of freedom that can reflect, e.g., differences in the electricity mix between peak and off-peak hours. Our online appendix lists robustness checks with additional covariates and a global model with de-seasonalization [35], resulting in similar findings.

Our key independent variable of interest is given by the price for EUA. We additionally incorporate the infeeds of wind farms and photovoltaic power plants, as well as the grid load, since they serve as standard control variables when studying electricity prices [22,23]. These also cover most factors that are specific to the demand and supply sides. In case of the latter, the short-term dynamics of prices are mostly affected not by power generation from fossil fuels but by variable sources of power generation, of which wind and solar power represent the largest shares in the German market. Moreover, we incorporate dummy variables for the weekday and the month in order to adjust for seasonal variations.

3.3. Asymmetric Influence Via Quantile Regressions

Previous research argues in favor of a nonlinear pass-through of emission costs [15]. Accordingly, the influence on electricity prices is stronger at the upper end of the merit order curve conditional on a large infeed from expensive sources of power generation. The influence of carbon prices on electricity prices is thus likely to be asymmetric [9,36]. For instance, we expect a large effect when all power plants produce electricity, since this might include carbon-intensive forms of power generation. As a consequence, increasing the carbon price is supposed to impact the electricity price more strongly than decreasing carbon prices. This suggests that the coefficient β is subject to variations and thus attains values that are dependent on the electricity price. In other words, the estimated free parameters of each quantile regression cover the quantiles of the distribution of electricity prices. Therefore, our later analysis also addresses the distributional influence of carbon prices on electricity prices.

For this purpose, we use quantile regressions, which differ from ordinary least squares (OLS). Whereas the latter measures the impact of predictor variables on the mean of an outcome variable, quantile regressions incorporate the entire distribution of electricity prices by estimating the influence of carbon prices at different percentiles of the dataset [37]. Therefore, we can shed light upon potential variations in the effect across different percentiles.

Mathematically, the OLS regression estimates global coefficients $\beta_1, \ldots, \beta_4$ that quantify the effect of regressors X on the electricity price $\tilde{P}_t$ at the mean. This is given by $E\left[P_t \parallel X\right] = X\left[\beta_1, \ldots, \beta_4\right]^T$. Quantile regression, by contrast, allows us to estimate the effect of predictor variables X on a selected quantile of outcome variable $\tilde{P}_t$ [37]. Hence, we obtain separate coefficients $\hat{\beta}_1^\tau, \ldots, \hat{\beta}_4^\tau$ for different quantiles τ of $\tilde{P}_t$, given by $Q_\tau\left[P_t \parallel X\right] = X\left[\hat{\beta}_1^\tau, \ldots, \hat{\beta}_4^\tau\right]^T$. This thus yields the full impact of carbon prices across the full (conditional) distribution of electricity prices. For mathematical details of this estimation procedure, we refer to Reference [37].

3.4. Dataset

An overview of our dependent variables and the covariates is presented in Table 1. The dependent variable is given by the electricity prices in the day-ahead, as well as intraday spot market. Here, delivery of electricity for a certain hour h of day d can be traded continuously from 3 p.m. on day $d-1$ until 30 min before hour h on day d. Hence, we follow previous research and perform our analysis based on the average hourly price [29]. The prices originate from the European Power Exchange (EPEX), which is a joint venture of the Energy Exchange (EEX) and the French Powernext.

Table 1. Overview of variables taken into account in the subsequent evaluation.

Variable, Unit	Symbol	Frequency	Description	Data Source
Day-ahead spot price, €/MWh	P_t^A	Hourly	Electricity price of day-ahead auction with delivery in Germany and Austria: The auction price is set at 12 a.m. for each hour of the next day	European Power Exchange
Intraday spot price, €/MWh	P_t^I	Hourly	Continuous intraday electricity price with delivery in Germany and (partially) Austria, where trading is possible up to 30 min before delivery	European Power Exchange
Wind infeed, MW	$Wind_t$	Hourly	Aggregated total wind infeed from the four transmission system operators (TransnetBW, Tennet, Amprion, and 50 Hertz) in Germany	Energy Exchange (EEX) Transparency
Solar infeed, MW	PV_t	Hourly	Aggregated total photovoltaic infeed from the four transmission system operators (TransnetBW, Tennet, Amprion, and 50 Hertz) in Germany	EEX Transparency
Load, MW	$Load_t$	Hourly	Total hourly electricity consumption in Germany	ENTSO-E
Price of EUA, €/tCO2	EUA_t	Daily	Setting price of EEX European Emission Allowance (EUA) future that is continuously traded on the Intercontinal Exchange (ICE): one EUA entitles its holder to emit one ton of carbon dioxide or its equivalents	Thomson Reuters Datastream

Note: ENTSO-E is the European Network of Transmission System Operations for Electricity.

Here, our main variable of interest is the price of European Emission Allowance (EUA). Table 2 reports the descriptive statistics. As part of the switch from phase II to III in the EU ETS, the price of EUA has undergone major changes. As such, the mean price per EUA dropped by −54.26 % from € 11.61 to € 5.23. Furthermore, the standard deviation declined by −71.35 % from 3.56 to 1.02.

Table 2. Descriptive statistics for the price of European Emission Allowance (in €/tCO_2) in phases II and III.

	Phase II	Phase III	Relative Change
Time Period	**Jan 2010–Dec 2012**	**Jan 2013–Dec 2014**	
Observations	**782**	**522**	
Mean	11.61	5.23	−54.26%
Median	12.71	5.13	−59.64%
Min.	5.74	2.70	−52.96%
Max.	17.03	7.37	−56.72%
Std. dev.	3.56	1.02	−71.35%
Skew.	−0.16	0.01	106.25%
Kurt.	−1.57	−0.67	57.32%

4. Results

This section analyzes the influence of the carbon prices on electricity prices. We first test the stationarity of our time series and, afterwards, estimate the autoregressive models and the quantile regressions for the hours $h = 8, 16, 24$.

4.1. Stationarity

We need to validate whether our time series are stationary in order to rule out a spurious regression [35]. Here, the augmented Dickey–Fuller (ADF) tests find stationary time series in levels for both electricity prices, the load, as well as for feed-ins from solar and wind power for each hour separately. The hourly EUA price is stationary in levels during phase II and phase III but not over the entire period. Therefore, in order to obtain stationary time series, we take the first differences of EUA when studying the entire sample. Due to the extensive nature of the ADF test for each hour and covariates (504 ADF tests), the values are omitted for brevity.

4.2. Influence of Carbon Price during the EU ETS Regimes

We now estimate the impact of the EUA price on both day-ahead and intraday electricity prices. Section 4.1 has already established that the price of EUA is integrated of order one during the entire sample. Therefore, we instead consider the first differences in order to ensure the absence of a spurious regression [35]. Detailed results are reported in the online appendix, while we restrict our presentation to key findings in the following.

We perform a series of diagnostic tests: First, we find only stationary residuals and thus eliminate the risk of analyzing a spurious regression. Second, the p-value belonging to the F-statistics of each model is zero and, thus, the combination of model variables has an influence on the dependent variable. Furthermore, we find autocorrelation of the residuals by using the Durbin–Watson test, while a Breusch–Pagan test reveals heteroscedastic residuals. To adjust the test values for both autocorrelation and heteroscedasticity, we use the Newey–West procedure (e.g., Reference [35,38]). Its advantage is that we yield t-statistics that are robust to a general form of serial correlation and heteroscedasticity but let the regression coefficients be unaffected [35].

According to Table 3, the influence of the EUA price on electricity prices is as follows. In the day-ahead market, the coefficient of the EUA time series (in the form of first differences) is non-significant at common statistical significance levels for each hour of the day (Table 3 exemplary reports the hours 8, 16, and 24). These findings change when we split the dataset into phase II and III. While in phase II, the EUA price has an impact in hours 13 to 16, in phase III the EUA price has an significant negative impact in each hour of the day, even in nighttime. Controversially, the impact remains negative over the entire day. We observe a quite similar picture with regard to the intraday market. In this case, the price of emission allowances shows a negative impact in phase III, being highly significant.

The relative impact on the dependent variable subsequent to a one standard deviation increase in the EUA price enables us to compare the strength of the effect independent of the scaling and thus across both the day-ahead and intraday markets. In the day-ahead market, the corresponding coefficient is negative ranging from −0.32 to −0.13. This means that a one standard deviation increase in the price of EUA results in a −0.32 to −0.13 standard deviation decrease in the electricity price variable depending on the hour of the day. In the intraday market, the significant negative impact lies between −0.36 to −0.28, revealing that the intraday market is more strongly affected by EUA prices.

Consistent with previous literature, we find that demand- and supply-side factors show a statistically significant impact on electricity prices, though the results differ across markets. In the day-ahead market, the impact of solar and wind feed-ins on the price of electricity is weaker than in the intraday market. Additionally, load governs the intraday market to a larger extent. Several other studies discuss the influence of external variables, such as load and feed-ins from renewables, in more detail (cf. References [23,29,33,39]).

According to the goodness-of-fit, all models in the day-ahead market as well as in the intraday market reveal a strong explanatory power with an adjusted R^2 above 0.94.

Table 3. Estimated coefficients of separate autoregressive models belonging to different hours ($h = 8, 16, 24$) of the day: These measure the influence of the EUA price on electricity prices of the EU ETS. Dependent variables are the hourly day-ahead and intraday electricity prices.

	Dependent Variable: Hourly Day-Ahead Electricity Price $\hat{P}_t^A$								
	Phase II (2010–2012)			Phase III (2013–2014)			Phases II & III (2010–2014)		
	Hour 8	Hour 16	Hour 24	Hour 8	Hour 16	Hour 24	Hour 8	Hour 16	Hour 24
$Wind_{t-1}$	**−0.13**	**−0.11**	**−0.08**	**−0.09**	**−0.08**	**−0.04**	**−0.11**	**−0.10**	**−0.08**
t-value	(−11.07)	(−12.97)	(−7.55)	(−4.74)	(−4.11)	(−3.22)	(−9.41)	(−9.60)	(−7.55)
PV_{t-1}	−0.01	**−0.02**		**−0.08**	**−0.11**		**−0.05**	**−0.07**	
t-value	(−1.86)	(−2.21)		(−3.11)	(−3.99)		(−4.28)	(−5.93)	
$Load_{t-1}$	**0.85**	**0.67**	**0.30**	**1.00**	**0.92**	**0.40**	**0.72**	**0.63**	**0.30**
t-value	(13.77)	(15.20)	(0.71)	(6.67)	(9.20)	(7.06)	(12.39)	(14.97)	(0.71)
ΔEUA_{t-1}	−0.05	**0.05**	0.00	**−0.32**	**−0.26**	**−0.13**	−0.01	−0.01	0.00
t-value	(−1.68)	(2.56)	(0.71)	(−3.45)	(−4.54)	(−3.00)	(−0.86)	(1.42)	(0.71)
P_{t-1}^A	**0.16**	**0.22**	**0.42**	**0.19**	**0.27**	**0.50**	**0.21**	**0.27**	**0.42**
t-value	(7.45)	(8.32)	(13.86)	(4.41)	(7.52)	(16.22)	(10.48)	(12.14)	(13.86)
P_{t-2}^A	**0.17**	**0.15**	**0.23**	**0.18**	**0.20**	**0.22**	**0.20**	**0.19**	**0.23**
t-value	(6.18)	(6.61)	(10.74)	(4.06)	(5.45)	(5.95)	(8.13)	(10.31)	(10.74)
P_{t-7}^A	**0.15**	**0.12**	**0.15**	**0.21**	**0.17**	**0.12**	**0.21**	**0.19**	**0.15**
t-value	(4.11)	(3.98)	(5.75)	(3.12)	(4.28)	(3.80)	(5.57)	(7.38)	(5.75)
Observations	1089	1089	1089	716	716	716	1819	1819	1819
Adjusted R^2	0.98	0.98	0.99	0.94	0.95	0.98	0.96	0.97	0.98
F-statistic	4661.56	7564.34	883.99	1148.79	1659.21	3809.93	4669.10	7051.32	22148.95
	Dependent Variable: Hourly Intraday Electricity Price $\hat{P}_t^I$								
	Phase II (2010–2012)			**Phase III (2013–2014)**			**Phases II & III (2010–2014)**		
	Hour 8	**Hour 16**	**Hour 24**	**Hour 8**	**Hour 16**	**Hour 24**	**Hour 8**	**Hour 16**	**Hour 24**
$Wind_t$	**−0.16**	**−0.14**	**−0.14**	**−0.18**	**−0.25**	**−0.24**	**−0.17**	**−0.18**	**−0.16**
t-value	(−8.94)	(−11.79)	(−7.08)	(−14.26)	(−13.94)	(−14.40)	(−13.02)	(−16.51)	(−11.07)
PV_t	−0.02	**−0.05**		**−0.05**	**−0.28**		**−0.04**	**−0.14**	
t-value	(−2.68)	(−4.00)		(−5.44)	(−11.06)		(−5.24)	(−9.07)	
$Load_t$	**1.04**	**0.86**	**0.49**	**1.16**	**1.54**	**0.97**	**0.87**	**0.88**	**0.43**
t-value	(18.20)	(16.10)	(8.12)	(20.42)	(19.70)	(13.46)	(20.38)	(20.84)	(9.55)
ΔEUA_t	−0.07	0.05	0.05	**−0.36**	**−0.42**	**−0.28**	−0.00	−0.00	0.00
t-value	(−1.85)	(1.74)	(1.88)	(−6.25)	(−5.86)	(−5.15)	(−0.48)	(−0.07)	(0.66)
P_{t-1}^I	**0.13**	**0.21**	**0.27**	**0.11**	**0.15**	**0.16**	**0.17**	**0.23**	**0.30**
t-value	(4.44)	(7.66)	(8.25)	(4.00)	(6.33)	(4.92)	(8.07)	(12.77)	(11.11)
P_{t-2}^I	**0.07**	**0.07**	**0.14**	**0.13**	0.01	**0.14**	**0.12**	**0.09**	**0.20**
t-value	(3.03)	(3.11)	(4.01)	(4.97)	(0.49)	(4.52)	(6.45)	(4.88)	(6.51)
P_{t-7}^I	0.05	0.04	**0.13**	**0.20**	**0.09**	**0.11**	**0.15**	**0.14**	**0.19**
t-value	(1.74)	(1.43)	(3.63)	(6.04)	(3.35)	(3.91)	(6.04)	(6.57)	(6.66)
Observations	1089	1089	1089	716	716	716	1819	1819	1819
Adjusted R^2	0.97	0.95	0.97	0.97	0.95	0.97	0.95	0.96	0.96
F-statistic	2359.96	3526.41	3940.32	2277.76	1485.25	2852.08	3959.73	4505.73	6164.12

Dummies: weekday, month; Stated: standardized coefficients (because of different units); robust t-statistics in parenthesis; $t-1$ refers to the same hour the day before, due to price-setting time points; Bold highlighting: coefficients with a p-value of below 0.05.

4.2.1. Robustness Checks

As part of our robustness checks, we estimate an analysis in accordance to Wolff and Feuerriegel [29] by incorporating additional control variables in order to ensure that their inclusion does not confound our results. These variables concern the daily coal price (Credit Suisse Commodity Benchmark for coal API 2 spot return price index at the Amsterdam-Rotterdam-Antwerp Hub in USD/t.) $Coal_t$, the gas price (Setting price of natural gas first near future at the virtual gas trading hub Title Transfer Facility (TTF) in EUR/MWh.) Gas_t, the oil price (Brent crude oil spot price in USD per barrel.) Oil_t, and the closing price of the USD-EUR exchange rate FX_t. We again take the first differences where time series are integrated of order one. As earlier, we only report the t-statistics that are robust to heteroscedasticity and serial correlation. The results are given in Table 4 for the impact of the EUA price on both the day-ahead and intraday electricity price.

The estimation results reveal the same picture as our earlier analysis: The EUA shows a significantly negative impact in the day-ahead market during phase III. Interestingly, the EUA price positively affects the electricity price for the hour 24 when the solar feed-in is zero, while in the main model, we find a small positive impact for hour 16. In the case of the intraday market, the price of emission allowances shows a negative impact in phase III that is statistically highly significant. Hence, a one standard deviation increase in the price of EUA results in a −0.79 to −2.53 standard deviation decrease in the electricity price variable.

Table 4. Estimated coefficients of separate autoregressive models belonging to different hours ($h = 8, 16, 24$) of the day as part of a robustness check: These measure the influence of the EUA price on electricity prices of the EU ETS incorporating control variables. Dependent variables are the hourly day-ahead and intraday electricity prices.

	Dependent Variable: Hourly Day-Ahead Electricity Price $\hat{P}_t^A$								
	Phase II (2010–2012)			Phase III (2013–2014)			Phases II & III (2010–2014)		
	Hour 8	**Hour 16**	**Hour 24**	**Hour 8**	**Hour 16**	**Hour 24**	**Hour 8**	**Hour 16**	**Hour 24**
$Wind_{t-1}$	**−0.28**	**−0.20**	**−0.21**	**−0.15**	**−0.12**	**−0.05**	**−0.22**	**−0.16**	**−0.13**
t-value	(−11.56)	(−14.51)	(−9.39)	(−5.09)	(−5.37)	(−3.36)	(−10.39)	(−11.75)	(−8.14)
PV_{t-1}	**−0.69**	**−0.07**		**−1.07**	**−0.10**		**−1.07**	**−0.11)**	
t-value	(−3.68)	(−3.59)		(−2.90)	(−4.66)		(−5.17)	(−7.81)	
$Load_{t-1}$	**0.68**	**0.51**	**0.32**	**0.66**	**0.47**	**0.20**	**0.55**	**0.41**	**0.21**
t-value	(14.09)	(12.86)	(7.88)	(5.59)	(5.60)	(3.20)	(12.09)	(13.65)	(8.16)
ΔEUA_{t-1}	−0.12	−0.15	**1.86**	**−2.20**	**−1.25**	−0.34	−1.03	−1.37	0.38
t-value	(−0.15)	(−0.20)	(2.38)	(−2.82)	(−2.36)	(−0.84)	(−0.97)	(−1.41)	(0.54)
$\Delta Coal_{t-1}$	−0.16	−0.13	−0.09	−0.57	−0.02	0.25	−0.14	−0.02	0.07
t-value	(−0.71)	(−0.73)	(−0.60)	(−0.83)	(−0.03)	(0.60)	(−0.72)	(−0.10)	(0.51)
ΔOil_{t-1}	0.03	−0.02	0.05	−0.08	−0.06	−0.04	0.05	0.04	0.02
t-value	(0.30)	(−0.22)	(0.68)	(−0.33)	(−0.32)	(−0.29)	(0.53)	(0.50)	(0.38)
Gas_{t-1}	**0.40**	**0.26**	**0.25**	**0.46**	**0.48**	**0.28**	**0.41**	**0.38**	**0.23**
t-value	(3.74)	(2.60)	(3.89)	(3.35)	(3.81)	(3.49)	(5.80)	(5.68)	(4.41)
ΔFX_{t-1}	16.74	−13.82	10.15	67.93	45.04	11.23	27.65	3.24	13.60
t-value	(26.84)	(−0.64)	0.69	(0.89)	(0.75)	0.26	(1.03)	(0.14)	(0.88)
P^A_{t-1}	**0.17**	**0.22**	**0.27**	**0.19**	**0.26**	**0.49**	**0.20**	**0.26**	**0.40**
t-value	(7.84)	(8.31)	(7.73)	(4.76)	(8.21)	(16.13)	(10.80)	(12.24)	(13.30)
P^A_{t-2}	**0.15**	**0.15**	**0.21**	**0.17**	**0.17**	**0.21**	**0.18**	**0.17**	**0.22**
t-value	(6.50)	(6.34)	(9.92)	(4.32)	(6.00)	(5.58)	(8.33)	(10.13)	(10.69)
P^A_{t-7}	**0.12**	**0.13**	**0.12**	**0.16**	**0.12**	**0.09**	**0.18**	**0.17**	**0.13**
t-value	(3.40)	(4.08)	(2.92)	(3.03)	(3.04)	(2.89)	(5.50)	(6.43)	(4.97)
Observations	1089	1089	1089	716	716	716	1819	1819	1819
Adjusted R^2	0.97	0.99	0.99	0.94	0.96	0.98	0.96	0.97	0.98
F-statistic	3610	6066	6919	857.3	1282	2739	3548	5430	7760

	Dependent Variable: Hourly Intraday Electricity Price $\hat{P}_t^I$								
	Phase II (2010–2012)			Phase III (2013–2014)			Phases II & III (2010–2014)		
	Hour 8	**Hour 16**	**Hour 24**	**Hour 8**	**Hour 16**	**Hour 24**	**Hour 8**	**Hour 16**	**Hour 24**
$Wind_t$	**−0.35**	**−0.26**	**−0.25**	**−0.29**	**−0.31**	**−0.26**	**−0.33**	**−0.28**	**−0.23**
t-value	(−9.44)	(−12.94)	(−7.41)	(−13.82)	(−14.25)	(−14.95)	(−13.81)	(−17.07)	(−11.14)
PV_t	**−0.96**	**−0.14**		**−0.64**	**−0.23**		**−0.95**	**−0.20**	
t-value	(−3.84)	(−5.35)		(−4.84)	(−11.16)		(−6.94)	(−10.53)	
$Load_t$	**0.86**	**0.65**	**0.42**	**0.78**	**0.87**	**0.45**	**0.70**	**0.59**	**0.32**
t-value	(17.16)	(15.26)	(7.62)	(15.75)	(14.44)	(8.95)	(18.06)	(17.30)	(8.01)
ΔEUA_t	−0.88	−0.69	1.25	**−2.53**	**−2.37**	**−0.79**	−0.58	−0.24	0.62
t-value	(−0.82)	(−0.59)	(1.13)	(−5.17)	(−4.68)	(−2.28)	(−0.62)	(−0.21)	(0.75)
$\Delta Coal_t$	0.01	−0.04	**−0.43**	−0.39	−0.59	−0.14	−0.03	−0.09	−0.30
t-value	(0.02)	(−0.16)	(−2.10)	(−0.78)	(−0.88)	(−0.39)	(−0.08)	(−0.39)	(−1.81)
ΔOil_t	−0.03	0.01	0.02	−0.07	0.29	0.05	0.12	0.18	0.11
t-value	(−0.26)	(0.06)	(0.27)	(−0.37)	(1.34)	(0.38)	(1.14)	(1.44)	(1.51)
Gas_t	**0.41**	**0.44**	**0.25**	**0.38**	**0.28**	**0.31**	**0.33**	**0.37**	**0.11**
t-value	(2.55)	(3.13)	(3.16)	(4.97)	(2.88)	(5.66)	(4.65)	(5.04)	(2.45)
ΔFX_t	−2.99	34.27	22.80	84.03	−10.93	−3.62	15.34	25.36	14.47
t-value	(−0.08)	(1.18)	(0.83)	(1.12)	(−0.14)	(−0.09)	(0.46)	(0.86)	(0.58)
P^A_{t-1}	**0.14**	**0.21**	**0.27**	**0.12**	**0.15**	**0.14**	**0.17**	**0.23**	**0.30**
t-value	(4.80)	(8.00)	(7.69)	(4.30)	(6.45)	(4.02)	(8.41)	(12.48)	(10.95)
P^A_{t-2}	**0.06**	**0.07**	**0.14**	**0.12**	0.01	**0.13**	**0.12**	**0.08**	**0.20**
t-value	(2.98)	(2.86)	(4.41)	(4.90)	(0.37)	(4.29)	(6.41)	(4.61)	(6.57)
P^A_{t-7}	0.04	0.04	**0.13**	**0.16**	**0.07**	**0.09**	**0.13**	**0.12**	**0.18**
t-value	1.19	(1.43)	(4.00)	(4.66)	(2.88)	(3.29)	(5.67)	(5.94)	(6.75)
Observations	1089	1089	1089	716	716	716	1819	1819	1819
Adjusted R^2	0.95	0.97	0.97	0.97	0.95	0.97	0.96	0.96	0.97
F-statistic	1719	2687	2923	1713	1084	2005	3034	3622	4816

Dummies: weekday, month; Stated: standardized coefficients (because of different units); robust t-statistics in parenthesis; $t-1$ refers to the same hour the day before, due to price-setting time points; Bold highlighting: coefficients with a p-value of below 0.05.

4.3. Asymmetric Influence of the Carbon Price on Electricity Prices

As previous research suggests [3,9,36], the carbon price may have an asymmetric impact on electricity prices. Therefore, we perform quantile regressions for the 25 %, 50 %, and 75 % quantiles. Figures 1 and 2 present the estimation results across different phases of the EU ETS.

Interestingly, we observe different results for the day-ahead and intraday markets compared to the results from Section 4.2. First of all, we find that the impact of the price of EUA on electricity prices appears predominantly in phase III of the EU ETS. However, we see a considerable impact in the day-ahead market in phase III, with a diverging influence for the 25 % and 75 % quantiles. In the 25 % quantile, a one standard deviation increase in the price of EUA results in an low impact with values around zero. By contrast, in the 75 % quantile, the price of EUA links to a decrease by up to −0.165 standard deviations in the day-ahead electricity price for hour 8. In the intraday market, the relationship remains stronger with statistically significant standardized coefficients up to 0.320 in the 75 % quantile for hour 24. All in all, our results indicate an asymmetric influence of the EUA price.

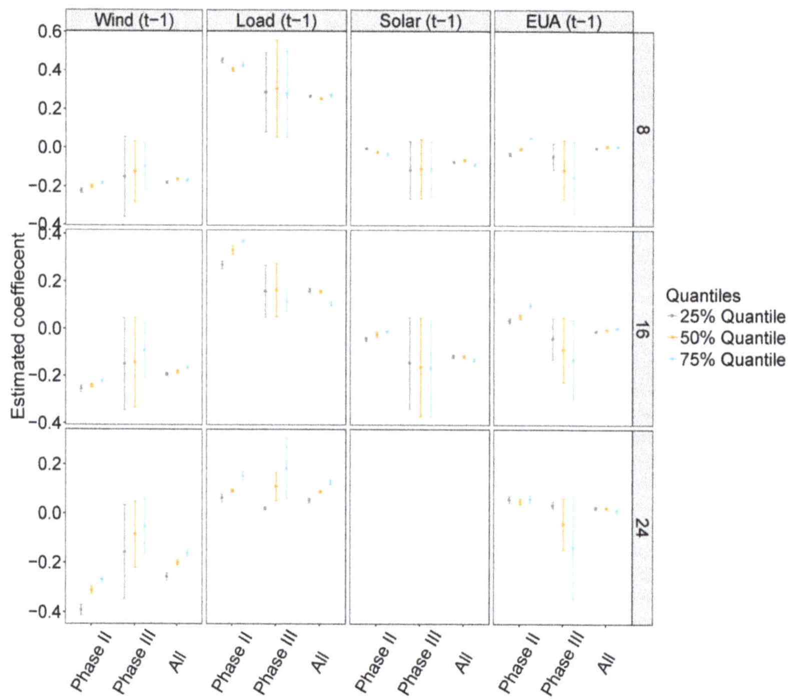

Figure 1. Results of the quantile regression for the day-ahead electricity market indicating an asymmetric influence of the EUA price on day-ahead electricity prices: Here, separate regressions are analyzed belonging to different hours ($h = 8, 16, 24$); see different rows. We first obtained difference EUA prices for the entire period to ensure stationary time series.

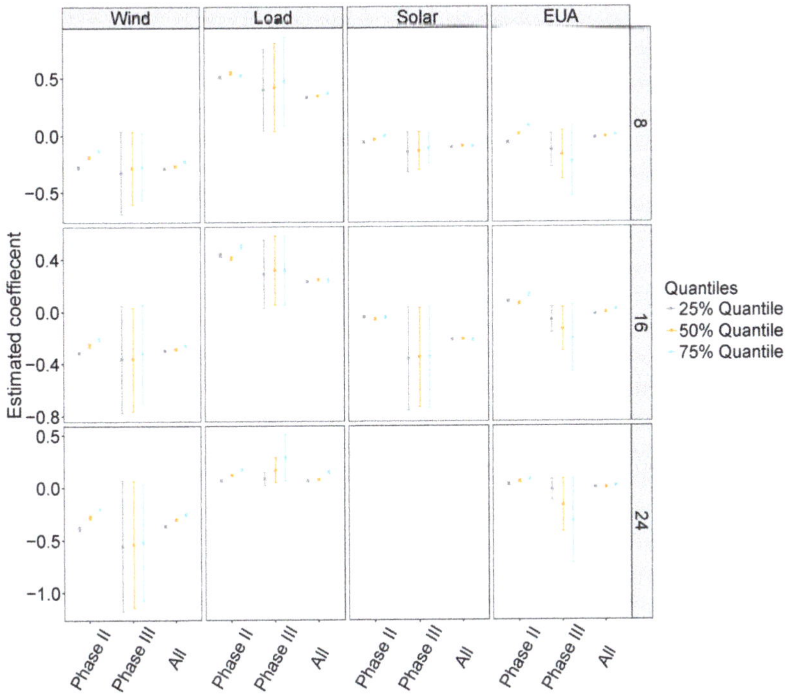

Figure 2. Results of the quantile regression for the intraday electricity market indicating an asymmetric influence of the EUA price on intraday electricity prices: Here, separate regressions are analyzed belonging to different hours ($h = 8, 16, 24$); see different rows.

5. Discussion of Findings

This section discusses our findings regarding the relationship between carbon and electricity prices. Our abovementioned findings from the autoregressive models deserve attention. According to expectation, the price of electricity is supposed to rise with a higher carbon price, thus giving rise to a positive relationship, since spending on emission allowances introduces an additional cost driver. We thus shed light on the potential reasons for the nature of the relationship:

- **Excess supply of emission allowances.** Since we find a weak and inconsistent influence of the emission allowance prices on the electricity price, the price must be too low to play a significant role in power generation. This is particularly evident in our autoregressive model during phase II, where we observe no direct effect. The same model evinces a statistically significant negative impact only for the day-ahead and intraday market in phase III. These findings are consistent with previous research, suggesting that the impact of low carbon prices is rather moderate [9,15] and becomes observable only above certain thresholds. Surprisingly, even though more industries are forced to engage in emissions trading in phase III of the EU ETS, actual power generation and the corresponding electricity price seem unaffected by carbon trading. Consequently, the presence of nonsignificant influences partially originate from an excess supply of emission allowances.
- **Changing energy mix.** The Emissions Trading System functions in a highly intricate interplay with other policies, especially the incentivized introduction of renewable energy sources. Renewables account for a growing portion of the total electricity supply. (Retrieved from http://www.bmwi-energiewende.de/EWD/Redaktion/Newsletter/2015/1/Meldung/infografik-strommix-2014-erneuerbare-auf-rekordhoch.html on 9 June 2019.) As these electricity sources replace fossil-fuel power plants, the demand for emission allowances (relative to the

total electricity demand) must decrease at the same pace. Otherwise, the burgeoning share of renewables inherently results in an excess supply of emission allowances, thus counteracting one of the main advantages of renewable energies.

- **Merit order effect.** Large carbon prices are also linked to larger marginal costs for carbon-intensive power plants as opposed to renewable energies. This explains the differential influence of carbon prices as revealed by our quantile regressions. An additional reason is given by support schemes for renewables that sometimes grant preferential treatment. That is, wind and solar power must be consumed before power is generated via other means. Hence, carbon prices have a less significant effect on electricity prices when renewables produce a high amount of electricity, i.e., when electricity prices are low.

6. Conclusions

The European Union has established a market for trading emission allowances of greenhouse gases. Its objective is to contribute to sustainability goals and to reduce emissions. The spending on emission allowances represents an integral cost driver of power generation affecting operational decision-making and thus price setting. It is therefore of interest to investigate the relationship between carbon and electricity prices across different phases of the Emissions Trading System.

6.1. Summary of the Findings

This paper contributes to the existing literature by analyzing the (asymmetric) impact of carbon prices on EPEX electricity prices, with a special focus on the intraday market. We thus use an autoregressive model with exogenous variables. The results show a behavior of the model contradictory to the intentions of policy-makers. We find a statistically highly significant negative impact in the intraday market during phase III of the European Union Emissions Trading System. Here, a one standard deviation change in the price of emission allowances decreases the price of electricity by up to −0.44 standard deviations. Moreover, the effect is weaker in the day-ahead market during phase III, while throughout phase II, we do not observe any measurable effect. Among the reasons are a growing share of renewable energy resources and an excess supply of emission allowances.

Most notably, we observe differences between the day-ahead and intraday markets. While our autoregressive model detects a weaker (negative) short-run impact of the EUA price in the day-ahead market, we find a stronger significant negative impact of the EUA price on the intraday electricity prices during phase III of the EU ETS. This outcome partially contradicts empirical studies into other markets, which mostly measure a positive impact [9]. Similarly, further research cannot find a short-run impact of the carbon prices on electricity prices [5]. Paraschiv et al. [18] provides evidence of both a positive and negative influence.

Altogether, various factors explain the relationship between emission allowances and electricity generation. Among them, we identify a combination of an excess supply of emission allowances, a growing share of renewable energy source, and the merit order effect. Since we are not aware of previous literature examining the price of emission allowances in the intraday market, we cannot compare these results to the findings of others.

To conclude, the transition from phase II to phase III of the EU ETS was motivated by policy-makers in order to reduce emissions from electricity generation. This is in correspondence to our results, according to which the price of European Emission Allowance was not linked to electricity prices at common statistical significance thresholds during phase II. It is also in line with expectations, as one was not required to hold EUA for carbon-intensive power generation during this phase. For phase III, however, we find evidence that counteracts the intention of policy-makers. Based on our model, we see that a higher EUA price is not reflected in higher electricity prices. Prior literature discussing the general design of the EU ETS has already suggested that the influence of carbon prices is fairly low, particularly due to a large supply. This is also seen in our analysis: The average price for EUA dropped extensively during the move from phase II to phase III of the EU ETS, i.e., from more than 11 €/tCO_2 to a little more of 5 €/tCO_2. This eases the pressure for electricity providers to incorporate carbon prices in their pricing models and to thus adapt their generation accordingly (e.g., they could be incentivized to use existing carbon-intensive capacities for electricity generation before an increase in carbon prices takes place).

6.2. Limitations and Call for Future Research

We focused on the short-run relationship of the EUA price and electricity prices since our aim is to investigate the price setting mechanism of electricity providers in both the day-ahead and intraday market. By following a short-run view, we were able to address the intraday variation of electricity prices for each hour of the day, which corresponds to the price setting mechanisms in the markets. As a results, future studies could built upon our research and analyze the impact of EUA prices on electricity prices, focusing on the long run. This would allow a deeper understand of how EUA prices are reflected in electricity prices. While a strength of our work is the focus on the price setting decisions, another limitation of this study originates from the circumstance that, as in other research, the actual trading models of energy firms are proprietary and thus not available for research. Nevertheless, our approach by studying ex post prices is able to shed light on the underlying price setting mechanism.

Our findings deserve attention by academics and policy-makers who should critically reflect whether this matches their intention. Hence, we regard this paper as a starting point for future research in analyzing the functioning of EU ETS with respect to electricity producers. The design of appropriate markets for emissions trading can considerably benefit from further research as many questions are still left unanswered. First of all, it is worthwhile to extensively investigate interactions between different regulations in order to derive policy implications. Here, one could even consider studying distinct price setting mechanisms, especially under increased shares of renewables in the system. These might better pass-through costs of emission allowances and thus help establish real incentives to reduce greenhouse gases. Second, further effort is necessary to understand how the price and trading volume of emission allowances impacts other commodities. With the upcoming advances in the Emissions Trading System, future research should continuously monitor the effect of carbon prices on electricity prices in order to evaluate all impending policy changes.

Supplementary Materials: The following are available online at http://www.mdpi.com/1996-1073/12/15/2894/s1.

Author Contributions: Both authors contributed equally to this work.

Funding: This research received no external funding.

Conflicts of Interest: The authors declare no conflict of interest.

References

1. European Commission. The EU Emissions Trading System. 2018. Available online: https://ec.europa.eu/clima/policies/ets (accessed on 29 March 2019).
2. Hintermann, B.; Peterson, S.; Rickels, W. Price and Market Behavior in Phase II of the EU ETS: A Review of the Literature. *Rev. Environ. Econ. Policy* **2015**, *10*, 108–128.
3. Sijm, J.; Neuhoff, K.; Chen, Y. CO_2 cost pass-through and windfall profits in the power sector. *Clim. Policy* **2006**, *6*, 49–72. [CrossRef]
4. Jouvet, P.A.; Solier, B. An overview of CO_2 cost pass-through to electricity prices in Europe. *Energy Policy* **2013**, *61*, 1370–1376. [CrossRef]
5. Cotton, D.; Mello, L.d. Econometric analysis of Australian emissions markets and electricity prices. *Energy Policy* **2014**, *74*, 475–485. [CrossRef]
6. Thoenes, S. Understanding the Determinants of Electricity Prices and the Impact of the German Nuclear Moratorium in 2011; *Energy J.* **2014**, *35*, 61–78. [CrossRef]
7. Bunn, D.; Fezzi, C. Interaction of European carbon trading and energy prices. *J. Energy Mark.* **2009**, *2*, 53–69. [CrossRef]

8. Bello, A.; Reneses, J. Electricity price forecasting in the Spanish market using cointegration techniques. In Proceedings of the 33rd Annual International Symposium on Forecasting (ISF 2013), Seoul, Korea, 23–26 June 2013; pp. 1–7.
9. Freitas, C.J.P.; da Silva, P.P. European Union emissions trading scheme impact on the Spanish electricity price during phase II and phase III implementation. *Util. Policy* **2015**, *33*, 54–62. [CrossRef]
10. Fell, H. EU-ETS and Nordic electricity: A CVAR analysis. *Energy J.* **2010**, *31*, 1–25. [CrossRef]
11. Jabłońska, M.; Viljainen, S.; Partanen, J.; Kauranne, T. The impact of emissions trading on electricity spot market price behavior. *Int. J. Energy Sect. Manag.* **2012**, *6*, 343–364. [CrossRef]
12. Bannör, K.; Kiesel, R.; Nazarova, A.; Scherer, M. Parametric model risk and power plant valuation. *Energy Econ.* **2016**, *59*, 423–434.
13. Chernyavs'ka, L.; Gullì, F. Marginal CO_2 cost pass-through under imperfect competition in power markets. *Ecol. Econ.* **2008**, *68*, 408–421.
14. Hintermann, B. Pass-Through of CO_2 Emission Costs to Hourly Electricity Prices in Germany. *J. Assoc. Environ. Resour. Econ.* **2016**, *3*, 857–891. [CrossRef]
15. Zachmann, G.; von Hirschhausen, C. First evidence of asymmetric cost pass-through of EU emissions allowances: Examining wholesale electricity prices in Germany. *Econ. Lett.* **2008**, *99*, 465–469. [CrossRef]
16. Aatola, P.; Ollikainen, M.; Toppinen, A. Impact of the carbon price on the integrating European electricity market. *Energy Policy* **2013**, *61*, 1236–1251. [CrossRef]
17. Woo, C.K.; Olson, A.; Chen, Y.; Moore, J.; Schlag, N.; Ong, A.; Ho, T. Does California's CO_2 price affect wholesale electricity prices in the Western U.S.A.? *Energy Policy* **2017**, *110*, 9–19.. [CrossRef]
18. Paraschiv, F.; Erni, D.; Pietsch, R. The impact of renewable energies on EEX day-ahead electricity prices. *Energy Policy* **2014**, *73*, 196–210. [CrossRef]
19. Bierbrauer, M.; Menn, C.; Rachev, S.T.; Trück, S. Spot and derivative pricing in the EEX power market. *J. Bank. Financ.* **2007**, *31*, 3462–3485.
20. Deng, S.; Oren, S.S. Electricity derivatives and risk management. *Energy* **2006**, *31*, 940–953.
21. Knittel, C.R.; Roberts, M.R. An empirical examination of restructured electricity prices. *Energy Econ.* **2005**, *27*, 791–817.
22. Weron, R. *Modeling and Forecasting Electricity Loads and Prices: A Statistical Approach*; John Wiley & Sons: Chichester, UK, 2007; Volume 403.
23. Paschen, M. Dynamic analysis of the German day-ahead electricity spot market. *Energy Econ.* **2016**, *59*, 118–128.
24. Contreras, J.; Espinola, R.; Nogales, F.J.; Conejo, A.J. ARIMA models to predict next-day electricity prices. *IEEE Trans. Power Syst.* **2003**, *18*, 1014–1020.
25. Fuglerud, M.; Vedahl, K.E.; Fleten, S.E. Equilibrium simulation of the Nordic electricity spot price. In Proceedings of the 2012 9th International Conference on the European Energy Market, Florence, Italy, 10–12 May 2012; pp. 1–10.
26. Ludwig, N.; Feuerriegel, S.; Neumann, D. Putting big data analytics to work: Feature selection for forecasting electricity prices using the LASSO and random forests. *J. Decis. Syst.* **2015**, *24*, 19–36.
27. Weron, R. Electricity price forecasting: A review of the state-of-the-art with a look into the future. *Int. J. Forecast.* **2014**, *30*, 1030–1081. [CrossRef]
28. Ferkingstad, E.; Løland, A.; Wilhelmsen, M. Causal modeling and inference for electricity markets. *Energy Econ.* **2011**, *33*, 404–412.
29. Wolff, G.; Feuerriegel, S. Short-term dynamics of day-ahead and intraday electricity prices. *Int. J. Energy Sect. Manag.* **2017**, *11*, 557–573.
30. Beran, P.; Pape, C.; Weber, C. Modelling German electricity wholesale spot prices with a parsimonious fundamental model—Validation and application. *Util. Policy* **2019**, *58*, 27–39. doi:10.1016/j.jup.2019.01.008. [CrossRef]
31. Benhmad, F.; Percebois, J. Photovoltaic and wind power feed-in impact on electricity prices: The case of Germany. *Energy Policy* **2018**, *119*, 317–326. [CrossRef]
32. Clò, S.; Cataldi, A.; Zoppoli, P. The merit-order effect in the Italian power market: The impact of solar and wind generation on national wholesale electricity prices. *Energy Policy* **2015**, *77*, 79–88.
33. Gelabert, L.; Labandeira, X.; Linares, P. An ex-post analysis of the effect of renewables and cogeneration on Spanish electricity prices. *Energy Econ.* **2011**, *33*, 59–65.

34. Mosquera-López, S.; Nursimulu, A. Drivers of electricity price dynamics: Comparative analysis of spot and futures markets. *Energy Policy* **2019**, *126*, 76–87. [CrossRef]
35. Wooldridge, J. *Introductory Econometrics: A Modern Approach*, 5th ed.; Cengage Learning: Boston, MA, USA, 2012.
36. Paraschiv, F.; Bunn, D.W.; Westgaard, S. *Estimation and Application of Fully Parametric Multifactor Quantile Regression with Dynamic Coefficients*; University of St. Gallen, School of Finance Research Paper No. 2016/07; 2016. Available online: https://ssrn.com/abstract=2741692 (accessed on 29 March 2019).
37. Koenker, R.; Bassett, G. Regression Quantiles. *Econometrica* **1978**, *46*, 33–50.
38. Pape, C.; Hagemann, S.; Weber, C. Are fundamentals enough? Explaining price variations in the German day-ahead and intraday power market. *Energy Econ.* **2016**, *54*, 376–387. [CrossRef]
39. Würzburg, K.; Labandeira, X.; Linares, P. Renewable generation and electricity prices: Taking stock and new evidence for Germany and Austria. *Energy Econ.* **2013**, *40*, 159–171.

© 2019 by the authors. Licensee MDPI, Basel, Switzerland. This article is an open access article distributed under the terms and conditions of the Creative Commons Attribution (CC BY) license (http://creativecommons.org/licenses/by/4.0/).

energies

MDPI

Article

Classification of Renewable Sources of Electricity in the Context of Sustainable Development of the New EU Member States

Jacek Brożyna [1,*], Grzegorz Mentel [1], Eva Ivanová [2] and Gennadii Sorokin [3]

1 Department of Quantitative Methods, The Faculty of Management, Rzeszow University of Technology, 35-959 Rzeszow, Poland; gmentel@prz.edu.pl

2 Department of Economics and Economy, Faculty of Social Economics Relationships, Alexander Dubcek University of Trencin, 911 50 Trencin, Slovakia; eva.ivanova@tnuni.sk

3 Department of Business Informatics and Mathematics, Tyumen Industrial University, 625000 Tyumen, Russia; sgenall@yandex.ru

* Correspondence: jacek.brozyna@prz.edu.pl; Tel.: +48-792-395-486

Received: 12 May 2019; Accepted: 10 June 2019; Published: 13 June 2019

Abstract: Climate change and awareness of the need to care for the environment have resulted in a global increase in the interest in renewable energy sources. The European Union (EU) is active in this respect and requires Member States to fulfill specific plans in the transformation of their energy systems. We employed hierarchical cluster analysis in an attempt to distinguish those countries among the new EU Member States that increased their electrical capacity from renewable energy sources to the greatest extent while paying attention to their energy intensity. The analyses were conducted in two scenarios for both 2004 and 2016. The first scenario assumed an analysis of all known renewable energy sources, whereas in the second scenario, only renewable energy sources from wind and solar power plants were included. The division of analyses into these two variants showed the importance of the differences in the energy assessment of individual countries, depending on classification of renewable energy sources. We identified groups of countries where electrical capacity from renewable energy sources increased the most. Conducting analyses using two variants allowed distinguishing countries that based most of their renewable energy on modern renewable energy sources, such as solar and wind power plants. The inclusion of gross domestic product in the analyses allowed us to identify countries with the worst energy efficiency value.

Keywords: renewable energy; sustainable development; gross domestic product; GDP; electrical capacity; energy intensity; hierarchical cluster analysis

JEL Classification: Q01; Q40; Q48; Q56; Q48; Q20; R11; O10

---

1. Introduction

Electricity is the basis for the functioning of the modern world, but its acquisition is often not environmentally friendly. Nearly 200 years have passed since the beginning of the industrial revolution, but many countries still base their energy on fossil fuels [1–3]. Along with an increase in ecological awareness, many countries are aiming to reduce human interference in the natural environment and obtain energy from renewable sources. Wind energy, solar radiation, precipitation, tides, sea waves, and geothermal energy are considered renewable energy sources [4,5]. Biofuels, biomass, and biogas are also considered renewable energy sources if their origin is ecological, but their conversion into energy through combustion is not [6,7]. Similarly, the qualification of hydroelectric power plants as renewable energy sources is controversial. Large hydroelectric plants have a negative impact on the environment [8–16], and for this reason, as in case of biofuels, they are often not included in studies on renewable energy sources [17–22].

Many factors impact the introduction of pro-ecological solutions into the energy policy of states. Apart from the most obvious, such as the level of economic development and the volume of energy production, there are also important geographical determinants [23,24] or social acceptance factors of renewable energy prices must be considered [25–27]. In the literature, these factors are often analyzed in pairs or groups, e.g., gross domestic product (GDP) and CO_2 emissions [28–33].

Regardless of the region of the world, the majority of energy demand forecasts show that demand will grow in the within several years or even a few decades [3,34–42]. Various studies show a relationship between economic development and energy demand. Most of the papers indicate that economic growth determines energy demand, which is referred to as the "conservation hypothesis" [43–49]. The "growth hypothesis" states that economic growth depends on energy production, but, as research shows, economic growth happens when a government policy leads to an excessive reduction of energy consumption [38,50–53]. Regardless of which of the listed factors is a determinant, individual countries need to implement a sustainable growth policy that maintains a balance between the development of countries and regions and the demand for energy [24,54–60]. International commitments and growing environmental awareness result in the same countries often deciding to subsidize investments in renewable energy sources to ensure uninterrupted energy supply to customers, and to simultaneously to provide green energy [2,61–65]. Notably, subsidies do not include hydroelectric power plants because they cannot always be considered environmentally friendly and, due to the need to diversify energy sources, do not fit into sustainable development policy [10,17,18,66].

In Europe, the need to reduce energy consumption and care for the environment started being widely discussed in the late 1960s [67–69]. The oil crises of 1973–1974 and 1979–1982 contributed to discussions on the common energy market, which was reflected in the Treaty of Lisbon in 1992. The four main assumptions of the common energy policy were contained in Article 194 of the Treaty on the Functioning of the European Union (TFEU) and they concerned: guaranteeing the functioning of the energy market; guaranteeing energy supplies to European Union (EU) countries; promoting energy efficiency and development of new, renewable energy sources of energy; and promoting inter-state energy connections.

In subsequent years, further documents and directives were published, e.g., Green Paper, White Paper, and Directives 96/92/EC and 98/30/EC, aimed at regulating the common energy market in the EU, considering the specific energy markets of individual Member States. The low effectiveness in the implementation of new laws in the Member States and the largest enlargement of the EU in history in 2004 created a need to develop a new law that would effectively regulate the common energy market. The Directive of 2009 imposed an obligation on Member States to reduce greenhouse gas emissions considering the structure of energy systems of individual countries and the level of their economic development.

The presented legal regulations are reflected in data on renewable energy sources in EU. The data show that at the beginning of the 21st century, the energy infrastructure was modernized to a greater extent so that it would be less harmful to the environment [1]. This is visible for the entire EU and even more so among its new Member States, as shown in Figure 1. This chart includes the aggregated capacity of renewable energy sources from wind and solar power plants.

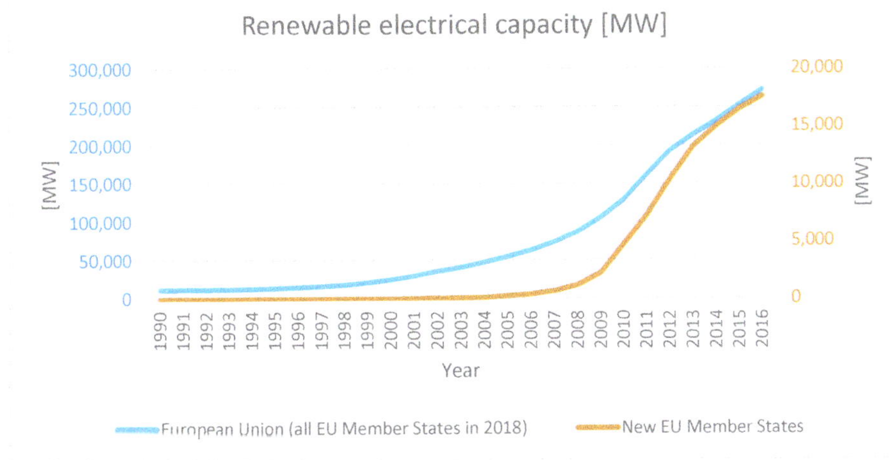

Figure 1. Joint electrical capacity in the European Union (EU). Source: Own visualizations based on [70].

As mentioned above, the EU, while regulating the common energy market, pays special attention to caring for the natural environment, but simultaneously understands that the energy markets and economies of individual countries differ. It is important to consider these differences in requirements for individual countries as, for example, in Central and Eastern European countries, (CEECs) rapid growth of energy production from renewable sources to the level in richer countries of Western Europe cannot be expected. Germany, for example, produces 42% of electricity from renewable sources [25]. Not only is the level of wealth of a given country significant, as measured by GDP, but also the initial structure of renewable energy shares in all types of energy sources. The ratio of energy to gross domestic product (E/GDP), i.e., energy intensity, is also important [35,54,71–76]. This factor describes the energy efficiency of the state's economy and indicates the cost of converting energy to GDP. The values of these factors differ significantly between the countries of Western Europe and Central and Eastern Europe and change dynamically over time. The best example for comparison is Germany and Romania. In 1992, energy intensity in Romania was four times higher than in Germany, and in 2013, it was only twice as much [35]. Countries that joined the EU in 2004 and subsequent years [77,78] are significantly different from the older Member States of the EU. Differences are related to many aspects, among which the most important from the point of view of this paper are the level of economic development and the structure of energy sources. The GDP of new Member States of the EU was much lower than that of the Member States of the EU from Western Europe [79,80], and with few exceptions, combustible fuels were the main sources of energy for new EU Member States [81–83]. For this reason, we focused on analyzing renewable energy sources against the background of economic growth only among new Member States of the EU. The sources of renewable energy were analyzed using two variants. In the first, all officially recognized renewable energy sources were considered as renewable energy sources. In the second variant, to remove the impact of the hydroelectric power stations built several dozens of years ago, only wind and solar power plants were accepted as renewable sources. Both options for the classification of renewable energy sources were analyzed for 2004 and 2016 to determine the level of change. The tests were performed with an application of cluster analysis [84–87], enabling us to create groups of countries similar to each other in terms of renewable energy sources and GDP in 2004 and 2016.

In the literature, many publications have separately analyzed renewable energy sources as a whole or individually. A novelty in our article is a comparative analysis that considered the different classifications of renewable energy sources. As such, we were able to determine the differences in the energy assessment of individual countries, depending on the classification of renewable energy sources. Separation of modern energy sources from all energy sources in comparison with the GDP also enabled an assessment of the energy efficiency from these renewable energy sources.

2. Materials and Methods

2.1. Data

We analyzed the changes in the capacity of the electrical infrastructure with a special focus on renewable energy sources and their relationship with GDP. The analysis included countries that have been Member States of the EU since 2004. The largest number of countries joined the EU on May 1, 2004: Cyprus, the Czech Republic, Estonia, Hungary, Latvia, Lithuania, Malta, Poland, Slovakia, and Slovenia. On January 1, 2007, Bulgaria and Romania joined, as did Croatia on July 1, 2013. Since 2004, the EU expanded by 13 countries; the main ones were from Central and Eastern Europe, as presented in Figure 2.

Figure 2. Map of the studied countries with the year they joined the EU.

The analyzed data included the capacity of the electrical infrastructure and the GDP of the aforementioned countries, and they were collected from Eurostat websites [70,88]. We are aware of possible errors during the process of data sampling [89]; therefore, the data were collected and verified again after a few weeks. The data have annual periodicity. Recent data on electrical infrastructure (accessed on June 14, 2018) were from 2016, and GDP data were from 2017. Therefore, the surveys included a 13-year period from 2004 to 2016. The data on the capacity of the electrical infrastructure are expressed in megawatts (MW), and GDP at current prices in million euro.

2.2. Methodology

For data sets that contain different features, clustering can be used to compare them [84]. A taxonomic analysis was used to designate groups of countries similar in terms of the capacity of the electrical infrastructure and share in the GDP [85] using Ward's clustering, which is a hierarchical method [86]. This method, which is an agglomerative clustering method, is one of the best, through which homogeneous aggregates can be obtained.

In this method, at the beginning, it is assumed that each observation vector is a separate cluster. Then, between all pairs of vectors in Equations (1) and (2), using the squared Euclidean distance (SED) in Equation (3), a distance matrix is determined, using Equation (4), which describes their similarity.

$$a = [a_1, \ldots, a_i] \tag{1}$$

$$b = [b_1, \ldots, b_i] \tag{2}$$

where a and b are the observation vector

$$d(a,b) = \sqrt{\sum_{i=1}^{p} (a_i - b_i)^2} \tag{3}$$

where p denotes the number of variables (vector length).

$$d(a,b) = \begin{bmatrix} 0 & d_{12} & \cdots & d_{1n} \\ d_{21} & 0 & \cdots & d_{2n} \\ \vdots & \vdots & \ddots & \vdots \\ d_{n1} & d_{n2} & \cdots & 0 \end{bmatrix} \tag{4}$$

where d_{ij} is the distance between the ith and the jth observation

The above distance matrix is based on physical space. This is reminiscent of the topological distance matrix based on network structures [90,91]. Clusters (groups) are created by applying one of several available grouping methods on the distance matrix [92–94]. In Ward's method, the distance between clusters is estimated by an analysis of variance. It is assumed that each cluster is represented by a centroid, as shown in Figure 3.

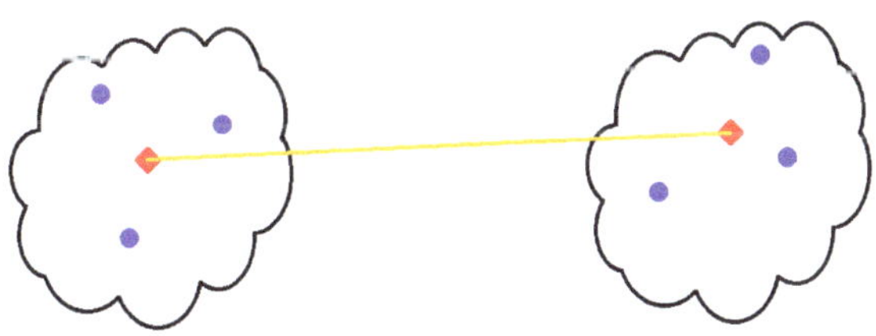

Figure 3. Centroids of clusters.

At each stage of the agglomeration hierarchical grouping process into a new cluster, the two most similar clusters are combined, e.g., A and B (Figure 4), for which there is the smallest increase in the sum of the squared error (SSE):

$$d(A,B) = SSE_{A \cup B} - (SSE_A + SSE_B) \tag{5}$$

$$SSE_{A \cup B} = \sum_{i=1}^{n_{AB}} (y_i - \overline{y_{AB}})'(y_i - \overline{y_{AB}}) \tag{6}$$

$$SSE_A = \sum_{i=1}^{n_A} (a_i - \overline{a})'(a_i - \overline{a}) \tag{7}$$

$$SSE_B = \sum_{i=1}^{n_B} \left(b_i - \overline{b}\right)'\left(b_i - \overline{b}\right) \tag{8}$$

where a_i represents the ith observation vector in cluster A, $\overline{a}$ is the centroid of cluster A, b_i represents the ith observation vector in cluster B, $\overline{b}$ the centroid of cluster B, y_i represents the ith observation vector in cluster AB, and $\overline{y_{AB}}$ the centroid of newly formed cluster AB.

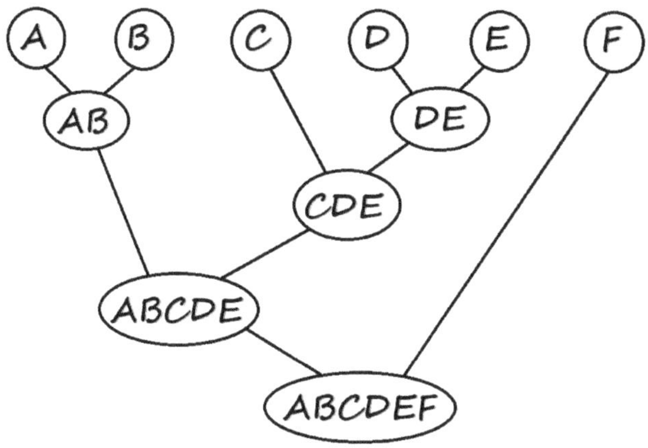

Figure 4. Hierarchical clustering.

The minimize function using the Ward minimal variance method can also be written as:

$$d(A,B) = \frac{n_A n_B}{n_A + n_B}\left(\overline{a} - \overline{b}\right)'\left(\overline{a} - \overline{b}\right) \tag{9}$$

where $\overline{a}$ and $\overline{b}$ represent the centroids of clusters A and B, respectively.

The process of determining the distance between clusters and joining them ends when all clusters are combined into one large cluster, e.g., ABCDEF in Figure 4.

Variables analyzed with the use of the Ward's method should have a coefficient of variation greater than 10% and should not be very strongly correlated. However, leaving variables out that do not meet these criteria for an analysis is allowed if these variables are significant from the point of view of the studied phenomenon.

3. Results

3.1. Cluster Analysis

As mentioned above, the data used for the analyses included the electrical capacity and GDP in the new EU Member States. Their short forms are introduced for the analysis: **EC** is the **E**lectrical **C**apacity, which is the sum of the capacity of all types of electricity sources; **ECR** is the **E**lectrical **C**apacity **R**enewable, which is the total capacity of the renewable energy sources by the most commonly used divisions, i.e., hydro, geothermal, wind, and solar; and **ECRN** denotes **E**lectrical **C**apacity **R**enewable **N**ew, which is the total capacity of only new types of renewable energy sources, i.e., wind and solar.

Cluster analysis was conducted for the data from the beginning and end of the analyzed period, i.e., 2004 and 2016. For this purpose, the variable designations introduced above were additionally determined for the relevant year: EC2004, ECR2004, ECRN2004, GDP2004, EC2016, ECR2016, ECRN2016, and GDP2016. These variables were used to create indicators describing the ratio of electrical capacity from renewable sources to total electric capacity and the ratio of electrical capacity from renewable sources and all sources in the ratio of GDP, which is referred to in the literature as energy intensity. The developed indicators were divided into those concerning the analysis of all renewable energy sources (indicators renewable energy; IRE) and those that concern only renewable energy sources of new type (indicators renewable energy new; IREN) for both 2004 and 2016, as presented in Tables 1 and 2.

Table 1. Indicators of renewable energy.

IRE 2004	IRE 2016
ECR2004/EC2004	ECR2016/EC2016
EC2004/GDP2004	EC2016/GDP2016
ECR2004/GDP2004	ECR2016/GDP2016

Table 2. Indicators of new renewable energy.

IREN 2004	IREN 2016
ECRN2004/EC2004	ECRN2016/EC2016
EC2004/GDP2004	EC2016/GDP2016
ECRN2004/GDP2004	ECRN2016/GDP2016

The above division of indicators allowed us to perform four cluster analyses (two for each of the analyzed years) to check how these groups changed over the period, but also to determine how the type of selected energy influenced the formation of these groups. The coefficients of variation of all indicators presented in Table 3 are 10% above the criterion, which means that they could be used for the cluster analysis.

Table 3. Coefficients of variation of the indicators.

Indicator	Coefficient of Variation (%)
ECR2004/EC2004	104.3
ECR2010/EC2010	78.8
ECR2016/EC2016	54.0
ECRN2004/EC2004	240.0
ECRN2010/EC2010	85.4
ECRN2016/EC2016	41.9
EC2004/GDP2004	65.2
EC2010/GDP2010	40.8
EC2016/GDP2016	36.8
ECR2004/GDP2004	104.4
ECR2010/GDP2010	90.4
ECR2016/GDP2016	74.7
ECRN2004/GDP2004	227.0
ECRN2010/GDP2010	100.7
ECRN2016/GDP2016	67.8

Correlation coefficients between indicators are presented in Table 4.

The coefficient of correlation only exceeded 90% for the pairs of ECRN2004/GDP2004—ECRN2004/EC2004. However, due to the high volatility of ECRN2004/GDP2004 of 227.0%, ECRN2004/EC2004 of 240.0%, and the need to examine this indicator, it was not rejected. The need to maintain the same set of variables to ensure comparability of results was also an argument for including these indicators. When analyzing the source data for 2004, we confirmed that the resulting correlation is apparent because it resulted from the lack of renewable energy sources of a new type in almost all studied countries, which further affected the almost zero value of the discussed factors. In subsequent years, the electrical capacity from new types of renewable energy sources increased, which confirms the need to retain all indicators to ensure the comparability of groups.

Table 4. Coefficients of variation of indicators.

Indicator *i*—Indicator *j*	Coefficient of Variation (%)
ECR2004/EC2004—EC2004/GDP2004	7.3
ECR2004/GDP2004—EC2004/GDP2004	63.0
ECR2004/GDP2004—ECR2004/EC2004	77.5
ECR2010/EC2010—EC2010/GDP2010	20.6
ECR2010/GDP2010—EC2010/GDP2010	62.2
ECR2010/GDP2010—ECR2010/EC2010	85.6
ECR2016/EC2016—EC2016/GDP2016	39.7
ECR2016/GDP2016—EC2016/GDP2016	80.9
ECR2016/GDP2016—ECR2016/EC2016	83.9
ECRN2004/EC2004—EC2004/GDP2004	−4.4
ECRN2004/GDP2004—EC2004/GDP2004	−0.8
ECRN2004/GDP2004—ECRN2004/EC2004	99.4
ECRN2010/EC2010—EC2010/GDP2010	25.4
ECRN2010/GDP2010—EC2010/GDP2010	69.5
ECRN2010/GDP2010—ECRN2010/EC2010	85.2
ECRN2016/EC2016—EC2016/GDP2016	23.9
ECRN2016/GDP2016—EC2016/GDP2016	82.3
ECRN2016/GDP2016—ECRN2016/EC2016	73.6

Countries were grouped separately for each year: once for the indicators including all renewable energy sources (IRE), and the second for the indicators where only wind and solar power plants (IRENs) were accepted as renewable energy sources. In total, four analyses were performed, where the division of the optimal number of clusters was determined [40,41]. Statistica 12.5 (TIBCO Software Inc., Palo Alto, CA, USA) was used as a tool to develop clusters.

3.1.1. Groups for IRE Indicators in 2004

In 2004 (and many years before), hydroelectric and geothermal power plants were the most frequently used renewable sources of electricity in the world, and the main source of energy in the Central and Eastern European countries (CEES) was hydroelectric power plants. While grouping such data in 2004, a tree diagram was developed, as shown in Figure 5.

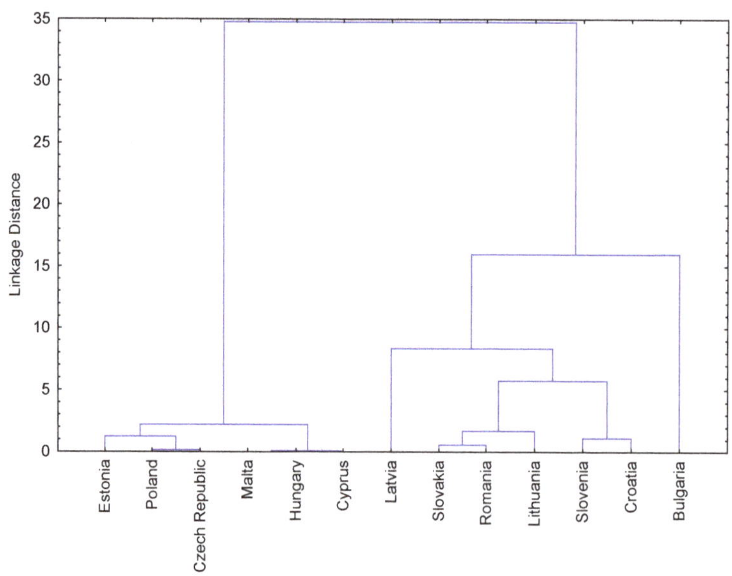

Figure 5. Tree Diagram for IRE in 2004.

Cutting off the tree diagram at a distance of 10 resulted in three clusters containing countries (Table 5) with the averages of the groups shown in Table 6. The analysis of variance (ANOVA) for the indicators from Table 6 was completed at the 0.05 significance level. The P-value for each of the indicators is smaller than the assumed level of significance, which means statistically significant differences exist between the groups of countries listed in Table 5.

Table 5. Clusters for IRE in 2004.

Group No.	A	B	C
1	Cyprus	Croatia	Bulgaria
2	Czech Republic	Latvia	
3	Estonia	Lithuania	
4	Hungary	Romania	
5	Malta	Slovakia	
6	Poland	Slovenia	

Table 6. Group averages for IRE in 2004.

Group\Indicator	ECR2004/EC2004	EC2004/GDP2004	ECR2004/GDP2004
A	0.034000	0.150845	0.005761
B	0.396833	0.210875	0.075654
C	0.232000	0.585706	0.136107
All groups	0.220944	0.212002	0.048046
p-value	*0.000742*	*0.002187*	*0.005618*

To compare groups, on the basis of Table 6, the values of indicators in each group were determined in relation to the general average, as shown in Figure 6.

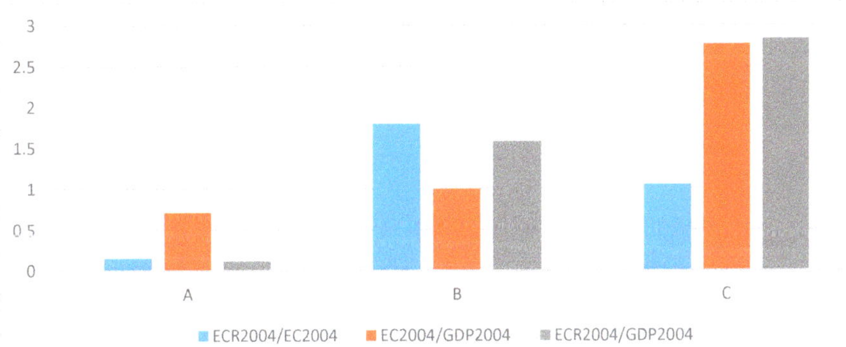

Figure 6. Comparison of group averages for IRE in 2004.

When using cluster analysis for traditional renewable energy sources, three clusters were created for the IRE data in 2004. In Table 5, groups A and B contain six countries, while group C only contains Bulgaria. When analyzing Figure 6, in this group (and the only country in this group), renewable energy sources (ECR2004/EC2004) mean that it is ranked in the middle of the surveyed countries. However, high energy intensity (EC2004/GDP2004 and ECR2004/GDP2004) was the main reason for the creation of this group, which means high energy costs are responsible for generating the GDP of Bulgaria. Groups A and B are opposites in terms of traditional, renewable energy sources. Group A included countries whose main sources of electricity were non-renewable energy sources, and renewable energy sources accounted for only a small percentage of all electrical capacity or none at all, as shown in Figure 7.

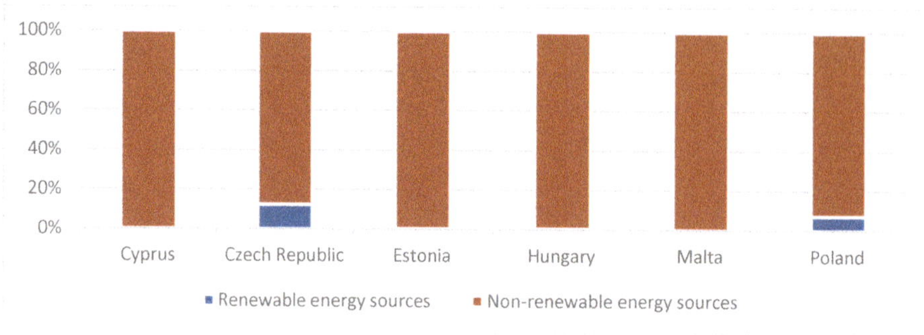

Figure 7. Energy sources of the countries in group A in 2004.

In the countries from group B, traditional renewable energy sources accounted for for several dozen percent of all electricity sources, and the cost of electricity converted into GDP was moderate. This means that group B is the most ecological group of countries according to the assumed criterion for 2004.

3.1.2. Groups for IRE Indicators in 2016

After joining the EU, new countries are obliged to implement a pro-ecological policy. For example, the Directive 2009/29/EC obliged Member States to reduce greenhouse gas emissions. With the increasing demand for electricity and restrictions resulting from EU directives, the most reasonable solution was to increase the electrical infrastructure capacity through investments in renewable energy sources. Treating all types of renewable energy sources in the same way and subjecting the countries to a re-analysis of clusters for 2016, a tree diagram was produced, as shown in Figure 8.

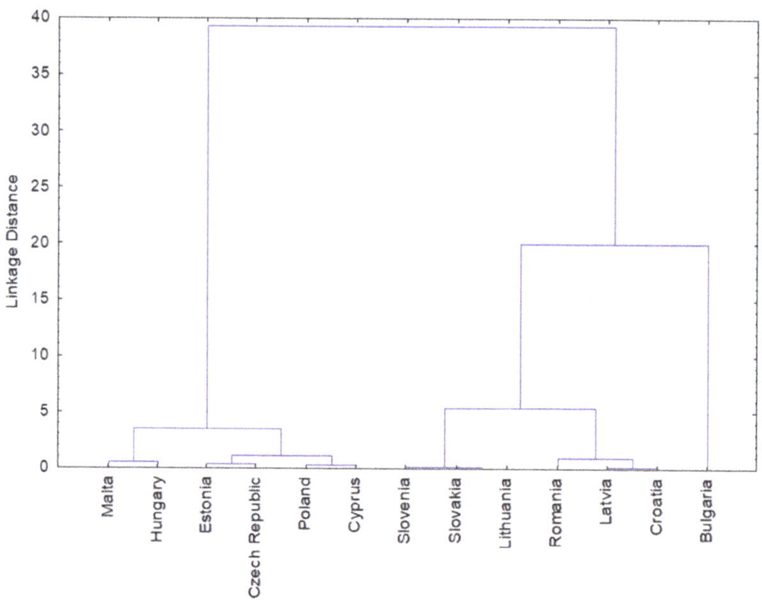

Figure 8. Tree diagram for IRE in 2016.

By cutting off the tree diagram at a distance of 10 for 2016 and 2004, three clusters re-emerged. Countries belonging to individual groups and average values in these groups are listed in Tables 7 and 8.

Table 7. Clusters for IRE in 2016.

Group No.	A	B	C
1	Cyprus	Croatia	Bulgaria
2	Czech Republic	Latvia	
3	Estonia	Lithuania	
4	Hungary	Romania	
5	Malta	Slovakia	
6	Poland	Slovenia	

Table 8. Group averages for IRE in 2004.

Group Indicator	ECR2016/EC2016	EC2016/GDP2016	ECR2016/GDP2016
A	0.153333	0.094233	0.014727
B	0.469833	0.106539	0.050602
C	0.461000	0.223131	0.102849
All Groups	0.323077	0.109828	0.038063
p-value	*0.000027*	*0.001363*	*0.000042*

The differences between groups are statistically significant, as demonstrated in the analysis of variance (*p*-value in Table 8). The values of the indicators in the groups in relation to the average of all groups are presented in Figure 9.

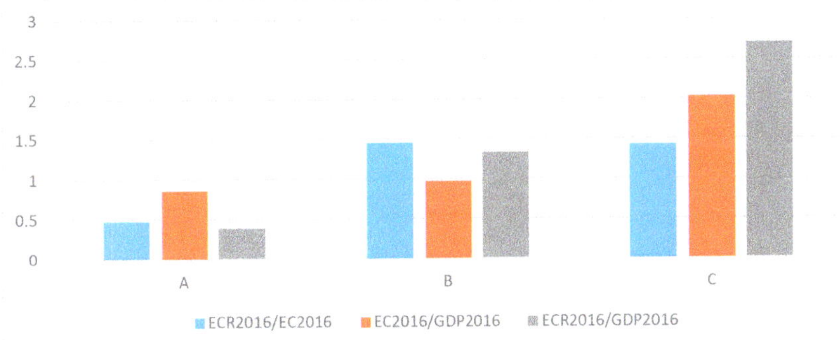

Figure 9. Comparison of group averages for IRE in 2016.

When comparing the results of the clustering analysis for 2016 with previously obtained results from 2004, no major changes are visible. The number of created groups is the same and their composition is identical. When comparing Figure 9 with its counterpart for data from 13 years ago (Figure 6), the similarities are noticeable. When analyzing Figure 9 more precisely, the biggest change is visible in group A, where the share of electrical capacity from renewable energy sources increased the most, both in relation to the total electrical capacity and the GDP. This conclusion is also confirmed by the analysis of the distribution of types of energy sources in the countries of group A, which is presented in Figure 10.

The traditional classification of renewable energy sources means that, as in the analyses presented above, the actual investment of countries in switching their economies to greener and more modern energy sources can be overlooked. This is due to the fact that hydroelectric power plants are also renewable energy sources, which in some countries have been a large part of electricity capacity for decades. This situation mean that with relatively young wind and solar energy infrastructure, expenditure on their development may be unnoticeable or misinterpreted if their electrical capacity in the analysis is included with the electric capacity from hydropower. Despite the classification of hydroelectric power plants as renewable energy sources, only small power plants with a capacity of up to several megawatts are considered as such. Larger hydropower plants have a negative impact on the environment, and thus should not be treated as renewable energy sources.

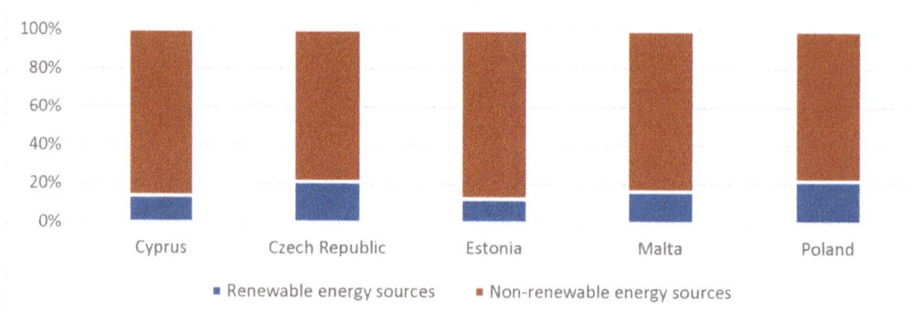

Figure 10. Energy sources in the countries of group A in 2016.

3.1.3. Groups for IREN Indicators in 2004

In the last dozen or so years when writing about renewable energy, we were rather thinking of dynamically developing wind and solar energy, not about hydroelectric power plants. This trend resulted from the global energy policy, which assumes that renewable energy is not enough—it needs to be sustainable. For this reason, the analyses for 2004 and 2016 were reconstructed for a comparison with the assumption that only wind and solar power plants are renewable energy sources. These analyses showed the extent to which the new type of renewable energy sources affect the classification of countries and the ratio of electrical capacity only from this type of energy in relation to GDP. These indicators designated for this type of energy were designated as IREN.

Assuming that only wind and solar power plants are renewable sources of energy, in 2004, they constituted only 0.08% of the total electrical capacity. For comparison, the electrical capacity of the hydroelectric plants alone was 18.60%. The introduced change fully altered the tree diagram (Figure 11) resulting from cluster analysis for IREN indicators compared to that presented earlier for IRE indicators in 2004.

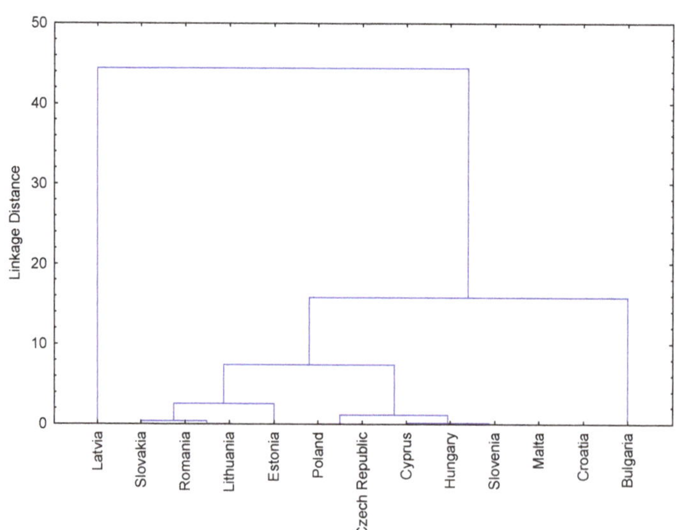

Figure 11. Tree diagram for IREN in 2004.

Cutting off the tree diagram as in the previous analyses at a distance of 10 resulted in the creation of three groups, but other than the C group, the groups included different countries (Table 9).

Table 9. Clusters for IREN in 2004.

No. Group	A	B	C
1	Latvia	Croatia	Bulgaria
2		Cyprus	
3		Czech Republic	
4		Estonia	
5		Hungary	
6		Lithuania	
7		Malta	
8		Poland	
9		Romania	
10		Slovakia	
11		Slovenia	

Group averages and their values in relation to the overall average are presented in Table 10 and Figure 12. The analysis of variance at the significance level of 0.05 showed statistically significant differences between the values of indicators in these groups, which means the groups have been correctly created and are significantly different from each other.

Table 10. Group averages for IREN in 2004.

Group\Indicator	ECRN2004/EC2004	EC2004/GDP2004	ECRN2004/GDP2004
A	0.012054	0.184781	0.002227
B	0.000520	0.180504	0.000115
C	0.000082	0.585706	0.000048
All Groups	0.001373	0.212002	0.000273
p-value	*0.000000*	*0.000009*	*0.004595*

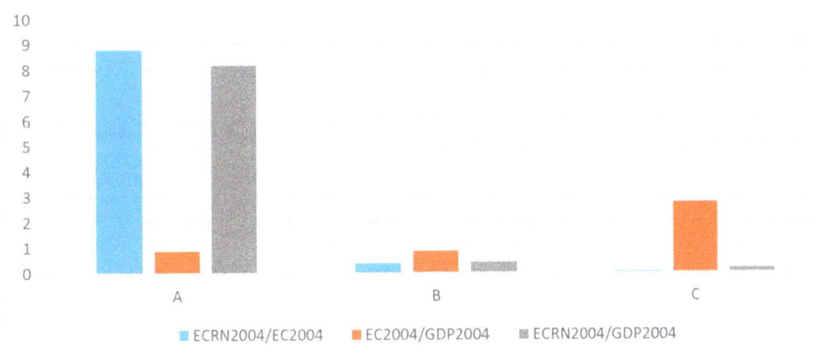

Figure 12. Comparison of group averages for IREN in 2004.

As mentioned above, the groups based on IREN indicators have different compositions, but, as shown in Figure 12, their nature is also different. Group A only contains Latvia, which stands out from the rest of the world in having the largest electrical capacity from wind energy. Group C, as in previous analyses, contains only Bulgaria. Again, the main reason for this situation is the high cost of energy in relation to GDP. The largest group B contains as many as 11 countries where the electrical capacity from wind and solar plants in relation to the total electrical capacity of each of these countries is negligible.

3.1.4. Groups for IREN Indicators in 2016

An increase in demand for electricity and the EU's climate policy has forced the Member States countries to invest in renewable energy sources. Wind and solar power plants belong to the most frequently developed investments in recent years. In most countries, the construction of new power plants for combustible fuels was practically discontinued due to their negative impact on the environment and long construction time and high costs. Similarly, the construction of nuclear power plants requires large financial outlays and building time. The security of these facilities and the use of radioactive waste are also debatable. Hydroelectric plants, although they are classified as renewable energy sources, but as mentioned before, have a negative impact on the natural environment if their power generated is greater than a dozen or so megawatts, and the vast majority of hydroelectric power plants in the studied countries produce much more power. The time required to design and build such plants is also quite long. Wind and solar power plants have become a natural choice as their construction time is shorter compared to power plants. The electrical capacity of wind and solar power plants depends mainly on the space they occupy, meaning smaller investors can also build them and create a dispersed network of small power plants. Considering only this type of power plants in cluster analysis allowed us to eliminate data disturbances caused by hydroelectric plants, and thus to more accurately group countries in terms of their investments in renewable energy sources. The tree diagram created for IREN indicators for 2016 is presented in Figure 13.

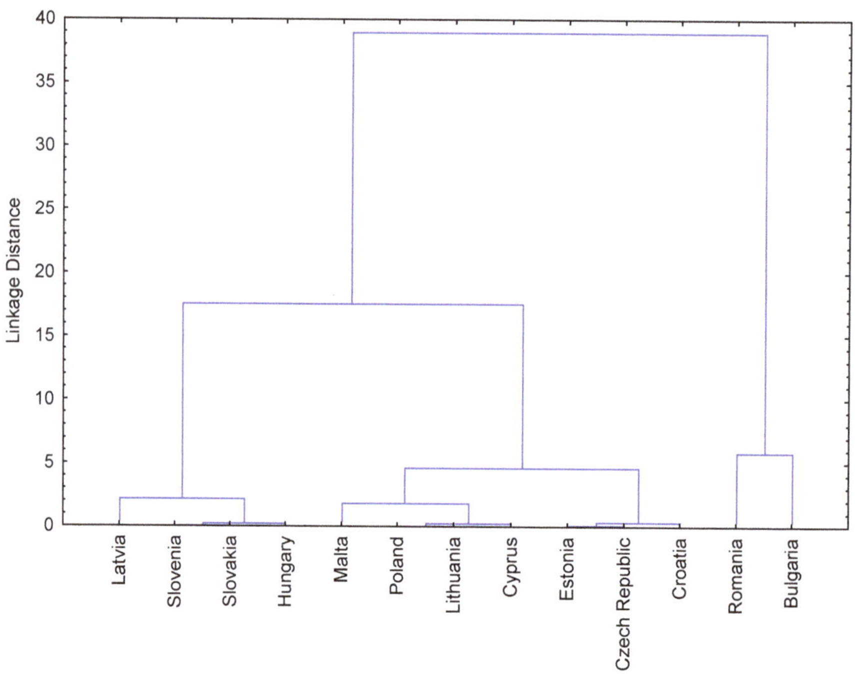

Figure 13. Tree diagram for IREN in 2016.

By cutting off the tree diagram, as in previous analyses, at a distance of 10, three clusters ere formed. Despite the same number of clusters as in the analysis for IRE 2016 indicators (Table 7), the clusters for IREN 2016 contain other countries (Table 11).

Table 11. Clusters for IREN in 2016.

No. Group	A	B	C
1	Hungary	Croatia	Bulgaria
2	Latvia	Cyprus	Romania
3	Slovakia	Czech Republic	
4	Slovenia	Estonia	
5		Lithuania	
6		Malta	
7		Poland	

Cluster averages were calculated to characterize clusters (Table 12) and the results are presented in relation to the average of all figures in Figure 14. As in all previous clusters, the analysis of variance of indicators in newly created groups was conducted. With the assumed significance level of 0.05, Table 12 shows that the p-value is always less than this value, indicating statistically significant differences between the clusters.

Table 12. Group averages for IREN in 2016.

Group/Indicator	ECRN2016/EC2016	EC2016/GDP2016	ECRN2016/GDP2016
A	0.055908	0.094332	0.005045
B	0.135949	0.098345	0.012990
C	0.173644	0.181012	0.030891
All Groups	0.117120	0.109828	0.013300
p-value	*0.000154*	*0.000004*	*0.008662*

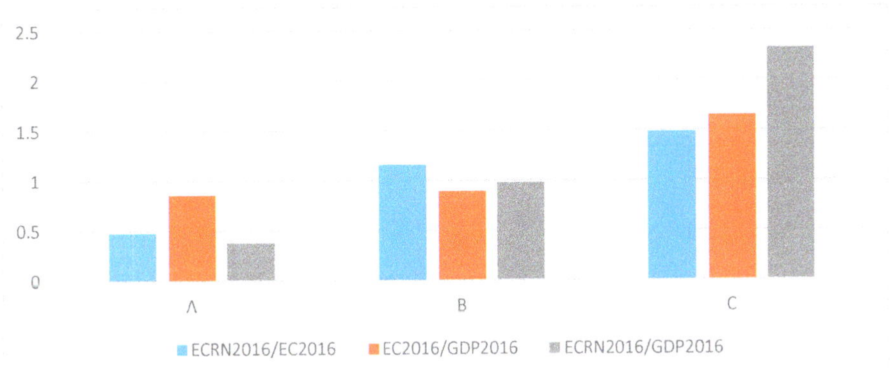

Figure 14. Comparison of group averages for IREN in 2016.

The average values in the groups of IREN indicators for 2016 (Figure 14) are similar for IRE indicators for 2016 (Figure 9). However, this similarity is only accidental as the countries in particular groups for IREN indicators (Table 11) only slightly overlap the countries in the groups for the IRE indicators (Table 7). Groups A and B contain countries whose cost of obtaining energy is moderate in relation to GDP, whereas Romania is also in group C in addition to Bulgaria. Group C, therefore, contains the poorest countries of the EU, where the energy cost is highest in relation to GDP, but the share of wind and solar power plants in the electrical capacity simultaneously increased the most in these countries (Figure 15).

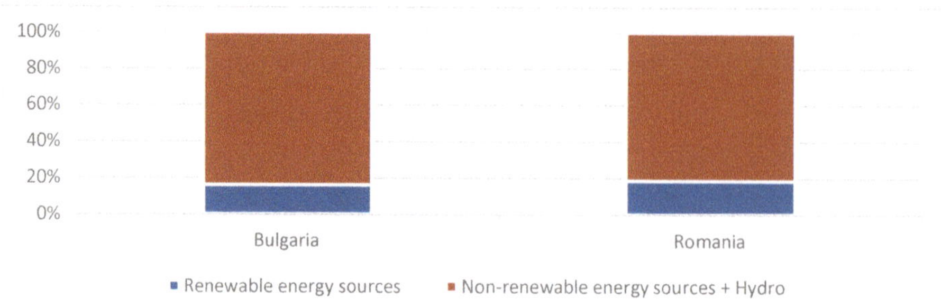

Figure 15. Energy sources in the countries of group C in 2016.

The structure of the electrical capacity divided into wind and solar power plants (renewable energy sources) and sum of non-renewable energy sources and hydro power is presented for groups A and B in Figures 16 and 17, respectively.

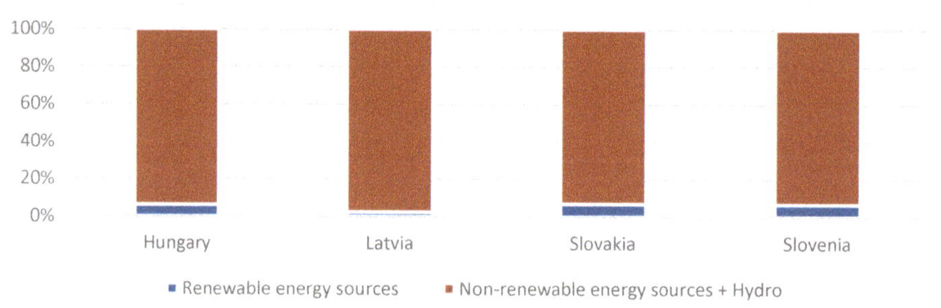

Figure 16. Energy sources in the countries of group A in 2016.

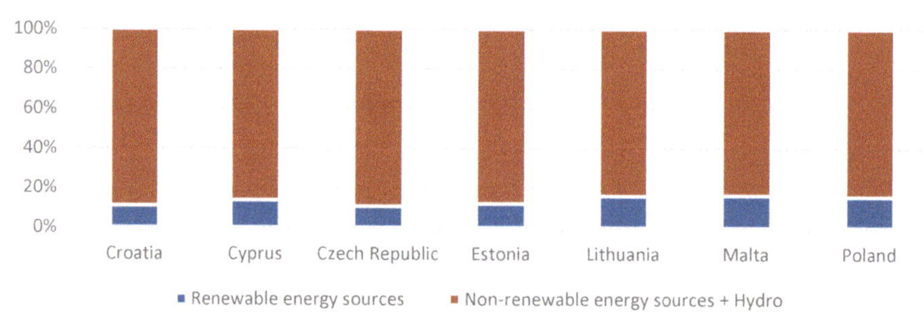

Figure 17. Energy sources in the countries of group B in 2016.

The exclusion of hydroelectric power plants from renewable energy sources led to the level of investments of individual countries in ecological sources of electricity being more visible in the conducted analyses. The countries of groups A and B in 2016 had a similar cost of energy conversion to GDP, but the level of investment in renewable energy sources was significantly different. Four countries from group A during the 13 years only slightly increased the production of electricity from wind and solar power plants, which means that this group can be considered the least ecological. Group B contains 7 of the 13 countries and is characterized by a much higher increase in energy production from wind and solar farms than the B group with a similar cost of energy conversion to GDP.

4. Discussion

In the context of the global development of renewable energy sources, we attempted to classify new EU Member States in terms of their progress in this area, and presented energy costs against their GDP. The electrical capacity selected for the analysis was divided into renewable and non-renewable energy sources as well as GDP. In order to compare the progress in the implementation of EU directives related to the reduction of greenhouse gas emissions through a change in the structure of energy sources, two variants of a cluster analysis were conducted for 2004 and 2016. In the first variant, the analyses were conducted for all known renewable energy sources. These analyses showed that all new EU countries in the analyzed period implemented a policy of increasing the electrical capacities from renewable sources to a similar level, which caused the groups of countries similar to each other in 2004 and 2016 to be identical. The distinctive group was group C, which contained only Bulgaria. Bulgaria is characterized by a large energy intensity, which is the cost of transforming energy into GDP. However, the significant reduction in energy intensity over the considered period can be regarded as a country success. In order to omit the influence of large hydropower plants on the results, analogous cluster analyses of only renewable energy sources of a new type, i.e., wind and solar power plants, were completed. The analysis for 2004 showed that the differences from previous analyses for 2004 were significant. The number of created groups was the same, but their characteristics and composition were different. The only group that remained was C, containing Bulgaria, again due to its high energy intensity. Group A only contained Latvia, which at the time was characterized by the possession of wind energy sources. The remaining 11 countries were so similar to each other that all were categorized into group B. This group was characterized in 2004 by zero electrical capacity from wind and solar power plants. Another analysis for 2016 showed which groups of countries increased their electrical capacity from modern renewable energy sources to the greatest extent. In this respect, the largest changes occurred in Group C, which included Bulgaria and Romania. In the two poorest countries of the EU, where the energy cost in relation to GDP was the highest, the electricity from hydropower and solar plants was already nearly 20% of the total electrical capacity. Group B, consisting of seven countries, where this type of energy constitute for over a dozen percent of electricity, can also be positively assessed. The smallest increase in electrical capacity from hydro and solar power plants occurred in group A, which included Hungary, Latvia, Slovakia, and Slovenia. It is particularly worth paying attention to Latvia, which was already distinguished in 2004 by having wind energy sources, and yet within 13 years it achieved the smallest increase (2%) in electrical capacity from wind and solar plants among all the surveyed countries. The division of analyses into the two options discussed above, apart from selecting groups of countries similar to each other in terms of electrical capacity and GDP, also showed the importance of the differences in the energy assessment of individual countries in the context of sustainable development, depending on whether their environmental achievements included current, often dozens-of-years-old and large hydropower plants, or whether only modern sources of energy from water and sun were analyzed.

By making analogous analyses, as in this article, the research can be extended to the entire EU. This would allow assessing whether a similar increase in investments in new types of renewable energy sources occurred countries that were members of the EU before 2004, as in the countries that joined in 2004 and later. Conducting analyses for all EU Member States would allow us to determine the differences between old and new members. Further questions could be posed about the composition of the groups in 2004 and 2016: Is there a division between old and new Member States in 2004? Are there any noticeable changes in the groups in 2016 among only old EU countries? Are there groups of mixed old and new Member States in the 2016 groups?

Questions of this type could help determine whether, for example, the new Member States are able to match the old EU Member States, or, despite investments in renewable energy sources, the old Member States do it even more efficiently. An analysis for the entire EU would also show differences in energy efficiency between old and new Member States.

Subsequent research could include Eastern European countries that are not members of the EU. Investigating the level of investments in various types of renewable energy sources in these countries, their development in relation to EU countries could be determined. Depending on the results obtained, it would be possible to analyze whether EU membership and legislation stimulates larger investments in renewable energy sources. Would the new Member States have developed these branches of energy if they did not become members of the EU?

In recent years, the high costs of producing energy from renewable sources have been widely discussed. The proposed classification of renewable energy sources could also be used to assess their impact on electricity prices, production costs, inflation, and consumer purchasing power, among others.

Author Contributions: Conceptualization, J.B.; Methodology, J.B., G.M. and G.S.; Formal Analysis, J.B., G.M., E.I. and G.S.; Investigation, J.B.; Resources, G.S.; Data Curation, J.B.; Writing—Original Draft Preparation, J.B. and G.S.; Writing—Review & Editing, J.B., G.M., G.S. and E.I.; Visualization, J.B.; Supervision, J.B., G.M. and G.S.; Funding Acquisition, G.M., E.I., and G.S.

Funding: This research received no external funding.

Conflicts of Interest: The authors declare no conflict of interest.

References

1. Smil, V. *Energy Transitions: History, Requirements, Prospects*; Praeger: Santa Barbara, CA, USA, 2010; ISBN 978-0-313-38177-5.
2. Sampedro, J.; Arto, I.; González-Eguino, M. Implications of switching fossil fuel subsidies to solar: A case study for the European Union. *Sustainability* **2017**, *10*, 50. [CrossRef]
3. Liao, H.; Liu, Y.; Gao, Y.; Hao, Y.; Ma, X.-W.; Wang, K. Forecasting residential electricity demand in provincial China. *Environ. Sci. Pollut. Res.* **2017**, *24*, 6414–6425. [CrossRef] [PubMed]
4. Ellabban, O.; Abu-Rub, H.; Blaabjerg, F. Renewable energy resources: Current status, future prospects and their enabling technology. *Renew. Sustain. Energy Rev.* **2014**, *39*, 748–764. [CrossRef]
5. Resch, G.; Held, A.; Faber, T.; Panzer, C.; Toro, F.; Haas, R. Potentials and prospects for renewable energies at global scale. *Energy Policy* **2008**, *36*, 4048–4056. [CrossRef]
6. Ketsetzi, A.; Capraro, M.M. Renewable energy sources. In *A Companion to Interdisciplinary STEM Project-Based Learning*; Sense Publishers: Rotterdam, The Netherlands, 2016; pp. 145–153. ISBN 978-94-6300-485-5.
7. Simionescu, M.; Albu, L.-L.; Szeles, M.R.; Bilan, Y. The impact of biofuels utilisation in transport on the sustainable development in the European Union. *Technol. Econ. Dev. Econ.* **2017**, *23*, 667–686. [CrossRef]
8. Auestad, I.; Nilsen, Y.; Rydgren, K. Environmental restoration in hydropower development—Lessons from norway. *Sustainability* **2018**, *10*, 3358. [CrossRef]
9. Nguyen, T.H.T.; Everaert, G.; Boets, P.; Forio, M.A.E.; Bennetsen, E.; Volk, M.; Hoang, T.H.T.; Goethals, P.L.M. Modelling tools to analyze and assess the ecological impact of hydropower dams. *Water* **2018**, *10*, 259. [CrossRef]
10. Frey, G.W.; Linke, D.M. Hydropower as a renewable and sustainable energy resource meeting global energy challenges in a reasonable way. *Energy Policy* **2002**, *30*, 1261–1265. [CrossRef]
11. Tromp, T.K.; Shia, R.-L.; Allen, M.; Eiler, J.M.; Yung, Y.L. Potential environmental impact of a hydrogen economy on the stratosphere. *Science* **2003**, *300*, 1740–1742. [CrossRef]
12. Galy-Lacaux, C.; Delmas, R.; Jambert, C.; Dumestre, J.-F.; Labroue, L.; Richard, S.; Gosse, P. Gaseous emissions and oxygen consumption in hydroelectric dams: A case study in French Guyana. *Glob. Biogeochem. Cycles* **1997**, *11*, 471–483. [CrossRef]
13. Rosa, L.P.; dos Santos, M.A.; Matvienko, B.; dos Santos, E.O.; Sikar, E. Greenhouse gas emissions from hydroelectric reservoirs in tropical regions. *Clim. Chang.* **2004**, *66*, 9–21. [CrossRef]
14. Fearnside, P.M. Greenhouse gas emissions from hydroelectric dams: Controversies provide a springboard for rethinking a supposedly 'clean' energy source. An editorial comment. *Clim. Chang.* **2004**, *66*, 1–8. [CrossRef]
15. Hidrovo, A.; Uche, J.; Martínez, A. Accounting for GHG net reservoir emissions of hydropower in Ecuador. *Renew. Energy* **2017**, *112*, 209–221. [CrossRef]
16. Kemenes, A.; Melack, J.; Forsberg, B. Downstream emissions of CH4 and CO2 from hydroelectric reservoirs (Tucuruí, Samuel, and Curuá-Una) in the Amazon basin. *Inland Waters* **2016**, *6*, 295–302. [CrossRef]

17. Citizens United for Renewable Energy and Sustainability Steps Toward a Sustainable Energy Future: Position Paper of the CURES Network for the CSD 2006.
18. The EU Environmental NGO Statement on CSD 15. Available online: http://www.wecf.eu/english/articles/2007/03/euro_csd15.php (accessed on 24 November 2018).
19. Zhang, H.; Zheng, Y.; Zhou, D.; Zhu, P. Which subsidy mode improves the financial performance of renewable energy firms? A panel data analysis of wind and solar energy companies between 2009 and 2014. *Sustainability* **2015**, *7*, 16548–16560. [CrossRef]
20. Abazaj, J.; Moen, Ø.; Ruud, A. Striking the balance between renewable energy generation and water status protection: Hydropower in the context of the European renewable energy directive and water framework Directive. *Environ. Policy Gov.* **2016**, *26*, 409–421. [CrossRef]
21. More Bang for the Buck: Record New Renewable Power Capacity Added at Lower Cost. Available online: https://about.bnef.com/blog/bang-buck-record-new-renewable-power-capacity-added-lower-cost/ (accessed on 24 May 2018).
22. Arıoğlu Akan, M.Ö.; Selam, A.A.; Oktay Fırat, S.Ü.; Er Kara, M.; Özel, S. A comparative analysis of renewable energy use and policies: Global and turkish perspectives. *Sustainability* **2015**, *7*, 16379–16407. [CrossRef]
23. Franzitta, V.; Curto, D.; Rao, D. Energetic sustainability using renewable energies in the Mediterranean Sea. *Sustainability* **2016**, *8*, 1164. [CrossRef]
24. Qoaider, L.; Steinbrecht, D. Photovoltaic systems: A cost competitive option to supply energy to off-grid agricultural communities in arid regions. *Appl. Energy* **2010**, *87*, 427–435. [CrossRef]
25. Ntanos, S.; Kyriakopoulos, G.; Chalikias, M.; Arabatzis, G.; Skordoulis, M. Public perceptions and willingness to pay for renewable energy: A case study from Greece. *Sustainability* **2018**, *10*, 687. [CrossRef]
26. Stigka, E.K.; Paravantis, J.A.; Mihalakakou, G.K. Social acceptance of renewable energy sources: A review of contingent valuation applications. *Renew. Sustain. Energy Rev.* **2014**, *32*, 100–106. [CrossRef]
27. Piao, M.; Piao, Y.; Lee, J.Y. Symmetrical uncertainty-based feature subset generation and ensemble learning for electricity customer classification. *Symmetry* **2019**, *11*, 498. [CrossRef]
28. Bengochea-Morancho, A.; Higón-Tamarit, F.; Martínez-Zarzoso, I. Economic growth and CO2 emissions in the European Union. *Environ. Resour. Econ.* **2001**, *19*, 165–172. [CrossRef]
29. Azomahou, T.; Laisney, F.; Nguyen Van, P. Economic development and CO2 emissions: A nonparametric panel approach. *J. Public Econ.* **2006**, *90*, 1347–1363. [CrossRef]
30. Narayan, P.K.; Narayan, S. Carbon dioxide emissions and economic growth: Panel data evidence from developing countries. *Energy Policy* **2010**, *38*, 661–666. [CrossRef]
31. Al-mulali, U. Factors affecting CO_2 emission in the Middle East: A panel data analysis. *Energy* **2012**, *44*, 564–569. [CrossRef]
32. Odhiambo, N. Economic growth and carbon emissions in South Africa: An empirical investigation. *Int. Bus. Econ. Res. J.* **2012**, *28*, 37–46.
33. Brożyna, J.; Mentel, G.; Szetela, B. Renewable energy and economic development in the European Union *Acta Polytech. Hung.* **2017**, *14*, 11–34.
34. Mehedintu, A.; Sterpu, M.; Soava, G. Estimation and forecasts for the share of renewable energy consumption in final energy consumption by 2020 in the European Union. *Sustainability* **2018**, *10*, 1515. [CrossRef]
35. Malla, S.; Timilsina, G.R. *Long-Term Energy Demand Forecasting in Romania: An End-Use Demand*; The World Bank: Washington, DC, USA, 2016; pp. 1–38.
36. Ardakani, F.J.; Ardehali, M.M. Long-term electrical energy consumption forecasting for developing and developed economies based on different optimized models and historical data types. *Energy* **2014**, *65*, 452–461. [CrossRef]
37. Hamedmoghadam, H.; Joorabloo, N.; Jalili, M. Australia's long-term electricity demand forecasting using deep neural networks. *arXiv* **2018**, arXiv:1801.02148.
38. Rodríguez-Caballero, C.V.; Ventosa-Santaulària, D. Energy-growth long-term relationship under structural breaks. Evidence from Canada, 17 Latin American economies and the USA. *Energy Econ.* **2017**, *61*, 121–134. [CrossRef]
39. Kishita, Y.; Yamaguchi, Y.; Umeda, Y.; Shimoda, Y.; Hara, M.; Sakurai, A.; Oka, H.; Tanaka, Y. Describing long-term electricity demand scenarios in the telecommunications industry: A case study of Japan. *Sustainability* **2016**, *8*, 52. [CrossRef]

40. McCarthy, R.; Yang, C.; Ogden, J.M. *California Energy Demand Scenario Projections to 2050*; Institute of Transportation Studies, University of California: Davis, CA, USA, 2008.
41. Brożyna, J.; Mentel, G.; Szetela, B. A mid-term forecast of maximum demand for electricity in Poland. *Montenegrin J. Econ.* **2016**, *12*, 73–88. [CrossRef]
42. Kharlamova, G.; Chernyak, O.; Nate, S. Renewable energy and security for Ukraine: Challenge or smart way? *J. Int. Stud.* **2016**, *9*, 88–115. [CrossRef] [PubMed]
43. Sadorsky, P. Renewable energy consumption and income in emerging economies. *Energy Policy* **2009**, *37*, 4021–4028. [CrossRef]
44. Sari, R.; Ewing, B.T.; Soytas, U. The relationship between disaggregate energy consumption and industrial production in the United States: An ARDL approach. *Energy Econ.* **2008**, *30*, 2302–2313. [CrossRef]
45. Lee, C.-C. The causality relationship between energy consumption and GDP in G-11 countries revisited. *Energy Policy* **2006**, *34*, 1086–1093. [CrossRef]
46. Shahbaz, M.; Tang, C.F.; Shahbaz Shabbir, M. Electricity consumption and economic growth nexus in Portugal using cointegration and causality approaches. *Energy Policy* **2011**, *39*, 3529–3536. [CrossRef]
47. Zhang, X.-P.; Cheng, X.-M. Energy consumption, carbon emissions, and economic growth in China. *Ecol. Econ.* **2009**, *68*, 2706–2712. [CrossRef]
48. Caraiani, C.; Lungu, C.I.; Dascălu, C. Energy consumption and GDP causality: A three-step analysis for emerging European countries. *Renew. Sustain. Energy Rev.* **2015**, *44*, 198–210. [CrossRef]
49. Li, H.; Lo, K.; Wang, M.; Zhang, P.; Xue, L. Industrial energy consumption in Northeast China under the revitalisation strategy: A decomposition and policy analysis. *Energies* **2016**, *9*, 549. [CrossRef]
50. Fang, Y. Economic welfare impacts from renewable energy consumption: The China experience. *Renew. Sustain. Energy Rev.* **2011**, *15*, 5120–5128. [CrossRef]
51. Inglesi-Lotz, R. The impact of renewable energy consumption to economic growth: A panel data application. *Energy Econ.* **2016**, *53*, 58–63. [CrossRef]
52. Menegaki, A.N. Growth and renewable energy in Europe: A random effect model with evidence for neutrality hypothesis. *Energy Econ.* **2011**, *33*, 257–263. [CrossRef]
53. Kayhan, S.; Adiguzel, U.; Bayat, T.; Lebe, F. Causality relationship between real GDP and electricity consumption in Romania (2001–2010). *Romanian J. Econ. Forecast.* **2010**, *169*, 169–183.
54. Peñalvo-López, E.; Cárcel-Carrasco, F.J.; Devece, C.; Morcillo, A.I. A methodology for analysing sustainability in energy scenarios. *Sustainability* **2017**, *9*, 1590. [CrossRef]
55. Katre, A.; Tozzi, A. Assessing the sustainability of decentralized renewable energy systems: A comprehensive framework with analytical methods. *Sustainability* **2018**, *10*, 1058. [CrossRef]
56. Cîrstea, S.D.; Moldovan-Teselios, C.; Cîrstea, A.; Turcu, A.C.; Darab, C.P. Evaluating renewable energy sustainability by composite index. *Sustainability* **2018**, *10*, 811. [CrossRef]
57. Zhao, H.; Guo, S. External benefit evaluation of renewable energy power in China for sustainability. *Sustainability* **2015**, *7*, 4783–4805. [CrossRef]
58. Saiah, S.B.D.; Stambouli, A.B. Prospective analysis for a long-term optimal energy mix planning in Algeria: Towards high electricity generation security in 2062. *Renew. Sustain. Energy Rev.* **2017**, *73*, 26–43. [CrossRef]
59. Zhang, Y.; Zhao, X.; Zuo, Y.; Ren, L.; Wang, L. The development of the renewable energy power industry under feed-in tariff and renewable portfolio standard: A case study of China's photovoltaic power industry. *Sustainability* **2017**, *9*, 532. [CrossRef]
60. Kasperowicz, R.; Pinczyński, M.; Khabdullin, A. Modeling the power of renewable energy sources in the context of classical electricity system transformation. *J. Int. Stud.* **2017**, *10*, 264–272. [CrossRef]
61. Nicolini, M.; Tavoni, M. Are renewable energy subsidies effective? Evidence from Europe. *Renew. Sustain. Energy Rev.* **2017**, *74*, 412–423. [CrossRef]
62. Zhang, L.; Xue, B.; Liu, X. Carbon emission reduction with regard to retailer's fairness concern and subsidies. *Sustainability* **2018**, *10*, 1209. [CrossRef]
63. Choi, Y. The Asian values of Guānxì as an economic model for transition toward green growth. *Sustainability* **2018**, *10*, 2150. [CrossRef]
64. Choi, G.; Heo, E.; Lee, C.-Y. Dynamic economic analysis of subsidies for new and renewable energy in South Korea. *Sustainability* **2018**, *10*, 1832. [CrossRef]
65. Bowden, N.; Payne, J.E. The causal relationship between U.S. energy consumption and real output: A disaggregated analysis. *J. Policy Model.* **2009**, *31*, 180–188. [CrossRef]

66. Li, X. Diversification and localization of energy systems for sustainable development and energy security. *Energy Policy* **2005**, *33*, 2237–2243. [CrossRef]
67. First Orientation for a Common Energy Policy, Communication from the EC to the Council 1968.
68. Jasiński, P.; Skoczny, T. *Polityka Energetyczna Wspólnot Europejskich—Tło Historyczne*; CE UW: Warszawa, Poland, 1996.
69. Immenga, U. The developement of European energy policy: From ECSC treaty to the internal market. In *Natural Gas in the Internal Market*; Mestmäcker, E.J., Ed.; Kluwer Law International: London, UK, 1992; ISBN 978-1-85333-795-6.
70. Eurostat, Infrastructure—Electricity. Available online: http://appsso.eurostat.ec.europa.eu/nui/show.do?dataset=nrg_113a (accessed on 14 June 2018).
71. Wurlod, J.-D.; Noailly, J. The impact of green innovation on energy intensity: An empirical analysis for 14 industrial sectors in OECD countries. *Energy Econ.* **2018**, *71*, 47–61. [CrossRef]
72. Voigt, S.; De Cian, E.; Schymura, M.; Verdolini, E. Energy intensity developments in 40 major economies: Structural change or technology improvement? *Energy Econ.* **2014**, *41*, 47–62. [CrossRef]
73. Tvaronavičienė, M.; Prakapienė, D.; Garškaitė-Milvydienė, K.; Prakapas, R.; Nawrot, Ł. Energy efficiency in the long-run in the selected European countries. *Econ. Sociol.* **2018**, *11*, 245–254. [CrossRef] [PubMed]
74. Bi, C., Jia, M.; Zeng, J. Nonlinear effect of public infrastructure on energy intensity in China: A panel smooth transition regression approach. *Sustainability* **2019**, *11*, 629. [CrossRef]
75. Sequeira, T.; Santos, M. Education and energy intensity: Simple economic modelling and preliminary empirical results. *Sustainability* **2018**, *10*, 1–17. [CrossRef]
76. Wiśnicki, B.; Chybowski, L.; Czarnecki, M. Analysis of the efficiency of port container terminals with the use of the data envelopment analysis method of relative productivity evaluation. *Manag. Syst. Prod. Eng.* **2017**, *25*, 9–15. [CrossRef]
77. European Union. The 28 Member Countries of the EU. Available online: https://europa.eu/european-union/about-eu/countries_en#tab-0-1 (accessed on 22 May 2018).
78. McCormick, J. *Understanding the European Union: A Concise Introduction*; Palgrave Macmillan: London, UK, 2017; ISBN 978-1-137-60776-8.
79. Pastor, J.M.; Peraita, C.; Serrano, L.; Soler, Á. Higher education institutions, economic growth and GDP per capita in European Union countries. *Eur. Plan. Stud.* **2018**, *26*, 1616–1637. [CrossRef]
80. Campos, N.F.; Coricelli, F.; Moretti, L. *Economic Growth and Political Integration: Estimating the Benefits from Membership in the European Union Using the Synthetic Counterfactuals Method*; Social Science Research Network: Rochester, NY, USA, 2014.
81. Gawlik, L. The polish power industry in energy transformation process. *Miner. Econ.* **2018**, *31*, 229–237. [CrossRef]
82. Dogan, E.; Seker, F. Determinants of CO2 emissions in the European Union: The role of renewable and non-renewable energy. *Renew. Energy* **2016**, *94*, 429–439. [CrossRef]
83. Scarlat, N.; Dallemand, J.-F.; Monforti-Ferrario, F.; Banja, M.; Motola, V. Renewable energy policy framework and bioenergy contribution in the European Union—An overview from national renewable energy action plans and progress reports. *Renew. Sustain. Energy Rev.* **2015**, *51*, 969–985. [CrossRef]
84. Everitt, B.S.; Landau, S.; Leese, M.; Stahl, D. *Cluster Analysis*, 5th ed.; John Wiley & Sons: New York, NY, USA, 2011; ISBN 978-0-470-97844-3.
85. Bailey, K.D. *Typologies and Taxonomies: An Introduction to Classification Techniques; Typologies and Taxonomies: An Introduction to Classification Techniques*; Sage Publications, Inc.: Thousand Oaks, CA, US, 1994; ISBN 978-0-8039-5259-1.
86. Ward, J.H.J. Hierarchical grouping to optimize an objective function. *J. Am. Stat. Assoc.* **1963**, *58*, 236–244. [CrossRef]
87. Pacesila, M.; Burcea, S.G.; Colesca, S.E. Analysis of renewable energies in European Union. *Renew. Sustain. Energy Rev.* **2016**, *56*, 156–170. [CrossRef]
88. Eurostat GDP and Main Components. Available online: http://appsso.eurostat.ec.europa.eu/nui/show.do?dataset=nama_10_gdp (accessed on 14 June 2018).
89. Shang, Y. Subgraph robustness of complex networks under attacks. IEEE Trans. *Syst. Man Cybern. Syst.* **2019**, *49*, 821–832. [CrossRef]
90. Shang, Y. Distance estrada index of random graphs. *Linear Multilinear Algebra* **2015**, *63*, 466–471. [CrossRef]

91. Shang, Y. Estimating the distance estrada index. *Kuwait J. Sci.* **2016**, *43*, 14–19.
92. Giudici, P. *Applied Data Mining: Statistical Methods for Business and Industry*; John Wiley & Sons: New York, NY, USA, 2005; ISBN 978-0-470-87139-3.
93. Hastie, T.; Tibshirani, R.; Friedman, J. *The Elements of Statistical Learning: Data Mining, Inference, and Prediction*, 2nd ed.; Springer: New York, NY, USA, 2009; ISBN 978-0-387-84857-0.
94. Szekely, G.J.; Rizzo, M.L. Hierarchical clustering via joint between-within distances: Extending ward's minimum variance method. *J. Classif.* **2005**, *22*, 151–183. [CrossRef]

© 2019 by the authors. Licensee MDPI, Basel, Switzerland. This article is an open access article distributed under the terms and conditions of the Creative Commons Attribution (CC BY) license (http://creativecommons.org/licenses/by/4.0/).

energies

MDPI

Article

Novel Conceptual Architecture for the Next-Generation Electricity Markets to Enhance a Large Penetration of Renewable Energy

Javier Rodríguez-García, David Ribó-Pérez *, Carlos Álvarez-Bel and Elisa Peñalvo-López

Universitat Politècnica de València, Institute for Energy Engineering, Camino de Vera s/n, edificio 8E, escalera F, 5ª planta, València 46022, Spain

* Correspondence: david.ribo@iie.upv.es; Tel.: +34-963879240

Received: 24 May 2019; Accepted: 4 July 2019; Published: 6 July 2019

Abstract: A transition to a sustainable energy system is essential. In this context, smart grids represent the future of power systems for efficiently integrating renewable energy sources and active consumer participation. Recently, different studies were performed that defined the conceptual architecture of power systems and their agents. However, these conceptual architectures do not overcome all issues for the development of new electricity markets. Thus, a novel conceptual architecture is proposed. The transactions of energy, operation services, and economic flows among the agents proposed are carefully analysed. In this regard, the results allow setting their activities' boundaries and state their relationships with electricity markets. The suitability of implementing local electricity markets is studied to enforce competition among distributed energy resources by unlocking all the potential that active consumers have. The proposed architecture is designed to offer flexibility and efficiency to the system thanks to a clearly defined way for the exploitation of flexible resources and distributed generation. This upgraded architecture hereby proposed establishes the characteristics of each agent in the forthcoming markets and studies to overcome the barriers to the large deployment of renewable energy sources.

Keywords: Electricity markets; power system; conceptual architecture; distributed generation; flexible resources; local electricity markets

1. Introduction

A transition from a fossil-fuel-based energy system to a decarbonized one is key to performing a cost-effective strategy to mitigate climate change [1] and achieve the 2 °C threshold aim of the Paris agreement. Within this context, renewable energy sources (RESs) represent the most promising technology for the transition and the future system. RESs are almost free-emission technologies and, during the last few years, RESs achieved economic competitiveness against conventional energy sources. However, their deployment in traditional power systems is not absent of challenges. The stochastic nature of renewable generation, the non-storable characteristic of electricity in a cost-effective way, and the low elasticity in demand associated with its difficulties to participate in electricity markets [2] make their variability a major issue with a wider impact on smaller systems. Moreover, the final energy consumption will tend to become electric in order to reduce emissions. Thus, future loads will impose new demands and challenges to the power system such as the massive penetration of electric vehicles (EV) to electrify transport.

In order to overcome this problem, the smart grid concept was an accepted solution for some time now. Smarts grids are electricity networks that intelligently integrate their users' actions to efficiently deliver economic, secure, and sustainable electricity [3]. The implementation of smart grids implies broad and sophisticated functionalities of electric transport and distribution systems, improving their flexibility, allowing bidirectional energy flows, and facilitating RES and demand response (DR) integration. The demand response is based on developing active participation of customers with new requirements that take into account technology and equipment for customer communications, relations, and services. However, just with the participation of demand, the security of supply will still be jeopardized with larger levels of stochastic production associated with renewable generation. Thus, storage systems will also be required to provide flexibility and ensure reliability to the system [4]. Moreover, the batteries' cost reductions make them a key component in the future power systems [5].

Currently, the electricity sector finds itself making three classes of transformations: firstly, the improvement of the current infrastructure; secondly, the addition of the digitalization of power systems, which is the essence of communications and data generation in smart grids; thirdly, business process transformation to perform, in addition to the traditional activities, new ones, or providing infrastructure and data to agents such as aggregators and virtual power plants (VPPs). These agents do new activities related to meeting customer needs and expectations in a more efficient way than the traditional centralized system. These three transformations were approached in several different ways, which were mainly described on a very abstract level [6] or focused on specific aspects such as just information and communication technology (ICT) [7]. Different standardization bodies developed specific concepts such as the American National Institute of Standards and Technology (NIST) framework and roadmap for smart grid standards [8] and the European Smart Grid Architecture Model (SGAM) [9]. However, the necessary new activities, agents, and interactions among them in the future electricity markets are not clearly defined and authors still characterize them in different ways. Therefore, it is necessary to align specific agents to established practical conceptual architectures as suggested by Neuriter et al., [10].

The functionality of the future power systems and markets may look quite different according to the local social, regulatory, or economic environment. Nevertheless, they have common applications and requirements for digital processing and communications to implement advanced control in all elements of the power system, allowing for bidirectional communication and energy flows [8], understanding the automation of processes and systems as digital processing to retrieve data and perform actions. According to this context, smart grids enable greater information management and efficiency compared to conventional power systems, thus allowing the exploitation of the benefits associated with RES, demand response, storage systems, and real-time competition and response in local markets. Local markets are arising as a new mechanism to provide an efficient allocation and pricing of the growing distributed generation (DG) and flexible demand [11,12].

Thus, smart grids are emerging as a solution for the future of power systems [13]. This broad concept that comprises many different agents, actors, and technology was approached in different ways. Its future faces different problems and sub-problems, which were widely studied. According to Reference [14], some of these are operation and management, energy storage, security, stability, and protection, demand control, or service restoration, among others.

For instance, some authors proposed multi-agent systems that optimize resource scheduling in smart grids [15,16]. These agents enable the system to behave in a more reliable and efficient way. However, the description of these agents does not follow any standardized premise. The authors of References [17,18] proposed energy management systems in smart grids. The agents as in Reference [15] did not include a clear definition of the agent boundaries of action or relationships and presented conflicts between them. A review of agent-based models was presented in Reference [19], where the necessity of harmonization between studies was highlighted.

In order to tackle the previously mentioned standardization problems, different meta-architectures were developed. These conceptual architectures provide a family of ontologies to map smart grids and guidelines on how to use standards [7]. The main two developments were the previously mentioned NIST work and SGAM.

In the United States of America (USA), the NIST created relevant conceptual models for the smart grid. NIST considered the approach that the smart grid can be divided into seven domains [8]. These domains and their sub-domains enclose the conceptual roles and services, including stakeholders, interactions, and types of services. On the other hand, the M/490 working group on reference architectures created the SGAM, which can be seen as a similar effort on the European level. SGAM is based on NIST and proposes a model with five interoperability layers, five domains, and six zones, as can be seen in Figure 1. Thus, every element in the model can be located in a three dimension grid according to its interoperability, domain, and zone characteristics [9]. As in the case of NIST, SGAM requires stronger integration between the design and the use cases and formal semantics [20], as it lacks of precise descriptions.

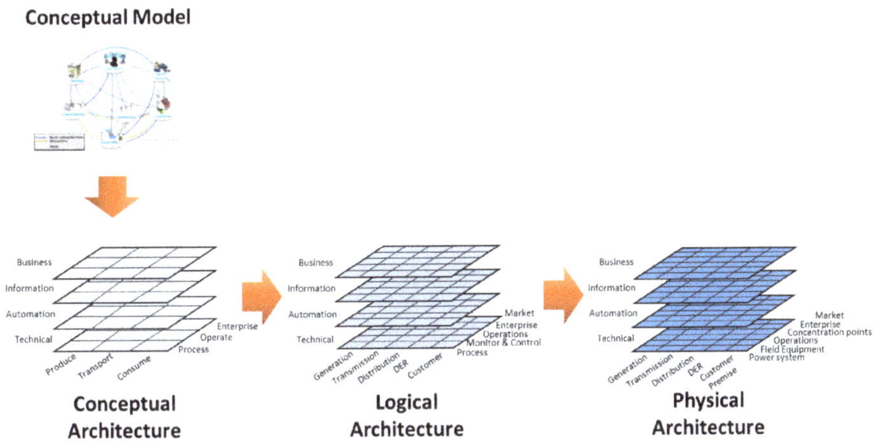

Figure 1. Smart Grid Architecture Model (SGAM) iterations, layers, and planes. Own elaboration based on Reference [8].

Highly correlated with smart grid development, the three novel agents of aggregator, storage, and virtual power plant (VPP) are being developed. In all these cases, several authors published studies on the topic. However, if the case of smart grids is still not clear and no standard definitions are used, VPP, storage, and aggregators offer an even wider range of variation and disagreement. The importance of these three agents is relevant for the conception of smart grids since these agents are crucial for the security and reliability of power systems with increasing levels of renewable penetration [21]. For instance, some authors optimized VPP bidding strategies [22–24], renewable energy integration [25,26], the use of demand response in smart grids [27], or the usage of RESs at the residential level [28,29]. However, there exists a lack of a standardized definition, interactions, and roles performed by a VPP.

Demand response is also stated to have an increasing role in power systems due to its potential capacity to help manage renewable variability [30]. Work was done in analyzing the cost of automated DR systems [31], the suitability of different customers [32], the evaluation of the action performance [33,34], or its optimization in smart grid programs [35]. Moreover, its role among active consumers at the distribution level is gaining importance [36]. Storage is seen as the key technology to enable RES integration in the future power systems [4,37]. Under this paradigm, storage systems are already a key agent in the power system as in the case of the Tesla Battery of South Australia [38]. However, the particularities and services that they provide are far from being homogeneous or clear among scholars and systems. Finally, in a similar line, aggregators were approached in different ways by authors and regulators, but also lack a clear common definition [39]. Moreover, authors do not share a common view on the size that optimal aggregation should have. For instance, while the authors of Reference [40] argued that aggregation is only profitable at large levels, the authors of Reference [41] defended that, even at low levels, aggregation offers benefits. In sum, agents are not clearly defined and the interactions between them vary among authors.

The conceptual architecture here developed is based on the NIST framework [8] and builds on providing the relationships and interaction design between the different agents. These agents can be performed by different entities or one entity, company, or organization that could hold more than one of the agents' responsibilities. Reference levels of power, voltage, and minimum bidding levels were parameterized to be chosen depending on the system, thus providing an easy way to implement the conceptual architecture to any power system. Thus, the proposed conceptual architecture can be applied to any type of power sector, independently of the level of decentralization and its size.

The main contributions of this paper are the following:

- A novel conceptual architecture for the development of the next-generation electricity markets to unlock all the hidden potential of flexible and distributed energy resources, taking into special consideration the potential benefits for active consumers, is proposed based on the analysis of the shortcomings of the current standardized models that can be found in the literature. This model provides a path that policy-makers can follow to eliminate barriers to integrate Distributed Energy Resources (DER) in a competitive way at distribution level.
- A complete description of the main roles/activities that should be assumed by the different agents in the proposed architecture is provided based on an ontological and a service-oriented analysis.
- A detailed proposal of the interactions that would occur among agents of the developed architecture is presented. These interactions were carefully analyzed from all points of view: energy flows, operation services, and economic transactions.
- The impacts on the performance of the conceptual model associated with the inclusion of local energy markets are analyzed and presented in this paper. This could help overcome the current flaws in real-time trading.

The rest of the paper is structured as follows: Section 2 outlines the NIST methodology used for building the proposed design to upgrade the current one. Then, the specific agents proposed for a standardized architecture are developed in Section 3. Finally, in Section 4, some conclusions are drawn.

2. Materials and Methods

The power system and market conceptual design methodology is described in this section. This method is framed under the framework of the NIST roadmap for smart grids [8]. The methodology proposed by the NIST was considered as a base to develop smart grid conceptual architectures by several authors and other standards [9,42]. In this regard, this methodology was selected as a meta architecture to develop the proposed upgrade of the existing architecture.

According to Reference [7], the first action is the specification of the roles/services that should be expected from the general implementation of smart grids. In addition to the traditional roles/services that are inherent in an electricity distribution system (i.e., generators and retailers), some additional agents should be expected from the combination of the new environmental requirements and advanced technology.

In this regard, the smart grid agents need to be designed to enable the system to successfully respond to the following needs:

- Providing a full technical and economic integration of distributed generation. This generation is generally difficult to integrate because of the low size, intermittent production, quality problems, and inability to provide operation services.
- Providing enhanced services and opportunities to the customers, allowing more tailored trading of their demand/generation resources, including interaction with retail energy and services markets/products.
- Providing an enhanced operation of the distribution system, both in normal conditions (such as reconfiguration for more efficient operation or for more secure supply) and in faulty conditions in order to allow a faster and more effective reaction to faults (fault location, reconfiguration, self-healing, etc.).

- Providing information services, based on measurements, to actors in the field of the energy supply such as aggregators, energy services companies (ESCOs), VPPs, etc.
- Providing the ability to accommodate and manage the presence of new loads at the customer level, such as the massive connection of electric vehicles.

It is important to highlight that the implementation of these agents can require the participation of new entities or the redesign of functions that will have to be performed by existing organizations.

A conceptual architecture is necessary to design a system capable of carrying out the roles/services that smart grids must perform according to the abovementioned needs. At this point, it is necessary to define a set of concepts that can be widely used along the description of the architecture:

- **Agent**: a specific function, capability, or sum of services played by an entity that cannot be split. In some systems, one entity can have in its business portfolio duties of several agents of this conceptual architecture.
- **Activities**: things that an agent does or has the capability to do.
- **Component**: a basic part from which something is made; the physical assets that are intrinsic to each agent.
- **Transaction**: agreement between two agents (one buys and the other one sells) to exchange goods, services, or financial instrument.

In order to align the architecture with the required services of the system, an ontological definition is required according to Reference [7]. For doing so, the methodology proposed in NIST, shown in Figure 2, was used.

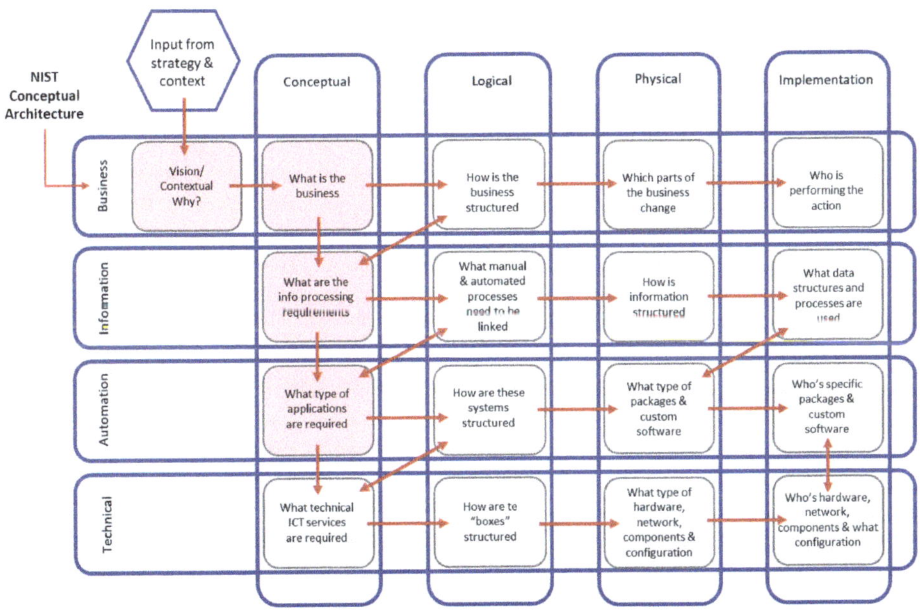

Figure 2. National Institute of Standards and Technology (NIST) conceptual architecture mapped onto the architecture matrix service orientation and ontology. Own elaboration based on Reference [8].

According to this procedure, four architectural levels must be considered to design the agents: business, information, automation, and technology. All these levels must be described to answer the four required layers: conceptual, logical, physical, and its implementation.

After this first context analysis, the interactions among the different agents were carefully studied to satisfy the required relationship needs among them. The entities required to implement a smart grid are, in general, quite standard; however, some agents' activities assigned to these entities may not be so established and, in some cases, can be a bit confusing in the literature, where different approaches to the same agents can be found.

The next section is devoted to presenting the novel conceptual architecture. Firstly, each agent is defined based on the existing knowledge and literature, and the activities expected for the agent are identified. According to these activities, the necessary physical components that each agent owns are described. This includes assets like physical generators, transmission lines, etc. Finally, the power flows, operating service, or economic transactions of each agent with the rest of them are described to fulfil the expected new requirements and functionalities of smart grids.

3. Discussion of Agent Conceptual Architecture for Market Implementation

The agents and nomenclature required for the upgraded conceptual architecture proposed in this paper are depicted in Table 1. The integration of different types of distributed generation, storage, and demand response resources to provide firm power production, as well as the active participation of the customers, were considered in detail.

Table 1. Summary of agents and elements considered in the future electricity market framework.

Agents	**Characteristics**
Active consumers	Self-generation, flexible demand, buying/selling electricity, and operation services with a proactive perspective [43]
Generators	Electricity generation and procurement of operation services
Virtual power plants (VPPs)	Buying/selling electricity and operation services from their distributed generation portfolio to different agents in a coordinated way
Aggregators	Buying and selling of small and medium demand resources to sell it to different agents in a coordinated way
Storage	Highly flexible elements that can consume, generate, and provide operation services
Transmission system operator (TSO)	Ensures power quality and security at a transmission level
Transmitter	Owns transmission grid and in charge of its maintenance
Distribution system operator (DSO)	Ensures power quality and security at a distribution level
Distributer	Owns distribution grid and in charge of its maintenance
Wholesale market operator (WMO)	Ensures independency and the good functioning of the wholesale market
Local market operator (LMO)	Ensures independency and correct functioning of the local market
Retailers	Provides electricity supply to consumers, buys excess of self-generated electricity
Key Concepts	**Definition**
Smart grids	An electric grid with an additional communication system that allows a better management of the system [44]
Demand response	Changes in electric usage by end-use customers from their normal consumption patterns [45]
Smart metering	All agents in the system have smart meters that provide data acquisition, transmission, processing, and interpretation [46].
Self-generation	Share of the total energy production directly consumed by the energy production system owner (based on Reference [47])
Distributed generation	Power generation within distribution networks [48]

Table 1. *Cont.*

Parameters	**Definitions**
V_{HV}	Voltage threshold defined as high voltage in the system parameters
$E_{W\text{-}S}$	Minimum energy required to sell electricity in the electricity market during a period of time unit (PTU)
$E_{W\text{-}B}$	Minimum energy required to buy electricity in the electricity market during a PTU
$E_{L\text{-}S}$	Minimum energy required to sell electricity in the local electricity market during a PTU
$E_{L\text{-}B}$	Minimum energy required to buy electricity in the local electricity market during a PTU
$P_{OS\text{-}T}$	Minimum power required to participate in operation services at transmission level
$P_{OS\text{-}D}$	Minimum power required to participate in operation services at distribution level

The conceptual architecture was completed with the transactions allowed between agents, as summarized in Table 2, where economic, energy, and operation service transactions between the different agents are proposed. A matrix representation of the allowed transactions among agents is shown in different colors in this table. The possible transactions from the agent in a row to the agent in the column are represented by triangles. For instance, position T_{12} shows the transactions from consumers to generators, which are only economic, as consumers just pay generators for consuming electricity. On the other hand, T_{21} shows how generators provide energy to consumers. Another example could be position T_{43}, where aggregators provide power flows and operating services to VPPs. In exchange for this, T_{34}, VPPs make economic payments to aggregators.

Table 2. Summary of the transactions among agents on the proposed smart grid framework (▲ [red]: economic transaction; ▲ [blue]: energy transaction; ▲ [green]: operation service transaction).

	Consumers	Generators	VPP	Aggregators	Storage	TSO
Consumers		▲ [red]	▲ [red] ▲ [blue] ▲ [green]	▲ [green]	▲ [red]	▲ [red] ▲ [green]
Generators	▲ [blue]		▲ [red] ▲ [blue] ▲ [green]		▲ [blue]	▲ [green]
VPP	▲ [red] ▲ [blue]	▲ [red] ▲ [green]		▲ [red]	▲ [red] ▲ [blue]	▲ [green]
Aggregators	▲ [red]		▲ [green]			▲ [green]
Storage	▲ [blue]	▲ [green]	▲ [red] ▲ [blue] ▲ [green]			▲ [green]
TSO	▲ [red]	▲ [red]	▲ [red]	▲ [red]	▲ [red]	
Transmitter						
DSO	▲ [red]		▲ [red]	▲ [red]	▲ [red]	
Distribution						
WMO		▲ [red]	▲ [red]		▲ [red]	
LMO	▲ [red]	▲ [red]	▲ [red]		▲ [red]	
Retailers	▲ [red]	▲ [red]	▲ [red]			

	Transmitter	DSO	Distribution	WMO	LMO	Retailers
Consumers	▲ [red]	▲ [red] ▲ [green]	▲ [red]	▲ [red]	▲ [red]	▲ [red] ▲ [blue]
Generators	▲ [red]	▲ [green]	▲ [red]	▲ [red]		
VPP		▲ [green]			▲ [red]	
Aggregators		▲ [green]				
Storage	▲ [red]	▲ [green]	▲ [red]	▲ [red]	▲ [red]	
TSO						
Transmitter						
DSO						
Distribution						
WMO						
LMO						
Retailers				▲ [red]		

The different agents must accomplish these transactions (economic, energy, or service) in a coordinated way, based on what is required to interchange information with the rest of the participants in the power system. Traditional and new entities coexist in the proposed model. Agents whose activities change from traditional models are described in more detail in this chapter, while traditional ones are described when some of their original characteristics change.

3.1. Active Consumers

Consumers are the end-users of electricity, and they use it to perform specific activities (industrial, commercial, or residential). Three different types of consumers are considered depending on their connection point to the grid as follows:

1. **Low voltage (LV):** Consumers. The voltage supply is lower than V_{HV} kV, and they are connected to the LV distribution network. They are usually residential or small commercial customers.
2. **High voltage (HV):** Consumers to distribution. Connected to the distribution power system with a voltage larger than VHV kV. They are typically medium industrial and commercial consumers.
3. **High voltage (HV):** Consumers to transmission. Connected to the transmission or sub-transmission power system level with a voltage larger than V_{HV} kV. They are typically large industrial and commercial consumers.

Consumers used to be a static agent that only consumed energy. Currently, this activity can be complemented with the production of electricity through self-generation, providing demand response resources, and being an active participant in electricity markets.

Consumers can be understood as a sum of loads that can own the metering equipment. Recently, it is becoming more and more common that customers may build their own generation resources, especially by using renewable resources. These generation facilities may range from a few kW to several MW. When generated electricity exceeds the demand, it can be sold to the main grid through retail companies that will be responsible for ensuring the economic compensation to small consumers by providing an electricity net balance with the system specified prices.

Regarding demand response resources (DRRs), they may exist in the customer facilities as a part of the demand that can be reduced/incremented according to the prices in the operation markets. Currently, it is becoming common that consumers own electric vehicles and small storage systems that can be operated in a smart way by aggregators or themselves [49] to have the possibility to offer operation services. Consumers should have the required communication systems to provide DRR in this case. Consequently, and depending on their size, consumers may require communication systems with other agents. For example, large flexible consumers will require direct communication with the TSO if they are connected to the transmission grid or direct communication with the distribution system operator (DSO) if they are connected to the distribution system. On the other hand, small and medium consumers will just interact with aggregators.

The consumer's main traditional transaction is to buy electricity from the grid and pay for it. Consumers can also now sell electricity to the grid and, eventually, may offer DRR directly to the DSO in a case where the size of the operable load is higher than the required $P_{OS\text{-}D}$. Additionally, these DRRs could also be offered directly to the transmission system operator (TSO) if they are larger than $P_{OS\text{-}T}$ or through the aggregator. Regarding the economic transactions, consumers pay for the electricity consumed to retailers if they do not directly access the markets. If they do, they pay for energy to the wholesale market operator or the local market operator, to whom they can also sell electricity for dynamic balancing. Additionally, they can also establish bilateral contracts with generators or VPPs. Regarding the operation services, consumers receive payments for the use of their flexible resources from the TSO, DSO, aggregators, VPPs, and generators, depending on who uses their flexibility. Finally, since consumers are the end-users of the system, they defray most of the incurred costs, such as transmission and distribution system usage, market and system operators, etc. They may pay them directly to the involved agents or, more commonly, they make a single payment to the retailer who divides it up with the rest of agents that receive payments from the consumer.

3.2. Generators

An electricity generator is an agent that owns the facilities to convert any type of primary energy into electricity.

The main activity of generators is to produce the electricity that is used by consumers. Moreover, generators have the capability to provide operation services (OSs), which are mandatory in some cases and optional in the rest. Optional OSs may be traded in markets or through contracts. Both energy and operation services can be provided to other agents via markets or bilateral contracts. Moreover, the regulation in most countries enforces the obligation to provide some type of primary (spinning) reserve to the TSO from any committed generator [21].

In addition to the generators and turbines, the generation plants have the control and communication systems to ensure the correct operation to supply the electricity to the grid in a reliable and secure way. New generators can also own the new assets regarding substations and transmission lines. Traditional generators were large centralized power plants, normally far away from consumers. Now, electricity generation also occurs at the distribution level and lower scales, which is known as DER [8]. Thus, electricity generators can be differentiated regarding their connection point with the grid (transmission or distribution), size, and dispatchability. Thus, generators can be bulk generators if they have large sizes and are connected to the transmission network, or they can be connected to the distributed network as DER. Moreover, a key characteristic of generating technologies is if they have the capability of varying their power output at will. Therefore, generating technologies can be differentiated in dispatchable and non-dispatchable technologies. It is common today for renewable generators to include batteries in their facilities to operate as conventional generators and provide operation services. Among all technologies, they can also be categorized as renewable (green), non-renewable (orange), nuclear (yellow), and renewable with storage (blue). The most common ones are the following: gas, coal, fuel, Combined Heat and Power (CHP), nuclear, hydroelectric, wind, solar photovoltaics (PV), solar thermal, and biomass. These classifications of technologies based on their connection point, dispatchability, and availability can be seen in Figure 3.

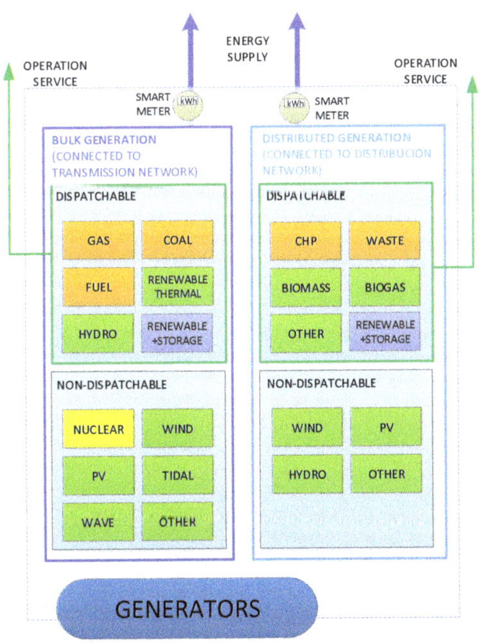

Figure 3. Generator technology types.

Generators mainly receive payments for the energy they produce and the operation services they offer. Generators provide electricity to the grid they are connected to (transmission or distribution), and this electricity can be managed by the generators or via a VPP that operates its assets. Regarding the operation services, they also provide them at the network level they are connected to. These services can be provided to the transmission and distribution operators if they meet the system operation service requirements ($P_{OS\text{-}T}$, $P_{OS\text{-}D}$). Thus, generators produce electricity that they sell in the wholesale market, local market (if connected to distribution), or via bilateral contracts to consumers, VPPs, and storage agents in exchange for economic transactions. Moreover, generators can also provide operation services via markets or contracts with the TSO, DSO, VPPs, and storage, receiving in exchange for them economic transactions. On the other hand, they can also purchase operation services from VPPs and storage agents. Finally, generators may pay fees for participating and using Wholesale Electricity Market (WEM), LMO, and the transmission and distribution grids (if connected to them).

3.3. Virtual Power Plants (VPPs)

VPPs are defined as an entity that integrates small and geographically distributed generators connected to the distribution system with the objective to provide firm and tradable generation.

VPPs integrate small and disperse generation to perform as a single entity in the wholesale market and power system [25]. Therefore, VPPs behave as a traditional generator in the system, providing energy but also operation services. VPPs help small generators, usually with no control capability, to become a viable and fully qualified generator in the market. The VPP provides this control capacity for them (primary and secondary reserve and voltage regulation) so that they can compete in energy and operation services and markets. The generation resources included in one VPP can easily be modified or switched on or off providing the required flexibility for operation purposes. This flexibility can also be obtained from the DRR by interacting directly with large consumers or through demand aggregators for small and medium-size demand resources. Energy storage may be also a key asset when providing VPP services.

The generators belonging to a VPP are usually spread out over a limited geographical area. The basic activities, relations, and minimum conditions in the framework of the proposed model are shown in Figure 4.

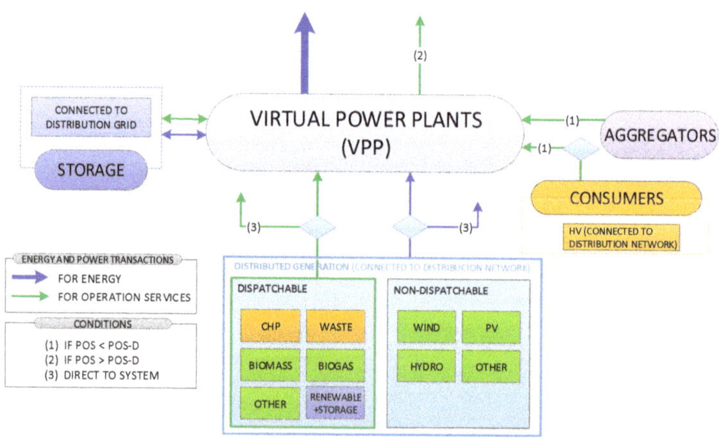

Figure 4. Virtual power plant (VPP) activities and relations.

VPPs agents may own or control generators such as renewables, cogeneration plants, traditional thermal generators, or storage systems. Moreover, VPPs need to have the same communication and control needs available as the traditional generators. These requirements should be more complex due to the fact that the VPP has to control large amounts of very distributed resources downstream and, in some cases, very small amounts according to their rated power. Therefore, their communication and computing systems have to be more complex to participate in energy markets.

Regarding its transactions, VPPs interact with many agents. VPPs buy electricity from DG generators connected to the distribution grid and storage agents, or from the local energy market. VPPs sell the electricity to the different markets (wholesale or local). Regarding operation services, VPPs purchase them from medium consumers connected to distribution, aggregators, and storage facilities. These are offered to DSO (if they are larger than $P_{OS\text{-}D}$), TSO (if they are larger than $P_{OS\text{-}T}$), or to other generators via bilateral contracts. Regarding their economic transactions, VPPs purchase electricity from generators and storage to sell them. Bilateral contracts can also be established between VPPs and consumers, retailers, or storage. Between storage and VPPs, bidirectional energy flows may exist. Finally, VPP agents receive payments from the TSO, DSO, and generators after providing the above-mentioned services. In order to obtain these services, VPPs have to purchase them from consumers, aggregators, and storage systems. The VPP is not supposed to pay any fee for participating in the market or using the transmission or distribution grids, as these costs will be translated to the generators that they operate or the consumers that buy electricity from them.

3.4. Aggregators

An aggregator is an entity that groups different consumer agents of a power system to represent and operate them as a single agent that participates in the operation service markets [40,50].

Its main activity is to put into value for the system the small customer demand response resources that, when independently considered, are not valuable for other network operators. Thus, they unlock potential resources based on economies of scale [39]. The aggregator manages the customer demand by clustering small (a few kW) demand resources with similar characteristics, or combining them to provide valuable resources to the operator, in terms of size, duration, advance notification time, etc. These products are able to compete in quality and price with those offered by other actors like generators. One special type of aggregator activity is the electric vehicle charging management, which manages the EV load charging process (and discharging) in a specific EV concentration point or area, with the objective to manage this special and flexible load and to provide additional storage to the system. Aggregators are also responsible for managing the small generation so that they can offer DRR products combining load and generation. The aggregator requires tools to evaluate the individual consumer response (or in low aggregation levels as in the case of residential customers) so that it may evaluate and foresee the main parameters of the customer response such as reduced power, duration, up and down ramps, etc. Then, it may proceed to the associated settlement when the transaction is completed. In addition, aggregators may also implement on/off control for small generators.

The basic activities, relations, and minimum conditions for the aggregator in the proposed model are shown in the Figure 5.

The aggregator's main components include an extensive communication facilities system and computational capability. The first has to provide fast and reliable performance, and the second needs to properly receive the requests from the network operators and respond to them using suitable resources without compromising the customer requirements and expectations.

The aggregator's main clients are VPPs, DSO, and TSO, to whom they provide operation services and power in exchange for economic payments. These operation services are provided according to the minimum required levels at distribution ($P_{OS\text{-}D}$) and transmission ($P_{OS\text{-}T}$). Moreover, they may also offer their services to other actors such as energy suppliers (retailers) and generators so that they may balance their buy/sell portfolio if necessary. Since all their resources come from consumers, aggregators have to pay consumers for their resources. These economic incentives that they have to provide are crucial for the seamless operation of this agent and to unlock the disaggregated opportunities.

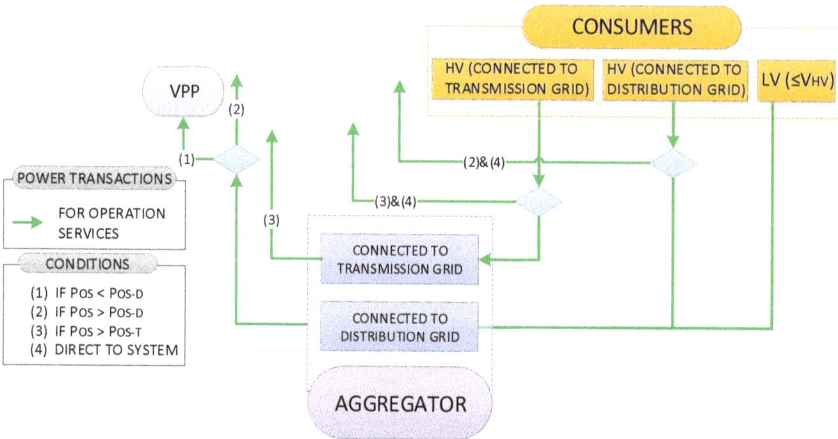

Figure 5. Aggregator activities and relations.

3.5. Storage

This agent consumes and generates electricity and has the ability to store it for using it afterward.

Storage is rapidly becoming a key technology in energy systems. Storage systems can help balance and flatten the electricity load profile. They are characterized by very fast responses, which provide storage with the capability to efficiently deliver operation services such as frequency response, black-start capability, load following, or capacity mechanisms [51]. Additionally, storage can participate in the wholesale market, leveling the load, competing with other peak power plants [52], and balancing short-term deviations. Storage was pointed out as one of the key factors to ensure reliable large renewable penetration in power systems [4], mainly because of its ability to balance the excess and deficit of renewable production, thereby avoiding curtailment and also helping the system operator.

This agent has the capability to store energy in other forms such as thermal, potential, mechanical, or chemical. This includes technologies such as pumped hydro, flywheels, molten salts, hydrogen, and electrochemical batteries [53]. The storage agent also has to have available information and control systems to be allowed to participate in the electricity market.

The storage agent implements power and energy transactions with the grid it is connected to (distribution or transmission). If connected to the transmission grid, storage injects and absorbs electricity from the grid to perform its activity and provide operation services with a minimum power ($P_{OS\text{-}T}$) to the TSO that manages the transmission grid. If connected to the distribution grid, the storage may exchange power and operation services with a minimum size ($P_{OS\text{-}D}$) not only in the distribution grid and the DSO but also through VPP. These could be implemented through bilateral contracts, which can occur for aggregating capacities to better participate in the markets. With respect to economic transactions, storage can receive payments from the wholesale market, local market, and VPPs for the energy sold. It can also receive payments from the TSO, DSO, VPP, and generators for operation services. Storage can also buy electricity from the wholesale market, local markets, and VPPs, and it may have to pay for the associated fees of markets and grid assets.

3.6. Transmission System Operator

This agent ensures the correct operation of the transmission system. Its main activities are to guarantee secure operation of the power system. This agent has to obtain the resources to operate the network not only from traditional generators but also eventually from VPPs, large customers, and storage as proposed in the architecture. To do so, the TSO needs information that is provided by the WMO, transmitter, and other agents connected to the transmission grid. The TSO is committed to balancing the system and identifying network restrictions, which requires a reliable monitoring and control capability either for committed generators or VPPs and, eventually, demand response resources, directly managed or through aggregators. These control signals require fast and reliable communication.

For doing so, the TSO needs to have assets to ensure the information and measurements flow is available, regarding the operation of the transmission network through a control center. The communication and cooperation between a DSO and TSO are essential in this new conceptual architecture. Furthermore, the TSO also has to manage exchanges with other power systems considering the capacity of the interconnections.

In the proposed model, the TSO has to also consider the use of resources to operate the transmission network not only from traditional generators but also eventually from VPPs, large customers connected to transmission, and storage. All these operation services require a minimum but homogeneous power ($P_{OS\text{-}T}$) for all participants that is determined according to the size of the system. Agents need to fulfil these requirements to compete in equal conditions. The TSO rewards economic payments in exchange for operation services to generators, VPPs, aggregators, storage systems, and consumers connected to the transmission network. As the main beneficiaries of the reliable and secure operation of the transmission grid are consumers, they pay the maintenance of the TSO via fees.

3.7. Transmitter

This agent is in charge of carrying the electricity from the bulk generation to the distribution system. The activity that it performs is to transport the electricity throughout the assets that it owns. Moreover, the transmitter has to plan and build (usually in a regulated framework) new lines, as well as reinforce the ones to account for future demand perspectives. It also verifies the connection procedure of new-generation capacity.

This agent has a physical infrastructure between the large generators and the distribution grid or large consumers. This includes high-voltage transformers and transmission lines.

This agent is highly regulated since it is a natural monopoly [54]. Therefore, the only transactions of this agent are the received fees from generators, storage, and consumers. The users of the transmission system bear the costs of its maintenance and modernization via taxes.

3.8. Distribution System Operator

This agent refers to the entity in charge of ensuring the operation of the distribution system. The DSO plays the important role of managing the distribution system. Moreover, since distributed generation is usually embedded in the distribution system, the system behavior increases in complexity (direction of energy flows, distribution operation constraints, etc.). To account for this situation, the DSO needs to have the necessary resources, which come from the customer resources directly operated or, if desirable, through aggregators. The following new roles that DSOs realize are of extreme importance:

- Enhancement of the competition and usage of different local resources to manage technical constraints at a distribution level, allowing the optimization of network planning and solving congestions at the distribution level [55].
- Provision of the forecast and availability of flexible resource to both TSO and local market operators, helping both to accurately predict and contrast the reliability of the resources [56].
- Improvement of power quality monitoring and control strategies associated with the inclusion of distributed energy generation at the distribution level [57].

Therefore, this agent needs to have assets to ensure the information and measurement flows are available regarding the operation of the distribution network, allowing the detection or prediction of undesirable conditions (current flows or voltages), and finding the resources to cope with the situation. According to this fact, fast and reliable communication channels with the TSO, aggregators, VPP, and generators connected to the distribution system are crucial. Moreover, they also own control centers to safeguard the operation of the system.

In the proposed model, the DSO has to also consider the use of resources to operate the network, not only from traditional generators but also eventually from VPPs, large customers, aggregators, and storage. All these operation services require a minimum but homogeneous power ($P_{OS\text{-}D}$) for all participants that is dictated by the size of the system. Agents need to fulfil these requirements to compete in equal conditions. Thus, the DSO is able to provide economic payments in exchange for operation services to generators, VPPs, aggregators, storage systems, and consumers connected to the distribution network. On the other hand, since the beneficiaries of the safe and secure operation of the distribution grid are consumers connected to the distribution, they pay for the maintenance of the DSO via fees.

3.9. Distributor

This agent is in charge of carrying the electricity at the final stage of the delivery, between the transmission grid and the final consumers connected to distribution.

Traditionally, the only objective of this agent was to provide the physical infrastructure between the transport grid and the final consumers. However, its activities are now larger due to the amount of information that they manage generated by smart meters. Therefore, it became an information provider too, since it manages all the telemetry and metering infrastructure. This agent as traditionally highly regulated since it was considered a natural monopoly [54]. Nevertheless, efforts to make the sector more competitive are arising [58].

A new critical activity for the distributor is as the "information provider", being responsible for gathering measurements and other information of the rest of the agents so that they may evaluate the response. For doing so, the distribution agent owns a large number of physical assets. Among them are medium- and low-voltage grids, transformers, and consumer's telemetry equipment; the distributor also owns a large advanced metering infrastructure (AMI) that collects large quantities of information. After this, thanks to a measured data management (MDM) system, all this information is filtered, processed, and organized in order to obtain valuable information for the correct functioning of the system.

The entities in charge of this agent have to maintain, monitor, and improve the physical assets and provide the collected information. Therefore, the only transactions of this agent are received fees from generators, storage, and consumers. The users of the distribution system bear the costs of its maintenance and modernization via taxes.

3.10. Wholesale Market Operator

This agent is an entity that provides a service, whereby the offers to sell electricity are matched with bids to buy electricity, ensuring the balance between them [59,60].

The main objective is to ensure the correct and transparent functioning of the economic transactions associated with the power sector, as well as organizing the different electricity markets, including wholesale, future markets, and the collection of all the bilateral contracts over the counter (OTC) that have an impact on the system. This information has to be provided to the TSO to ensure the correct functioning of the system.

The WMO is an independent actor in liberalized frameworks, strictly regulated. The WMO is characterized by a trading platform that it controls in order to manage all the bids to buy and sell products. One of its main tasks is to couple the market by matching the sell and buy offers.

Regarding transactions among agents, the generators, storage, and the consumers bear the costs associated with the WMO, paying the fees directly or via a third party. Regarding energy transactions, a minimum level for buying ($E_{W\text{-}B}$) and selling ($E_{W\text{-}S}$) electricity in this market is established depending on its size. Generators and VPPs offer electricity in the market and are compensated with cash flows. These come from the retailers and consumers that participate in the market. Storage has the capacity to buy and sell electricity to obtain benefits. Thus, cash flows between storage and the WMO are bidirectional.

3.11. Local Market Operator

Currently, local electricity markets (LEMs) are probably the least developed component of smart grids. The implementation of electricity markets in the last 20 years did not result in a significant reduction in the price ties of the energy or the increment of opportunities for most of the final consumers. Local markets are being designed to bring competitive advantages to these consumers, by implementing local trading (peer-to-peer) either directly or through aggregators and VPPs [43].

LEMs need to be reliably established to enhance the fair trading for customer-owned renewable generation and flexible resources.

This requires the development and implementation of dynamic and automatic trading platforms, for the negotiation of energy for short periods of time (shorter than the ones applied to wholesale markets) and probably closing a minute before delivery. LME platforms have to offer consumers, aggregators, and VPPs the chance to virtually trade energy services in a geographically constrained area [61]. These markets complement wholesale markets and bilateral contracts that do not have the capability to react in real time to the myriad of small demand resources and distributed generation [62]. The LMO manages and operates the LEM from an independent perspective, enabling a more dynamic trading of electricity.

Its main activity is to promote the diversity and competitiveness of the market, while ensuring the correct functioning of it by matching buying and selling bids. Furthermore, they have to monitor all the energy transactions to communicate them to the DSO to ensure a reliable operation under the technical limits. This information is provided according to the geographic control area of the DSO associated with the LEM.

The main components that characterize the LMO are the trading platforms that it controls to manage all the bids to buy and sell products. All these agents have to be in a local area and interconnected in a distribution grid. This allows a fast negotiation process and a dynamic response to prices.

Due to its role of market operator, the LMO receives payments from all the agents participating in this market. The local market manages payments among the participating agents; to do so, a minimum level for buying ($E_{L\text{-}B}$) and selling ($E_{L\text{-}S}$) electricity in these markets is established depending on its size. While generators and VPP agents receive payments for the energy traded, consumers and retailers pay for it. As in other markets, the storage has bidirectional energy flows, having the capacity to buy and sell electricity. Finally, consumers, storage, and generators pay an established fee for participating in the market directly or throughout a third party.

3.12. Retailer

Electricity retailers are entities that bridge the gap between consumers and the wholesale markets [60]. The activities of this agent do not change significantly from the traditional one. They buy the electricity in the market and sell it to their customers. Nevertheless, in the proposed model, the self-generation becomes a common possibility for small customers, being the interaction for these customers directly handled by retailers. These interactions translate in contracts with consumers to absorb the self-generation excess and economically compensate them afterward.

The retailers do not have specific components on their assets. They play a role of intermediary, thus owning strong communication and prediction systems for optimizing their performance.

This agent needs to interact for energy trading with wholesale and local markets. They can also sign these transactions through bilateral contracts with generators and VPPs. For these reasons, they need reliable and secure communication and information channels. Moreover, according to the proposed architecture, retailers are also allowed to interact with customers and aggregators for portfolio balancing purposes, needing for that the capability to interact through dynamic pricing (not control capabilities) with the customers. They are also responsible for implementing the self-consumption or net balance contracting, needing for that information about the customer buying and selling electricity. Retailers are also responsible for paying the fees in representation of consumers to the different market operators. In sum, interactions between retailers are with customers, aggregators, VPPs, generators, and market operators.

3.13. Conceptual Architecture and Interactions among Agents

The above-described agents establish a series of relationships among them as summarized in Table 2. More specifically, the figures below map the different interactions that take place in the newly proposed conceptual architecture. Thus, these figures explicitly depict each of the transactions above explained.

Figure 6 shows the transactions among agents associated with the physical commodity (electricity), which can be due to power, operation services, or balancing requirements. The blue arrows show transactions among agents related to energy; for instance, generators can supply power to the grid if they are connected to generation. In contrast, if they are connected to the distribution grid, they can supply its energy to the grid or through a VPP if their capacity is small. Another example can be storage, which has the capability to provide or purchase electricity from the grid. Depending on which grid (transmission or distribution) it is connected to, the energy fluxes will vary. The green arrows represent the operation service transactions. These are related to frequency and voltage control, energy imbalance, or system protection [21]. It can be seen that these transactions are applied to the transmission or distribution grid, depending on which grid the resources are connected to. Afterward, these operation resources at the distribution level can be managed at higher levels by the TSO thanks to the communication between DSO and TSO.

Figure 7 shows the economic transactions among agents, differentiated depending on if they are associated with an energy supply, bilateral contracts, operation services, balancing of own assets, fees, and grid usage transactions. Thus, blue arrows refer to an economic payment associated with a power exchange, dashed blue arrows show energy bilateral contracts, green arrows represent payments related with operation services, green dashed arrows represent payments for balancing portfolios or demands, orange arrows represent fees, and gray arrows represent taxes for the usage of the grid. For instance, aggregators receive payments for operation services from the DSO, TSO, and VPP, but they pay these operation services to consumers. Retailers buy energy from wholesale and local markets and bilateral contracts with VPPs and generators. Afterward, this energy is sold to consumers that pay for it. On the other hand, the transmitter agent and the distributor agent only receive payments associated with taxes, which are only paid by consumers, storage, and generators, the agents that are considered the final users of the infrastructure. Specifically, only agents connected to the distribution grid pay to the distributor.

Finally, some agents associated with energy services can also balance their own portfolio to optimize their performance in the market. These last arrows can be seen as green dashed lines.

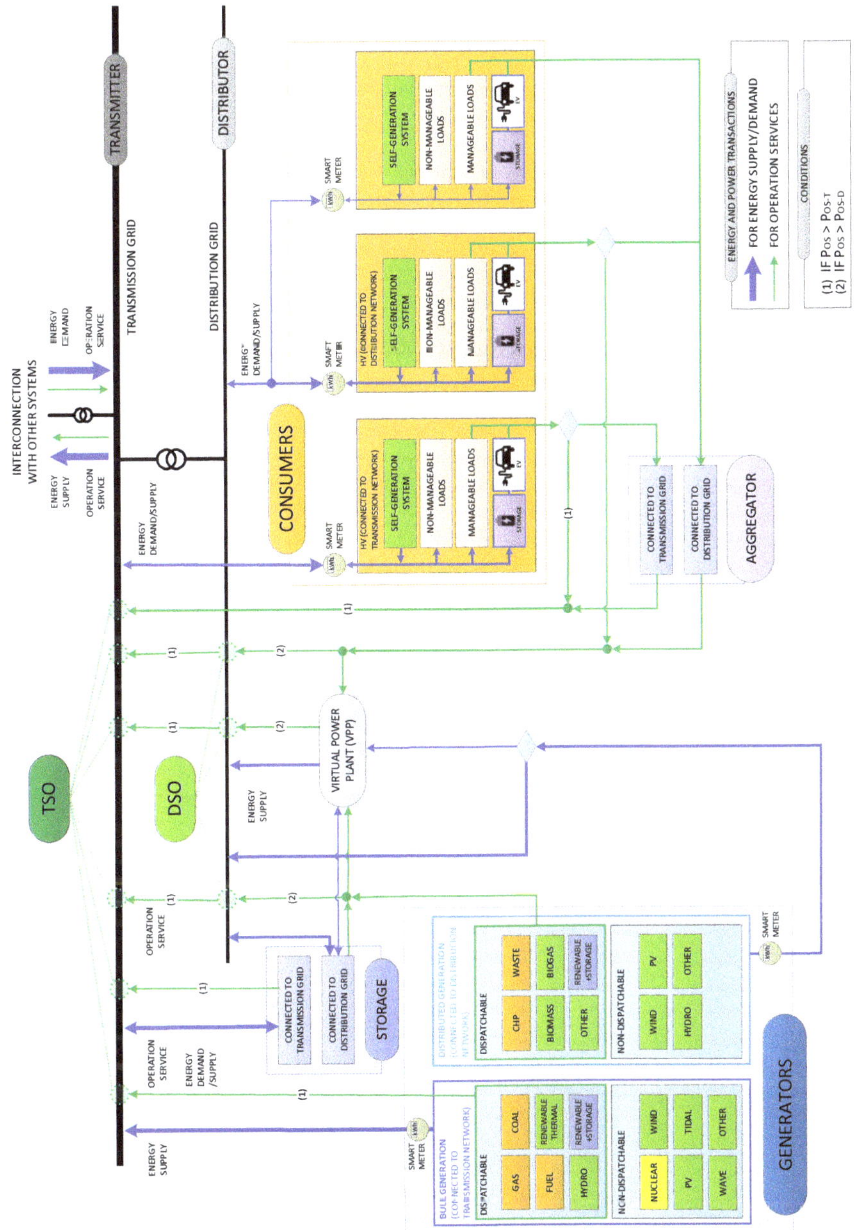

Figure 6. Conceptual architecture with physical transactions

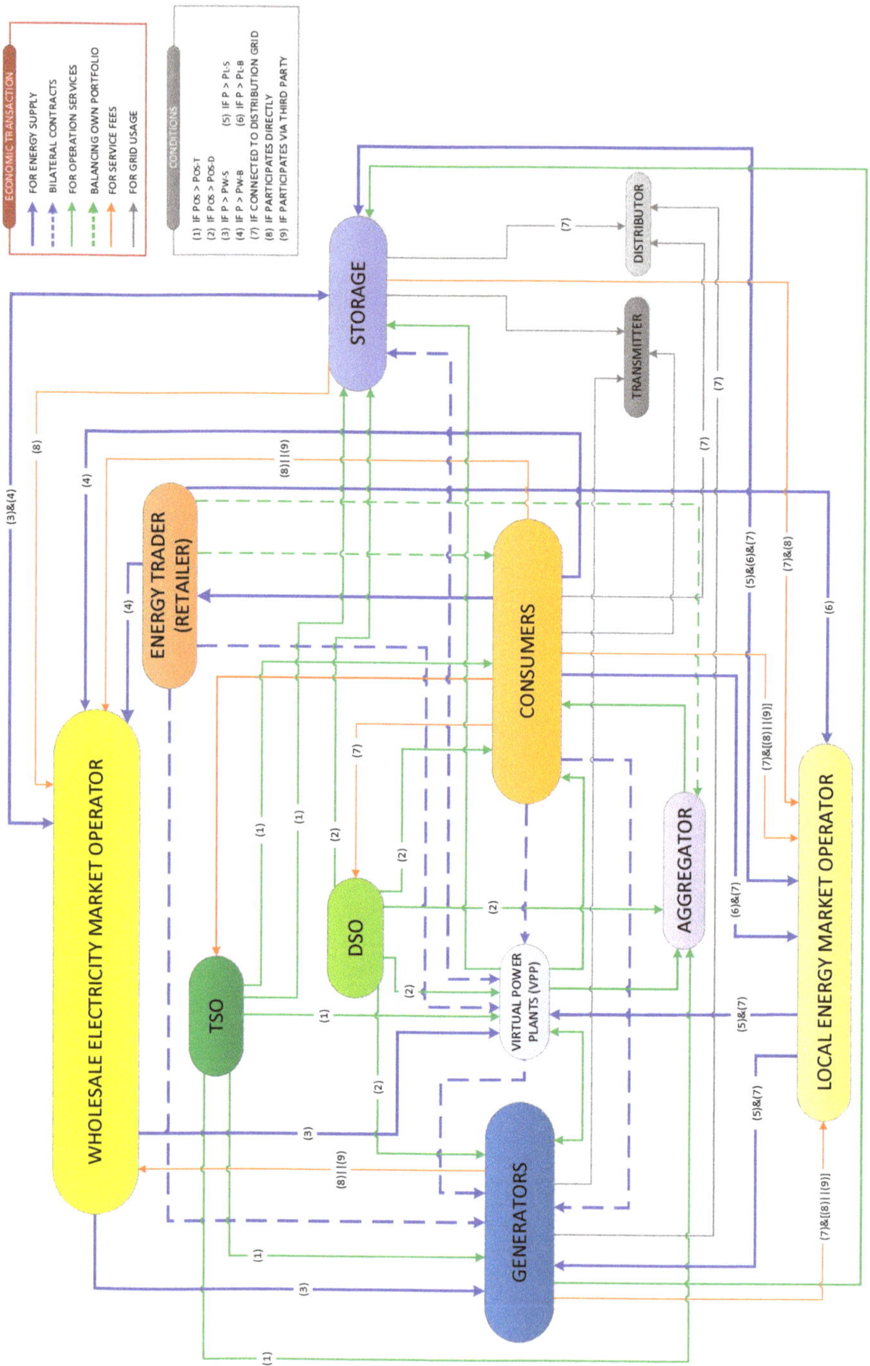

Figure 7. Conceptual architecture with economic transactions.

4. Conclusions

This paper presented a novel conceptual architecture for the development of the next-generation electricity markets. The architecture helps unlock all the hidden potential of flexible and distributed energy resources, taking into special consideration the potential benefits for active consumers. The novel architecture was proposed based on the analysis of the shortcomings of the existing models that can be found in the literature. This model provides a path that policy-makers can follow to eliminate barriers to integrate DER in a competitive way at the distribution level.

In this new paradigm with a massive integration of renewables, the need for electricity storage and for enhancing the value of demand response resources forces agents' services and transactions to appear. The proposed new architecture focuses on agents who enable flexible resources to be exploited such as storage, virtual power plants, and aggregators. These agents are already operating in some systems and emerging in others. However, the model includes the transactions among them based on an ontological analysis. Furthermore, the transactions among the presented agents are separated in energy, operating services, and economic transactions, which were clearly analyzed and described regarding the offered services, taking into account the technical restrictions. This results in a clear proposal of how the future electricity markets could be implemented.

This architecture also presents and characterizes the flexible resources available in the next-generation electricity markets, paving the way for its transactions. This flexibility can be available for two functions: to provide operation services, and the fast and dynamic balancing of electricity consumption and generation at different network levels. Three types of flexibility were shown in the proposed conceptual architecture. Similar to traditional generators, intermittent renewables with batteries are also able to provide flexibility. Consumers with self-generation and batteries can also become a flexible resource for the systems. This also helps them optimize their electricity cost by unlocking resources and allowing them to use their flexibility with an economic purpose. Finally, electric vehicles will also become a major source of flexibility in the system. Even though they are a concrete application, the massive electrification of transport gives as an opportunity to provide flexibility to the system. EVs can be described as consumers with self-generation and batteries if vehicle-to-grid chargers are implemented, or just as flexible consumers if only grid-to-vehicle chargers are installed.

Another novel element is the inclusion of local electricity markets in the conceptual architecture. Currently, these markets are gaining importance and interest due to their capability of reacting to the novel scenario of larger intermittency and decentralized generation at the distribution level. However, their relationships with other agents of the system were not previously studied from an ontological perspective. These relationships were carefully studied and stated. LEMs represent a valuable tool to exchange energy locally in a more dynamic and cost-efficient way for the power system (grid loss reduction). Furthermore, they also present an opportunity for decentralization and enhancement of competition in real time. It is important to highlight the need to have a fast and reliable communication channel between the local market operator and the DSO. The latter provides the technical restrictions that determine under what limits energy can be traded in these LEMs.

Finally, future work should assess the implementation of a case study with the proposed architecture to assess how the model enhances a more competitive electricity market and how agents are integrated in existing systems. It is also necessary to develop a clear cost–benefit analysis of the implemented model to gain knowledge of it. Moreover, simulations of the market behavior under different time domains also remain as a future objective.

Author Contributions: Conceptualization, J.R.-G., D.R.-P., E.-P.L., and C.Á.-B.; data curation, J.-M.C.; formal analysis, J.R.-G. and D.R.-P.; investigation, J.R.-G. and D.R.-P.; methodology, J.R.-G., D.R.-P., and E.-P.L.; supervision, J.R.-G. and C.Á.-B.; validation, J.R.-G., D.R.-P., E.-P.L., and C.Á.-B.; writing—original draft and editing, J.R.-G. and D.R.-P.; writing—review, J.R.-G., D.R.-P., E.-P.L., and C.Á.-B.

Funding: This work was supported by the Ministerio de Economía, Industria, y Competitividad (Spanish Government) under research project ENE-2016-78509-C3-1-P, and EU FEDER funds. The authors received funds from these grants for covering the costs to publish in open access. This work was also supported by the Spanish Ministry of Education under the scholarship FPU16/00962.

Conflicts of Interest: The authors declare no conflicts of interest.

Abbreviations

AMI	Advanced metering infrastructure
CHP	Combined Heat and Power
DER	Distributed Energy Resources
DG	Distributed generation
DR	Demand response
DRR	Demand response resources
DSO	Distribution system operator
ESCO	Energy services company
EV	Electric vehicle
HV	High voltage
ICT	Information and communication technology
LEM	Local electricity markets
LMO	Local market operator
LV	Low voltage
MDM	Measured data management
NIST	National Institute of Standards and Technology
OS	Operation services
OTC	Over the counter
PV	Photovoltaics
RES	Renewable energy sources
SGAM	Smart Grid Architecture Model
TSO	Transmission system operator
VPP	Virtual power plants
WEM	Wholesale Electricity Market
WMO	Wholesale market operator

References

1. IPCC. *Fifth Assesment Report: Chapter 7 Energy Systems*; IPCC: Geneva, Switzerland, 2014.
2. Gabaldón, A.; Guillamón, A.; Ruiz, M.C.; Valero, S.; Álvarez, C.; Ortiz, M.; Senabre, C. Development of a methodology for clustering electricity-price series to improve customer response initiatives. *IET Gener. Transm. Distrib.* **2010**, *4*, 706–715. [CrossRef]
3. European Technology Platform Smart Grids. *Smart Grids: Strategic Deployment Document for Europe's Electricity Network of the Future | 2degrees COMMUNITY | 2degrees*; European Technology Platform Smart Grids: Brussels, Belgium, 2010.
4. Weitemeyer, S.; Kleinhans, D.; Vogt, T.; Agert, C. Integration of renewable energy sources in future power systems: The role of storage. *Renew. Energy* **2015**, *75*, 14–20. [CrossRef]
5. Lebedeva, N.; Tarvydas, D.; Tsiropoulos, I. *Li-Ion Batteries for Mobility and Stationary Storage Applications—Scenarios for Costs and Market Growth*; Publications Office of the European Union: Luxembourg, 2018.
6. Kaitovic, I.; Lukovic, S. Adoption of model-driven methodology to aggregations design in Smart Grid. In Proceedings of the 2011 9th IEEE International Conference on Industrial Informatics (IEEE), Lisbon, Portugal, 26–29 July 2011; pp. 533–538.
7. Albano, M.; Ferreira, L.L.; Pinho, L.M. Convergence of Smart Grid ICT architectures for the last mile. *IEEE Trans. Ind. Inf.* **2015**, *11*, 187–197. [CrossRef]
8. NIST. *Framework and Roadmap for Smart Grid Interoperability Standards, Release 3.0*; NIST: Gaithersburg, MD, USA, 2014.
9. Smart Grid Coordination Group. *Smart Grid Reference Architecture. CEN-CENELEC-ETSI*; Smart Grid Reference Architecture: Brussels, Belgium, 2012.

10. Neureiter, C.; Uslar, M.; Engel, D.; Lastro, G. A standards-based approach for domain specific modelling of smart grid system architectures. In Proceedings of the 2016 11th System of Systems Engineering Conference (SoSE), IEEE, Kongsberg, Norway, 12–16 June 2016; pp. 1–6.
11. Goncalves Da Silva, P.; Ilic, D.; Karnouskos, S. The impact of smart grid prosumer grouping on forecasting accuracy and its benefits for local electricity market trading. *IEEE Trans. Smart Grid* **2014**, *5*, 402–410. [CrossRef]
12. Ampatzis, M.; Nguyen, P.H.; Kling, W. Local electricity market design for the coordination of distributed energy resources at district level. In Proceedings of the IEEE PES Innovative Smart Grid Technologies, Europe (IEEE), Istanbul, Turkey, 12–15 October 2014; pp. 1–6.
13. Ipakchi, A.; Albuyeh, F. Grid of the future. *IEEE Power Energy Mag.* **2009**, *7*, 52–62. [CrossRef]
14. Coelho, V.N.; Weiss Cohen, M.; Coelho, I.M.; Liu, N.; Guimarães, F.G. Multi-agent systems applied for energy systems integration: State-of-the-art applications and trends in microgrids. *Appl. Energy* **2017**, *187*, 820–832. [CrossRef]
15. Logenthiran, T.; Srinivasan, D.; Khambadkone, A.M. Multi-agent system for energy resource scheduling of integrated microgrids in a distributed system. *Electr. Power Syst. Res.* **2011**, *81*, 138–148. [CrossRef]
16. Radhakrishnan, B.M.; Srinivasan, D. A multi-agent based distributed energy management scheme for smart grid applications. *Energy* **2016**, *103*, 192–204. [CrossRef]
17. Yoo, C.-H.; Chung, I.-Y.; Lee, H.-J.; Hong, S.-S. Intelligent control of battery energy storage for multi-agent based microgrid energy management. *Energies* **2013**, *6*, 4956–4979. [CrossRef]
18. Zhao, B.; Xue, M.; Zhang, X.; Wang, C.; Zhao, J. An MAS based energy management system for a stand-alone microgrid at high altitude. *Appl. Energy* **2015**, *143*, 251–261. [CrossRef]
19. Ringler, P.; Keles, D.; Fichtner, W. Agent-based modelling and simulation of smart electricity grids and markets—A literature review. *Renew. Sustain. Energy Rev.* **2016**, *57*, 205–215. [CrossRef]
20. Dänekas, C.; Neureiter, C.; Rohjans, S.; Uslar, M.; Engel, D. *Towards a Model-Driven-Architecture Process for Smart Grid Projects*; Springer: Cham, Switzerland, 2014; pp. 47–58.
21. Wang, Q.; Zhang, C.; Ding, Y.; Xydis, G.; Wang, J.; Østergaard, J. Review of real-time electricity markets for integrating Distributed energy resources and demand response. *Appl. Energy* **2015**, *138*, 695–706. [CrossRef]
22. Pandžić, H.; Kuzle, I.; Capuder, T. Virtual power plant mid-term dispatch optimization. *Appl. Energy* **2013**, *101*, 134–141. [CrossRef]
23. Pandzic, H.; Morales, J.M.; Conejo, A.J.; Kuzle, I. Offering model for a virtual power plant based on stochastic programming. *Appl. Energy* **2013**, *105*, 282–292. [CrossRef]
24. Rahimiyan, M.; Baringo, L. Strategic bidding for a virtual power plant in the day-ahead and real-time markets: A price-taker robust optimization approach. *IEEE Trans. Power Syst.* **2016**, *31*, 2676–2687. [CrossRef]
25. Pudjianto, D.; Ramsay, C.; Strbac, G. Virtual power plant and system integration of distributed energy resources. *IET Renew. Power Gener.* **2007**, *1*, 10. [CrossRef]
26. Pudjianto, D.; Ramsay, C.; Strbac, G. Microgrids and virtual power plants: Concepts to support the integration of distributed energy resources. *Proc. Inst. Mech. Eng. Part A J. Power Energy* **2008**, *222*, 731–741. [CrossRef]
27. Mnatsakanyan, A.; Kennedy, S.W. A novel demand response model with an application for a virtual power plant. *IEEE Trans. Smart Grid* **2015**, *6*, 230–237. [CrossRef]
28. Bartolucci, L.; Cordiner, S.; Mulone, V.; Santarelli, M.; Bartolucci, L.; Cordiner, S.; Mulone, V.; Santarelli, M. Ancillary services provided by hybrid residential renewable energy systems through thermal and electrochemical storage systems. *Energies* **2019**, *12*, 2429. [CrossRef]
29. Cucchiella, F.; D'Adamo, I.; Gastaldi, M.; Stornelli, V. Solar photovoltaic panels combined with energy storage in a residential building: An economic analysis. *Sustainability* **2018**, *10*, 3117. [CrossRef]
30. Dupont, B.; De Jonghe, C.; Olmos, L.; Belmans, R. Demand response with locational dynamic pricing to support the integration of renewables. *Energy Policy* **2014**, *67*, 344–354. [CrossRef]
31. Piette, M.A.; Black, D.; Yin, R. Comparison of Actual Costs to Integrate Commercial Buildings with the Grid; Jun. 2016. Available online: https://www.semanticscholar.org/paper/Comparison-of-Actual-Costs-to-Integrate-Commercial-Piette-Black/b953cfef9716b1f87c759048ef714e8c70e19869/ (accessed on 9 July 2019).
32. Alfonso, D.; Pérez-Navarro, A.; Encinas, N.; Álvarez, C.; Rodríguez, J.; Alcázar, M. Methodology for ranking customer segments by their suitability for distributed energy resources applications. *Energy Convers. Manag.* **2007**, *48*, 1615–1623. [CrossRef]

33. Rodríguez-García, J.; Álvarez-Bel, C.; Carbonell-Carretero, J.-F.; Alcázar-Ortega, M.; Peñalvo-López, E. A novel tool for the evaluation and assessment of demand response activities in the industrial sector. *Energy* **2016**, *113*, 1136–1146. [CrossRef]
34. Morales, D.X.; Besanger, Y.; Sami, S.; Alvarez Bel, C. Assessment of the impact of intelligent DSM methods in the Galapagos Islands toward a Smart Grid. *Electr. Power Syst. Res.* **2017**, *146*, 308–320. [CrossRef]
35. Derakhshan, G.; Shayanfar, H.A.; Kazemi, A. The optimization of demand response programs in smart grids. *Energy Policy* **2016**, *94*, 295–306. [CrossRef]
36. Söyrinki, S.; Heiskanen, E.; Matschoss, K. Piloting demand response in retailing: Lessons learned in real-life context. *Sustainability* **2018**, *10*, 3790. [CrossRef]
37. McPherson, M.; Tahseen, S. Deploying storage assets to facilitate variable renewable energy integration: The impacts of grid flexibility, renewable penetration, and market structure. *Energy* **2018**, *145*, 856–870. [CrossRef]
38. Electrek. Hornsdale Power Reserve, Year 1 Technical and Market Impact Case Study. Available online: https://www.aurecongroup.com/markets/energy/hornsdale-power-reserve-impact-study/ (accessed on 9 July 2019).
39. Burger, S.; Chaves-Ávila, J.P.; Batlle, C.; Pérez-Arriaga, I.J. A review of the value of aggregators in electricity systems. *Renew. Sustain. Energy Rev.* **2017**, *77*, 395–405. [CrossRef]
40. Niesten, E.; Alkemade, F. How is value created and captured in smart grids? A review of the literature and an analysis of pilot projects. *Renew. Sustain. Energy Rev.* **2016**, *53*, 629–638. [CrossRef]
41. Calvillo, C.F.; Sánchez-Miralles, A.; Villar, J.; Martín, F. Optimal planning and operation of aggregated distributed energy resources with market participation. *Appl. Energy* **2016**, *182*, 340–357. [CrossRef]
42. Lopes, A.J.; Lezama, R.; Pineda, R. Model based systems engineering for smart grids as systems of systems. *Procedia Comput. Sci.* **2011**, *6*, 441–450. [CrossRef]
43. Lüth, A.; Zepter, J.M.; Crespo del Granado, P.; Egging, R. Local electricity market designs for peer-to-peer trading: The role of battery flexibility. *Appl. Energy* **2018**, *229*, 1233–1243. [CrossRef]
44. Kabalci, Y. A survey on smart metering and smart grid communication. *Renew. Sustain. Energy Rev.* **2016**, *57*, 302–318. [CrossRef]
45. U.S. Department of Energy. *Benefits of Demand Response in Electricity Markets and Recommendations for Achieving them a Report to the United States Congress Pursuant to Section 1252 of the Energy Policy Act of 2005*; U.S. Department of Energy: Washington, DC, USA, 2006.
46. Alahakoon, D.; Yu, X. Smart electricity meter data intelligence for future energy systems: A survey. *IEEE Trans. Ind. Inf.* **2016**, *12*, 425. [CrossRef]
47. Luthander, R.; Widén, J.; Nilsson, D.; Palm, J. Photovoltaic self-consumption in buildings: A review. *Appl. Energy* **2015**, *142*, 80–94. [CrossRef]
48. Ackermann, T.; Andersson, G.; Söder, L. Distributed generation: A definition. *Electr. Power Syst. Res.* **2001**, *57*, 195–204. [CrossRef]
49. Jha, M.; Blaabjerg, F.; Khan, M.A.; Bharath Kurukuru, V.S.; Haque, A.; Jha, M.; Blaabjerg, F.; Khan, M.A.; Bharath Kurukuru, V.S.; Haque, A. Intelligent control of converter for electric vehicles charging station. *Energies* **2019**, *12*, 2334. [CrossRef]
50. Ikäheimo, J.; Evens, C.; Kärkkäinen, S. *DER Aggregator Business: The Finnish Case*; CiteSeer: Princeton, NJ, USA, 2010.
51. Günter, N.; Marinopoulos, A. Energy storage for grid services and applications: Classification, market review, metrics, and methodology for evaluation of deployment cases. *J. Energy Storage* **2016**, *8*, 226–234. [CrossRef]
52. Akhil, A.A.; Huff, G.; Currier, A.B.; Kaun, B.C.; Rastler, D.M.; Chen, S.B.; Cotter, A.L.; Bradshaw, D.T.; Gauntlett, W.D. *Sandia Report DOE/EPRI Electricity Storage Handbook in Collaboration with NRECA*; U.S. Department of Energy Office of Scientific and Technical Information: Oak Ridge, TN, USA, 2018.
53. Market Analysis. Full Report Australian Energy Storage. Available online: https://www.smartenergy.org.au/resources/australian-energy-storage-market-analysis/ (accessed on 9 July 2019).
54. Künneke, R.W. Electricity networks: How 'natural' is the monopoly? *Util. Policy* **1999**, *8*, 99–108. [CrossRef]
55. Rivero Puente, E.I.; Gerard, H.; Six, D. A set of roles for the evolving business of electricity distribution. *Util. Policy* **2018**, *55*, 178–188. [CrossRef]
56. Silva, J.; Sumaili, J.; Bessa, R.J.; Seca, L.; Matos, M.A.; Miranda, V.; Caujolle, M.; Goncer, B.; Sebastian-Viana, M. Estimating the active and reactive power flexibility area at the TSO-DSO interface. *IEEE Trans. Power Syst.* **2018**, *33*, 4741–4750. [CrossRef]

57. Armendariz, M.; Babazadeh, D.; Brodén, D.; Nordström, L. Strategies to improve the voltage quality in active low-voltage distribution networks using DSO's assets. *IET Gener. Transm. Distrib.* **2017**, *11*, 73–81. [CrossRef]
58. Helm, D. *Burn Out, The Endgame for Fossil Fuels*; IngentaConnect: New Haven, CT, USA, 2017.
59. European Commission. *Proposal for a Directive of the European Parliament and of the Council on Common Rules for the Internal Market in Electricity*; European Commission: Brussels, Belgium, 2016.
60. Kirschen, D.; Strbac, G. *Fundamentals of Power System Economics*; John Wiley & Sons, Ltd: Chichester, UK, 2004; ISBN 9780470020593.
61. Mengelkamp, E.; Notheisen, B.; Beer, C.; Dauer, D.; Weinhardt, C. A blockchain-based smart grid: Towards sustainable local energy markets. *Comput. Sci. Res. Dev.* **2018**, *33*, 207–214. [CrossRef]
62. Monacchi, A.; Elmenreich, W. Assisted energy management in smart microgrids. *J. Ambient Intell. Humaniz. Comput.* **2016**, *7*, 901–913. [CrossRef]

© 2019 by the authors. Licensee MDPI, Basel, Switzerland. This article is an open access article distributed under the terms and conditions of the Creative Commons Attribution (CC BY) license (http://creativecommons.org/licenses/by/4.0/).

energies

MDPI

Article

Electricity Markets for DC Distribution Systems: Design Options

Longjian Piao [1,2,*], Laurens de Vries [1], Mathijs de Weerdt [2] and Neil Yorke-Smith [2,3]

1. Technology, Policy and Management, Delft University of Technology, 2628 BX Delft, The Netherlands
2. Electrical Engineering, Mathematics and Computer Science, Delft University of Technology, 2628 XE Delft, The Netherlands
3. Olayan School of Business, American University of Beirut, Beirut 1107 2020, Lebanon

* Correspondence: L.Piao@tudelft.nl; Tel.: +31-15-278-9295

Received: 28 May 2019; Accepted: 6 July 2019; Published: 10 July 2019

Abstract: DC distribution systems (DCDSs) are a promising alternative to AC systems because they remove AC-DC conversions between renewable sources and loads. Their unique features compared to AC include low system inertia, strict power limits and power-voltage coupling. In a liberalised electricity market, merely applying an AC market design to a DCDS cannot guarantee the latter's supply security and voltage stability; new markets must be designed to meet DC challenges. This article identifies the key design options of DCDS electricity markets. To identify these options, we develop a comprehensive design framework for local electricity markets; to our knowledge, we provide the first such analysis. Whereas previous studies focus on separate aspects of DCDS markets, we widen the scope to include the role of market architecture and investigate the arrangements of sub-markets. As an illustration, we demonstrate three promising DCDS market designs that can be defined in our framework, and provide a first assessment of their performance.

Keywords: electricity market design; direct current; distribution system; local market; flexibility

1. Introduction

A high proportion of future electric power will be generated by direct current (DC) renewable sources [1–3] and consumed or stored locally by DC or DC-ready devices [4,5]. For instance, micro wind turbines, flywheels, and the motors and heating/cooling devices with variable-speed drives have a DC link (AC-DC or AC-DC-AC conversion). The rise of DC generation and consumption—characterised as prosumption—brings challenges. For instance, on the one hand, more rooftop PVs inject volatile power into distribution networks; on the other hand, vehicle electrification and the deployment of heat pumps may create new load peaks [1] that are an order of magnitude higher than conventional residential load peaks. Energy storage systems (especially batteries) are typically DC by nature, but the need for twice AC-DC conversions has reduced their energy efficiency. These changes pose challenges to the legacy alternating current (AC) distribution system, which typically has low power capacity, high energy losses and complex control due to synchronisation and AC-DC conversions. DC distribution systems (DCDSs), by contrast, facilitate the integration of renewable sources, loads and storage systems by removing such conversions. Compared to AC, a DCDS does not need complex control of synchronisation, inrush current, three-phase imbalances and reactive power [6,7]. Technically, it is also feasible to upgrade existing AC lines into DC lines with remarkably higher power capacity; such upgrades only demand simple changes in tower heads and insulation [8]. While AC networks have simpler voltage transformation and protection mechanisms, a DCDS has higher power capacity, energy efficiency, reliability and simpler control and is a potential competitor to AC systems [6,9,10].

Although regulations empower prosumer participation in electricity markets [11,12], the existing AC markets cannot be applied to a DCDS. The latter's unique technical features, including low system inertia, strict power limits and power–voltage coupling [13], pose new challenges to the market design. First, DCDS substations, either connected to AC or DC transmission systems, are typically converters with much stricter power limits than AC transformers [14]. While the latter have a higher tolerance to temporary overloading, the precision of converter design and manufacturing leaves little room for DC converters to be overloaded. However, rapid electrification and large-scale renewable integration may soon push these substations to congestion. Second, a DCDS mainly consists of non-spinning devices, and its system inertia is much lower [15] than interconnected AC systems with large inertia [1]. Hence, substation congestion management is crucial to a DCDS, because the latter may suffer from severe voltage disturbances once the match between local supply and demand is broken. Third, DC nodal voltage is solely linked to power flow [16]; this is different from AC in which voltage magnitude and power flow can be controlled separately. To sum up, a DCDS is a local system by nature: its network issues, including voltage deviation and network congestion [17], highlight the local value of flexibility and call for energy exchange among flexible prosumers. Merely applying AC market designs to DC may cause voltage stability issues, which motivates the design of new markets tailored to DCDS. Researchers proposed pricing mechanisms to resolve DC congestion and voltage deviations [14,18,19], but few have investigated the economic DCDS operation in a liberalised electricity market. This article is inspired by the overlooked potential of DC at the distribution level and focuses on DCDS markets' short-term economic efficiency, namely minimising system operational costs.

Studies on local electricity markets have focused on prosumer-friendly energy trading [20,21], distribution congestion management [22,23], local ancillary services [24,25] and market implementation [25,26]. However, the broad scope of electricity market research has resulted in market designs with the following negative consequences. First, market designs that ignore crucial design goals are doubtful in terms of credibility and feasibility of implementation. Second, markets aiming at one specific challenge cannot be applied directly to the real world, in which multiple interrelated challenges exist. Third, researchers who study a limited set of design variables have not thoroughly justified this choice of scope. Finally, previous works aimed at single sub-markets did not investigate the strong linkage among the sub-markets, which crucially affect the overall market performance [27,28]. All the above calls for a systematic design framework and specified design options for local electricity markets, yet, to date, there is no consensus on such a framework to our knowledge.

This article provides such a comprehensive market design framework based on an engineering design process (Section 2), and, with it, identifies the key variables that determine a DCDS market's performance. First, we enumerate the common goals of local electricity markets (Section 3). Second, we recognise the design variables that crucially impact market efficiency, and then evaluate the consequences of the choice of each design option (Sections 4 and 5). Whereas previous studies focus on separate markets, we widen the scope to include the role of market architecture and investigate the arrangement of sub-markets. As an illustration, we demonstrate three promising DCDS market designs within our framework (Section 6): *integrated market design*, *locational energy market design*, and *locational Flex market design*. The latter two pay prosumer flexibility (Flex) directly via Flex contracts. We introduce each market's principle and organisation, and then briefly discuss its advantages and challenges. As we conclude in Section 7, this article represents the first step towards a comprehensive DCDS market design and is a preparatory step towards a quantitative study of DCDS markets.

2. Market Design Framework

As we have seen, to date, there is no consensus on a general design framework for local electricity markets. This article develops such a framework based on an engineering design process. We adopt qualitative methods such as literature review and systematic analysis.

Figure 1 illustrates our design framework for local electricity markets, where each block corresponds to a section of this paper. It is based on an engineering design process of identifying goals, determining the design space, testing and evaluation [29]. This article focuses on the first two stages. Whereas previous studies focus on separate markets, we widen the scope to include the role of architecture and investigate the arrangement of sub-markets, as suggested by Stoft [27]. Accordingly, we divide the space into *architecture design*—choice and arrangement of sub-markets—and *sub-market design* that decides detailed trading rules.

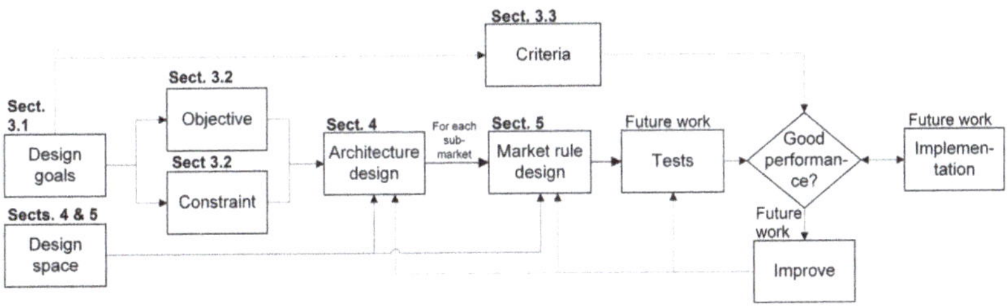

Figure 1. Design framework for local electricity markets.

Notably, as shown in Figure 1, we include a feedback loop and allow step-by-step improvements along with the test and implementation, inspired by the concept of agile design [30]. An electricity market involves complex systems and multiple stakeholders, thus the market design should be done in several iterations, starting with a minimum level of testing [31]. Since both technical systems and prosumers change rapidly, there is no single best market per se but one should improve the designs continuously during the test and implementation.

Electricity market design is an interdisciplinary study involving power systems, economics, computer science and social–environmental issues. Each discipline sets unique and sometimes contradictory requirements. As the *World Energy Trilemma* [32] suggests, the goals of energy security, energy equity and environmental sustainability challenge each other, thus the design must balance them carefully. Section 3 elaborates some unique goals of local markets, such as open access, transparency and simplicity. The negotiable goals become *objectives*, whereas the others become *constraints*. To validate a market design, we need *criteria* that define the minimum required level for each of these goals.

The design space describes the freedom to adjust design variables [33] and represents the feasible region of a design problem, where each variable represents one dimension with a set of design options. Although a large design space allows for diversified markets, it also complicates the choice and validation. We aim to limit the space and focus on those variables crucial to market efficiency.

Market architecture design: The market architecture describes the choice and arrangement of sub-markets [27], each serving a technical function required for system operation. Stoft [27] suggests that the choice of sub-markets, their types, the linkages between sub-markets are three key design variables for market architecture. For local markets, however, we identify the linkage to wholesale markets as the fourth design variable.

Sub-market design: In each sub-market, properly designed rules yield competitive prices and prevent gaming [31] by regulating information and prosumer behaviour. The selection of the design variables is based on a literature review over general electricity markets, balancing markets and flexibility markets. Based on the stages of market operation [34], we categorise the design variables into the general organisation, bid format, allocation and payment, and settlement.

Market designs without comprehensive tests may contain serious flaws that lead to failures [31]. Before implementation, a market must be thoroughly tested against uncertainty and complex prosumer behaviour, by agent simulations or rigorous field tests for instance. One should start with bottom-line tests to identify fundamental design flaws before bringing them into further studies [31]. A set of criteria, unbiased and preferably quantitative, should be used to judge if the design goals are met. We briefly discuss the role of criteria in Section 3.3 and leave the test and implementation for future work.

Due to our focus on short-term economic dispatch, we make the following assumptions. First, for globally efficient market operation, we assume that a converter connects a DCDS to the utility grid, and the price fluctuations of the wholesale markets are passed to final customers. Second, we adopt the general microeconomics assumption that prosumers are self-interested and operate their devices to maximise utility. Third, since a DCDS market requires high-frequency trading, we assume that automatic agents control devices and trade on prosumers' behalf. Fourth, since national energy policies decide taxes and levies, we assume the latter to be outside the scope of local market design.

3. Design Goals

Adopting the design framework of Section 2, this section commences the DCDS market design by stating the goals. Section 3.1 categorises the common market design goals of energy policy documents and technical reports. Section 3.2 divides the goals into objectives and constraints. Section 3.3 briefly discusses the criteria that decide whether the above goals are met or not.

3.1. Listing of the Design Goals

Energy policy documents and technical reports have revealed the goals of electricity markets, as categorised in Table 1 [12,35]. The primary goal is *productive and allocative efficiency*, where efficient prices coordinate efficient prosumption [1]. Next, an efficient market requires *reliable system operation*. Another crucial goal is to *involve prosumers* into the market. Finally, markets should be *practical to implement* in real life. Some goals are inevitably contradictory and require a balance.

Table 1. Design goals of local electricity markets.

Category	Goal	Role
Economic efficiency	efficient production	objective
	efficient allocation	objective
	completeness	constraint
	incentive-compatibility	constraint
	complete risk-hedging	constraint
	cost recovery	constraint
	liquidity & competitiveness	objective
System reliability	sufficient network capacity	constraint
	voltage regulation	constraint
	power balance	constraint
Prosumer involvement	non-discriminatory access	constraint
	information transparency	objective
	privacy	objective
	fairness	objective
	simplicity	objective
Implementability	technical feasibility	constraint
	scalability	objective
	stakeholder agreement	objective
	compatibility with wholesale markets	objective
	consistency with regulations	objective

The market's primary goal is to *produce and allocate resources efficiently* [12]. It should be *complete* so that each tradeable commodity (for which universal participation, exchangeability and cost causation of a service is guaranteed) is exchanged at low transaction costs [27]. Incentive-compatible prices should let prosumers support DCDS [34] as they reflect a resource's scarcity in time [36] and space [37]. A market should provide complete risk-hedging tools [38] and pay off investments in the long run [1]. Given the few players, it should also improve market liquidity [34] and competitiveness [39].

Efficient market operation depends on *system reliability* [40]. The power prosumption of a community-level grid is highly stochastic and hardly predictable, leading to network congestion [37] and voltage deviations [41]. Such issues must be solved immediately in a DCDS, especially if a DC substation cannot be overloaded; otherwise, a low-inertia DCDS must balance local prosumption immediately by unplanned curtailments.

Another goal is *prosumer involvement* [23]: a market should grant prosumers non-discriminatory access [42]. Information transparency [12] facilitates optimal allocation at the cost of prosumer privacy [43]. The allocation and pricing should be fair [35] so that prosumers pay for their actual contribution [23]. The trading rules should be simple enough for prosumers to master [23].

Finally, a market should be *implementable* [37] regarding technical feasibility, scalability, existing stakeholders and regulations. Market clearing mechanisms should be tractable and scalable [44,45]. The market should respect existing stakeholders [46,47], be compatible with wholesale markets [24] and consistent with regulations [42], thereby removing implementation barriers.

3.2. Objectives and Constraints

We further divide the design goals into objectives and constraints, as listed in Table 1 on the right side. A *constraint* limits the design space and lists feasible options, whereas an *objective* evaluates them in order to select design options that meet the goals. Economic efficiency is the fundamental goal and our primary objective. Since wrong incentives reduce economic efficiency, market completeness and incentive-compatibility become constraints. A market should offer stakeholders complete risk-hedging tools and steady revenue to recover investments; hence, they are also considered constraints. Reliability is crucial to power systems and is a constraint: A market should mitigate substation congestion and voltage deviations by matching supply and demand immediately. Prosumer involvement and implementability also play a key role, where the two constraints are non-discriminatory access (in order to support small prosumers) and the technical feasibility (regarding computational and communication complexity). The other goals, by contrast, become the objectives of the market design.

3.3. Criteria

To conclude whether a market design meets the goals, we need unbiased *criteria* that define the minimum required level for each goal. Criteria assist our design choices by: (1) excluding markets that violate design constraints; (2) suggesting the most promising designs with the help of objectives; and (3) indicating the direction of future improvements. This article does not discuss the full set of criteria but gives two examples. As discussed in Section 3.2, reliability is a key concern of power system operation and is a crucial constraint for DCDS market operation. For instance, a DCDS requires immediate power balancing due to strict converter power limits; a violation of this requirement will either lead to unplanned curtailments or a system-wide voltage collapse. Thus, we propose two quantitative criteria, namely a maximum substation congestion ratio (such as 10%) and a maximum nodal voltage deviation (such as $\pm$30V), to verify different market designs for a DCDS. Such verification demands detailed modelling of a DCDS's power network and market players.

4. Market Architecture Design Variables

Sections 4 and 5 investigate the design space of DCDS markets, namely a set of design variables and their options. For each variable, we aim to answer: How is the variable defined? What is its role in the overall market design? Which options are there and what does each option imply?

This section identifies the design variables for market architecture—the choice and arrangement of sub-markets—then lists different options and evaluates their features. Table 2 lists the four design variables on the left, i.e., the choice of sub-markets, their types, the linkages between sub-markets, and the linkage to wholesale markets. The first three are identified by Stoft [27], whereas the fourth one is from our analysis. For each design variable, Table 2 lists the options on the right.

Table 2. Electricity market architecture: design variables and their options.

Design Variable	Design Options
Choice of sub-markets	energy/substation capacity/voltage regulation
Market type	bilateral/organised
Linkage between sub-markets	explicit/implicit
Linkage to wholesale markets	complete/partial

4.1. Choice of Sub-Markets

The choice of sub-markets determines the commodities a market remunerates. It lays the foundation for the incentive scheme. To avoid *missing market* problems [1], a market design should reward all tradeable commodities; a commodity still plays a role even if it is not paid directly [27].

The DCDS operation relies on power dispatch, congestion management, plus various ancillary services regarding voltage regulation, contingency supply, safety, protection and power quality [13]. When deciding which commodities to reward, one should consider non-discriminatory access, completeness (and no repeated remuneration), transaction costs and transparent operation [27]. According to these criteria, (electrical) energy, network capacity (substation capacity in particular) and voltage regulation are qualified for a sub-market [13]. By contrast, the services for contingency supply, safety, protection and power quality have either high entry barriers (technical requirements for instance) or low tradeability (challenging to measure for instance). Therefore, such services should be provided by a distribution system operator (DSO) or regulated by DC network codes. To sum up, energy, network capacity and voltage regulation are the three candidate sub-markets of a DCDS.

4.2. Market Type

The market type describes the arrangement of trading, and it affects the available information in the market. An organised market, such as a pool (with side payments) or an exchange (without these), adopts central clearing and facilitates information exchange [48]. It uses standardised contracts to lower transaction costs but has high requirements for computation and communication infrastructure. Since a DCDS requires small-amount, high-frequency trading, organised markets are advantageous in efficiency, transparency and transaction costs. A bilateral market (based on bulletin boards or brokers) allows peer-to-peer trading and diversified contracts [49], but the information exchange is less efficient and transparent, thus reducing the market efficiency and DCDS security.

At the first stage of implementation, one may choose not to set up a sub-market but instead create a pricing scheme for substation capacity or voltage regulation. If market players are not familiar with such markets, an incentive-compatible pricing signal could still guide them to use resources efficiently.

4.3. Linkage Between Sub-Markets

The linkage between sub-markets is "the heart of market architecture", which naturally arises because of time, location and financial arbitrage [27]. Implicit linkages are common between sub-markets: in a DCDS, for instance, energy and voltage regulation markets are closely linked due to power–voltage coupling. Implicit linkages lead to information exchange and arbitrages between sub-markets. An explicitly-linked market [50], by contrast, integrates various commodities into one. Figure 2 lists all candidate sub-markets—energy, substation capacity and voltage regulation—and five possible linkages between them (solid lines for explicit linkages and dash lines for implicit ones).

The linkages should contribute to economic efficiency and reduced market complexity [27]. For instance, if the linkages between all sub-markets are explicit, we obtain an *integrated* market that merges various commodities into one (Figure 2d). Explicit linkages may increase market efficiency thanks to improved coordination, but may not if it the value of both sub-markets is not correctly represented [27]. Otherwise, we obtain multi-commodity electricity markets with separate prices for each commodity. Figure 2a represents a locational energy market (hereinafter, a sub-market is referred to as a market if it is clear from the context), which links the substation capacity to energy market via locational energy prices. Figure 2b represents a locational voltage regulation market, where the local flexibility (Flex) for voltage regulation is priced differently at each node of the DCDS. Figure 2c represents an energy market with voltage-based pricing. Finally, Figure 2e represents a market where three sub-markets are organised separately. Further study should balance economic efficiency and the extra complexity an explicit linkage brings.

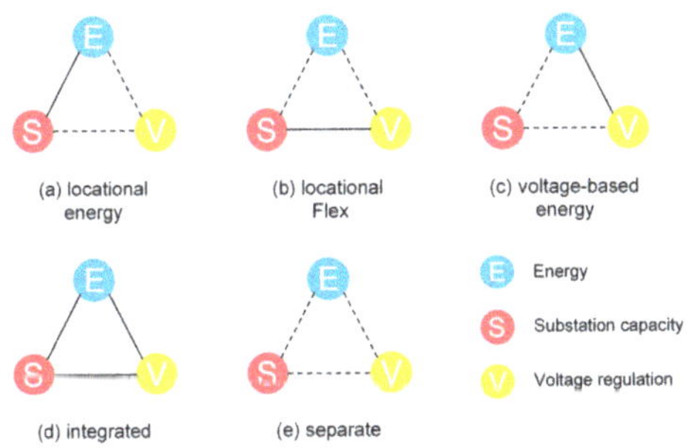

Figure 2. Choice of sub-markets and their linkages.

4.4. Linkage to Wholesale Markets

The above design variables are identified in wholesale markets [27]. For local markets, we identify the linkage to wholesale markets as the fourth design variable, indicating how a local sub-market connects to a corresponding wholesale market [24]. Our motivation is twofold. First, a local market should facilitate prosumer participation in the wholesale market. Second, local resource allocation should aim at the global optimum. Prosumers should be exposed to wholesale market prices so that they share local resources efficiently in a broad marketplace. Here, the design criterion is the completeness [51], i.e., if each sub-market in a DCDS is linked to a wholesale market. A partial linkage hinders globally efficient resource allocation and separates prosumers from the wholesale market. Readers may refer to Tohidi et al. [52] for a more comprehensive review of such linkages.

4.5. Summary

This section identifies some critical design variables of DCDS market architecture and analyses their options. The market architecture sets the foundation for a market design, based on which we set rules for each sub-market. Its design variables are the choice of sub-markets, market type, linkages between sub-markets, and linkage to wholesale markets. Further study should investigate the linkage between sub-markets and its impact on the overall market performance.

5. Sub-Market Design Variables

Section 4 lists the sub-markets of a DCDS and discusses their arrangement. For each sub-market, this section identifies the crucial design variables that affect its efficiency and competitiveness. To the best of our knowledge, Table 3 lists some critical design variables; for each identified variable, the table shows the options on the right. The selection of the variables is based on literature review of general electricity markets [34,53,54], balancing markets [55,56] and flexibility markets [24,25,57].

Based on different stages of market operation [34], we further categorise the design variables into four groups: general organisation, bid format, allocation and payment, and settlement. The *general organisation* decides buyers and sellers. The *bid format* regulates the information gathered from prosumers. The *allocation rules* determine the economic efficiency of the allocation, while the *pricing rules* sets monetary incentives. Finally, the *settlement rules* guarantee the delivery of commodities.

Table 3. Electricity sub-markets: design variables and their options.

Category	Design Variable	Design Options
General organisation	buyer and seller	one-sided/double-sided
	entry requirements	universal/tech-specific, voluntary/mandatory
	info disclosure	fully transparent–fully hidden
Bid format	bid content	simple/complex
	time resolution	1 s–15 min
	gate closure time	1 s–24 h
	locational info	global/zonal/nodal
Allocation & payment	objective	economic efficiency/renewables/self-sufficiency/...
	pricing mechanism	uniform/discriminatory
	price cap	yes/no (or sufficiently high)
Settlement	method	physical/financial
	pricing directions	one-price/two-price
	risk-hedging tools	no/forward market/options/stochastic clearing/...

5.1. General Organisation

The general organisation decides buyers, sellers and the available market information. The design variables are: (1) *the arrangement of buyers and sellers*; (2) *entry requirements*, the conditions for prosumer participation; and (3) *information disclosure policy* related to prosumers' privacy.

Arrangement of buyers and sellers: The arrangement of buyers and sellers defines the supply and demand side of a market. It has a major influence on the market structure, namely, different parties' market share and their competition. The design variable is the bidding sides [58]: one-sided or double-sided. A one-sided market has either a monopoly (for instance, in substation capacity auctions) or a monopsony (for instance, in frequency regulation markets), whose significant market power reduces economic efficiency. By contrast, a double-sided market promotes competitions on both sides and is preferred when possible.

Entry requirements: Entry requirements are the conditions (or obligations) for a prosumer to enter a market. An entry barrier can be a minimum size of bidding quantity or qualification of performance; such barriers prevent non-discriminatory access and thus reduce market liquidity. If open access is a major consideration, we should remove technology-specific entry requirements, so that flexible generation, flexible loads and storage systems are equally treated [12]. Mandatory participation yields more predictable market volume and prices, but all the prosumers should accept it.

Information disclosure policy: The information disclosure policy decides to which detail prosumers should reveal private information. While public information (local prosumption forecasts and wholesale prices) should be fully transparent to support prosumers' decisions, bids and allocation results contain sensitive, private information [59]. Disclosing truthful information may yield more efficient allocation [57], yet it should be safe and beneficial to prosumers (one option is to publish anonymous or aggregated bids) [60]. Hence, one should balance information transparency and privacy.

5.2. Bid Format

The bid format determines the information gathered for allocation. The design variables are: (1) the *bid content*, the information a prosumer's bid contains; (2) *time resolution* of allocation; (3) *gate closure time*, the deadline for bid submission; and (4) the inclusion of prosumer's *locational information*.

Bid content: The bid content is the information a prosumer's bid contains. More information potentially increases market efficiency but challenges computational tractability. Simple bids with price and quantity are commonly used in power exchanges, whereas complex bids with additional costs, constraints and location [61] are used in power pools. In energy and substation capacity markets, simple bids may be sufficient because the services are identical. In flexibility markets, however, players are much different in operational constraints so complex bids may be necessary.

Time resolution: A bid resolution is the fineness of allocation or payment in time [56], price [62], or quantity [63]. A low-inertia DCDS is vulnerable to real-time substation congestion, so the market needs small-amount, high-frequency trading. Regarding this, the price and quantity resolutions can be set high to facilitate prosumer participation. However, the time resolution, which is bound by 1 s (DCDS response speed) and 15 min (wholesale market response speed), should be chosen carefully. Although a higher time resolution matches local supply–demand more accurately [53], it increases the computational and communication burden of the market clearing.

Gate closure time: The gate closure time is the deadline for bid updates. Its lower bound is set at the acceptable uncertainty level, and the upper bound is limited to the system response time. Both bounds are much lower in a DCDS market than in wholesale markets. Variable renewables push up the upper bound to one day to address uncertainty; DC converters and flexible devices push down the lower bound to 1 s thanks to their prompt response. A later gate closure allows the use of more accurate, updated information [53], whereas an earlier one provides more flexibility.

Locational information: The locational information, included in prosumer bids, indicates the spacial scarcity of a resource [64]. A DCDS relies on locational information for congestion management and voltage regulation. Nodal pricing ameliorates this reliance through incentive-compatible prices, but it has challenges with large numbers of nodes. Zonal pricing is sufficient if congestion only occurs at some critical points (such as substations) that divide the DCDS into several price zones.

5.3. Allocation and Payment

The allocation rules decide to whom and how a market allocates resources. The payment rules, on the other hand, reward the accepted bids adequately, thereby setting the bidding incentives. Both rules affect market efficiency and prosumers' welfare. The design variables are: (1) the *objective*, the desired direction of resource allocation; (2) the *pricing mechanism* for the allocation; and (3) the *price cap* that limits a commodity's price.

The objective: The objective quantitatively describes the desired direction of resource allocation. The primary objective of a DCDS market is economic efficiency under reliability constraints. Other objectives such as integration of local renewables or community energy self-sufficiency may be considered as well.

Pricing mechanism: The pricing mechanism defines at which price a deal is closed [65]; it lays the basis of the incentive scheme. Payment is either universal (such as in uniform price auctions) or discriminatory (such as in pay-as-bid auctions) among market parties [53]. Universal pricing schemes are incentive-compatible and more predictable. However, marginal pricing may yield high prices; in such cases, we may consider discriminatory pricing, although it can be vulnerable to strategic bidding.

Price cap: A price cap (or floor) sets the maximum (or minimum) price of a commodity. In European wholesale markets, the energy price cap ranges from 150 to 3000 Euro/MWh [53]. Although it is meant to protect consumers against extreme prices, it limits prosumer's scarcity rents and affects incentive-compatibility. To avoid *missing money* problems [1], we suggest avoiding price caps or keeping them sufficiently high [46], for instance to the value of the lost load.

5.4. Settlement

Finally, a market operator should settle transactions to guarantee the delivery of commodities. The design variables are: (1) the *method* to deliver a commodity; (2) the *pricing direction* when settling deviations; and (3) *risk-hedging tools* to deal with market uncertainties.

Settlement method: The settlement method defines the way a commodity is delivered. It is: (1) physical, if the commodity must be delivered in real time; or (2) financial, when cash payments are sufficient [66]. A physical settlement guarantees supply security (typically with penalties for non-delivery), but the limited market liquidity may invite market power. A financial settlement yields higher liquidity thanks to arbitrageurs and is preferable in forward markets for risk hedging [34].

Settlement pricing directions: The settlement pricing direction defines whether the deviation of a contract is settled at different prices for long and short positions [67]. It affects incentive-compatibility and investment incentives. The one-price settlement acknowledges the equal position of flexible generation, demand response and storage. However, their dispatching costs are different in real time, so we may consider a two-price settlement to make payments incentive-compatible.

Risk-hedging tools: A DCDS has high operational uncertainty that risks the reliability and market efficiency. Such uncertainty stems from generation availability, load fluctuation, wholesale markets, bidding behaviour, among others [68]. Since high uncertainty distorts market efficiency and prosumer welfare, a DCDS market should offer risk-hedging tools, such as forward markets [69], options [70], or stochastic clearing with risk measures [61].

5.5. Summary

This section lists the design variables of local electricity market rules and analyses their options. For each sub-market, we must set rules for general organisation, bid format, allocation and payment, and settlement. The choice of a design variable must carefully balance conflicting design goals; further quantitative studies might be warranted. Variables for which this is relevant include the information disclosure policy, time resolution, gate closure time and allocation pricing rules.

Table 4. A brief comparison of three market designs.

Market Design	IM	LEM	LFM
Explicit linkage	all sub-markets	energy–network capacity	Flex–network capacity
Commodity	integrated product	locational energy + Flex	energy + locational Flex
Flex payment	implicit	explicit, non-location-specific	explicit, location-specific
Advantages	optimal dispatch in theory, incentive-compatible price	promoting Flex deployment, liquid Flex market	promoting free energy trading and Flex deployment, Flex at right places
Challenges	privacy issue, sophisticated clearing algorithm, unpredictable price	standard Flex contract, Flex pricing, Flex at wrong places; if a DSO sells Flex: distorted incentive, tariff fairness	standard Flex contract, Flex pricing, less liquid Flex market

6. Three Promising Market Designs

This section demonstrates three illustrative examples of DCDS market designs and provides a first qualitative verification of our design method in Section 2. Table 4 compares the three promising designs that fit into our framework, i.e., *integrated market (IM) design*, *locational energy market (LEM) design* and *locational Flex market (LFM) design*. Regarding market architecture (Section 4), we chose designs with all the required sub-markets. Regarding sub-markets (Section 5), we chose simple, fast and efficient mechanisms that facilitate prosumer participation. Whereas the architecture distinguishes these market designs, the sub-market rules also affect their overall performance. In Table 4, Rows 2–4 list the market features and the last two rows compare their advantages and disadvantages.

6.1. Integrated Market (IM) Design

An integrated market design explicitly links all the three sub-markets to create an integrated product, which remunerates energy as well as substation capacity and voltage regulation. The principle of this design is illustrated in Figure 2d. The only commodity is the electrical energy available at a specific time and location. The real-time price reflects the temporal and locational scarcity of energy, whereas the price fluctuation implicitly remunerates prosumers for providing flexibility.

Such design represents a centrally organised market based on security-constrained economic dispatch, where the objective is economic welfare maximisation. For global market efficiency, the opportunity for trade between the local market and the wholesale market should be maximised. All prosumers are involved in the mandatory real-time market. They submit complex bids, including their devices' state, constraints and additional costs. Prosumers are charged (paid) by their marginal contribution to the system, resulting in real-time locational marginal prices.

This design provides incentive-compatible prices, but challenges are privacy and the need for sophisticated market clearing algorithms. While prosumers are not familiar with the integrated product, they need to submit private information; hence, the market requires their trust. Meanwhile, the sophisticated market clearing requires considerable computation and communication infrastructure. If flexibility is needed and present, this market design is theoretically optimal, unlike the next two designs that we discuss. Further, since local energy prosumption is volatile, the local energy price may be unpredictable, which could be mitigated by the introduction of a voluntary forward market.

6.2. Locational Energy Market (LEM) Design

The second design, as shown in Figure 2a, explicitly links energy and network capacity markets into a locational energy market (LEM) while leaving voltage regulation in a standalone market. The LEM optimally allocates energy under network constraints; an example is locational marginal (energy) pricing [71], which is widely adopted in the US wholesale markets.

As stated in Section 1, voltage regulation of a DCDS requires local changes in energy prosumption and is therefore dependent on local flexibility (Flex). The DSO, who is responsible for voltage regulation, can provide this as a system service or contract it from prosumers in an explicit Flex market. Although LEMs have been studied for DC [14,18,19], few researchers have discussed the use of Flex trading for DC voltage regulation. Below, we discuss a case with and one without an explicit Flex market.

6.2.1. Flex Market for Voltage Regulation

In this case, Flex is an explicit, standard commodity which the DSO purchases from prosumers. It is defined as an option to adjust prosumers' power in real time. Flex contracts directly remunerate prosumer flexibility in addition to their revenues from energy trading. Other parties who may purchase Flex include wholesale market players such as balance responsible parties or aggregators [72]. A Flex market creates new business models for storage systems and demand response. In this market design, Flex payments are universal across the DCDS and are not location-specific; compared to the next design, this one has higher liquidity thanks to larger supply. However, as the Flex market is not location-specific, there is no guarantee that Flex will be deployed where necessary. Since the LEM takes care of power matching, the Flex market can be settled less frequently to improve scalability.

6.2.2. The DSO Provides Voltage Regulation

This case represents the current DSO model: Flex is a service provided by the DSO, who passes the costs along to prosumers in its tariffs [72]. A DSO may own or rent flexible devices and use them for voltage regulation [10]. One challenge is that voltage deviations may increase because prosumers are not incentivised to limit them. Another challenge is to set distribution tariffs fairly: instead of maximising prosumer welfare, a DSO may overcharge prosumers or deploy Flex for extra profit.

6.3. Locational Flex Market (LFM) Design

The third design, depicted in Figure 2b, explicitly links Flex and network capacity markets in a *locational Flex market* (LFM), while keeping the energy market standalone. An LFM aims to bring prosumers into wholesale energy markets by resolving local network issues. It acknowledges the locational value of flexibility [73] and aims to attract Flex investments to where they are needed.

The organisation of an LFM is similar to a standalone Flex market in Section 6.2.1, except that the Flex prices vary by location. The market must strictly respect DC network constraints; as real-time Flex dispatch requires extensive information from prosumers, the LFM should be centrally organised and will be less scalable than the Flex market in Section 6.2.1. As the number of providers may be very limited, we adopt pay-as-bid auctions to mitigate gaming and to lower DSOs' Flex procurement costs.

Flex markets, including LFMs [74,75], are not well studied and may generate new challenges. First, Flex products and contracts are difficult to standardise due to their complex constraints. A Flex contract may set requirements for ramping speed, energy capacity, response delay and tracking accuracy. Notably, some Flex providers, such as storage systems and flexible loads, have strong inter-temporal constraints. Second, Flex pricing is challenging because it depends on both the condition of the DCDS and the state of each Flex device. Third, Flex markets may be susceptible to market power because of their low liquidity.

7. Conclusions

This article identifies the key design options of electricity markets for DC distribution systems (DCDSs). Compared to AC systems, a DCDS has higher power capacity, energy efficiency, reliability and simpler control—anticipating the future where a large amount of renewable power is generated and consumed locally in DC. We develop a comprehensive design framework for local electricity markets to structure alternative options. To our knowledge, we provide the first such analysis.

The unique features of DCDS, such as low system inertia, strict power limits and power–voltage coupling, make a DCDS market fundamentally different from AC: it requires short response times, precise congestion management (as DC converters cannot be overloaded) and a different approach to voltage regulation. A DCDS is a local system by nature where flexibility has a high local value and needs to be exchanged for economically efficient DCDS operation.

The major elements of a DCDS *market architecture* are *energy* delivery, the provision of *substation capacity*, and *voltage regulation*. It is possible to provide all three services by creating a sub-market for each, such as a local energy exchange, a substation capacity auction and a payment scheme for voltage regulation. However, we found that DC energy and voltage regulation markets are interlinked due to power–voltage coupling: DC nodal voltage is a function of flexible power generation and consumption. Compared to the case with a DSO regulating voltage, the inclusion of a prosumer-oriented Flex market may provide the same service with better price incentives and higher economic efficiency.

For each selected *sub-market*, we analysed the design options for the general organisation, bid format, allocation and payment, and settlement. However, the choice of some design variables must trade off conflicting design goals. The degree of *information disclosure* should balance information transparency and prosumer privacy. The *time resolution* should balance a DCDS's need for short response time (efficient prosumption) and the computational burden (technical feasibility). The *gate closure time* should balance a lower power matching error (efficient prosumption) and higher flexibility for DC voltage regulation (system reliability). The *allocation pricing* rules should balance incentive-compatibility and market competitiveness (few players).

Our systematic analysis of the design options led to three promising DCDS markets. First, the *integrated market design* explicitly links three sub-markets (for energy, substation capacity and voltage regulation) to create a single commodity—an integrated product. It aims at incentive-compatible, volatile price signals that encourage prosumers to resolve congestion and voltage issues, but the challenges are privacy concerns and the need for sophisticated market clearing algorithms. Second, the *locational energy market design* links energy and substation capacity markets but leaves voltage regulation separate. Although a DSO may provide the latter as a system service, the introduction of a Flex market may offer the same service with better prosumer incentives. Third, the *locational Flex market design* links Flex and network capacity markets, thereby encouraging prosumers to help regulate DC voltage at the most critical nodes. However, further study should resolve issues regarding product definition, pricing and market power prevention.

Building on our design framework, the next step is to analyse the design options using quantitative criteria, each corresponding to a design goal in Section 3. An important direction for future work is the development of quantitative models to compare the performance of different market designs. For market architecture, further studies should balance economic efficiency and the extra complexity an explicit linkage brings. For sub-markets, researchers should balance conflicting goals by adjusting four design variables, namely the information disclosure policy, time resolution, gate closure time and allocation pricing rules. This analysis could be, for example, based on the IEEE European Low-Voltage Test Feeder (upgraded to DC). Lastly, to develop DCDS markets that are technically feasible and economically efficient, researchers should test these market designs against uncertainty and strategic behaviour.

Author Contributions: L.P. performed technical analysis, identified design options and wrote the article; L.d.V. contributed to the application of the market design framework; and all authors performed analysis and contributed to conceptualisation, supervision (except L.P.), review and editing. The authors would like to thank the anonymous reviewers for their suggestions, and Ilija Pecelj for his valuable and constructive feedback.

Funding: This work received funding in the framework of the joint programming initiative ERA-Net Smart Grids Plus under the European Union Horizon 2020 programme (Grant No. 646039).

Conflicts of Interest: The authors declare no conflict of interest.

Abbreviations

AC	Alternating Current
DC	Direct Current
DCDS	Direct Current Distribution Systems
DSO	Distribution System Operator
Flex	(local) flexibility
IM	integrated market
LEM	locational energy market
LFM	locational flexibility market
PV	photovoltaics

References

1. Newbery, D.; Pollitt, M.G.; Ritz, R.A.; Strielkowski, W. Market design for a high-renewables European electricity system. *Renew. Sustain. Energy Rev.* **2018**, *91*, 695–707. [CrossRef]
2. Cetin, E.; Yilanci, A.; Ozturk, H.K.; Colak, M.; Kasikci, I.; Iplikci, S. A micro-DC power distribution system for a residential application energized by photovoltaic–wind/fuel cell hybrid energy systems. *Energy Build.* **2010**, *42*, 1344–1352. [CrossRef]
3. Kumar, M.; Singh, S.N.; Srivastava, S.C. Design and control of smart DC microgrid for integration of renewable energy sources. In Proceedings of the 2012 IEEE Power and Energy Society General Meeting, San Diego, CA, USA, 22–26 July 2012; pp. 1–7. [CrossRef]
4. Torreglosa, J.P.; García-Triviño, P.; Fernández-Ramirez, L.M.; Jurado, F. Control strategies for DC networks: A systematic literature review. *Renew. Sustain. Energy Rev.* **2016**, *58*, 319–330. [CrossRef]
5. Mackay, L.; Vandeventer, E.; Ramirez-Elizondo, L.; Bauer, P. Capacitive Grounding for DC Distribution Grids with Multiple Grounding Points. In Proceedings of the 2017 IEEE Second International Conference on DC Microgrids (ICDCM), Nuremburg, Germany, 27–29 June 2017; pp. 76–80.
6. Justo, J.J.; Mwasilu, F.; Lee, J.; Jung, J.W. AC-microgrids versus DC-microgrids with distributed energy resources: A review. *Renew. Sustain. Energy Rev.* **2013**, *24*, 387–405. [CrossRef]
7. Planas, E.; Andreu, J.; Gárate, J.I.; Martínez De Alegría, I.; Ibarra, E. AC and DC technology in microgrids: A review. *Renew. Sustain. Energy Rev.* **2015**, *43*, 726–749. [CrossRef]
8. Larruskain, D.M.; Zamora, I.; Abarrategui, O.; Aginako, Z. Conversion of AC distribution lines into DC lines to upgrade transmission capacity. *Electr. Power Syst. Res.* **2011**, *81*, 1341–1348. [CrossRef]
9. Hammerstrom, D.J. AC versus DC distribution systems-did we get it right? In Proceedings of the 2007 IEEE Power Engineering Society General Meeting, Tampa, FL, USA, 24–28 June 2007; pp. 1–5. [CrossRef]
10. Mackay, L. Steps Towards the Universal Direct Current Distribution System. Ph.D. Thesis, Delft University of Technology, Delft, The Netherlands, 2018. [CrossRef]

11. Federal Energy Regulatory Commission (FERC). *Wholesale Competition in Regions with Organized Electric Markets: Final Rule (RM07-19-000, AD07-7-000)*; Federal Energy Regulatory Commission (FERC): Washington, DC, USA, 2008.
12. European Commission. *Proposal for a Directive of the European Parliament and of the Council on Common Rules for the Internal Market in Electricity (Recast)*; COM(2016) 864 Final/2; European Commission: Brussels, Belgium, 2016.
13. Piao, L.; de Weerdt, M.; de Vries, L. Electricity Market Design Requirements for DC Distribution Systems. In Proceedings of the 2017 IEEE Second International Conference on DC Microgrids (ICDCM), Nuremburg, Germany, 27–29 June 2017; pp. 95–101. [CrossRef]
14. Mackay, L.; van der Blij, N.H.; Ramirez-Elizondo, L.; Bauer, P. Towards the Universal DC Distribution System. *Elect. Power Compon. Syst.* **2017**, *45*, 1032–1042. [CrossRef]
15. Mackay, L.; Hailu, T.G.; Mouli, G.C.; Ramirez-Elizondo, L.; Ferreira, J.A.; Bauer, P. From DC nano- and microgrids towards the universal DC distribution system—A plea to think further into the future. In Proceedings of the 2015 IEEE Power & Energy Society General Meeting, Denver, CO, USA, 26–30 July 2015. [CrossRef]
16. Yang, N.; Paire, D.; Gao, F.; Miraoui, A.; Liu, W. Compensation of droop control using common load condition in DC microgrids to improve voltage regulation and load sharing. *Int. J. Electr. Power Energy Syst.* **2015**, *64*, 752–760. [CrossRef]
17. Ilieva, I.; Bremdal, B.; Olivella, P. D6.1 Market Design; Technical Report; EMPOWER: Local Electricity Retail Markets for Prosumer Smart Grid Power Services. 2015. Available online: http://empowerh2020.eu/wp-content/uploads/2016/05/D6.1_Market-design.pdf (accessed on 8 July 2019).
18. Disfani, V.R.; Fan, L.; Miao, Z. Distributed DC Optimal Power Flow for radial networks through partial Primal Dual algorithm. In Proceedings of the 2015 IEEE Power & Energy Society General Meeting, Denver, CO, USA, 26–30 July 2015. [CrossRef]
19. Li, C.; de Bosio, F.; Chen, F.; Chaudhary, S.K.; Vasquez, J.C.; Guerrero, J.M. Economic Dispatch for Operating Cost Minimization Under Real-Time Pricing in Droop-Controlled DC Microgrid. *IEEE J. Emerg. Sel. Top. Power Electron.* **2017**, *5*, 587–595. [CrossRef]
20. Sousa, T.; Soares, T.; Pinson, P.; Moret, F.; Baroche, T.; Sorin, E. Peer-to-peer and community-based markets: A comprehensive review. *Renew. Sustain. Energy Rev.* **2019**, *104*, 367–378. [CrossRef]
21. Lüth, A.; Zepter, J.M.; Crespo del Granado, P.; Egging, R. Local electricity market designs for peer-to-peer trading: The role of battery flexibility. *Appl. Energy* **2018**, *229*, 1233–1243. [CrossRef]
22. Huang, S.; Wu, Q.; Liu, Z.; Nielsen, A.H. Review of congestion management methods for distribution networks with high penetration of distributed energy resources. In Proceedings of the IEEE PES Innovative Smart Grid Technologies, Europe, Istanbul, Turkey, 12–15 October 2014; pp. 1–6. [CrossRef]
23. Picciariello, A.; Reneses, J.; Frias, P.; Söder, L. Distributed generation and distribution pricing: Why do we need new tariff design methodologies? *Electr. Power Syst. Res.* **2015**, *119*, 370–376. [CrossRef]
24. Ramos, A. Coordination of Flexibility Contracting in Wholesale and Local Electricity Markets. Ph.D. Thesis, Katholieke Universiteit Leuven, Leuven, Belgium, 2017.
25. Minniti, S.; Haque, N.; Nguyen, P.; Pemen, G. Local Markets for Flexibility Trading: Key Stages and Enablers. *Energies* **2018**, *11*, 3074. [CrossRef]
26. Mendes, G.; Nylund, J.; Annala, S.; Honkapuro, S.; Kilkki, O.; Segerstam, J. *Local Energy Markets: Opportunities, Benefits, and Barriers*; CIRED: Ljubljana, Slovenia, 2018.
27. Stoft, S. Market Architecture. In *Power System Economics: Designing Markets for Electricity*; Wiley-IEEE Press: Piscataway, NJ, USA, 2002; pp. 82–92.
28. Wu, T.; Rothleder, M.; Alaywan, Z.; Papalexopoulos, A.D. Pricing Energy and Ancillary Services in Integrated Market Systems by an Optimal Power Flow. *IEEE Trans. Power Syst.* **2004**, *19*, 339–347. [CrossRef]
29. Herder, P.M.; Stikkelman, R.M. Methanol-based industrial cluster design: A study of design options and the design process. *Ind. Eng. Chem. Res.* **2004**, *43*, 3879–3885. [CrossRef]
30. Brhel, M.; Meth, H.; Maedche, A.; Werder, K. Exploring principles of user-centered agile software development: A literature review. *Inf. Softw. Technol.* **2015**, *61*, 163–181. [CrossRef]
31. Stoft, S. Designing and Testing Market Rules. In *Power System Economics: Designing Markets for Electricity*; Wiley-IEEE Press: Piscataway, NJ, USA, 2002; pp. 93–106.
32. World Energy Council. *World Energy Trilemma Index 2018*; Technical Report; World Energy Council: London, UK, 2018.

33. Parkinson, A.R.; Balling, R.J.; Hedengren, J.D. *Optimization Methods for Engineering Design: Applications and Theory*; Brigham Young University: Provo, UT, USA, 2013.
34. Morey, M.J. *Power Market Auction Design: Rules and Lessons in Market Based Control for the New Electricity Industry*; Edison Electric Institute: Washington, DC, USA, 2001.
35. Federal Energy Regulatory Commission (FERC). *Working Paper on Standardized Transmission Service and Wholesale Electric Market Design*; Technical Report; FERC: Washington, DC, USA, 2002.
36. Ampatzis, M.; Nguyen, P.H.; Kling, W. Local electricity market design for the coordination of distributed energy resources at district level. In Proceedings of the IEEE PES Innovative Smart Grid Technologies, Europe, Istanbul, Turkey, 12–15 October 2014.
37. Strbac, G.; Mutale, J. Framework and Methodology for Pricing of Distribution Networks with Distributed Generation (A Report to OFGEM). Centre for Distributed Generation and Sustainable Electrical Energy, UK. 2005. Available online: https://www.ofgem.gov.uk/ofgem-publications/44458/10147-strbacmutalepdf (accessed on 10 July 2019).
38. European Commission. *Proposal for a Regulation of the European Parliament and of the Council on Risk-Preparedness in the Electricity Sector and Repealing Directive 2005/89/EC*; European Commission: Brussels, Belgium, 2016.
39. Cramton, P. Electricity market design. *Oxf. Rev. Econ. Policy* **2017**, *33*, 589–612. [CrossRef]
40. Divshali, P.H.; Choi, B.J. Electrical market management considering power system constraints in smart distribution grids. *Energies* **2016**, *9*, 405. [CrossRef]
41. Pérez-Arriaga, I.J.; Ruester, S.; Schwenen, S.; Battle, C.; Glachant, J.M. *From Distribution Networks to Smart Distribution Systems: Rethinking the Regulation of European Electricity DSOs*; Technical Report; European University Institute: Fiesole, Italy, 2013. [CrossRef]
42. European Commission. *Proposal for a Regulation of the European Parliament and of the Council on the Internal Market for Electricity (Recast). COM(2016) 861 Final*; European Commission: Brussels, Belgium, 2016.
43. Cuijpers, C.; Koops, B.J. Smart Metering and Privacy in Europe: Lessons from the Dutch Case. In *European Data Protection: Coming of Age*; Springer: Dordrecht, The Netherlands, 2013; Chapter 12, pp. 269–293. [CrossRef]
44. Conejo, A.J.; Sioshansi, R. Rethinking restructured electricity market design: Lessons learned and future needs. *Int. J. Electr. Power Energy Syst.* **2018**, *98*, 520–530. [CrossRef]
45. USEF. *The Framework Explained*; Technical Report; Universal Smart Energy Framework (USEF): Arnhem, The Netherlands, 2015. Available online: https://www.usef.energy/download-the-framework/ (accesssed on 10 July 2019).
46. Jong, J.D.; Genoese, F.; Egenhofer, C. *Reforming the Market Design of EU Electricity Markets: Addressing the Challenges of a Low-Carbon Power Sector*; Technical Report; Centre for European Policy Studies (CEPS): Brussels, Belgium, 2015. Avaailable inline: https://www.ceps.eu/ceps-publications/reforming-market-design-eu-electricity-markets-addressing-challenges-low-carbon-power/ (accessed on 10 July 2019).
47. Reinders, A.; Übermasser, S.; van Sark, W.; Gercek, C.; Schram, W.; Obinna, U.; Lehfuss, F.; van Mierlo, B.; Robledo, C.; van Wijk, A. An Exploration of the Three-Layer Model Including Stakeholders, Markets and Technologies for Assessments of Residential Smart Grids. *Appl. Sci.* **2018**, *8*, 2363. [CrossRef]
48. Biskas, P.N.; Chatzigiannis, D.I.; Bakirtzis, A.G. Market coupling feasibility between a power pool and a power exchange. *Electr. Power Syst. Res.* **2013**, *104*, 116–128. [CrossRef]
49. Mengelkamp, E.; Gärttner, J.; Rock, K.; Kessler, S.; Orsini, L.; Weinhardt, C. Designing microgrid energy markets: A case study: The Brooklyn Microgrid. *Appl. Energy* **2018**, *210*, 870–880. [CrossRef]
50. Ries, S.; Neumann, C.; Glismann, S.; Schoepf, M.; Fridgen, G. Rethinking short-term electricity market design: Options for market segment integration. In Proceedings of the 2017 14th International Conference on the European Energy Market (EEM), Dresden, Germany, 6–9 June 2017. [CrossRef]
51. O'Neill, R.P.; Fisher, E.B.; Hobbs, B.F.; Baldick, R. Towards a complete real-time electricity market design. *J. Regul. Econ.* **2008**, *34*, 220–250. [CrossRef]
52. Tohidi, Y.; Farrokhseresht, M.; Gibescu, M. A review on coordination schemes between local and central electricity markets. In Proceedings of the 2018 15th International Conference on the European Energy Market (EEM), Lodz, Poland, 27–29 June 2018; pp. 1–5. [CrossRef]
53. Hu, J.; Harmsen, R.; Crijns-Graus, W.; Worrell, E.; van den Broek, M. Identifying barriers to large-scale integration of variable renewable electricity into the electricity market: A literature review of market design. *Renew. Sustain. Energy Rev.* **2018**, *81*, 2181–2195. [CrossRef]

54. Ela, E.; Helman, U. *Wholesale Electricity Market Design Initiatives in the United States: Survey and Research Needs*; Technical Report; Electric Power Research Institute (EPRI): Palo Alto, CA, USA, 2016.
55. Abbasy, A.; Hakvoort, R.A. Exploring the design space of balancing services markets—A theoretical framework. In Proceedings of the 2009 Second International Conference on Infrastructure Systems and Services: Developing 21st Century Infrastructure Networks (INFRA), Chennai, India, 9–11 December 2009.
56. Van der Veen, R.A.; Hakvoort, R.A. The electricity balancing market: Exploring the design challenge. *Utilities Policy* **2016**, *43*, 186–194. [CrossRef]
57. Rosen, C. Design Considerations and Functional Analysis of Local Reserve Energy Markets for Distributed Generation. Ph.D. Thesis, RWTH Aachen University, Aachen, Germany, 2014.
58. Chao, H.P. Demand response in wholesale electricity markets: The choice of customer baseline. *J. Regul. Econ.* **2011**, *39*, 68–88. [CrossRef]
59. Layton, B. *The Markets for Electricity in New Zealand (Report to the Electricity Commission)*; Technical Report; The New Zealand Institute of Economic Research: Wellington, New Zealand, 2007.
60. Roth, A.E. The art of designing markets. *Harv. Bus. Rev.* **2007**, *85*, 118–126. [CrossRef]
61. Zheng, Q.P.; Wang, J.; Liu, A.L. Stochastic Optimization for Unit Commitment—A Review. *IEEE Trans. Power Syst.* **2015**, *30*, 1913–1924. [CrossRef]
62. Albadi, M.H.; El-Saadany, E.F. A summary of demand response in electricity markets. *Electr. Power Syst. Res.* **2008**, *78*, 1989–1996. [CrossRef]
63. Alagna, V.; Cauret, L.; Entem, M.; Evens, C.; Fritz, W.; Hashmi, M.; Mutale, J.; Linares, P.; Lombardi, M.; Melin, S.; et al. D5.1 Description of Market Mechanisms Which Enable Active Demand Participation in the Power System; Technical Report; ADDRESS Project. 2011. Available online: http://www.addressfp7.org/config/files/ADD-WP5-ContractsMarketsandRegulation.pdf (accessed on 10 July 2019).
64. Roques, F.; Perekhodtsev, D.; Hirth, L. *Electricity Market Design and RE Deployment*; Technical Report; IEA Renewable Energy Technology Deployment: Utrecht, The Netherlands, 2016.
65. Maurer, L.; Barroso, L. *Electricity Auctions: An Overview of Efficient Practices*; The World Bank: Washington, DC, USA, 2011. [CrossRef]
66. Chao, H.P.; Wilson, R. Design of Wholesale Electricity Markets. 2001. Available online: http://web.mit.edu/esd.126/www/StdMkt/ChaoWilson.pdf (accessed on 10 July 2019).
67. Scharff, R.; Amelin, M. Trading behaviour on the continuous intraday market Elbas. *Energy Policy* **2016**, *88*, 544–557. [CrossRef]
68. Aien, M.; Hajebrahimi, A.; Fotuhi-Firuzabad, M. A comprehensive review on uncertainty modeling techniques in power system studies. *Renew. Sustain. Energy Rev.* **2016**, 1077–1089.
69. Stoft, S. The Two-Settlement System. In *Power System Economics: Designing Markets for Electricity*; Wiley-IEEE Press: Piscataway, NJ, USA, 2002.
70. Pineda, S.; Conejo, A.J. Using electricity options to hedge against financial risks of power producers. *J. Mod. Power Syst. Clean Energy* **2013**, *1*, 101–109. [CrossRef]
71. Huang, S.; Wu, Q.; Oren, S.S.; Li, R.; Liu, Z. Distribution Locational Marginal Pricing Through Quadratic Programming for Congestion Management in Distribution Networks. *IEEE Trans. Power Syst.* **2015**, *30*, 2170–2178. [CrossRef]
72. Olivella-Rosell, P.; Lloret-Gallego, P.; Munne-Collado, I.; Villafafila-Robles, R.; Sumper, A.; Odegaard Ottessen, S.; Rajasekharan, J.; Bremdal, B.A. Local Flexibility Market Design for Aggregators Providing Multiple Flexibility Services at Distribution Network Level. *Energies* **2018**, *11*, 822. [CrossRef]
73. Eid, C.; Koliou, E.; Valles, M.; Reneses, J.; Hakvoort, R. Time-based pricing and electricity demand response: Existing barriers and next steps. *Utilities Policy* **2016**, *40*, 15–25. [CrossRef]
74. Kim, D.; Kwon, H.; Kim, M.K.; Park, J.K.; Park, H. Determining the flexible ramping capacity of electric vehicles to enhance locational flexibility. *Energies* **2017**, *10*, 2028. [CrossRef]
75. Bucher, M.A.; Delikaraoglou, S.; Heussen, K.; Pinson, P.; Andersson, G. On quantification of flexibility in power systems. In Proceedings of the 2015 IEEE Eindhoven PowerTech, Eindhoven, The Netherlands, 29 June–2 July 2015. [CrossRef]

© 2019 by the authors. Licensee MDPI, Basel, Switzerland. This article is an open access article distributed under the terms and conditions of the Creative Commons Attribution (CC BY) license (http://creativecommons.org/licenses/by/4.0/).

energies
MDPI
Article

A Q-Cube Framework of Reinforcement Learning Algorithm for Continuous Double Auction among Microgrids

Ning Wang [1], Weisheng Xu [1,*] and Weihui Shao [2] and Zhiyu Xu [1]

1 School of Electronics and Information Engineering, Tongji University, Shanghai 201804, China
2 Education Technology and Computing Center, Tongji University, Shanghai 200092, China
* Correspondence: xuweisheng@tongji.edu.cn; Tel.: +86-21-6598-1061

Received: 9 July 2019; Accepted: 23 July 2019; Published: 26 July 2019

Abstract: Decision-making of microgrids in the condition of a dynamic uncertain bidding environment has always been a significant subject of interest in the context of energy markets. The emerging application of reinforcement learning algorithms in energy markets provides solutions to this problem. In this paper, we investigate the potential of applying a Q-learning algorithm into a continuous double auction mechanism. By choosing a global supply and demand relationship as states and considering both bidding price and quantity as actions, a new Q-learning architecture is proposed to better reflect personalized bidding preferences and response to real-time market conditions. The application of battery energy storage system performs an alternative form of demand response by exerting potential capacity. A Q-cube framework is designed to describe the Q-value distribution iteration. Results from a case study on 14 microgrids in Guizhou Province, China indicate that the proposed Q-cube framework is capable of making rational bidding decisions and raising the microgrids' profits.

Keywords: microgrids; continuous double auction; Q-learning algorithm; battery energy storage system, Q-cube framework; bidding strategy

1. Introduction

The power system has experienced the evolution from a traditional power grid to the smart grid and then to the Energy Internet (EI), driven by economic, technological and environment incentives. Distributed energy resources (DERs) including distributed generation (DG), battery energy storage system (BESS), electric vehicle (EV), dispatchable load (DL), etc. are emerging and reconstructing the structure of power systems. In future EI, renewable energy sources (RESs) are regarded as the main primary energy owing to the wide application of solar panels, wind turbines and other new energy technologies [1]. According to a recent report from the U.S. Energy Information Administration (EIA), the U.S. electricity generation from RESs surpassed coal this April for the first time in history, providing 23% of the total electricity generation compared to coal's 20%. Meanwhile, the proportion of RES generation in Germany has already reached 40% in 2018. The considerable increase of RESs encourages a significant decrease in energy prices, which drives the reform of energy trading patterns and behaviors in the power system. In addition, flexible location and bi-direction energy trading ability of DERs lead to the transformation of management mode from centralized to decentralized [2]. In this process, the introduction of economic models to this decentralized system makes the power grid truly equipped with market characteristics [3].

As the aggregators of DERs in certain geographical regions, microgrids are important participants in the power market [4]. By implementing internal dispatch, microgrids can provide economic benefits through applying demand response projects and avoiding long distance energy transmission [5].

Moreover, microgrids give solutions for emergencies when power from the grid is disrupted. Energy trading among networked microgrids in the distribution network form the local energy market [6]. In early research and realistic practice, cooperative energy trading mechanisms have been proposed to achieve better performances on profit and management for the overall market. Models and algorithms have been investigated to describe features of the multi-microgrid system [7,8] and solve the optimization problem [9,10]. However, given a diverse internal network topology and device configurations, the microgrids' willingnesses of joining this cooperative energy trading market differ from each other. Though some mix-strategy Nash equilibrium points have been found by theoretical proofs, the freedom of energy trading have to be sacrificed in exchange for global optimum, as the solutions to this NP-hard problem often fail to satisfy everyone. In addition, a cooperative energy trading mechanism requires detailed information on power prediction and operating data of every device in the microgrids. This will expose residential energy consumption habits and behavioral preferences, causing privacy protection issues. Non-cooperative energy trading mechanisms are urgently needed. The development of information and communication technologies (ICTs) provides ideas for solving the above problems.

With the application of advanced ICT in the energy market, the degree of informatization has been greatly improved. Smart meter, mobile internet, blockchain and 5G, etc. help to extend the traditional power system to a three-layer architecture [11,12]: the bottom layer is the network of power devices and transmission lines. The middle layer is the network of information nodes, in which the ICTs play a very important role. Software-based negotiation agents participate in the energy trading market in the top layer [13]. In the energy trading market of networked microgrids, microgrid operators (MGOs) are set to trade energy with each other and the grid under the regulations formulated by distribution network operators (DNOs). Different economic models are implemented in this layer based on personalized behaviors of the participants, which is an emerging topic in both academic and practical fields. As a common method for allocating resources, continuous double auction (CDA) is frequently used to address the bidding problem in energy markets among multi-buyers and multi-sellers [14]. The authors in [15] discussed the efficiency of applying CDA in a computational electricity market with the midpoint price method. In [16], an adaptive aggressiveness strategy was presented in the CDA market to adjust bidding price according to market change. A stable CDA mechanism was proposed in [17], which alleviated the unnecessarily volatile behavior of normal CDA. Furthermore, peer-to-peer (P2P) energy trading mechanisms are drawing attention as ICTs like blockchain are making P2P energy trading in real time possible [18]. Wang et al. [19] proposed a parallel P2P energy trading framework with multidimensional willingness, mimicking the personalized behaviors of microgrids. In [20], a canonical coalition game was utilized to propose a P2P energy trading scheme, which proves the potential to corroborate sustainable prosumer participation in P2P energy trading. To summarize, the literature mentioned above is mainly concerned with the bidding price in the energy trading market, as the intersection of price sequences decides whether to close a deal or not. However, with the wide use of advanced ICT in the power grid, the uncertainty of DERs can be compensated by real-time behavior adjustment; meanwhile, DER responses to price signal become faster than ever before. Not only does the bidding price have impacts on the bidding results, but bidding quantity also simultaneously affects the real-time supply and demand relationship. Meanwhile, the capacity of BESS is only taken into consideration in the internal scheduling of each microgrid, neglecting the potential of BESS to participate in energy market dispatching.

At the same time as research on energy trading mechanisms, significant efforts have been devoted to model the complex bidding behaviors of negotiation agents in energy trading markets, among which the interest of applying reinforcement learning (RL) algorithms to solve power grid problems is emerging [21]. Reinforcement learning is a formal framework to study sequential decision-making problems, particularly relevant for modeling the behavior of financial agents [22]. The authors in [23] made a comprehensive review on the application of RL algorithms on electric power system decision and control. A few research works have begun to pay attention to this problem and made an effort to establish better bidding mechanisms [15,24–28]. Nicolaisen et al. [15] applied a modified Roth–Erev RL algorithm to determine the bidding price and quantity offers in each auction round. The authors in [25] presented an exact characterization of the design of adaptive learning rules for contained energy trading game concerning privacy policy. Cai et al. [26] analyzed the performance of evolutionary game-theory based trading strategies in the CDA market, which highlighted the practicability of the Roth–Erev algorithm. The authors in [27] presented a general methodology for searching CDA equilibrium strategies through the RL algorithm. Residential demand response enhanced by the RL algorithm was studied in [28] by a consumer automated energy management system. Particularly, Q-learning (QL) stands out because it is a model-free algorithm and easy to implement. The authors in [29] considered the application of QL with temperature variation for bidding strategies. Rahimiyan's work [30,31] concentrated on the adaptive adjustment of QL parameters with the energy market environment. Salehizadeh et al. [32] proposed a fuzzy QL approach in the presence of renewable resources under both normal and stressful cases. The authors in [33] introduced the concept of scenario extraction into a QL-based energy trading model for decision support.

The existing literature shows the potential of combining QL algorithms and energy trading mechanisms in obtaining better market performance. However, suitable answers to the following three issues are still unsettled, which are the motivations for this paper's research:

(1) How the QL algorithm could be combined to fit better with energy trading mechanisms to describe the characteristics of the future energy market. Bidding in the future energy market is close to real-time enhanced by ICTs, and the iteration of Q-values should be round-based rather than hour-based or day-based, whereas the time scale of updating Q-values in [29] couldn't reflect the latest market status. In addition, for a multi-microgrid system, the QL algorithm should be carried out separately by each microgrid. The authors in [34] provided the thought of applying a fitted-Q iteration algorithm in the electric power system, and more appropriate methods need to be proposed.

(2) How the coupling relationship of bidding price and quantity should be modeled and reflected by the Q-values of the Q-learning algorithm. Little research has been made about the impact of bidding quantity on bidding results in the above literature. Wang's work referred to the issue of bidding quantity [25], but only the bidding strategies of sellers in the market are discussed. In addition, the energy trading game presented in this paper adopted a discontinuous pricing rule. The impact of BESS on adjusting bidding quantity was mentioned in [35] without considering the ramping restriction, which is not practical in realistic scenes. The authors in [36] applied the extended fitted-Q iteration algorithm to control the operation modes of battery storage devices in a microgrid; however, only three actions were taken into consideration in this paper and the (dis)charge rate constraints were ignored.

(3) How the QL parameters should be decided by each microgrid, considering real-time energy market status, microgrid preferences, historical trading records and other time-varying factors. In QL algorithms, the risk characteristic of one microgrid is reflected by the values of QL parameters. However, in the existing literature, those QL-based models try to identify the bidding status according to the experiences gained from a series of trials in the current bidding rounds, ignoring the importance of historical trading records. The authors in [30,31] had noticed this issue, but the relationship between QL parameters and bidding performances were not analyzed in detail. In addition, the progress of QL research in other areas [37] hasn't been introduced into the energy trading market.

To tackle the above issues, we formulate the energy trading problem among microgrids as a Markov Decision Process (MDP) and investigate the potential of applying a Q-learning algorithm into a continuous double auction mechanism. Taking inspiration from related research on P2P trading and heuristic algorithms, a Q-cube framework of Q-learning algorithm is proposed to describe the Q-value distribution of microgrids, which is updated in each bidding round iteratively. To the best of the authors' knowledge, none of the previous work has proposed a non-tabular formation of Q-values for decision-making of the power grid.

The contributions of this paper are summarized as follows:

(1) The energy trading problem among microgrids in the distribution network is framed as a sequential decision problem. The non-cooperative energy market operation and bidding behaviors are modeled with a continuous double auction mechanism, which decreases the need for centralized control and suits the weakly-centralized nature of this distribution network.
(2) The high dimensional continuous problem is tackled by the Q-learning algorithm. Except for the bidding price, the bidding quantity of microgrids is considered as the second dimension of bidding action space and could be adjusted during the bidding process with the assistance of BESS, by which the coupling relationship between energy trading price and quantity during bidding process is handled. Related parameter setting and sharing mechanisms are designed.
(3) A non-tabular solution of Q-values considering two dimensions of action space is designed as a Q-cube. The Q-value distribution in the proposed Q-cube is in accordance with the behavior preferences of the microgrids.
(4) The real-time supply and demand relationship is highlighted as the state in the proposed Q-learning algorithm. A normal probability density distribution is divided into eight equal parts as eight states for all the microgrids. In addition, the idea of 'local search' in heuristic algorithms is applied in the proposed Q-learning algorithm for generating the action space. This approach not only takes the characteristics of power grids into consideration, but also achieves the compromise between exploitation and exploration in the action space.
(5) The proposed continuous double auction mechanism and Q-learning algorithm are validated by a realistic case from Hongfeng Lake, Guizhou Province, China. Profit comparison with traditional and P2P energy trading mechanisms highlights the doability and efficiency of the proposed method. A 65.7% and 10.9% increase in the overall profit of the distribution network could be achieved by applying a Q-learning based continuous double auction mechanism compared with the two mechanisms mentioned above.

The rest of this paper is organized as follows. In Section 2, the overview of a non-cooperative energy trading market is presented, along with a description of the proposed Q-learning based continuous double auction mechanism. A Q-cube framework of the Q-learning algorithm is introduced in Section 3. Case studies and analyses are demonstrated in Section 4 to verify the efficiency of the proposed Q-cube framework for a Q-learning algorithm and continuous double auction mechanism. Finally, we draw the conclusions and future works in Section 5.

2. Mechanism Design for Continuous Double Auction Energy Trading Market

In this section, we provide the overview of non-cooperative energy trading market and the analytical description of Q-learning based continuous double auction mechanism.

2.1. Non-Cooperative Energy Trading Market Overview

In a future distribution network, the DNO is the regulator of local energy trading market as it provides related ancillary services for market participants: (1) By gathering and analyzing the operation data from ICT, the DNO monitors and regulates the operation status of distribution network; (2) By carrying out centralized safety check and congestion management, the DNO guarantees the power flow in every transmission line is under limitation; (3) By adopting reasonable economic models, the DNO affects energy trading patterns and preferences of market participants. With the reform of the traditional energy market, along with the application of advanced metrology and ICT, the trend of peer-to-peer energy trading pattern is emerging. As peers in this energy market, we assume that MGOs have no information on their peers' energy trading preferences and internal configurations, which addresses the concern on privacy protection. In addition, each peer in this energy market is blind about the bidding target, it joins this energy trading market to satisfy its own needs for energy to the greatest extent rather than seeking cooperation. Each MGO can adjust its bidding price and quantity according to public real-time market information and private historical trading records. Accordingly, the energy trading among microgrids in the distribution network could be formulated as a non-cooperative peer-to-peer energy trading problem. Figure 1 shows the process of the non-cooperative energy trading market discussed in this paper.

Consider a distribution network containing a number of networked microgrids in a certain area. In the hour-ahead energy trading market before Time Slot N, each MGO deals with the internal coordinated dispatch (ICD) of local DERs and residents based on DERs' power prediction and BESS's state of charge (SOC) restriction information. Meanwhile, the DNO makes the distribution network scheduling for further procedures. A Q-learning based continuous double auction among microgrids is implemented according to ICD results and BESS's SOC status; detailed descriptions are presented in the following chapter. After the safety check and congestion management made by DNO, energy trading commands are confirmed and transmitted to each MGO. As the MGOs are empowered to set real-time price for regional energy, internal pricing for DER power and charge and discharge scheduling for BESS are completed in this period.

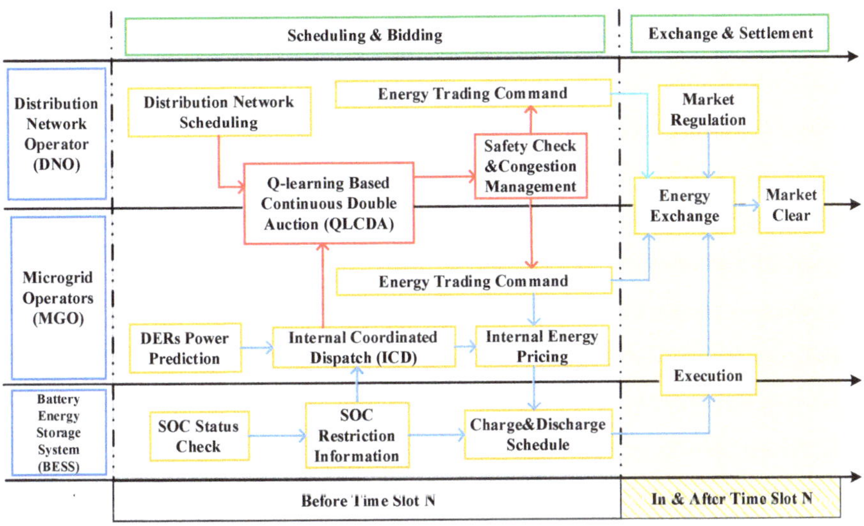

Figure 1. The process of the proposed non-cooperative energy trading market.

Energy is exchanged according to the pre-agreed trading contracts in Time Slot N under the regulation of DNO. Sufficient back-up energy supply and storage capacity are provided in case of the impact of extreme weather and dishonesty behaviors of the market participants. A market clear process is carried out after Time Slot N to ensure the accurate and timely settlement of energy transactions. Punishments are also made for the above abnormal market behaviors. Security and timeliness of the market clear process could be guaranteed by advanced ICT such as blockchain, smart meters, 5G, etc.

2.2. *Q-Learning Based Continuous Double Auction Mechanism*

This paper proposes a Q-learning based continuous double auction (QLCDA) mechanism for the energy trading market. Figure 2 presents the process of the proposed QLCDA in one time slot.

Before the QLCDA start in one time slot, each MGO tackles the ICD problem and generates the initial bidding information. The SOC check and charge and discharge restriction of the BESS are also completed in this initialization stage. In each round of CDA (indexed by n), each MGO reports its energy trading price and quantity to the DNO. Note that the trading quantity would be updated in each round; it is possible that one MGO changes its role as buyer or seller in the bidding process. Thus, an identity confirmation is made as the first step in CDA and the number of buyers (nb)/ sellers (ns) are obtained. Then, the DNO calculates and releases the overall supply and demand relationship (SDR) to these networked microgrids. Meanwhile, the reference prices for buyer and seller microgrids are calculated and released, which are the average price of selling and buying energy in the real-time market. MGOs update their bidding price and quantity according to real-time SDR and historical trading records based on the Q-Learning algorithm; the SOC restrictions are also taken into consideration to limit the behaviors of BESS in each microgrid. The bidding price of sellers and buyers are sorted in increasing order by the DNO; we have $price^b_{nb} < price^b_{nb-1} < \cdots < price^b_1$ and $price^s_1 < price_2 < \cdots < price^s_{ns}$. Once the price sequences of seller and buyers are intersected, i.e., $price^s_1 < price^b_1$, MGOs whose bidding prices are in this interval are chosen to join the energy sharing process. Actual trading price and quantity are decided in this step and the bidding quantity of each microgrid is updated based on the sharing results. If there is still untraded energy in the market, the QLCDA will repeat until the deadline of bidding rounds (N represents the maximum bidding round in one time slot). If energy demand or supply are fully satisfied before the deadline, QLCDA will be stopped in the current round. Results of QLCDA are confirmed by MGOs and sent to the DNO for further energy allocation and safety check. Detailed descriptions on initialization and the energy sharing mechanism are presented in the following chapters.

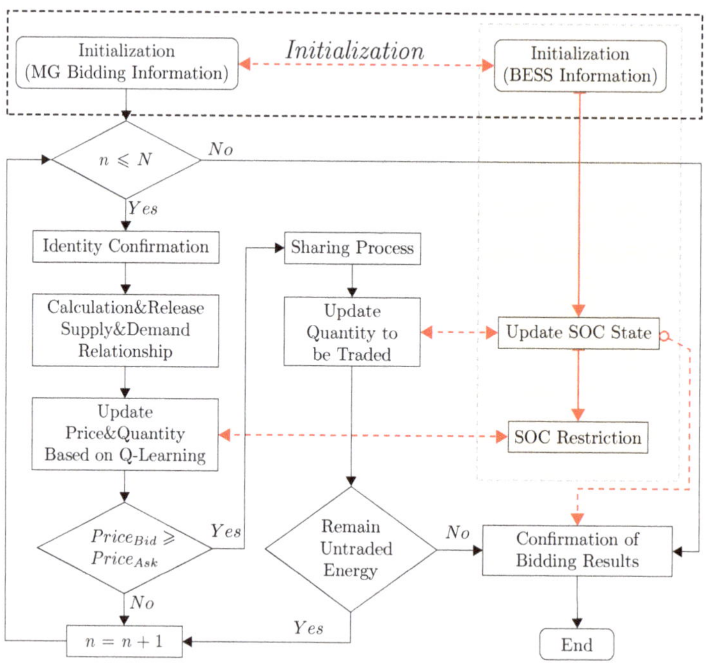

Figure 2. The process of Q-learning based continuous double auction in one time slot.

2.2.1. Initialization Setups

As the ICD of each microgrid is completed before QLCDA by each MGO, the scheduling plans are assumed to be fixed during one time slot, therefore the initial bidding quantity of QLCDA are set as the results of ICD. For seller microgrid i, the initial bidding price in time slot T is calculated as follows:

$$p_i^{initial} = p_{grid,buy}^t + \left(p\_hl^t - p\_ll^t\right) \cdot rand_i. \quad (1)$$

Similar to the seller, buyer microgrid j submits the bidding price as follows:

$$p_j^{initial} = p_{grid,sell}^t - \left(p\_hl^t - p\_ll^t\right) \cdot rand_j, \quad (2)$$

where $p_{grid,buy}^t$ and $p_{grid,sell}^t$ represent energy purchase and sell price of the grid in time slot t respectively. p_hl^t and p_ll^t are the highest/lowest bidding price limitation of this market in time slot t. $rand_i$ and $rand_j$ are random real numbers generated from the range of [0.95,1] to obtain higher/lower initial bidding price for sellers/buyers.

As each microgrid is equipped with BESS already, it is essential to consider the application of BESS in QLCDA and make full use of its charging and discharging capacity to improve real-time SDR inside the distribution network. The charging ability of microgrid i's BESS is given by:

$$P_{bess,i}^{t,charge} = \frac{C_i \cdot \min\left(SOC_i^{\Delta,charge}, (1 - SOC_i^t)\right)}{\Delta t \cdot \eta_i^{charge}}. \quad (3)$$

Similarly, the discharging ability of microgrid i's BESS is calculated as follows:

$$P_{bess,i}^{t,discharge} = \frac{C_i \cdot \min\left(SOC_i^{\Delta,discharge}, SOC_i^t\right) \cdot \eta_i^{discharge}}{\Delta t}, \quad (4)$$

where C_i is the capacity of microgrid i's BESS, SOC_i^t is the initial SOC of BESS in time slot t. Due to the limitation of material technology, charging and discharging behaviors of BESS are constrained, $SOC_i^{\Delta,charge}$ and $SOC_i^{\Delta,discharge}$ represent the ramp constraints on charging and discharging of microgrid i's BESS, respectively. Practical operations of BESS will cause energy loss, therefore we set η_i^{charge} and $\eta_i^{discharge}$ as the charging and discharging efficiency of BESS, respectively. During the QLCDA process, updated bidding quantity can't exceed the restrictions on these two parameters. Δt is the bidding cycle in this energy trading market.

2.2.2. Energy Sharing Mechanism

Once the price sequences of buyers and sellers are intersected, i.e., $price_1^s < price_1^b$, the microgrids whose bidding price are within the interval will be chosen to enter the energy sharing process. Due to the uncertainty and complexity of price intersections, a layering method and a price-prioritized quantity-weighted sharing rule are combined to solve the energy sharing problem.

The number of selected buyer and seller microgrids are nb_{share} and ns_{share}, respectively. Starting from the highest bidding price of sellers, the buyer microgrids whose bidding prices are higher than $p_{bs_{share}}^s$ and all of the seller microgrids are selected to be combined into a sharing layer. These buyer microgrids have the priority to trade with seller microgrids as they would like to pay the higher price for each unit of energy. Deals are made in this layer and related microgrids are removed from the sharing list depending on different situations. The layering method is applied repeatedly until there is no buyer microgrid in the sharing list or all the energy of seller microgrids is sold out. The detailed layering process is presented below:

- (1) Form a bidding layer according to the above-mentioned method and proceed to (2).

- (2) Allocate the energy in this layer. If energy demand exceeds supply in this layer, the sharing process is over after allocation. If energy supply exceeds demand in this layer, proceed to (3).
- (3) Remove the buyer microgrids in this layer from the optional sharing list as their energy demands are satisfied. Remove the sell microgrids whose selling prices are higher than the current highest price of buyer microgrids as there are no potential buyers for them. Return to (1) to form a new bidding layer.

Take the situation in Figure 3 as an example. Two buyer microgrids (p^b_1 and p^b_2) and three seller microgrids (p^s_1, p^s_2 and p^s_3) are selected to form Layer 1 as shown in Figure 3a. After energy allocation in Layer 1, all of the seller microgrids have surplus energy, therefore p^b_1 and p^b_2 are removed from the sharing list as their energy demands are satisfied. p^s_3 is also removed from the list as no buyer microgrid's bidding price is higher than his. Afterwords, Layer 2 is formed containing one buyer microgrid (p^b_3) and two seller microgrids (p^s_1 and p^s_2), as shown in Figure 3b. The sharing process ends after the energy allocation in this layer.

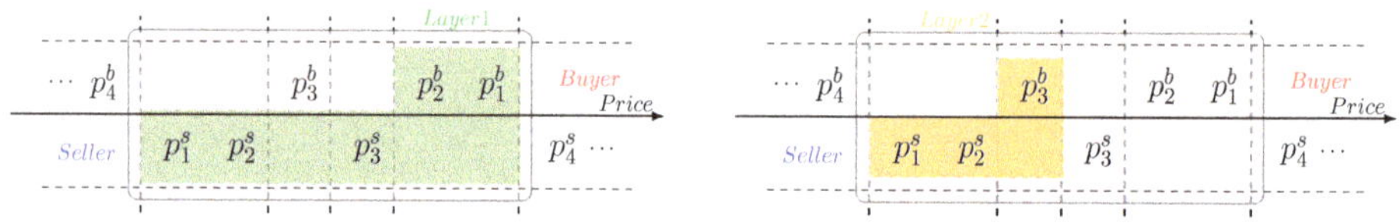

(a) Layer 1 **(b)** Layer 2

Figure 3. Layering methods in the proposed sharing mechanism.

For each layer in the energy sharing process, without loss of generality, we propose a price-prioritized quantity-weighted sharing rule for two situations. Figure 4 gives the sharing results of examples on these two situations, in which the bidding price/quantity of each deal is given below/above the figure. Energy quantity of buyers in a layer is sorted based on their quoted prices in descending order, while for sellers the quantity are sorted in ascending order. This rule ensures buyers with higher bid prices give priority to lower-priced energy. In Figure 4a, for the sharing process in round n, when $\sum q^b_i \geq \sum q^s_j$, every seller will sell out its energy, the exceeded part of demand will be cut and participate in the next round of bidding in the energy market. However, when $\sum q^b_i < \sum q^s_j$ as shown in Figure 4b, the sellers will have to fairly share the exceeded part of supply. A seller microgrid j's trading quantity is calculated as follows:

$$q^n_j = \begin{cases} q^n_j & if\ \sum q^b_i \geq \sum q^s_j, \\ q^n_j - q^n_{cut,j} & if\ \sum q^b_i < \sum q^s_j, \end{cases} \tag{5}$$

$$q^n_{cut,j} = \left(\sum q^s_j - \sum q^b_i\right) \cdot \frac{q^n_j}{\sum q^s_j}. \tag{6}$$

In Equation (6), $q^m_{cut,j}$ represents the cut quantity for microgrid j in round n. The oversupply burden is weighted shared to each seller microgrid and cut from their energy supply. This sharing rule guarantees that each seller microgrid could sell a non-negative quantity, which is more fair than the equally sharing mechanism. After the determination of sharing layers and trading quantity, the DNO can choose any suitable price within the interval $[p^s_i, p^b_j]$ as trading price at this time slot for microgrid i and j. We assume both sides of this transaction agree to trade at a price $p_{ij} = \theta \cdot (p^b_i + p^s_j)$, where $\theta \in (0, 1)$ is a predefined constant. Without loss of fairness, θ is set as 0.5 in this paper.

The proposed energy sharing mechanism ensures that buyer microgrids with higher bidding price and seller microgrids with lower bidding price have the priority in reaching a deal. In addition, the fairness of energy trading quantity is accomplished by a weighted sharing rule.

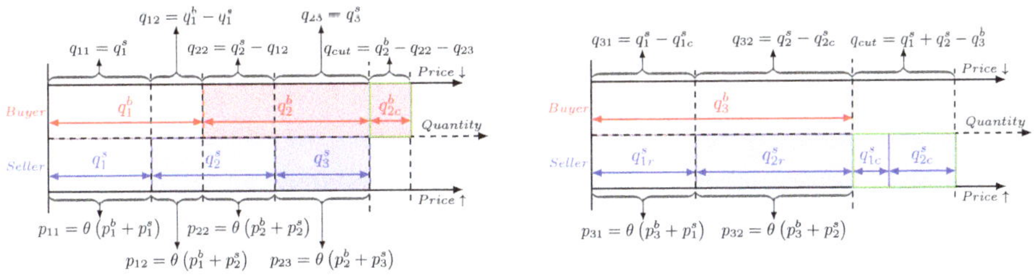

(a) Sharing results when $\sum q_{i,b}^{n} \geq \sum q_{j,s}^{n}$ **(b)** Sharing results when $\sum q_{i,b}^{n} < \sum q_{j,s}^{n}$

Figure 4. Sharing price and quantity under two situations.

3. A Q-Cube Framework for Q-Learning in a Continuous Double Auction Energy Trading Market

In a normal Q-learning algorithm, an agent learns from the environment information and interacts with relevant agents. By observing states and collecting rewards, an agent selects appropriate actions to maximize future profits. The agents are independent from one other both in terms of acting as well as learning. However, the particularity of the energy trading market creates a complex energy economic system. Non-cooperative trading pattern, personalized MGO preferences and time-varying market conditions bring difficulties to the selection of bidding strategies for market participants. As a model-free algorithm, Q-learning is capable of modeling the MGOs' bidding behaviors in a continuous double auction energy trading market. In this paper, a Q-cube framework of Q-learning algorithm is proposed especially for this multi-microgrid non-cooperative bidding problem, which addressed the exploitation–exploration issue.

3.1. Basic Definitions for Q-Learning

We base the Q-cube framework on an MDP consisting of a tuple $\langle S, A, S', r \rangle$. Detailed introductions of these variables are given as follows.

3.1.1. State Space

S represents the state space, which describes the state of MGOs in a real-time energy market. As a multi-agent system, it is impossible and senseless to select different state descriptions for each agent, whereas a common formulation is preferred. We propose to choose the real-time supply and demand relationship to form the state space for the following reasons: (1) the SDR has a decisive impact on bidding results. When the energy supply exceeds demand in a distribution network, seller microgrids are more willing to cut their bidding prices to make more deals, and exceeded supply is preferred to be stored in the BESS rather than selling to the grid at lower prices. In the meantime, buyer microgrids are not eager to raise their bidding prices quickly, but they tend to buy more energy for later use as the trading prices are much cheaper than those of the grid. The interactions between price and quantity on two roles of the energy market participants still exist when the energy demand exceeds supply. (2) The SDR reflects external energy transactions status of the networked microgrids. The more balanced the supply and demand relationship is, the less energy networked microgrids interacted with the distribution network. (3) The SDR describes the bidding situation as a public information of the energy trading market, which addresses the issue of privacy protection.

In this paper, the real-time SDR of round n in time slot T is formulated as a normal distribution with $\mu = 0$ and $\delta = 0.3$, whose value is extended to the interval of [0,2].

$$SDR^{n} = 2 \cdot \frac{1}{\sqrt{2\pi}\delta} \exp\left(-\frac{(CP^{n} - \mu)^{2}}{2\delta^{2}}\right), \tag{7}$$

$$CP^n = \frac{\sum q^n_{seller} - \sum q^n_{buyer}}{A}, \tag{8}$$

where CP^n is the clear power index, representing the clear power of the energy market in round n divided by a pre-defined constant A.

A pre-selection on the value of δ is performed and the results are shown in Figure 5a. A small choice of δ value ($\delta = 0.1$) will cause a sharp increase of SDR during the interval of $[-0.25, 0.25]$, which makes the SDR meaningless in a large clear power index range. Meanwhile, a large δ value ($\delta = 0.5$) will reduce the sensitivity of SDR when the energy supply and demand are close to equilibrium. Therefore, a compromise choice of δ value ($\delta = 0.3$) is preferred.

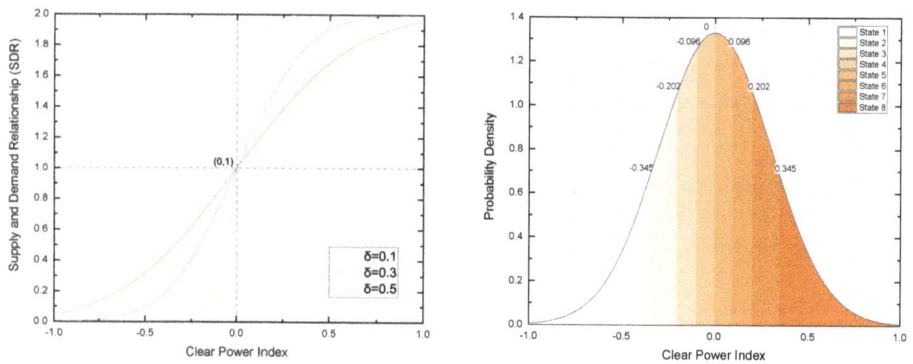

(**a**) SDR function based on clear power (**b**) State division based on probability density distribution

Figure 5. Supply and demand relationship function and state division of the proposed Q-cube framework.

The blue curve in Figure 5a shows the SDR under a different clear power index. When $\sum q^n_{seller} = \sum q^n_{buyer}$, $SDR^n = 1$, the energy supply and demand attain equilibrium. When $\sum q^n_{seller} \geq \sum q^n_{buyer}$, $SDR^n \geq 1$, vice versa. The SDR is sensitive in the interval close to 1 as the equilibrium between energy supply and demand is dynamic and in real time. In view of the fact that the SDR of energy trading market is a continuous variable, it is impossible to consider this MDP problem in an infinite space. In addition, it is impractical to model and simulate the energy trading market with limitless state descriptions. As a common method of applying Q-learning algorithm in practical problems, the state space should be divided into limited pieces for a better characterization of the SDR. For the Q-learning algorithm proposed in this paper, the number of states should be even-numbered as the SDR function is symmetrical. In addition, the probability of falling into each state should be equal. Without loss of fairness, the probability density distribution of the SDR function is divided into eight blocks with equal integral areas as shown in Figure 5b. These eight SDR intervals are defined as eight states in the proposed state space S for all the MGOs. The clear power index is also divided into eight intervals, corresponding to eight intervals of the SDR. When the clear power index is close to 0 (the market is near the equilibrium between energy supply and demand), the interval length of state is small as in most time slots the SDR experiences minor oscillation in the bidding rounds near the deadline. However, for the states whose clear power index is far from 0, the interval length is large as the SDR isn't sensitive, which means that the microgrids in the distribution network want to escape from these states.

3.1.2. Action

A represents the set of eligible actions of MGOs, which are the variation of bidding price and quantity in each bidding round. As most of the previous works aim at increasing market participants' profits via the dynamic adjustment of bidding pricing, we propose a two-dimensional formulation of action for Q-learning. By covering both bidding price and quantity, the action space is extended to a two-dimensional space rather than a set of single price actions, formulated as Equation (9):

$$a^n = (p^n, q^n) \qquad n = 1, 2, \cdots, N. \tag{9}$$

The basic idea on actions in this paper is that each MGO always optimistically assumes that all other MGOs behave optimally (though they often will not, due to their exploration and exploitation nature). In addition, all the MGOs play fair competition in the bidding process. Considering the particularity of energy trading market and agent-based simulation environment, the concept of 'Basic Action' is created to describe the rational and conventional action of each MGO. One point needs to be emphasized is that 'Basic Action' is just a point in the action space, showing the general choices of bidding price and quantity for MGOs. The mathematical expressions of basic price action are presented as follows:

$$p_i^{n,basic} = p_{i,step}^T \cdot (1 + TP^n) \cdot SDRF^n, \tag{10}$$

$$p_{i,step}^T = \frac{\left| price_i^{initial} - price_i^{history,T} \right|}{\beta \cdot N}, \tag{11}$$

$$TP^n = 1 - (1 - \frac{n}{N})^{e^{-1}}, \tag{12}$$

where $p_{i,step}^T$ represents the price changing step of MGO i, determined by MGO i's initial bidding price $price_i^{initial}$ and historical trading price $price_i^{history,T}$ in time slot T as shown in Equation (11). β is a regulation factor for the price changing step. As the QLCDA reaches the time deadline, both buyer and seller MGOs are willing to make a concession on the bidding price to make more deals. The setup of time pressure TP^n as presented in Equation (12) describes the degree of urgency over bidding rounds. Discussions on the choice of time pressure function have already been made in previous research [19]. In this paper, we adopt a simplified form in which the time pressure of each microgrid is only related to the bidding round index. The historical trading records of each microgrid are ignored in the description of time pressure. $SDRF^n$ is a modified factor based on real-time SDR. Different calculation expressions are adopted for buyer and seller MGOs as follows, inside which π is an adjustment coefficient in the range of [0.3,0.5]. The setting of π measures the influence of SDR on the basic bidding price:

$$SDRF_i^n = \begin{cases} \pi \cdot (1 - SDR^n) + 1 & \text{for buyers,} \\ \pi \cdot (SDR^n - 1) + 1 & \text{for sellers.} \end{cases} \tag{13}$$

Accordingly, the basic quantity action is calculated as follows:

$$q_i^{n,basic} = \begin{cases} q_i^n \cdot (SDR^n \cdot PS_i^n - 1) & \text{for buyers,} \\ q_i^n \cdot ((2 - SDR^n) \cdot PS_i^n - 1) & \text{for sellers,} \end{cases} \tag{14}$$

$$PS_i^n = \rho + 2 \cdot (1 - \rho) \cdot N(PR_i^n, \mu, \sigma), \tag{15}$$

$$PR_i^n = \lambda \cdot \frac{p_i^{history,T} - p_i^{reference,n}}{p_{hl}^t - p_{ll}^t}, \tag{16}$$

where the PR_i^n is a reference price factor calculated as a parameter of normal distribution, $\lambda = 1$ when MGO i is a buyer, while $\lambda = -1$ when MGO i is a seller. $p_i^{reference,n}$ is the reference price of MGO i in round n, calculated as the average price of potential transactions in the market. The values of μ and σ in Equation (15) are the same as those in Equation (7). ρ is a pre-defined adjustment coefficient located in the range of [0.95,1] for coordination with the change rate of SDR in Equation (14).

Since the action space is a continuous one, it is impossible to explore the whole action space in this problem. The idea of 'local search' in heuristic algorithms is applied in the proposed Q-learning algorithm: we intend to explore the neighborhood space of basic action on price and quantity dimensions for better bidding performance in the QLCDA process. Based on the basic action obtained in the former process, we search two directions of the price and quantity dimensions symmetrically, therefore the number of actions in each dimension is odd. Supposing that we choose more than two neighborhoods of the basic action in one direction, the total number of actions in this problem will be at least 25 actions, which is impractical and meaningless in both modeling and simulation. To limit the number of bidding actions and reduce computational complexity, only the closest neighborhoods are taken into account. The neighborhood actions are calculated as follows, where ξ and τ indicate the proximity of bidding price and quantity according to bidding experiences, respectively. ξ and τ are independent variables that only describe the neighborhood relationship of bidding price and quantity. Thus, a 3×3 action matrix is created as alternative behaviors of one MGO under a certain state. One factor, in particular, needs highlighting: the nine actions under a certain state represents nine bidding preferences and tendencies of each microgrid. Given that the SDR in one state might be different, the nine actions are also SDR-based and not totally the same for one state:

$$p_i^{n,-} = p_i^{n,basic} - \xi \cdot p_{i,step}^{n}, \tag{17}$$

$$p_i^{n,+} = p_i^{n,basic} + \xi \cdot p_{i,step}^{n}, \tag{18}$$

$$q_i^{n,-} = q_i^{n,basic} \cdot (1 - \tau), \tag{19}$$

$$q_i^{n,+} = q_i^{n,basic} \cdot (1 + \tau). \tag{20}$$

3.1.3. Q-Values and Rewards

The goal of the Q-learning algorithm for bidding strategy optimization is to choose the appropriate actions under different states for each MGO, and the Q-Values indicate the long-term values of state-action pairs. In the former Q-learning process, the Q-values for state-action pairs are arranged in the so-called Q-table. However, based on the action space, we mention, in the former chapter, a Q-cube framework of Q-learning algorithm is proposed as shown in Figure 6, in which the colors of state slices are corresponding to Figure 5.

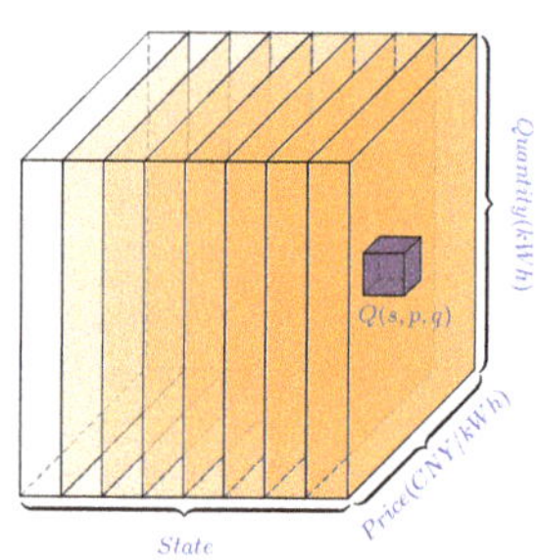

Figure 6. A Q-cube framework designed for the Q-learning algorithm.

The Q-value of taking one bidding action under one certain state is distributed in this Q-cube as shown with a small blue cube inside. Generally speaking, the proposed Q-cube is a continuous three-dimensional space, but, for practical purposes, we discrete the problem domain by taking eight states, three bidding prices and three bidding quantities under consideration in this paper.

Each MGO has a unique Q-cube showing the Q-value distribution in the proposed problem domain. The Q-values in the Q-cube are not cleared to zero at the end of each time slot but will be applied as initial configuration of the next time slot. The rolling iteration of Q-cube accumulates bidding experience in the energy trading market.

$r(s,a)$ is the reward function for adopting action a in state s. The selection of reward function is crucial, since it induces the behavior of MGOs. Seeing that we consider the dual effects of bidding price and quantity in QLCDA, both contributions of adopting one certain action should be taken into account in the reward function. The mathematical expression of reward function is presented in Equation (21). ω represents the weighted factor on bidding price and quantity. As price is the decisive factor in deciding whether a deal is closed, we pay more attention to the bidding price, therefore ω is usually set to be greater than 0.5:

$$r(s,a) = \omega \cdot r_p(s,a) + (1-\omega) \cdot r_q(s,a). \tag{21}$$

$r_p(s,a)$ and $r_q(s,a)$ represent the contributions of bidding price and quantity update on the reward function, which are calculated as follows. All of the variable definitions are the same as those in Equations (10)–(16):

$$r_p(s,a) = \frac{\left|\left|p_i^n - p_i^{history,T}\right| - \left|p_i^{reference,n} - p_i^{history,T}\right|\right|}{p_{i,step}^T}, \tag{22}$$

$$r_q(s,a) = \frac{\lambda \cdot (q_i^n - q_i^{initial,T})}{q_i^{initial,T} \cdot (SDR^n - 1)}. \tag{23}$$

3.2. Update Mechanism of the Proposed Q-Cube Framework

In the proposed Q-learning-based continuous double auction energy trading market, as two dimensions of MGOs' action, bidding price is the key factor in deciding whether to close a deal or not, bidding quantity affects the real-time SDR of the overall market. Meanwhile, the SDR (as the MGOs' states) has a decisive influence on MGOs' actions by updating Q-Values. The coupling relationship between MGOs' actions and market SDR is modeled in this chapter, as shown in Figure 7. One MGO takes a^{n-1} in round $n-1$ and the state transfers from s^{n-1} to s^n. After calculating rewards and updating Q-value, the probability of choosing any action in the action space is modified. Afterwards, given the new Q-cube and market SDR, the MGO might choose a^n as the action in round n and repeat the above process. Therefore, the state-action pair of one MGO in each bidding round is formulated in a spiral iteration way, considering both local private information and public environment. The Q-cube framework is a connector of the state perception process and a decision-making process, which is the core innovation of this paper.

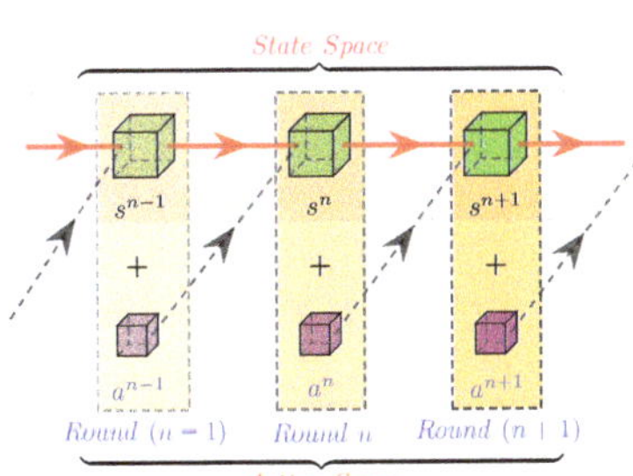

Figure 7. The coupling relationship between microgrid operators' actions and market supply demand relationship.

3.2.1. Q-Value Update

The common Q-Value update rule for the model-free Q-learning algorithm with learning rate α and discount factor γ is given as follows:

$$Q^{n+1}(s_i^n, a_i^n) = (1-\alpha) \cdot Q^n(s_i^n, a_i^n) + \alpha \cdot \left[r(s_i^n, a_i^n) - \gamma \cdot \max_{\forall a_i} Q^n(s_i^{n+1}, a_i^{n+1}) \right], \tag{24}$$

where $Q^{n+1}(s_i^n, a_i^n)$ represents the updated Q-value for MGO i adopting action a_i^n under state s_i^n in the nth bidding round. When observing the subsequent state s_i^{n+1} and reward $r(s_i^n, a_i^n)$, the Q-value is immediately updated. We adopt this common Q-value update rule for Q-learning in this paper.

The learning rate α and discount factor γ are two critical parameters of MGOs as they reflect each MGO's bidding preference. The learning rate defines how much the updated Q-value learns from the new state-action pair. $\alpha = 0$ means the MGO will learning nothing from new market bidding information, while $\alpha = 1$ indicates that the Q-value of a new state-action pair is the only important information. The discount factor defines the importance of future revenues. The MGOs whose γ near 0 are regarded as a short-sighted agent as it only cares about gaining short-term profits, but, for the MGOs whose γ is close to 1, they tend to wait until the proper time for more future revenues.

3.2.2. Action Update

In each round of QLCDA, for each MGO, firstly the basic action is calculated based on market SDR and historical trading records as shown in Figure 8 with red balls in the action space. The colors of action space slices represent the market state. The neighborhood actions are formed in the action space as shown with blue blocks. A selection process is carried out by creating the probability matrices of nine optional actions.

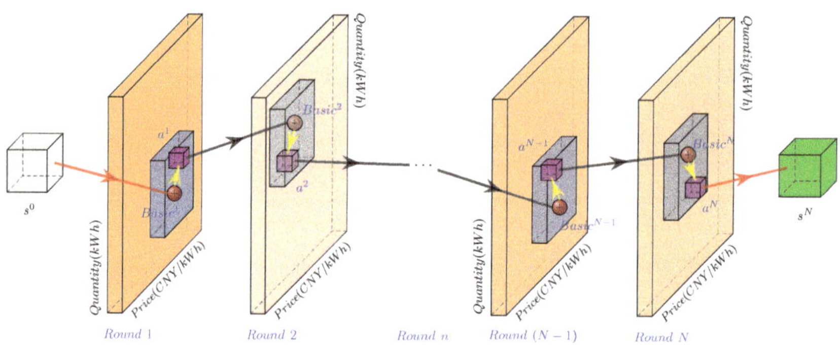

Figure 8. The update process of bidding action in the proposed Q-learning algorithm.

In Equation (25), the action matrix A_i^n is composed by combining the optional bidding price and quantity. Correspondingly, the elements the probability matrix Pro_i^n are formed according to 'ε-greedy' strategy. The probability (x^{bb}) of the basic action ($a_i^{bb} = (p_i^{basic}, q_i^{basic})$) is given preferential treatment and equals ε. For each microgrid, the setting of ε represents its degree of attention on optimal bidding action choice in theory, which is diverse from each other. The probability of other neighborhood actions are calculated by weighted sharing of the remaining probability according to their Q-value (as Equation (26)). The sum of nine probabilities on actions equal to 1:

$$\mathrm{A}_i^n = \begin{bmatrix} (p_i^-, q_i^+) & (p_i^{basic}, q_i^+) & (p_i^+, q_i^+) \\ (p_i^-, q_i^{basic}) & (p_i^{basic}, q_i^{basic}) & (p_i^+, q_i^{basic}) \\ (p_i^-, q_i^-) & (p_i^{basic}, q_i^-) & (p_i^+, q_i^-) \end{bmatrix} \Rightarrow \mathrm{Pro}_i^n = \begin{bmatrix} x^{-+} & x^{b+} & x^{++} \\ x^{-b} & x^{bb} & x^{+b} \\ x^{--} & x^{b-} & x^{+-} \end{bmatrix}. \tag{25}$$

For example, x^{-+} represents the probability of choosing action $a_i^{-+} = (p_i^-, q_i^+)$ under the current state, which is calculated as follows:

$$x^{-+} = (1-\varepsilon)\cdot\frac{Q^n(s_i^n, a_i^{-+})}{\sum\limits_{\forall a/a^{bb}} Q^n(s_i^n, a_i)}. \tag{26}$$

This selection mechanism means that all MGOs have a higher possibility of choosing actions with higher Q-values in each round of QLCDA. By putting the MGOs' best possible local actions together, the most suitable actions for the current global state are generated in a distributed non-cooperative way.

4. Case Studies and Simulation Results

In this section, we investigate the performance of the Q-learning algorithm for continuous double auction among microgrids by Monte Carlo simulation. The proposed algorithm is tested on the realistic case in Guizhou Province of China. The distribution network near Hongfeng Lake consists of 14 microgrids with different scales and internal configurations. Detailed topology of the networked microgrids are given in Figure 9. As power flow calculation and safety check are not the focus of this paper, distance information and transmission price in this distribution network are not provided here. The interested reader may refer to [24] for more details.

We simulate this non-cooperative energy trading market within a scheduling cycle of 24 h. The QLCDA is performed every $\Delta t = 0.5$ h. A scheduling cycle starts at 9:00 a.m. The internal coordinated dispatch of each microgrid is accomplished in advance, from which the dispatch results are treated as initial bidding information in QLCDA. BESS properties of the 14 microgrids are provided in Table A1, including capacity, initial SOC, charge and discharge restriction and charge and discharge efficiency. Guizhou Grid adopts the peak/flat/valley pricing strategy for energy trading, which divides a 24-hour scheduling cycle into three types of time intervals. The surplus energy injected to the grid is paid at 0.300 CNY for each kWh in the whole day. In addition, buying energy from the grid is charged at the price 1.197/0.744/0.356 CNY, respectively (see Table A2).

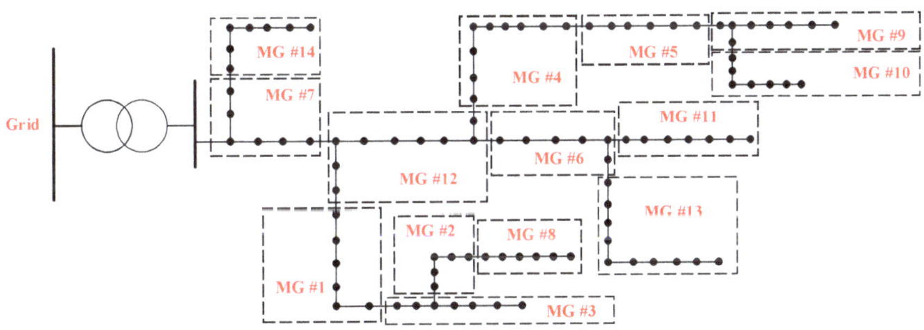

Figure 9. Network topology of 14 microgrids in Guizhou Province, China.

In order to simulate the microgrids' preferences in decision-making, different risk strategies are adopted by setting diverse Q-learning parameters. Fourteen microgrids' values of learning rate, discount rate and greedy degree are given in Table A3. Three risk strategies are defined and discussed according to different Q-learning parameter choices:

- Conservative Strategy: the high value of α (ranged in [0.6,1]) indicates that the MGO is greedy about new bidding information, but the higher choice of ε (ranged in [0.6,1]) indicates that he is conservative on basic QLCDA bidding actions. In addition, he's satisfied with a lower value of γ (ranged in [0,0.4]) as future revenue is not important for him.
- In-Between Strategy: Ordinary choices of three parameters ranged in [0.4,0.6].

- Risk-Taking Strategy: the MGO is not greedy about new bidding information (low value of α ranged in [0,0.4]) but likes to explore more potential actions (low value of ε ranged in [0,0.4]) as a risk-taker. In the meantime, he is eager for more future profits (high value of γ ranged in [0.6,1]).

Other hyper parameters in the proposed Q-learning algorithm are given in Table A4.

The proposed energy trading market model and QLCDA algorithm are implemented and simulated using MATLAB R2019a on an Intel Core i7-4790 CPU, 3.60GHz. Three case studies on bidding performances and profit comparisons are discussed in this section. All the three case studies are simulated repeatedly for 30 times, among which the bidding result of one certain Monte Carlo simulation is analyzed in detail in Case Studies 1 and 2, and the average values of bidding profits are adopted to compare with the profits of two other energy trading mechanisms in Case Study 3.

4.1. Case Study 1: Bidding Performance of the Proposed Continuous Double Auction Mechanism

4.1.1. Bidding Performance of the Overall Energy Trading Market

The proposed continuous double auction energy trading mechanism achieves significant effects on the energy trading among microgrids. Figure 10 shows the bidding process of price in Time Slot 12. In Figure 10a, the bidding price of all microgrids in the whole time slot is presented. Starting with different initial bidding prices, the slopes of price curves indicate different bidding strategies of the MGOs. Due to the fact that bidding price is the key factor in deciding whether to close a deal or not, different intersection points of the pricing curves represent deals under various market conditions. Buyer/Seller MGOs with stronger willingness of reaching deals prefer to raise/drop their prices quickly, expecting that their energy demand/supply is satisfied in the early stage of a time slot. Although patient MGOs would like to wait until the deadline for a better trading price, they have to experience fierce price competition near the deadline and face the possibility of no energy to trade.

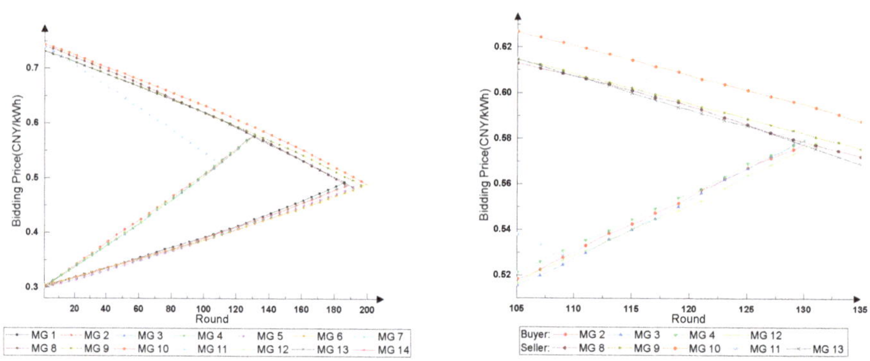

(**a**) Bidding price in the whole time slot (**b**) Bidding price details in round 105-135

Figure 10. The bidding process of price in time slot 12.

Figure 10b shows the bidding price details in rounds 105–135 of time slot 12. MG 11 hadn't traded energy with other microgrids for a long time according to historical records. With stronger willingness of selling energy, MG 11 drops its bidding price quickly and reaches a deal with MG 4 at the price 0.530 CNY/kWh. Unmet energy demand of MG 4 is satisfied by MG 13 with a higher price (0.579 CNY/kWh). MG 6 raises its bidding price slowly and closes deals with MG 9 and MG 10 at the price of 0.481 CNY/kWh and 0.489 CNY/kWh, respectively. However, 27.016 kWh of energy demand has to be bought from the grid with a higher price (0.744 CNY/kWh) as all the energy supply from other microgrids is sold out. This shows a trade-off between price and trading opportunity: one MGO might be eager for closing a deal, but the trading price might not be satisfactory. On the other hand, the energy trading market follows the principle of 'First, Bid, First, Deal', which means the closer the time to the deadline, the less energy one is able to trade.

Comparison on clear power curves before and after CDA is presented in Figure 11. Enhanced by the proposed CDA mechanism, the distribution network achieves better performance on the balance of energy supply and demand. As a result of more balanced energy trading market conditions, more energy is transacted within the distribution network rather than trading with the grid, which reduces long-distance energy transmission loss. With the help of BESS, an alternative form of 'demand response' is performed among microgrids by exerting the potential capacity of elastic loads, which expands the concept of demand response from time-slot-based to multi-agent-based by CDA. In addition, trading prices are more reasonable and profitable, taking care of each MGO's personal preferences.

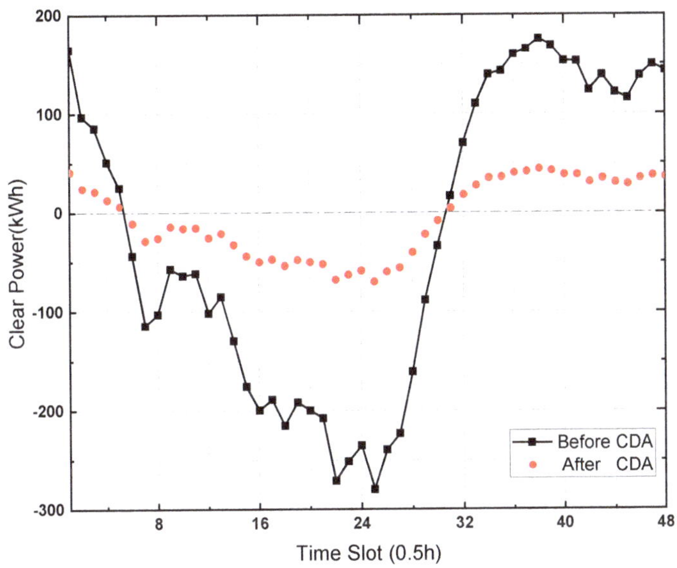

Figure 11. Clear power of the overall energy trading market before/after continuous double auction.

The comparison of trading quantity before and after the proposed CDA is given in Table 1. A significant effect could be obtained by adopting CDA as the trading quantity with grid decrease by different degrees. For example, only 10.8% of the energy demand of MG 3 is provided by the grid, while microgrids with heavy demand like MG 4 and MG 6 still depend on the grid to a large extent, holding 65.5% and 57.1%, respectively. Seller microgrids' dependency of the grid is obviously less than that of buyer microgrids with an average percentage of 26.1% as they prefer to sell energy within the distribution network. The BESS storage change and (dis)charge energy loss are also presented in Table 1, from which we could find that most of the microgrids' BESS obtain higher SOC at the end of one scheduling cycle. The larger BESS capacity and the more active the participation in the trading market, the more BESS (dis)charge energy loss will be caused.

Table 1. Comparison of trading quantity before/after continuous double auction.

Trading Quantity	MG 1	MG 2	MG 3	MG 4	MG 5	MG 6	MG 7
With Grid(Before)(kWh)	530.0	351.3	1540.7	1747.2	2840.3	6787.7	2167.5
With Grid(After)(kWh)	125.0 (23.6%)	85.5 (24.4%)	166.5 (10.8%)	1145.2 (65.5%)	741.1 (26.1%)	3876.4 (57.1%)	595.7 (27.5%)
BESS Storage Change (kWh)	21.7	13.3	55.5	31.2	32.0	151.1	36.1
BESS (Dis)Charge Loss(kWh)	7.5	8.1	14.6	10.9	26.6	51.9	10.4
Trading Quantity	**MG 8**	**MG 9**	**MG 10**	**MG 11**	**MG 12**	**MG 13**	**MG 14**
With Grid(Before)(kWh)	3754.1	1640.0	1275.4	1209.2	1616.9	4427.7	2230.6
With Grid(After)(kWh)	1131.1 (30.1%)	386.2 (23.5%)	287.9 (22.6%)	243.4 (20.1%)	622.7 (38.5%)	705.5 (15.9%)	720.9 (32.3%)
BESS Storage Change (kWh)	−5.9	15.2	14.2	3.9	26.4	−26.1	−9.7
BESS (Dis)Charge Loss(kWh)	16.5	12.7	7.7	12.5	20.1	33.9	34.8

4.1.2. Bidding Results of Specific Microgrids with Different Roles

The bidding results of specific microgrids with different roles are presented in this chapter, including bidding price and quantity. Figure 12 gives the energy trading price of MG 4 and MG 12. MG 4 plays the role of buyer in the whole scheduling cycle, and it successfully reaches deals with other microgrids in most of the time slots as shown in Figure 12a. On no-deal time slots, it buys energy from the grid at higher prices. During the valley interval, although the grid purchase price is low enough (0.356 CNY/kWh), there are still plenty of opportunities to trade with other microgrids in consideration of the real-time SDR. MG 4 succeeds at buying energy at lower prices in almost all the time slots in this interval. Different from MG 4, MG 12 plays two roles in different time slots. The detailed trading prices of MG 12 in time slots 9 to 32 are presented in Figure 12b. Good performance is obtained in both roles that MG 12 plays: during buyer intervals, it reaches deals with other microgrids at prices lower than the grid's, while, in seller intervals, it sells energy in every time slot for higher profits. The overall profit of MG 12 raised by 33.9% after joining the CDA energy trading market.

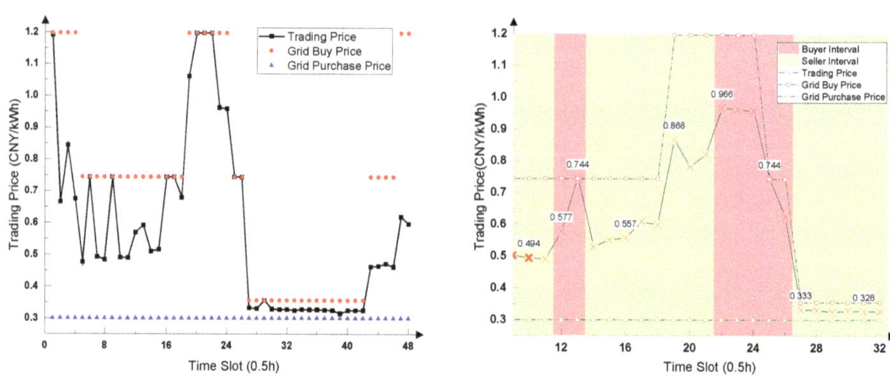

(**a**) Trading price of MG 4 in a 24h scheduling cycle (**b**) Trading price of MG 12 between time slot 9 and 32

Figure 12. Energy trading price of MG 4 and MG 12.

For MG 7, the bidding performance on quantity is presented in Figure 13a. As a buyer microgrid in the whole scheduling cycle, the gaps between original bidding quantity curve and actual trading quantity curve correspond to the real-time SDR. When $SDR \geq 1$ (the original clear power ≥ 0 as shown in Figure 13a above the blue horizontal line) in former and later time slots, MG 7 raises its trading quantity and stores more energy into its BESS to absorb the surplus energy in the market. During the middle time slots when $SDR < 1$ (the original clear power < 0), part of the energy demand is provided by its own BESS, which helps to balance the excessive energy demand in market. The two curves coincide at the end of the scheduling cycle as the BESS stores enough energy in time slot 32 to 38 and SOC is near 1. The same characteristics could be found in the bidding performance of MG 12. In Figure 13b, when energy demand exceeds supply as shown below the purple horizontal line, the BESS of MG 12 discharges to satisfy the energy demand. More energy is sold in these time slots to reach a better market SDR performance, while, during the nighttime, MG 12 charges the surplus energy to its BESS rather than selling to the grid. It is obvious that the actual trading quantity curves cohere better with the real-time SDR than the original bidding quantity curves in both the standpoints of buyer and seller microgrids.

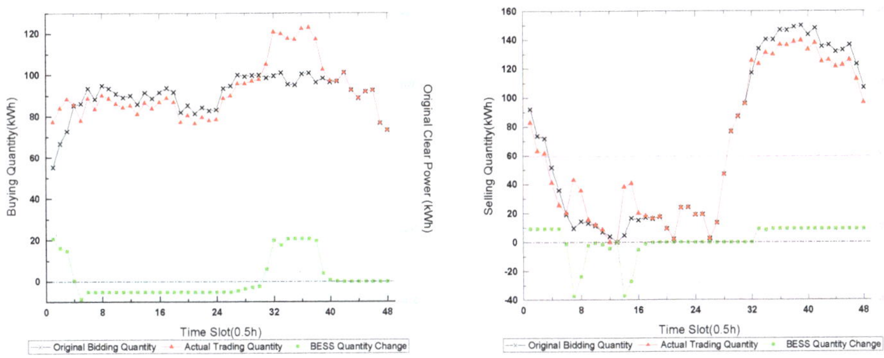

(**a**) Bidding quantity comparison of MG 7 (**b**) Bidding quantity comparison of MG 12

Figure 13. Bidding performance on quantity of MG 7 and MG 12.

The BESS SOC of MG 7 and MG 12 is presented in Figure 14, from which we could find the trend of SOC curves coheres with that of the SDR. When $SDR < 1$, both MG 7 and 12 discharge their BESS to compensate the lack of energy supply. The BESS of MG 12 releases all its energy and the SOC reaches 0 since time slot 16. However, when the energy supply exceeds demand during the nighttime, the BESS starts to charge and save surplus energy for later use. The SOC of MG 7 reaches 100% since time slot 40. Different from former research by [25], the charge and discharge behaviors of BESS are restricted by ramp constraints, which makes the simulation results closer to reality. Due to BESS capacity and (dis)charge energy loss, the regulatory ability of BESS on the energy trading market is limited. When $SOC = 0$ or $SOC = 1$, internal re-scheduling of each microgrid could be developed for greater bidding potential.

4.2. Case Study 2: Effectiveness Verification of the Proposed Q-Cube Framework

The Q-cube of a MGO is updated in each round of the whole scheduling cycle. Q-values are iteratively accumulated following the proposed update rules. In order to display this distribution in the three-dimension space, bidding actions are abstracted to nine actions. MG 6 and MG 13 are chosen as the examples of risk-taking strategy and conservative strategy, respectively. The Q-value distributions of these two microgrids are shown in Figure 15.

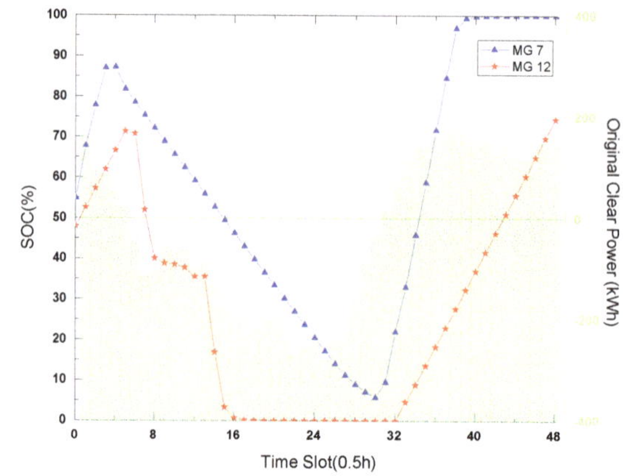

Figure 14. Battery energy storage system's SOC on MG 7 and MG 12.

As a risk-taker, the Q-value distribution in MG 6's Q-cube is a non-uniform distribution with a slight trench in the middle of action dimension as shown in Figure 15a. According to the Q-cube framework proposed in this paper, the low value of MG 6's greedy degree ($\varepsilon = 0.1680$) results in its curiosity on the neighborhood actions of basic action (action 5) for all the states. Neighborhood actions are given more opportunities to accumulate Q-values based on the action selection mechanism. The eagerness of obtaining more future profits aggravates this phenomenon as the discount factor ($\gamma = 0.6721$) is high. A low value of learning rate ($\alpha = 0.2617$) indicates that new bidding information in the real-time market has little impact on the choice of actions.

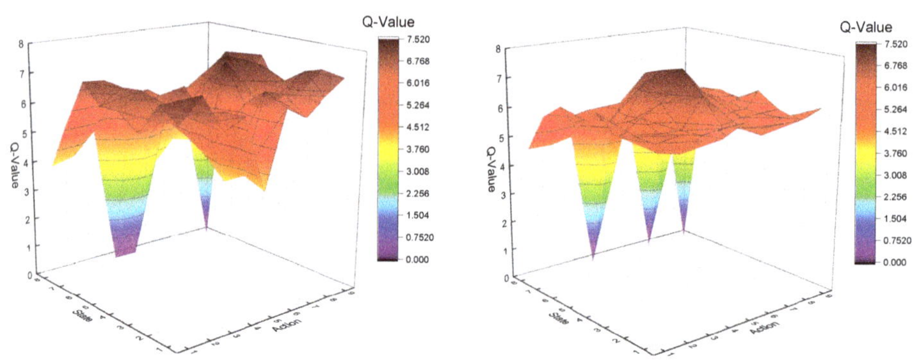

(a) MG 6 (Risk-Taking Strategy).

(b) MG 13 (Conservative Strategy).

Figure 15. The Q-value distribution of microgrids adopting two risk strategies.

On the contrary, MG 13 chooses to be conservative in the QLCDA process, whose Q-value distribution in the Q-cube is presented in Figure 15b. MG 13 likes to keep in touch with the latest market information and prefers to choose the basic action under states near $SDR = 1$, which leads to high values of learning rate ($\alpha = 0.6812$) and greedy degree ($\varepsilon = 0.8462$). He is satisfied with current revenues and doesn't have much interest in exploring new actions, so the discount factor of MG 13 is at a low level ($\gamma = 0.333$). Therefore, there is an obvious hump on the surface of Q-value plane around the middle part (near Q (state 4, action 5) and Q (state 5, action 5)), showing that MG 13 is rational and not greedy.

The iteration results of Q-values of different microgrids prove that the proposed Q-Cube framework for Q-learning algorithm is capable and effective in reflecting the microgrids' characteristics.

4.3. Case Study 3: Profit Analysis on Different Energy Trading Mechanisms

To verify the performance of the proposed QLCDA, a profit analysis on different energy trading mechanisms is carried out. Previous work of [19] on peer-to-peer energy trading mechanism is introduced here for comparison. As shown in Table 2, three energy trading mechanisms are simulated on the same case from Guizhou Grid for 30 times and the average values of energy trading profits are calculated and analyzed for statistically significance. Negative values indicate the cost paid to peer microgrids and the DNO. The proposed QLCDA mechanism is proved to have superior performance over tradition energy trading mechanism as expected. In addition, for most microgrids, a certain degree of increase on profits could be obtained compared to P2P mechanisms. The profits of seller microgrids are commonly raised as clean energy generated during valley intervals could be stored until the needed time rather than selling to the grid at lower prices. A 65.7% and 10.9% rise in the overall profits of the distribution network can be achieved by the QLCDA mechanism compared with that of the tradition energy trading mechanism and P2P mechanism, respectively.

However, for some buyer microgrids (particularly for MG 6), the profits by adopting the QLCDA mechanism is less than that of the P2P mechanism. This could be explained by the following reasons: (1) as presented in Table 1, the trading quantity is adjustable in the QLCDA mechanism, most of the microgrids obtain higher BESS SOC at the end of one scheduling cycle, inside which MG 6 stores the largest quantity of energy (151.1kWh). The profits by selling this part of stored energy are not calculated in Table 2, while, in a P2P mechanism, the effect of applying BESS and changes in bidding quantity is not taken into consideration. (2) MG 6 is a risk-taker based on its Q-learning parameters. The low value of greedy degree ($\varepsilon = 0.1680$) and learning rate ($\alpha = 0.2617$) indicate that MG 6 cares less about new bidding information and wants to explore more potential actions rather than sticking to the basic action. A high value of discount rate ($\gamma = 0.6721$) proves his eagerness for more future profits, therefore it prefers to keep its BESS at a high SOC and seek deals with lower trading prices near the deadline. From another standpoint of view, the profits analysis proves the effectiveness of the proposed Q-Cube framework for the Q-learning algorithm on energy trading problems.

Table 2. Contrast of profits among three energy trading mechanisms [1].

	MG 1	MG 2	MG 3	MG 4	MG 5	MG 6	MG 7
Profit in TM (CNY)	−428.3	−281.5	−1224.7	−1429.4	−1989.0	−4542.4	−1516.6
Profit in P2PM (CNY)	−340.9	−230.0	−993.5	−1090.6	−1483.0	−3539.1	−1199.6
Profit in QLCDAM (CNY)	313.7	217.4	−892.1	−1154.5	−1494.1	−3989.2	−1143.2
Growth Rate(Over P2PM)	8.0%	5.5%	10.2%	−5.9%	−0.7%	−12.7%	4.7%
	MG 8	**MG 9**	**MG 10**	**MG 11**	**MG 12**	**MG 13**	**MG 14**
Profit in TM (CNY)	1126.2	491.9	295.4	−110.1	421.2	1328.3	515.6
Profit in P2PM (CNY)	1687.4	808.4	459.3	−70.5	581.0	1874.2	708.4
Profit in QLCDAM (CNY)	1875.4	897.0	471.3	29.3	564.2	2105.8	743.1
Growth Rate(Over P2PM)	11.1%	11.0%	2.6%	141.6%	−2.9%	12.4%	4.9%

[1] TM: Traditional Mechanism; P2PM: Peer-to-Peer Mechanism; QLCDAM: Q-learning based Continuous Double Auction Mechanism.

Considering the equipment and operation costs of BESS, the proposed QLCDA mechanism might not be the best choice for energy trading among microgrids, but the simulation results prove its potential in increasing profits for microgrids with different configurations and preferences.

5. Conclusions

To better describe the characteristics of future electricity market, a non-cooperative continuous double auction mechanism, considering the coupling relationship of bidding price and quantity, was developed in this paper to facilitate energy trading among microgrids in the distribution network. An alternative form of 'demand response' is performed in the proposed energy trading mechanism by exerting the potential capacity of BESS, which expands the concept of demand response from time-based to multi-agent-based. The Q-learning algorithm was introduced to CDA mechanism as a decision-making method for each microgrid. To solve the existing defects on the application of Q-learning algorithm in power system, a non-tabular framework of Q-values considering two dimensions of the bidding action is proposed as a Q-cube. In addition, corresponding parameter setting and state-action architecture are designed to better reflect the microgrids' personalized bidding preferences and make rational decisions according to real-time status of the networked microgrids. Simulations on a realistic case from Hongfeng Lake, Guizhou Province, China prove the efficiency and applicability of the proposed CDA mechanism and Q-cube framework. All of the microgrids are able to make an appropriate negotiation response to the global real-time supply and demand relationship without disclosing personal privacy. A 65.7% and 10.9% increase in the overall profit of the distribution network could be achieved by applying a QLCDA mechanism compared with the traditional energy trading mechanism and P2P energy trading mechanism, respectively. In addition, the Q-value distribution in the proposed Q-cube gives a good response to microgrid's bidding behaviors and preferences on both theoretical analysis and simulation results. As has been demonstrated in this paper, the proposed Q-cube framework of a Q-learning algorithm for a continuous double auction mechanism can be applied to more energy trading markets in future EI.

There are still some limitations of the proposed Q-cube framework to be discussed: the interaction between bidding price and quantity should be better described as many other factors could have an influence on this coupling relationship, and it is still difficult to summarize the microgrids' energy bidding preferences with these existing parameters. Moreover, the power flow calculation should be considered synchronously as the energy trading quantity might cause safety issues in the distribution network. In future works, a two-layer energy bidding architecture could be discussed considering both QLCDA among microgrids and internal coordinated dispatch inside microgrids. The interaction of these two layers is worth studying. The power transmission limitations should be considered to ensure the safety of energy market. In addition, further extensions are to be carried out on the time-varying setting of QL parameters and a more appropriate description of the reward function.

Author Contributions: Conceptualization, N.W. and W.X.; Methodology, N.W.; Software, N.W.; Validation, N.W.; Formal Analysis, N.W.; Investigation, N.W. and Z.X.; Data Curation, Z.X.; Writing—original draft preparation, N.W.; Writing—review and editing, W.X., Z.X. and W.S.; Visualization, N.W.; Supervision, W.X.; Project Administration, W.X.; Funding Acquisition, W.X. All of the authors have read and approved the final manuscript.

Funding: This research was funded by the National Natural Science Foundation of China under Grant No. 61773292 and 71841036.

Acknowledgments: The authors thank Ke Sun and Yifan Cheng for careful reading and many helpful suggestions to improve the presentation of this paper.

Conflicts of Interest: The authors declare no conflict of interest.

Abbreviations

The following abbreviations are used in this manuscript:

EI	Energy Internet
DER	Distributed Energy Resource
DG	Distributed Generation
BESS	Battery Energy Storage System
EV	Electric Vehicle
DL	Dispatchable Load
RES	Renewable Energy Source
ICT	Information and Communication Technology

MGO	Microgrid Operator
DNO	Distribution Network Operator
CDA	Continuous Double Auction
P2P	Peer-to-Peer
RL	Reinforcement Learning
QL	Q-learning
MDP	Markov Decision Process
ICD	Internal Coordination Dispatch
SOC	State of Charge
QLCDA	Q-learning based Continuous Double Auction
SDR	Supply and Demand Relationship

Appendix A. Supplementary Case Data from the Guizhou Grid, China

Table A1 shows the BESS properties of the 14 microgrids in the Guizhou Grid, including capacity, initial SOC, charge and discharge restriction and charge and discharge efficiency.

Table A1. Battery energy storage system properties of 14 microgrids in the Guizhou Grid.

	MG 1	MG 2	MG 3	MG 4	MG 5	MG 6	MG 7
Capacity(kWh)	40	20	80	100	100	300	80
Initial SOC (%)	33.74	30.60	68.78	68.05	49.65	54.93	60.59
Charge & Discharge Restriction(%)	24.05	16.18	16.39	15.55	15.98	16.15	18.23
Charge & Discharge Efficiency	0.8543	0.9207	0.9465	0.8756	0.9156	0.9400	0.9309
	MG 8	**MG 9**	**MG 10**	**MG 11**	**MG 12**	**MG 13**	**MG 14**
Capacity(kWh)	150	70	50	60	100	200	150
Initial SOC (%)	60.59	44.99	32.84	59.05	48.00	69.58	62.65
Charge & Discharge Restriction(%)	18.23	20.91	19.93	20.91	23.39	18.96	18.87
Charge & Discharge Efficiency	0.9309	0.8899	0.8933	0.8791	0.8891	0.9004	0.8622

The peak/flat/valley electricity price formulated by Guizhou Grid, China is presented in Table A2, which divides a day into three types of time internals.

Table A2. Peak/flat/valley electricity price formulated by the Guizhou Grid.

Time Interval	Interval Type	Price (CNY/kWh)
8 a.m.–11 a.m., 6 p.m.–9 p.m.	Peak	1.197
6 a.m.–8 a.m., 11 a.m.–6 p.m., 9 p.m.–10 p.m.	Flat	0.744
10 p.m.–6 a.m.	Valley	0.356

The learning rate α, discount factor γ and greedy degree ε parameters of the 14 microgrids are given in Table A3.

Table A3. Q-Learning Parameters of 14 Microgrids.

	MG 1	MG 2	MG 3	MG 4	MG 5	MG 6	MG 7
Learning Rate α	0.5107	0.3205	0.7124	0.7969	0.7169	0.2617	0.6210
Discount Factor γ	0.5240	0.6423	0.7569	0.3373	0.7781	0.6721	0.4231
Greedy Degree ε	0.6564	0.3564	0.8156	0.4961	0.3485	0.1680	0.4894
	MG 8	**MG 9**	**MG 10**	**MG 11**	**MG 12**	**MG 13**	**MG 14**
Learning Rate α	0.3546	0.5083	0.3291	0.5566	0.2371	0.6812	0.2996
Discount Factor γ	0.4670	0.7065	0.2570	0.4146	0.6397	0.3330	0.4062
Greedy Degree ε	0.3648	0.2658	0.6173	0.3173	0.4983	0.8462	0.7943

The values of hyper parameters that appear in this paper are given in Table A4.

Table A4. Hyper parameters settings for the proposed Q-learning algorithm.

Parameter	θ	μ	δ	β	π	ρ	ξ	τ	ω
Value	0.50	0	0.30	1.20	0.40	0.98	0.30	0.10	0.70

References

1. Ampatzis, M.; Nguyen, P.H.; Kling, W. Local electricity market design for the coordination of distributed energy resources at district level. In *IEEE PES Innovative Smart Grid Technologies, Europe*; IEEE: Piscataway, NJ, USA, 2014; pp. 1–6, doi:10.1109/ISGTEurope.2014.7028888.
2. Al-Awami, A.T.; Amleh, N.A.; Muqbel, A.M. Optimal Demand Response Bidding and Pricing Mechanism With Fuzzy Optimization: Application for a Virtual Power Plant. *IEEE Trans. Ind. Appl.* **2017**, *53*, 5051–5061, doi:10.1109/TIA.2017.2723338. [CrossRef]
3. Dehghanpour, K.; Nehrir, M.H.; Sheppard, J.W.; Kelly, N.C. Agent-Based Modeling of Retail Electrical Energy Markets With Demand Response. *IEEE Trans. Smart Grid* **2018**, *9*, 3465–3475, doi:10.1109/TSG.2016.2631453. [CrossRef]
4. Jeong, G.; Park, S.; Lee, J.; Hwang, G. Energy Trading System in Microgrids With Future Forecasting and Forecasting Errors. *IEEE Access* **2018**, *6*, 44094–44106, doi:10.1109/ACCESS.2018.2861993. [CrossRef]
5. Wei, W.; Liu, F.; Mei, S. Energy Pricing and Dispatch for Smart Grid Retailers Under Demand Response and Market Price Uncertainty. *IEEE Trans. Smart Grid* **2015**, *6*, 1364–1374, doi:10.1109/TSG.2014.2376522. [CrossRef]
6. Olson, M.; Rassenti, S.; Rigdon, M.; Smith, V. Market Design and Human Trading Behavior in Electricity Markets. *IIE Trans.* **2003**, *35*, 833–849, doi:10.1080/07408170304406. [CrossRef]
7. Nunna, H.S.V.S.K.; Doolla, S. Energy Management in Microgrids Using Demand Response and Distributed Storage—A Multiagent Approach. *IEEE Trans. Power Deliv.* **2013**, *28*, 939–947, doi:10.1109/TPWRD.2013.2239665. [CrossRef]
8. Nunna, H.S.V.S.K.; Doolla, S. Demand Response in Smart Distribution System With Multiple Microgrids. *IEEE Trans. Smart Grid* **2012**, *3*, 1641–1649, doi:10.1109/TSG.2012.2208658. [CrossRef]
9. Sueyoshi, T.; Tadiparthi, G. Wholesale Power Price Dynamics Under Transmission Line Limits: A Use of an Agent-Based Intelligent Simulator. *IEEE Trans. Syst. Man Cybern. Part C (Appl. Rev.)* **2008**, *38*, 229–241, doi:10.1109/TSMCC.2007.913909. [CrossRef]
10. Kasbekar, G.S.; Sarkar, S. Pricing games among interconnected microgrids. In Proceedings of the 2012 IEEE Power and Energy Society General Meeting, San Diego, CA, USA, 22–26 July 2012; pp. 1–8, doi:10.1109/PESGM.2012.6344881. [CrossRef]
11. Zhang, C.; Wu, J.; Zhou, Y.; Cheng, M.; Long, C. Peer-to-Peer energy trading in a Microgrid. *Appl. Energy* **2018**, *220*, 1–12, doi:10.1016/j.apenergy.2018.03.010. [CrossRef]
12. Devine, M.T.; Cuffe, P. Blockchain Electricity Trading Under Demurrage. *IEEE Trans. Smart Grid* **2019**, *10*, 2323–2325, doi:10.1109/TSG.2019.2892554. [CrossRef]
13. Bunn, D.W.; Oliveira, F.S. Agent-based simulation-an application to the new electricity trading arrangements of England and Wales. *IEEE Trans. Evol. Comput.* **2001**, *5*, 493–503, doi:10.1109/4235.956713. [CrossRef]
14. Vytelingum, P.; Cliff, D.; Jennings, N.R. Strategic bidding in continuous double auctions. *Artif. Intell.* **2008**, *172*, 1700–1729, doi:10.1016/j.artint.2008.06.001. [CrossRef]
15. Nicolaisen, J.; Petrov, V.; Tesfatsion, L. Market power and efficiency in a computational electricity market with discriminatory double-auction pricing. *IEEE Trans. Evol. Comput.* **2001**, *5*, 504–523, doi:10.1109/4235.956714. [CrossRef]
16. Wang, J.; Wang, Q.; Zhou, N.; Chi, Y. A Novel Electricity Transaction Mode of Microgrids Based on Blockchain and Continuous Double Auction. *Energies* **2017**, *10*, 1971, doi:10.3390/en10121971. [CrossRef]
17. Tan, Z.; Gurd, J.R. Market-based grid resource allocation using a stable continuous double auction. In Proceedings of the 2007 8th IEEE/ACM International Conference on Grid Computing, Austin, TX, USA, 19–21 September 2007; pp. 283–290, doi:10.1109/GRID.2007.4354144. [CrossRef]

18. Long, C.; Wu, J.; Zhou, Y.; Jenkins, N. Peer-to-peer energy sharing through a two-stage aggregated battery control in a community Microgrid. *Appl. Energy* **2018**, *226*, 261–276, doi:10.1016/j.apenergy.2018.05.097. [CrossRef]
19. Wang, N.; Xu, W.; Xu, Z.; Shao, W. Peer-to-Peer Energy Trading among Microgrids with Multidimensional Willingness. *Energies* **2018**, *11*, 3312, doi:10.3390/en11123312. [CrossRef]
20. Tushar, W.; Saha, T.K.; Yuen, C.; Liddell, P.; Bean, R.; Poor, H.V. Peer-to-Peer Energy Trading With Sustainable User Participation: A Game Theoretic Approach. *IEEE Access* **2018**, *6*, 62932–62943, doi:10.1109/ACCESS.2018.2875405. [CrossRef]
21. Rocchetta, R.; Bellani, L.; Compare, M.; Zio, E.; Patelli, E. A reinforcement learning framework for optimal operation and maintenance of power grids. *Appl. Energy* **2019**, *241*, 291–301, doi:10.1016/j.apenergy.2019.03.027. [CrossRef]
22. Sutton, R.S.; Barto, A.G. *Reinforcement learning: An introduction*; MIT Press: Cambridge, MA, USA, 2018.
23. Glavic, M.; Fonteneau, R.; Ernst, D. Reinforcement Learning for Electric Power System Decision and Control: Past Considerations and Perspectives. *IFAC-PapersOnLine* **2017**, *50*, 6918–6927, doi:10.1016/j.ifacol.2017.08.1217. [CrossRef]
24. Foruzan, E.; Soh, L.; Asgarpoor, S. Reinforcement Learning Approach for Optimal Distributed Energy Management in a Microgrid. *IEEE Trans. Power Syst.* **2018**, *33*, 5749–5758, doi:10.1109/TPWRS.2018.2823641. [CrossRef]
25. Wang, H.; Huang, T.; Liao, X.; Abu-Rub, H.; Chen, G. Reinforcement Learning for Constrained Energy Trading Games With Incomplete Information. *IEEE Trans. Cybern.* **2017**, *47*, 3404–3416, doi:10.1109/TCYB.2016.2539300. [CrossRef] [PubMed]
26. Cai, K.; Niu, J.; Parsons, S. Using Evolutionary Game-Theory to Analyse the Performance of Trading Strategies in a Continuous Double Auction Market. In *Adaptive Agents and Multi-Agent Systems III. Adaptation and Multi-Agent Learning*; Tuyls, K., Nowe, A., Guessoum, Z., Kudenko, D., Eds.; Lecture Notes in Computer Science; Springer: Berlin/Heidelberg, Germany, 2008; pp. 44–59.
27. Schvartzman, L.J.; Wellman, M.P. Stronger CDA Strategies Through Empirical Game-theoretic Analysis and Reinforcement Learning. In Proceedings of the 8th International Conference on Autonomous Agents and Multiagent Systems (AAMAS '09), Budapest, Hungary, 10–15 May 2009; Volume 1; International Foundation for Autonomous Agents and Multiagent Systems: Richland, SC, USA, 2009; pp. 249–256.
28. O'Neill, D.; Levorato, M.; Goldsmith, A.; Mitra, U. Residential Demand Response Using Reinforcement Learning. In Proceedings of the 2010 First IEEE International Conference on Smart Grid Communications, Gaithersburg, MD, USA, 4–6 October 2010; pp. 409–414, doi:10.1109/SMARTGRID.2010.5622078. [CrossRef]
29. Naghibi-Sistani, M.B.; Akbarzadeh-Tootoonchi, M.R.; Javidi-Dashte Bayaz, M.H.; Rajabi-Mashhadi, H. Application of Q-learning with temperature variation for bidding strategies in market based power systems. *Energy Convers. Manag.* **2006**, *47*, 1529–1538, doi:10.1016/j.enconman.2005.08.012. [CrossRef]
30. Rahimiyan, M.; Rajabi Mashhadi, H. Supplier's optimal bidding strategy in electricity pay-as-bid auction: Comparison of the Q-learning and a model-based approach. *Electr. Power Syst. Res.* **2008**, *78*, 165–175, doi:10.1016/j.epsr.2007.01.009. [CrossRef]
31. Rahimiyan, M.; Mashhadi, H.R. An Adaptive Q-Learning Algorithm Developed for Agent-Based Computational Modeling of Electricity Market. *IEEE Trans. Syst. Man Cybern. Part C (Appl. Rev.)* **2010**, *40*, 547–556, doi:10.1109/TSMCC.2010.2044174. [CrossRef]
32. Salehizadeh, M.R.; Soltaniyan, S. Application of fuzzy Q-learning for electricity market modeling by considering renewable power penetration. *Renew. Sustain. Energy Rev.* **2016**, *56*, 1172–1181, doi:10.1016/j.rser.2015.12.020. [CrossRef]
33. Rodriguez-Fernandez, J.; Pinto, T.; Silva, F.; Praça, I.; Vale, Z.; Corchado, J.M. Context aware Q-Learning-based model for decision support in the negotiation of energy contracts. *Int. J. Electr. Power Energy Syst.* **2019**, *104*, 489–501, doi:10.1016/j.ijepes.2018.06.050. [CrossRef]
34. Ernst, D.; Glavic, M.; Geurts, P.; Wehenkel, L. Approximate Value Iteration in the Reinforcement Learning Context. Application to Electrical Power System Control. *Int. J. Emerg. Electr. Power Syst.* **2005**, *3*, doi:10.2202/1553-779X.1066. [CrossRef]

35. Bui, V.H.; Hussain, A.; Im, Y.H.; Kim, H.M. An internal trading strategy for optimal energy management of combined cooling, heat and power in building microgrids. *Appl. Energy* **2019**, *239*, 536–548, doi:10.1016/j.apenergy.2019.01.160. [CrossRef]
36. Mbuwir, B.V.; Ruelens, F.; Spiessens, F.; Deconinck, G. Battery Energy Management in a Microgrid Using Batch Reinforcement Learning. *Energies* **2017**, *10*, 1846, doi:10.3390/en10111846. [CrossRef]
37. Manchin, A.; Abbasnejad, E.; Hengel, A.V. Reinforcement Learning with Attention that Works: A Self-Supervised Approach. *arXiv* **2019**, arXiv: 1904.03367.

© 2019 by the authors. Licensee MDPI, Basel, Switzerland. This article is an open access article distributed under the terms and conditions of the Creative Commons Attribution (CC BY) license (http://creativecommons.org/licenses/by/4.0/).

energies

MDPI

Article

A Speculative Trading Model for the Electricity Market: Based on Japan Electric Power Exchange

Jun Maekawa [1,*] and Koji Shimada [2]

1 Ritsumeikan Global Innovation Research Organization (R-GIRO), Ritsumeikan University, 1-1-1 Noji-higashi, Kusatsu, Shiga 525-8577, Japan
2 Faculty of Economics, Ritsumeikan University, 1-1-1 Noji-higashi, Kusatsu, Shiga 525-8577, Japan
* Correspondence: jmaekawa@fc.ritsumei.ac.jp

Received: 30 June 2019; Accepted: 28 July 2019; Published: 31 July 2019

Abstract: Renewable energy sources produce less environmental impact and have little marginal cost. Thus, because of these characteristics, it is desirable to disseminate it for the purpose of economic efficiency. Because of the uncertainty in the supply of renewable energy and the special feature of electricity as a good, such as merit order curve, introducing forward markets is an essential factor in a liberalized market. In European countries, which have already established several mechanisms for managing liquidity including markets with several timelines, the market liquidity invites the investor to perform some speculative action. We present a simple electric power market model to analyze the speculative actions of electricity suppliers and the price effect of such actions. Moreover, we found that the speculative action improves the inelasticity of the demand in electricity market.

Keywords: electricity market; speculative trading; forward market

1. Introduction

In Japan, power-related administrative reforms have been progressing since the 2011 earthquake. The liberalization of the electricity market is a major pillar of this policy. From the ministry of economy, trade and industry (METI) report [1], until now, the power market has been monopolized by regional corporations, and a single large power company has provided almost all of the power in a given region. The Tokyo Electric Power Company, which caused a nuclear accident, is one such company. The public reaction on the accident has led to a demand for a system that will allow consumers to select a preferred power company. In fact, the liberalization of electricity retail began in 2016, and it is expected to continue being liberalized going forward.

The Japanese government is promoting the liberalization of electricity concurrently with the introduction of renewable energy; furthermore, it is trying to promote power trading through one market player, namely the Japan Electric Power Exchange (JEPX). In the spot market, all bids are matched every 30 min; thus, 48 products are traded every 30 min per day. The minimum volume that can be traded during a 30-min bid is 1 MW (equivalent to 500 kWh). The bid supply and demand are matched by using the price auction method. In the case of congestion of electricity, the exchange is split by a regional hub, and the transaction is carried out in each split market. When there is no intersection between the supply and demand caused by oversupply, the spot price is deemed to be zero (see JEPX website [2]).

However, as we explain in Section 3, the electricity market possesses some special properties (such as demand price-in-elasticity and merit order curve) that are different from the normal goods handled in economics. For this reason, electric power industries in regions such as Europe also supplement the electricity market through various mechanisms. The forward market (that is, the opening up of an electricity market before electric power is actually supplied) is one of the mechanisms for increasing the liquidity of electric power.

Of course, the forward market has already been opened by using JEPX. There are four types of forward products, including monthly 24-h products, weekly 24-h products, monthly daytime products, and weekly daytime products. Bid supply and demand is matched through continuous sessions.

The development of forward markets is also important from the perspective of renewable energy propagation. The growing popularity of renewable energy is, of course, a natural consequence of nuclear power plant accidents. Nuclear power has a lower environmental impact compared to thermal power, and it has been introduced for that purpose in Japan. Furthermore, the policy for cleaner renewable energy achieved national attention after the earthquake. However, renewable energy is produce through a variable energy system in which the amount of power generation is affected by various environmental factors such as sunshine hours, wind speed, and rainfall. Because weather information changes from moment to moment, the existence of markets with various timelines that correspond to such information is also essential for considering business operator's risk aversion.

Previous research has pointed out that increasing the liquidity of renewable energy is also an important factor in the penetration of renewable energy; thus, based on this aspect as well, the forwards market is an essential system. For example, researchers investigating wind power generation in Germany pointed out the importance of increasing the liquidity of electricity (Holttinen (2005) [3], Ummels et al. (2006)) [4]. Markets with several timelines raise the liquidity of electricity and it becomes easy to trade electricity produced by renewable energy.

However, increasing the amount of trading opportunities includes one other aspect: allowing market participants to dynamically perform speculative actions. Electric power is difficult to save, but if the liquidity of the market increases, and the electricity transaction becomes easier, it is natural that some market players make profits by using price differences. In this paper, we use a simple model based on JEPX to analyze the speculative behavior in the dynamic power market. Moreover, we show that the demand inelasticity is improved by the speculative trading. The inelasticity of demand in the electricity market is well known, and it is one of the causes of inefficiencies, such as price manipulation by suppliers. However, our findings show that speculative action may improve the inelasticity and increase market welfare. Our model is based on JEPX; however, the basic factor of the model is not specific to Japan. This model can be applied as a more general model in electricity markets.

The rest of the paper is organized as follows. Section 2 introduces related literature, and Section 3 explains the standard electricity market features. Section 4 then introduces the heterogeneous belief model, and it includes many of the electricity market features noted in Section 3. Section 5 presents the theoretical results and discusses some policy implications. Finally, Section 6 provides some concluding remarks.

2. Related Literature

Following the full-scale liberalization of electricity retailing, constructing forward market has become an urgent agenda for the further vitalization of the electricity wholesale market, which is an important source of power procurement for electricity retailers. It is crucial to develop an electricity forwards market that can effectively aid the formation of fair and transparent price indicators and help to hedge against the risk of fluctuations in electricity wholesale prices.

Lucia et al. (2002) [5] and Pilipovic (1998) [6] examined the importance of the regular pattern in the behavior of electricity prices and its implications for the purposes of forward pricing. Other empirical papers, such as those by Escribano et al. (2002) [7], Eydeland et al. (2003) [8], Huisman et al. (2003) [9], and Maekawa et al. (2018) [10], have introduced a panel model for determining hourly electricity prices in forward markets and thus examined their characteristics. These models consider several factors: seasonality, regime switching, or price spikes.

Research in this field has been actively conducted recently. For example, Botterud et al. (2010) [11] analyzed 11 years of historical spot and forward prices from the hydro-dominated electricity market and found that forward prices tended to be higher than spot prices.

By using multivariate models, Raviv et al. (2015) [12] demonstrated that the disaggregated hourly prices contained useful predictive information of the daily average price in the Nord Pool market.

All these papers utilized an empirical perspective. However, to understand the relationship between spot prices and forward prices, it is essential to construct a microeconomic theoretical model. The power market has many special conditions, and it is difficult to simply adopt any particular economic point of view.

In response to these empirical studies, power market prices are being modeled in many studies. There is a long list of papers for wholesale power prices and electricity derivatives (e.g., Cartea et al. (2005) [13], Weron (2007) [14], Hikspoors et al. (2007) [15], Benth et al. (2008) [16], and Jaimungal et al. (2011) [17]). Their models are very sophisticated and their relevance to the empirical data is deep. However, because they emphasize the relationship between investors and the market, they are not an economically closed model. To understand the relationship between spot prices and forward prices, it is essential to construct a microeconomic theoretical model. However, the power market has many special conditions, and it is difficult to simply adopt any particular economic point of view. The construction of theoretical model is expected to facilitate easier utilization of economic findings. We introduce a speculative model with heterogeneous beliefs to the electricity market.

Similar to our research, Cartea et al. (2018) [18] derived an investor's optimal trading strategy of electricity contracts traded in two locations. Their strategy was based on ambiguity averse to price spikes. Our model is to analyze speculative behavior of the more essential power market. We add more fundamental condition of electricity market features to a speculative market model and analyze speculative trading by electric suppliers, not investors.

The purpose of this paper is to provide a new theoretical foundation to understand the factors that drive the electricity markets. In our model, speculative behavior has a strong influence on the price of the electricity market. In addition, it is due to the strong presence of market participant heterogeneity.

The field of speculative model construction along with heterogeneous belief has been the target of study for a long time. If all investors possess a common belief, there is no incentive for trading assets. However, as some heterogeneity exists in investors' belief, each investor assigns a different value to the asset, and a trade can occur.

At first, Miller (1977) [19] and Harrison et al. (1978) [20] created speculative models by using the heterogeneous belief agent. Townsend (1983) [21] and Singleton (1987) [22] pointed out the importance of heterogeneous beliefs in economics. Market prices are influenced by fundamentals, but players' beliefs are also essential factors for determining prices. Sheinkman et al. (2003) [23] evolved these models to adapt to more general asset models. Thus, heterogeneity is treated in the theoretical model as a large driving force that drives the market.

For instance, in the energy market, as discussed by Gabriel et al. (2009) [24], investors can exert both strategic and hedging behaviors by utilizing heterogeneous expectations. Utilizing significant evidence, Joets (2015) [25] found that energy markets are composed of heterogeneous traders who exhibit different behaviors depending on the intensity of the price fluctuations and the uncertainty.

This paper uses the heterogeneity of market participants to create a new price model of speculative electricity market. We found that the speculative trading relaxes the price spikes. Electric price spikes are discussed in many papers (Huisman et al. (2003) [9], Weron et al. (2004) [26], Cartea et al. (2005) [13], Escribano et al. (2002) [7], Knittel et al. (2005) [27] and Chan et al. (2008) [28]). In these papers, some quantitative models are embedded with price spikes by using several factors such as seasonality, risks and events. Although the relationship between our speculative behavior and the price spike is simple, it can also contribute to the development of these studies.

3. Model Bases and Materials: Electricity Market Features

Our model was created based on the features of these JEPX market attributions. The following conditions are assumed in our model.

- Cost function follows the "merit order curve".

- There are two types of suppliers: normal and renewable suppliers.
- There are two markets: forward market and spot market.
- Consumers' demand curve is inelastic to the price.
- Only non-renewable suppliers have budgets for speculative trading.
- All suppliers and consumers are price-takers.

Since the marginal cost is almost constant under the same power generation method, the power generation company efficiently generates power to meet the limits such that marginal costs remain low in accordance with the demand.

In a power market, "merit order effect" is used as a term to describe the mechanism by which the market price is determined. The electric power supply is determined based on the "merit order"; among these, the sources with the cheapest marginal costs (mostly renewable energy sources such as wind power, solar energy, hydroelectric power, and nuclear power) will be sold more quickly. Renewable energy sources such as photovoltaic solar power, wind power, and hydroelectric power are located at the left end in the merit order curve because they have little marginal costs. Based on this, nuclear power, coal, oil, and natural gas follow in order (see Figure 1).

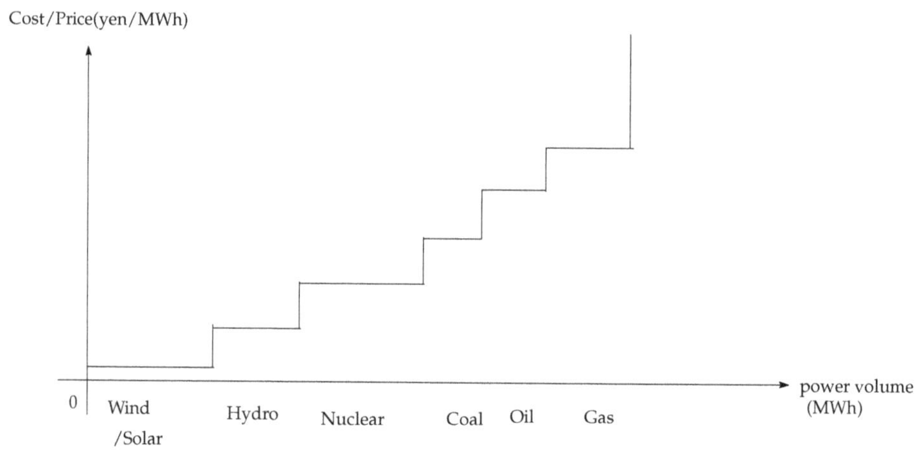

Figure 1. Merit order curve.

For the purposes of this study, we assumed that there are two types of suppliers: normal suppliers and renewable suppliers. Although regional monopolistic companies participate in the electricity market, there are no regional differences in power prices in areas other than Hokkaido, and, as a result of inter-regional competition, a single company cannot wield exclusive market power in JEPX. Recently, new suppliers who deal with renewable energy sources have begun to participate in JEPX, but most of them are small and local companies.

This paper's major results include determining the relationship between forward price and spot price. Two types of products are currently being traded on JEPX: spot market products and forward market products. Moreover, there are four types of forward products: monthly 24-h products, weekly 24-h products, monthly daytime products, and weekly daytime products. For simplicity, we assume that there are two market types in the model: spot market and forward market.

Power demand also has specific characteristics in economic theories. It is well-known that electricity demand is very inelastic to price. Naturally, a major reason is that electricity is an essential item. Electric power is one of the essential infrastructures of modern society, and it is very difficult to live without electric power in countries where urbanization has advanced. The difficulty of saving electricity is another factor of the inelasticity. If an appropriate storage system is not in place, it is inflexible with respect to price, as it is impossible to buy electricity when its price is low and to consume it when its price is high.

In our model, electricity was supplied through two resources: renewable resources where the marginal costs are zero and non-renewable energy resources, such as thermal power, where the marginal costs can be denoted with a positive real number a.

Our model is based on the particular situation of JEPX. In Japan, JEPX, the only power trading market, was established during the trend of electricity liberalization. JEPX was established through investments from electric power companies and new electric power companies, and started trading from 2005 onwards. Since only members of JEPX can trade in the market, general consumers cannot buy electricity directly. Therefore, we assumed that all market players were price-takers. No one has the market power required for controlling electricity prices.

4. Theoretical Method

4.1. Model Settings

We formulated the following equations as an original model. Suppliers can produce electricity at $t = 2$, and they can trade them at $t = 1$. The trading good for $t = 1$ is not electricity; rather, it is the right to sell electricity at $t = 2$. Therefore, at date 1, the suppliers who expect higher prices at date 2 have an incentive to buy them, and those who expect lower prices at date 2 have an incentive to sell them.

There are two types of electricity suppliers: renewable suppliers and normal suppliers.

Renewable suppliers can produce electricity by utilizing renewable resources, and their marginal costs thus tend to be zero. They are small firms and have no budgets for speculation.

Normal suppliers are conventional suppliers. They produce electricity at marginal costs $a > 0$. They have budgets $I > 0$ for speculative trading.

At $t = 1$, the electricity supplied at $t = 2$ is traded in the market. p_1 is $t = 1$ price, and p_2 is $t = 2$ price.

At $t = 1$, all firms can watch p_1 and determine the trading volume. All suppliers can sell electricity, but only normal suppliers can buy electricity in the market (renewable suppliers have no budgets for this activity).

All suppliers have beliefs about renewable supply at $t = 2$, R_2. If some firms expect high volumes of renewable supply at $t = 2$, they must also expect low electricity prices at $t = 2$, $E_i[p_2]$.

They determine their trading strategy based on their beliefs about $E_i[p_2]$.

Profit maximization at $t = 1$ is as follows:

$$Max_{x_1,x^b} \ (p_1 - a)x_1 + E_i[v(x_1, x^b)] \quad s.t.\ x_1 \le X,\ p_1 x^b \le I,\ x_1, x_b \ge 0 \tag{1}$$

x is the selling volume, x^b is the purchase volume, and $v(x_1, x^b)$ is the value function for $t = 2$.

$$v(x_1, x^b) = Max_{x_2} \ (p_2 - a)x_2 + p_2 x^b \quad s.t.\ x_2 \le X - x_1,\ x_2 \ge 0 \tag{2}$$

They determine the appropriate date for selling their electricity by comparing the prices and marginal costs. For example, if p_1 is higher than their expected forward price $E_i[p_2]$ and higher than marginal cost a, they sell the electricity.

Renewable suppliers are authorized to sell R_1 units of electricity. R_1 is interpreted as the minimum supply value for renewable energy sources (R_1 can be zero). The main problem of renewable suppliers can be expressed as follows:

$$\begin{aligned} &Max_{x_1} \;\; p_1 x_1 + E_i[v(x_1)] \\ &\qquad\qquad s.t.\; x_1 \leq R_1,\; x_1 \geq 0 \end{aligned} \tag{3}$$

x is the selling volume, and $v(x_1)$ is the value function for $t = 2$.

At $t = 2$, renewable suppliers can sell an additional R_2 units of electricity. R_2 is a random variable, and it is realized at date $t = 2$.

$$\begin{aligned} &v(x_1) = Max_{x_2} \;\; p_2 x_2 \\ &\qquad\qquad s.t.\; x_2 \leq R_1 + R_2 - x_1,\; x_2 \geq 0 \end{aligned} \tag{4}$$

Because their marginal costs amounts to 0, their strategy is simpler than that of the normal suppliers. If $p_1 > E_i[p_2]$, they usually sell electricity at $t = 1$; otherwise, they tend to wait until $t = 2$.

The main way to solve the problem can be expressed as follows:

Normal suppliers' strategy at $t = 1$ can be expressed as follows:

$$x_1 = X,\; x^b = 0 \qquad \text{if } p_1 > E_i[p_2] \text{ and } p_1 \geq a \tag{5}$$

$$x_1 = 0, x^b = 0 \qquad \text{if } p_1 > E_i[p_2] \text{ and } p_1 < a \tag{6}$$

$$x_1 = 0, x^b = \frac{I}{p_1} \qquad \text{if } p_1 \leq E_i[p_2] \tag{7}$$

Renewable suppliers' strategy at $t = 1$ is:

$$x_1 = R_1 \qquad \text{if } p_1 > E_i[p_2] \tag{8}$$

$$x_1 = 0 \qquad \text{if } p_1 \leq E_i[p_2] \tag{9}$$

Normal suppliers' strategy at $t = 2$ can be expressed as follows:

$$x_2 = X - x_1 \qquad \text{if } p_2 \geq a \tag{10}$$

$$x_2 = 0 \qquad \text{if } p_2 < a \tag{11}$$

Renewable suppliers' strategy at $t = 2$ can be expressed as follows:

$$x_2 = R_1 + R_2 - x_1 \tag{12}$$

The consumers' demand is $d_1(p_1)$ at $t = 1$. This is the industrial firm's demand. The consumers' demand at $t = 2$ is $d_2(p_2)$. We assumed that little elasticity would exist. That is, $d_1'(p_1) < 0$, and $d_2(p_2) < 0$. We assumed $lim_{p_1 \to \infty} d_1(p_1) = 0$, and $lim_{p_2 \to \infty} d_2(p_2) = 0$.

4.2. Model Equilibrium

The market equilibrium is determined by the intersection between the supply and demand. We can solve this model by conducting backward induction.

The equilibrium at $t = 2$ matches the normal electricity noted in Section 3.

To explain the equilibrium, we note the total electricity sold at date 1 as X^b.

$$X^b = \int_i x_i^b di \tag{13}$$

The market equilibrium is determined with some price p_1 (see Figure 2).

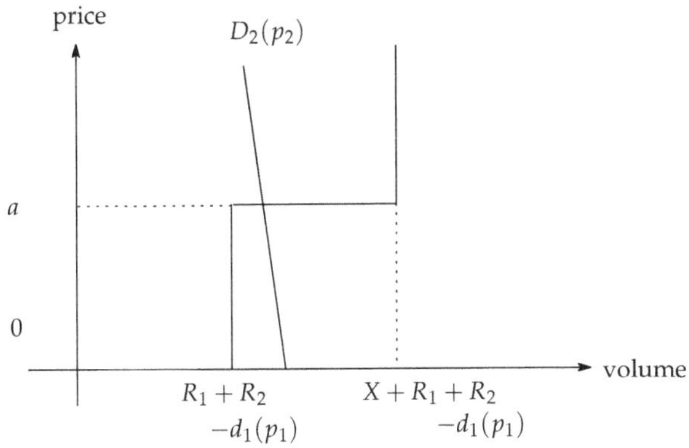

Figure 2. Date 2 market equilibrium.

Total demand $D_2(p_2)$ is expressed as follows:

$$D_2(p_2) = d_2(p_2) \tag{14}$$

Total supply $S_2(p_2)$ is expressed as follows:

$$0 \leq S_2(0) \leq R_1 + R_2 + X^b - d_1(p_1) \qquad \text{if } p_2 = 0 \tag{15}$$

$$S_2(p_2) = R_1 + R_2 + X^b - d_1(p_1) \qquad \text{if } 0 \leq p_2 \leq a \tag{16}$$

$$R_1 + R_2 + X^b - d_1(p_1) \leq S_2(p_2) \leq R_1 + R_2 + X - d_1(p_1) \qquad \text{if } p_2 \geq a \tag{17}$$

The market equilibrium is determined through the intersection of the demand and supply functions:

$$D_2(p_2) = S_2(p_2) \tag{18}$$

The electricity price at $t = 2$, p_2 is expressed as follows:

$$p_2 = 0 \qquad \text{if } d_2(0) + d_1(p_1) \leq R_1 + R_2 + X^b \tag{19}$$

$$0 < p_2 < a \qquad \text{if } d_2(0) + d_1(p_1) > X^b + R_1 + R_2 \text{ and } d_2(a) + d_1(p_1) \leq X^b + R_1 + R_2 \tag{20}$$

$$p_2 = a \qquad \text{if } d_2(a) + d_1(p_1) > X^b + R_1 + R_2 \text{ and } d_2(a) + d_1(p_1) \leq X + R_1 + R_2 \tag{21}$$

$$p_2 > a \qquad \text{if } d_2(a) + d_1(p_1) > X + R_1 + R_2 \tag{22}$$

p_2 depends on the renewable supply R_2 and the price at $t = 1$, p_1.
The market equilibrium is determined by the intersection of the demand and supply functions:

$$D_2(p_2) = S_2(p_2) \tag{23}$$

The belief of p_2 depends on the belief of R_2 and p_1.

Let $F_{p_1}(E_i[p_2])$ be the distribution of the expected price at $t = 2$ at price p_1. For solving the equilibrium, we need the following assumption.

Assumption 1. *If $p \leq p'$, for all $E_i[p_2]$, $F_p(E_i[p_2]) \leq F_{p'}(E_i[p_2])$.*

Although this is an assumption, we check whether this assumption is satisfied through the market equilibrium.

At $t = 1$, the population of suppliers who have beliefs $p_1 > E_i[p_2]$ is $1 - F_{p_1}(p_1)$.

Therefore, at $t = 1$, if $p_1 \leq a$, $1 - F_{p_1}(p_1)$ normal suppliers buy electricity, and $F_{p_1}(p_1)$ renewable suppliers sell electricity.

$p_1 > a$, $1 - F_{p_1}(p_1)$ normal suppliers buy electricity, and $F_{p_1}(p_1)$ renewable and normal suppliers sell electricity.

Total demand $D_1(p_1)$ is expressed as follows:

$$D_1(p_1) = d_1(p_1) + (1 - F_{p_1}(p_1))\frac{I}{p_1} \tag{24}$$

An analysis of Assumption 2 showed that the demand function is decreasing: $lim_{p_1 \to 0} D_1(p_1) = \infty$, and $lim_{p_1 \to \infty} D_1(p_1) = lim_{p_1 \to \infty} d_1(\infty) = 0$. Total supply $S_2(p_2)$ is expressed as follows:

$$F_{p_1}(p_1)R_1 \qquad \text{if } p_1 \leq 0 \tag{25}$$

$$F_a(a) \leq S_1(a) \leq F_a(a)(R_1 + X) \qquad \text{if } p_1 = a \tag{26}$$

$$F_{p_1}(p_1)(R_1 + X) \qquad \text{if } p_2 \geq a \tag{27}$$

An analysis of Assumption 2 showed that the supply function is increasing in p_1, and $S(0) = 0$ $lim_{p_1 \to \infty} S_1(p_1) = R_1 + X$. The market equilibrium is determined by the intersection of the demand and supply functions (see Figure 3):

$$D_1(p_1) = S_1(p_1) \tag{28}$$

The electricity price at $t = 1$, p_1 is expressed as follows:

$$d_1(p_1) + (1 - F_{p_1}(p_1))\frac{I}{p_1} = F_{p_1}(p_1)R_1 \qquad \text{if } p_1 \leq a \tag{29}$$

$$F_a(a)R_1 < d_1(p_1) + (1 - F_a(a))\frac{I}{p_1} < F_a(a)R_1 + F_a(a)X \qquad \text{if } p_1 = a \tag{30}$$

$$d_1(p_1) + (1 - F_{p_1}(p_1))\frac{I}{p_1} = F_{p_1}(p_1)R_1 + F_{p_1}(p_1)X \qquad \text{if } p_1 > a \tag{31}$$

$D_1(p_1)$ is decreasing, and $S_1(p_1)$ is increasing.

$$lim_{p_1 \to 0} D_1(p_1) = \infty \tag{32}$$

$$lim_{p_1 \to \infty} D_1(p_1) = lim_{p_1 \to \infty} d_1(p_1) = 0 \tag{33}$$

$$lim_{p_1 \to 0} S(p_1) = S(0) = 0 \tag{34}$$

$$lim_{p_1 \to \infty} S_1(p_1) = R_1 + X \tag{35}$$

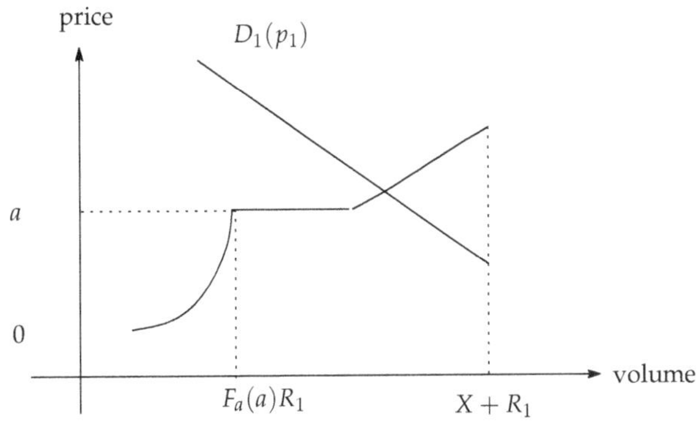

Figure 3. Date 1 market equilibrium.

5. Results and Discussion

This model has a mixed structure; it includes aspects of the heterogeneous belief model as well as the electricity market. Therefore, it includes many special features that are not available in other models. As noted, the shape of supply curve at date 1 is one of them. This shape is derived from the electricity market model. Electricity suppliers have ladder cost-related functions, and their strategy depends on the shape of that functions. We analyzed three major points of this model: price relation, speculative trading effect, and belief effect.

5.1. Price Relation

The market at date 2 is simpler than the market at date 1. However, the date 2 price is influenced by the date 1 price p_1. The next lemma shows the relation between p_1 and p_2.

The market equilibrium at date 2 is simpler than the date 1 equilibrium. As noted, total supply $S_2(p_2)$ is expressed as follows:

$$0 \leq S_2(0) \leq R_1 + R_2 + X^b - d_1(p_1) \qquad \text{if } p_2 = 0 \tag{36}$$

$$S_2(p_2) = R_1 + R_2 + X^b - d_1(p_1) \qquad \text{if } 0 \leq p_2 \leq a \tag{37}$$

$$R_1 + R_2 + X^b - d_1(p_1) \leq S_2(p_2) \leq R_1 + R_2 + X - d_1(p_1) \qquad \text{if } p_2 \geq a \tag{38}$$

X^b electricity units are sold by normal suppliers at date 1. Therefore, based on the date 1 equilibrium, the following can be expressed:

$$X^b = F_{p_1}(p_1)X \tag{39}$$

Therefore, X^b is an increasing function of p_1. Moreover, $d_1(p_1)$ is a decreasing function of p_1. Thus, if p_1 gets higher (and other conditions at date 2 remain equal), total supply moves towards the right. As a result, p_2 decreases (see Figure 4).

We can examine Assumption 2 by this price relation. Higher p_1 implies lower p_2 for all suppliers; therefore, $p_1 \leq p_1'$ for all $E_i[p_2]$.

$$F_{p_1}(E_i[p_2]) \leq F_{p_1'}(E_i[p_2]) \tag{40}$$

This implies that Assumption 2 is satisfied by this equilibrium.

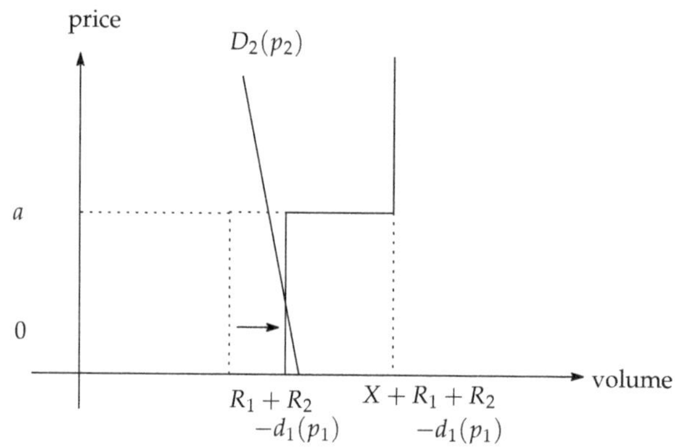

Figure 4. Date 2 market equilibrium with higher p_1.

5.2. Speculative Trading

Interestingly, the date 1 supply spikes at price $p_1 = a$. The optimistic normal suppliers have the incentive to sell electricity at date 1, but their marginal cost is a. Therefore, they choose to sell electricity only if p_1 exceeds a.

If the speculative trade is prohibited, the market structure become very simple. Suppliers cannot buy electricity at $t = 1$; that is, x^b must be 0. Therefore, normal suppliers' profit maximization at $t = 1$ changes as follows:

$$Max_{x_1, x^b} \ (p_1 - a)x_1 + E_i[v(x_1)] \quad s.t.\ x_1 \leq X,\ p_1 x^b \leq I,\ x_1 \geq 0 \tag{41}$$

x_1 is the selling volume, and $v(x_1)$ is the value function for $t = 2$.

$$v(x_1, x^b) = Max_{x_2} \ (p_2 - a)x_2 \quad s.t.\ x_2 \leq X - x_1,\ x_2 \geq 0 \tag{42}$$

The renewable suppliers face the same problem. Thus, the market without speculative trading equilibrium is as follows:

$$d_1(p_1) = F_{p_1}(p_1)R_1 \qquad \text{if } p_1 \leq a \tag{43}$$

$$F_a(a)R_1 < d_1(p_1) < F_a(a)R_1 + F_a(a)X \qquad \text{if } p_1 = a \tag{44}$$

$$d_1(p_1) = F_{p_1}(p_1)R_1 + F_{p_1}(p_1)X \qquad \text{if } p_1 > a \tag{45}$$

$$p_2 = 0 \qquad \text{if } d_2(0) + d_1(p_1) \leq R_1 + R_2 \tag{46}$$

$$0 < p_2 < a \qquad \text{if } d_2(0) + d_1(p_1) > R_1 + R_2 \text{ and } d_2(a) + d_1(p_1) \leq X^b + R_1 + R_2 \tag{47}$$

$$p_2 = a \qquad \text{if } d_2(a) + d_1(p_1) > R_1 + R_2 \text{ and } d_2(a) + d_1(p_1) \leq X + R_1 + R_2 \tag{48}$$

$$p_2 > a \qquad \text{if } d_2(a) + d_1(p_1) > X + R_1 + R_2 \tag{49}$$

Because suppliers cannot buy the electricity, the demand on date 1 is only $d_1(p_1)$.

Because optimistic suppliers expect that renewable supplies are low and electricity prices at $t = 2$ is high, they buy date 1 electricity for the purpose of making profits through resale. The date 1 price

p_1 is affected by the elasticity of the demand. Even if the consumers' demand $d_1(p_1)$ is inelastic with regard to the price, total demand is not inelastic with regard to the price.

Lemma 1. *The total demand becomes more elastic with regard to the price through speculative trading at $t = 1$.*

Proof. In the speculative trading model, the date 1 total demand is elastic.

$$D_1(p_1) = d_1(p_1) + (1 - F_{p_1}(p_1))\frac{I}{p_1} \tag{50}$$

Even if the consumers' demand $d_1(p_1)$ is perfectly inelastic with regard to the price, the total demand is not inelastic with regard to the price. □

This simple lemma implies that the trade of the suppliers has a role in establishing stability in the electricity market. If speculative trade is prohibited, the price p_1 jumps from 0 to a, with demand for d_1 shifting (see Figure 5).

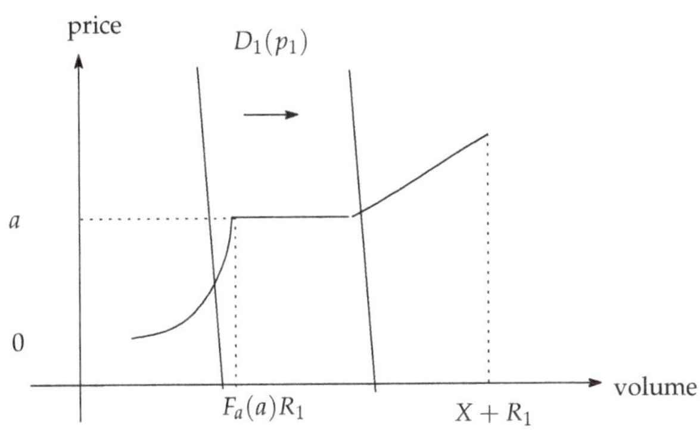

Figure 5. Date 1 market equilibrium without speculative trading.

Governments may often worry about consumers' surplus in the speculative electricity market. Interestingly, consumers' surplus is not always lower than that of the market without resale.

5.3. Heterogeneous Belief

The market equilibrium is heavily influenced by the suppliers' beliefs. It is essential to compare some measures of optimistic beliefs.

Assumption 2. *$G_{p_1}(p_2) > F_{p_1}(p_2)$ for all p_1, p_2.*

This means that, with distribution F_{p_1}, for all p_1, the population that expects to encounter a date 2 price that is higher than p_2, which is larger than that with G_{p_1}. Therefore, F_{p_1} is a more optimistic distribution than G_{p_1}.

Lemma 2. *The market price p_1 is higher under F_{p_1} compared to that under G_{p_1}.*

Proof. This implies $D_1(p)$ under $G(d)$ is lower than that under $F(d)$ for all d. Similarly, $S_1(p)$ is higher than that under $F(d)$. Then, the date 1 price p is lower than that under $F(d)$. This is because the date 2 price p_2 is expressed as follows:

$$d_1(p_1) + (1 - F_{p_1}(p_1))\frac{I}{p_1} = F_{p_1}(p_1)R_1 \qquad \text{if } p_1 \leq a \tag{51}$$

$$F_a(a)R_1 < d_1(p_1) + (1 - F_a(a))\frac{I}{a} < F_a(a)R_1 + F_a(a)X \qquad \text{if } p_1 = a \tag{52}$$

$$d_1(p_1) + (1 - F_{p_1}(p_1))\frac{I}{p_1} = F_{p_1}(p_1)R_1 + F_{p_1}(p_1)X \qquad \text{if } p_1 > a \tag{53}$$

Based on the assumption, for all, p_1 is expressed as follows:

$$d_1(p_1) + (1 - F_{p_1}(p_1))\frac{I}{p_1} \geq d_1(p_1) + (1 - G_{p_1}(p_1))\frac{I}{p_1} \tag{54}$$

$$F_{p_1}(p_1)R_1 \leq G_{p_1}(p_1)R_1 \tag{55}$$

$$F_{p_1}(p_1)R_1 + F_{p_1}(p_1)X \leq G_{p_1}(p_1)R_1 + G_{p_1}(p_1)X \tag{56}$$

Therefore, the total demand under F_{p_1} is larger than that under G_{p_1}, and the total supply under F_{p_1} is less than that under G_{p_1}. As a result, the market price increases under $F_{p_1}(p_2)$. □

Because of heterogeneous beliefs, the price of the electricity is higher compared to the no-resale market price. Firms have the incentive to buy the electricity for the speculative resale, and the consumer suffers the higher prices.

This model involves a heterogeneous bubble. Heterogeneous beliefs among investors create such bubbles. In such models, investors' beliefs differ because they have different prior belief distributions. Agents' heterogenous beliefs can occur because of many factors. For example, overconfidence about the precision of signals among investors can lead to different prior distributions (with lower variance) regarding the signals' noise term. Investors without common prior beliefs can agree to disagree even after they share all their information with each other. In the heterogeneous beliefs model with short-sale constraints, the asset price can result in the creation of bubbles. Optimistic agents buy the asset, and the price rises. Under the conditions of a short-sale constraint, pessimistic traders cannot make use of the high asset prices (Miller (1977) [19], Harrison et al. (1978) [20]). In a dynamic model, the asset price can even exceed the valuation of the most optimistic investor's expectation regarding the economy. In the model, firms with pessimistic beliefs about the demand on the next day sell the electricity immediately; this implies that the supply on the next day increases. Therefore, a pessimistic belief distribution, such as G(x), is very beneficial for the consumers' surplus.

As noted in Section 4.1, higher p_1 implies lower p_2 in this model. Therefore, the next lemma shows some negative correlations between the expected price and the realized price.

Lemma 3. *The market price p_2 is lower under F_{p_1} than that under G_{p_1}.*

Therefore, if the suppliers' belief is optimistic (that is, they expect higher $E[p_2]$), p_2 tends to be lower. This can be interpreted as a heterogeneous belief bubble, and this bubble burst on date 2.

6. Policy Implications and Conclusions

6.1. Policy Implications

By utilizing the three lemmas discussed in Sections 5.2 and 5.3, we can propose two policy implications. First, through speculative trading, electricity price spikes can be reduced. Pricing is one of the signals for market conditions. Therefore, in terms of efficiency, price jumping is not desirable. Price spikes are considered in several research papers (Huisman et al. (2003) [9], Weron et al. (2004) [26] and others as noted in Section 2). Price spikes occur because the electricity market is characterized by inelastic demand and a stair-like supply curve. However, speculative trading allows suppliers to be electricity buyers on date 1. This increases the elasticity of the total demand, and the price change becomes relaxed.

Abrupt price fluctuations are undesirable because they increase the market risk. The presence of forward markets is motivated by speculation, so they become more sensitive to price differences. As a result, forward market will be effective for reducing price fluctuations.

Second, the forward market is also an important factor for making policy decisions that may cause prices to move in the opposite direction. If the price on date 1 is large, the price on date 2 tends to be lower. As the government's main policy target includes consumers who lack market power, it is important to keep the price of the second quarter (that is, the real-time price) low.

The main risk posed by forward markets may be the speculative price hikes. In the case of other asset bubbles, high prices are factors that can hurt efficiency. However, the situation is different in the case of the electricity market. The power market cannot save power because of the characteristics of power; therefore, the power capacity within a certain time zone loses the opportunity cost if it is not sold. Therefore, prices do not remain high, as is the case for ordinary assets. As a result, it is unlikely that the consumer's utility will be impaired by the high electricity prices. Conversely, as Lemma 2 shows, price increases in the forward market are likely to increase consumer utility. From this viewpoint as well, the effectiveness of the forward market for electricity can be demonstrated.

6.2. Conclusions

Increasing market liquidity is an indispensable factor for introducing renewable energy. However, attempts to raise liquidity in economics can generally lead to speculative behavior. In this paper, we outline such problems with a simple model. Based on the results of the model's speculative behavior, we determined that, even if the consumer demand for electricity is inelastic with regard to the price, the price elasticity of the demand is newly born; in this way, price change becomes stable. However, the forward market creates speculative trading. Speculative trading naturally accompanies a decrease in consumer surplus. Policymakers, especially the Japanese government, have a strong tendency to vigorously construct speculative actions in the electricity market and strengthen regulations. However, this model shows that some regulation, in terms of optimistic prospects to the future electricity price, can increase consumers' surplus.

In addition, it has often been pointed out that, in the existing power market model, the market price is dominated by several corporations. In this situation, the market price can be easily raised by these corporations because of the special nature of electric power. However, in this model, the company is assumed to be a price-taker; this also shows that there are cases in which consumer surplus can increase. Thus, inviting new corporations is an important policy for the liberalization of the power market. Our main findings is that the speculative tradings can improve the inelasticity of the demand. This finding of our theory is the first one that points out this effect. This finding is also a very useful result when considering the market design of the electricity market. Speculative trading is an important factor in introducing the forward market to electricity. The results of this paper show that speculative trading has a positive externality and is likely to be a desirable policy in the electricity market.

Our model is based on JEPX. However, the basic elements are common to electricity markets in other countries, and can be applied as a more general electricity model. It would be necessary to comprehensively analyze the linkage between dynamic markets rather than individual markets. We propose a basic model as one of the attempts to describe these two types of markets.

Author Contributions: Methodology, J.M.; Formal Analysis, J.M.; Writing—Original Draft Preparation, J.M.; and Supervision, K.S.

Funding: This research was funded by JSPS KAKENHI Grant NO. 15K00645.

Acknowledgments: We are grateful to Izumi Inasawa, Hiroshi Kinokuni, and Ai Takeuchi for their helpful discussions and comments with regard to our model. All blame for any errors is attributed to us.

Conflicts of Interest: The authors declare no conflict of interest.

Abbreviations

The following abbreviations/nomenclature are used in this manuscript

JEPX	Japan Electric Power Exchange
p_1	date 1 price
p_2	date 2 price
X	maximum supply of normal suppliers
R_1	minimum supply of renewable suppliers
R_2	maximum supply of renewable suppliers
$d_1(p_1)$	consumers' demand at date 1
$d_2(p_2)$	consumers' demand at date 2
a	marginal cost of normal suppliers
$F_{p_1}(p)$	the population of suppliers who has beliefs $E_i[p_2] < p$ under p_1
x_1	suppliers' selling volume at date 1
x_2	suppliers' selling volume at date 1
x^b	suppliers' purchase volume at date 1
X^b	total suppliers' purchase volume at date 1
$S_1(p_1)$	total supply volume at date 1
$S_2(p_2)$	total supply volume at date 2
$D_1(p_1)$	total demand volume at date 2
$D_2(p_2)$	total demand volume at date 2

References

1. METI. Report about Progress of Total Liberalization of Electricity Retail in 2018. Available online: www.meti.go.jp/committee/sougouenergy/denryoku_gas/denryoku_gas_kihon/pdf/009_03_01.pdf (accessed on 1 July 2018).
2. Japan Electric Power Exchange. Available online: http://www.jepx.org (accessed on 30 May 2019).
3. Holttinen, H. Optimal Electricity Market for Wind Power. *Energy Policy* **2005**, *33*, 2052–2063. [CrossRef]
4. Ummels, B.C.; Gibescu, M.; Kling, W.L.; Paap, G.C. Integration of Wind Power in the Liberalized Dutch Electricity Market. *Wind Energy* **2006**, *9*, 579–590. [CrossRef]
5. Lucia, J.J.; Schwartz, E.S. Electricity Prices and Power Derivatives: Evidence from the Nordic Power Exchange. *Rev. Delivatives Res.* **2002**, *5*, 5–50. [CrossRef]
6. Pilipovic, D. *Energy Risk: Valuing and Managing Energy Derivatives*; McGraw Hill: New York, NY, USA, 1998.
7. Escribano, A.; Pena, J.; Villapana, P. *Modeling Electricity Prices: International Evidence*; Working Paper; Universidad Carlos III de Madris: Madrid, Spain, 2002; Volume 02–27, pp. 79–817.
8. Eydeland, A.; Wolyniec, K. *Energy and Power Risk Management; New Developments in Modeling, Pricing and Hedging*; John Wiley & Sons: Hoboken, NJ, USA, 2003.
9. Huisman, R.; Mahieu, R. Regime jumps in electricity prices. *Energy Econ.* **2003**, *25*, 425–434. [CrossRef]
10. Maekawa, J.; Hai, B.H.; Shinkuma, S.; Shimada, K. The Effect of Renewable Energy Generation on the Electric Power Spot Price of the Japan Electric Power Exchange. *Energies* **2018**, *11*, 2215. [CrossRef]
11. Audun, B.; Tarjei, K.; Marija, D. The relationship between spot and forward prices in the Nord Pool electricity market. *Energy Econ.* **2010**, *32*, 967–978.
12. Raviv, E.; Bouwman, K.E.; van Dijk, D. Forecasting day-ahead electricity prices: Utilizing hourly prices. *Energy Econ.* **2015**, *50*, 227–239. [CrossRef]
13. Cartea, Á.; Figueroa, M. Pricing in electricity markets: A mean reverting jump diffusion model with seasonality. *Appl. Math. Financ.* **2005**, *12*, 313–335. [CrossRef]
14. Weron, R. *Modeling and Forecasting Electricity Loads and Prices: A Statistical Approach*; John Wiley & Sons: New York, NY, USA, 2007; Volume 403.
15. Hikspoors, S.; Jaimungal, S. Energy spot price models and spread options pricing. *Int. J. Theor. Appl. Financ.* **2007**, *10*, 1111–1135. [CrossRef]
16. Benth, F.E.; Saltyte-Benth, J.; Koekebakker, S. *Stochastic Modeling of Electricity and Related Markets*; World Scientific: Hackensack, NJ, USA, 2008.

17. Jaimungal, S.; Surkov, V. Lévy-based cross-commodity models and derivative valuation. *SIAM J. Financ. Math.* **2011**, *2*, 464–487. [CrossRef]
18. Carteaa, Á.; Jaimungalc, S.; Qinc, Z. Speculative trading of electricity contracts in interconnected locations. *Energy Econ.* **2018**, *79*, 3–20. [CrossRef]
19. Miller, E.M. Risk, Uncertainty, and Divergence of Opinion. *J. Financ.* **1977**, *32*, 1151–1168. [CrossRef]
20. Harrison, J.M.; Kreps, D. Speculative Investor Behavior in a Stock Market with Heterogeneous Expectations. *Q. J. Econ.* **1978**, *89*, 323–326. [CrossRef]
21. Townsend, R. Forecasting theforecasts of others. *J. Political Econ.* **1983**, *91*, 546–588. [CrossRef]
22. Singleton, K.J. Asset prices in a time series model with disparately informed, competitive traders. In *New Approaches to Monetary Economics*; Burnett, W., Singleton, K., Eds; Cambridge University Press: Cambridge, UK, 1987.
23. Sheinkman, J.; Xiong, W. Overconfidence and Speculative Bubbles. *J. Political Econ.* **2003**, *111*, 1183–1219. [CrossRef]
24. Gabriel, S.A.; Zhuang, J.; Egging, R. Solving stochastic complementary problems in energy market modeling using scenario reduction. *Eur. J. Oper. Res.* **2009**, *197*, 1028–1040. [CrossRef]
25. Joets, M. Heterogeneous beliefs, regret, and uncertainty: The role of speculation in energy price dynamics. *Eur. J. Oper. Res.* **2015**, *247*, 204–215. [CrossRef]
26. Weron, R.; Bierbrauer, M.; Trueck, S. Modeling electricity prices with regime switching models. *Phys. Stat. Mech. Its Appl.* **2004**, *336*, 39–48. [CrossRef]
27. Knittel, C.R.; Roberts, M.R. An empirical examination of restructured electricity prices. *Energy Econ.* **2005**, *27*, 791–817. [CrossRef]
28. Chan, K.F.; Gray, P. Using extreme value theory to measure value-at-risk for daily electricity spot prices. *Int. J. Forecast.* **2006**, *22*, 283–300. [CrossRef]

© 2019 by the authors. Licensee MDPI, Basel, Switzerland. This article is an open access article distributed under the terms and conditions of the Creative Commons Attribution (CC BY) license (http://creativecommons.org/licenses/by/4.0/).

energies

MDPI

Article

The Uncertain Bidder Pays Principle and Its Implementation in a Simple Integrated Portfolio-Bidding Energy-Reserve Market Model

Dávid Csercsik [1,*], Ádám Sleisz [2] and Péter Márk Sőrés [2]

1 Faculty of Information Technology and Bionics, Pázmány Péter Catholic University, 1083 Budapest, Práter str. 50/A, Hungary
2 Department of Electric Power Engineering, Budapest University of Technology and Economics, 1111 Budapest, Egry J. u. 18., Hungary
* Correspondence: csercsik@itk.ppke.hu; Tel.: +36-1-886-4706

Received: 21 June 2019; Accepted: 26 July 2019; Published: 1 August 2019

Abstract: One reason for the allocation of reserves in electricity markets is the uncertainty of demand and supply. If the bias of the generation portfolio shifts from controllable generators to renewable sources with significantly higher uncertainty, it is natural to assume that more reserve has to be allocated. The price of reserve allocation in European models is dominantly paid by the independent system operator in the form of long-term paid reserve capacities and reserve demand bids submitted to various reserve markets. However, if we consider a scenario where the significant part of generation is allocated in day-ahead auctions, the power mix is not known in advance, so the required reserves can not be efficiently curtailed for the ratio of renewables. In the current paper we analyze an integrated European-type, portfolio-bidding energy-reserve market model, which aims to (at least partially) put the burden of reserve allocation costs to the uncertain energy bidders who are partially responsible for the amount of reserves needed. The proposed method in addition proposes a more dynamic and adaptive reserve curtailment method compared to the current practice, while it is formulated in a computationally efficient way.

Keywords: integration of renewable sources; integrated markets; co-optimization; reserve allocation

1. Introduction

Although modern renewable generators do have improved controllability properties compared to earlier solutions, they still exhibit a higher level of supply uncertainty compared to non-renewable generators. In addition, uncertainty is present in several forms in the power grid. First of all, in addition to renewable sources with fundamental characteristics of production uncertainty (of some level), conventional power plants are also naturally subject to failures and technological issues, which may limit their output from time to time. Furthermore, significant part of the demand corresponds to domestic consumers, the schedule of whom may be predicted only with limited accuracy [1,2].

As supply-demand imbalance causes frequency shift in the power network, certain forms of ancillary services (or 'reserves' to put it shorter) are needed for frequency stability. Activating these reserves in the appropriate time restores power balance and thus network frequency. Ancillary markets [3] are specialized energy-economical platforms, in which commodities connected to ancillary services are traded. Although there is a broad spectrum of ancillary service and reserve types, in this paper we consider reserves used for frequency control. In a typical ancillary market, reserve providers are paid for allocating the required reserves and an additional fee is paid if the reserve is activated as well. According to this, in the current paper (regarding reserves) we consider capacity-allocation payment (and we are not interested if later the reserve is activated or not and do not consider these activation payments).

Reserves may be classified furthermore into non-event and event driven resources. While non-event driven resources are used to compensate production-consumption imbalances, event driven resources are used in reaction to contingencies and power plant or line outages. In the current article we focus on non-event driven resources and their allocation.

As maintaining the frequency stability of the power grid is the responsibility of the system operator (The terminology may differ as the expressions TSO (transmission system operator) and ISO (independent system operator) are also used for this player. In our case, as reserves will be allocated in the integrated power-reserve auction, we will assume that the operator is in charge not only of the transmission system but also of the auction, so we will use ISO.), it is its task to ensure the allocation of the required amount of reserve via long-term contracts or from the ancillary markets. Ensuring the necessary reserves via long term contracts is by nature more inflexible, as the actual power mix resulting from day ahead power exchanges and its uncertainty properties can not be taken into account. Either the system operator takes a conservative (and costly) approach and allocates a high amount of reserves in long term contracts or uses a mix of long term and short term solutions to account for the actual reserve requirements.

Regarding the general issues of reserve allocation, Reference [4] formulates two general questions: How much reserve should be allocated and who should pay for it? The first problem, in addition to the simple approach presented in the same paper [4], has been a subject to several articles. Following the lead idea of Reference [4], namely that reserve should be purchased up to the point where the marginal cost of providing reserve is equal to the marginal value of this reserve, the article [5] considers the customer outage cost to determine the marginal value of reserve. Regarding more recent approaches, motivated by the increasing market share of renewable sources, stochastic unit commitment based reserve procurement procedure for power systems including wind farms is described in Reference [6], while more or less the same problem is approached by a different solution in Reference [7]. Reference [8] contributes to the same topic by applying a chance-constrained optimization to determine the required amounts of reserve capacity. A robust optimization based method of joint determination of day-ahead energy and reserve dispatch is described in Reference [9]. Allocation of reserve-related costs is however not discussed in detail in these articles.

Regarding the allocation of these costs, which is discussed less in the literature, the original paper [4] gives two approaches: In addition to the most simple solution, namely *'all consumers should pay a share of the cost of reserve on the basis of their consumption'*, it also adds that on the other hand *'the cost of reserve should be shared among the generators on the basis of their contribution to the need for reserve'*. It also discusses the possible scenario when generators forward these cost to their consumers. Regarding the allocation of reserve-related balancing and ramping costs, Reference [10] proposes a unit commitment-based approach, applying the principle of pareto-optimality for the problem. Reference [11] aims to distribute the reserve cost among the most appropriate consumers, applying agent-based modelling and simulation approach. A somewhat similar agent-based approach combined with stochastic unit commitment for the reserve cost allocation problem is presented in Reference [12]. The two latter papers both use the concept of *demand factors* (first defined in Reference [13]) to characterize the reliability level of customers.

In the current article we follow the second principle formulated in Reference [4] and aim to put the burden of reserve allocation costs to market participants whose activity significantly contributes to the need of reserve allocation. In the context of portfolio-bidding markets, the *'who pays for'* question is typically answered by the principle that the accepted demand bids cover the costs of products, thus we need to introduce any cost-allocation policy in the form of demand bids. This principle can be considered promising as it fits into the official market development plans of the European Union. It is stated in the Clean Energy Package that *'all market participants shall be responsible for the imbalances they cause in the system'* and this imposition includes variable renewable energy producers [14]. However, the explicit regulation concerns only balancing markets while the more liquid day-ahead trading platform may remain free of uncertainties. This gap could be filled using a model developed in the current article.

In order to facilitate European market development, the created clearing formulation has to be adjusted to the usual European approach. Thus, in contrast to previous results, which analyzed the problem in a unit commitment framework [15], we consider self-scheduling generators and a purely portfolio-bidding market based framework, consisting energy and reserve markets. We define the concept of supplementary reserve demand bids, which ensure that the owner of any uncertain energy bid, in the case of acceptance, will also automatically contribute to the costs of reserve allocation as well. While uncertainty characterization in the majority of previous literature was focussing to either power plants or customers, in the proposed approach we consider the potential uncertainty of both side of the energy market—both uncertain supply and demand energy bids are considered. Furthermore while previous methods use computationally demanding agent based modelling (as References [11,12]) or include quadratic constraints [10] (or their semidefinite relaxation), we formulate the suggested method as a simple mixed integer linear problem (MILP), which can be efficiently solved, for example, via Benders-decomposition [16] and/or the branch-and-bound algorithm [17]. (Both techniques are widely used to solve problems in the power sector [18–20].) In addition, the method suggested in the current paper uses a single scalar parameter according to which the set of uncertain/not uncertain energy bids are defined. Decreasing this parameter from a sufficiently large value, the market implementation of the method presented may be introduced to the market incrementally.

For the aim of simplicity and clarity, we introduce the proposed concept assuming a period decoupled market, in which no multiperiod block orders or minimum income condition (MIC) orders are present, thus every period may be dispatched independently of the others. This means that it is enough for us to define the framework for a single period and introduce the concepts in this context. Later, in Section 4, we discuss the principles according to which the proposed concepts may be implemented in markets using multi-period block orders or MIC orders.

2. Materials and Methods

2.1. Uncertainty Quantification

We assume that every bid of the market can be connected to a bidder. We assume that there are K bidders in the energy market. Similar to the approach of demand factors [12,13], we use simple scalar quantities to characterize the uncertainty of market participants. While in the case of the demand factor the characterizing scalar is formulated as the quotient of the expected energy not supplied (EENS) and the current load, we also account for deviations in the positive direction and define two uncertainty measures corresponding respectively to the positive and negative part of the deviation in question. Based on previous bidding and market behavior of the bidder (or if no data present, then based on the applied technology), we assign the uncertainty quantifiers (u_k^+ and u_k^-) for each energy-bidder. The previous market behaviour in this case means that we may analyze how many times and how much the bidder deviated from its previously fixed schedule in relative terms (%) in the positive or in the negative direction, weighted by the quantity of the bid in question. Let us denote the expected relative value in the positive direction of bidder k by u_k^+, while the expected relative value in the negative direction of bidder k by u_k^-. By nature, both u_k^+ and u_k^- are non-negative quantities.

For example, let us suppose a bidder corresponding to a renewable source with significant uncertainty. If we have for example, a bid/schedule realization history (H) for bidder k (For the indexing of bidders we use the variable k, while j and i are used for the indexing of bids.) as

$$H = \begin{pmatrix} 50 & 70 & 100 & 80 & 65 & 65 \\ 41 & 73 & 92 & 80 & 63 & 69 \end{pmatrix}, \tag{1}$$

Each column of H corresponds to a previous auction, where the bid of the bidder has been accepted. The top element of each column holds the nominal bid quantity, while the bottom row corresponds to the realized schedule. In this particular case we have the following signed relative deviation vector D (in %).

$$D_i = 100\frac{H_{2,i} - H_{1,i}}{H_{1,i}} \quad \rightarrow \quad D = \begin{pmatrix} -18 & 4.29 & -8 & 0 & -3.08 & 6.15 \end{pmatrix} \tag{2}$$

Taking the positive and the negative part of this vector and weighing with the first row of (1), we get the expected values

$$u_k^+ = \frac{\sum_i \frac{D_i+|D_i|}{2} H_{1,i}}{\sum_i H_{1,i}} = 1.63\% \quad u_k^- = \frac{\sum_i \frac{D_i-|D_i|}{2} H_{1,i}}{\sum_i H_{1,i}} = 4.42\% \tag{3}$$

for bidder k. Let us define the positive and negative uncertainty upper bounds $\overline{u}^+$ and $\overline{u}^-$. Bids belonging to bidders with $u_k^+ \geq \overline{u}^+$ and $u_k^- < \overline{u}^-$ will be called positively uncertain (U+) bids, while bids belonging to bidders with $u_k^+ < \overline{u}^+$ and $u_k^- \geq \overline{u}^-$ will be called negatively uncertain (U-) bids. Bids belonging to bidders with both $u_k^+ \geq \overline{u}^+$ and $u_k^- \geq \overline{u}^-$ will be called bi-uncertain (Ub) bids.

To consider a simple example, if we assume the above bid with $u_k^+ = 1.63\%$, $u_k^- = 4.42\%$ and $\overline{u}^+ = \overline{u}^- = 2\%$, the bid will be considered as a negatively uncertain bid. In contrast, if $\overline{u}^+ = \overline{u}^- = 1\%$, the bid will be taken into account as a bi-uncertain bid.

While the above example was demonstrating the case of a supply bid, we apply the same approach for demand bids as well in the proposed framework (domestic consumers may be for example, considered as uncertain demand bidders).

Let us note that uncertainty upper bounds in general may be different in the case of supply and demand bids, however in this paper we will not distinguish between uncertainty bounds of supply and demand bids. As we will see later, we will use these values to account for reserve allocation needed for the coverage of this uncertainty.

2.2. Market Model of the Single Period Case

We consider a basic portfolio bidding scenario, where participants capable of delivering a certain product (energy or reserve in this case) are represented by supply bids, while entities who are ready to pay for it are submitting demand bids. The market clearing aims to balance the supply with the demand in the terms of the traded quantity and the price.

As in the first step we do not consider multi-period block bids, which define interdependencies over time periods, the calculations for each period may be carried out independently. For this reason, to make the notation more simple, in the first step we describe the calculations regarding only a single time period. Later we discuss how the proposed approach may be generalized for multi-period cases including block orders. Regarding the bid format, the two generally used bid types in portfolio-bidding electricity markets are the step bid and the linear bid. In the case of the step bid the price per unit (PPU) of the bid is independent of the acceptance rate, while in the case of linear bid the price depends on the acceptance rate linearly. In other words, while step bids are parametrized by two values (the quantity (q) and the bid PPU), linear bids are parametrized by the quantity, a starting price and a final price. If a linear bid is partially accepted the resulting PPU may be derived as a linear interpolation of the two prices: If for example, the acceptance rate is 0.5, the resulting PPU is the average of the starting price and the final price. In the proposed framework, for the aim of simplicity and computational efficiency, we do not allow linear bids, only step bids.

In the proposed model, there are 3 sub-markets: the energy sub-market and the reserve sub-markets corresponding to positive and negative reserve. The term 'sub-market' is used to emphasize that interdependencies between these markets will be defined and thus they have to be cleared together—in this case the 'market' is composed of the sub-markets and the sub-markets are not independent entities anymore.

We do not consider fill-or-kill type bids in the market model, in other words partial acceptance is allowed for all energy bids. Regarding energy bids with uncertainty levels below the thresholds $\overline{u}^+$ and $\overline{u}^-$, the variable $y_j^{ES} \in [0,1]$ denotes the acceptance variable of j-th energy supply bid, while $y_j^{ED} \in [0,1]$ denotes the acceptance variable of j-th energy demand bid. In the case of uncertain bids, $y_j^{ESU+} \in [0,1]$, $y_j^{ESU-} \in [0,1]$ and $y_j^{ESUb} \in [0,1]$ denote the acceptance variables of energy supply bids with (respectively positive, negative or both) uncertainty, while $y_j^{EDU+} \in [0,1]$, $y_j^{EDU-} \in [0,1]$ and $y_j^{EDUb} \in [0,1]$ denote the acceptance variables of energy demand bids with uncertainty.

As we will see later, these acceptance indicators will be included in variable vector of the problem. In addition to the acceptance indicators, the variable vector will also hold the income variables, logical integer variables used in the formulation of logical constraints and the market clearing prices (MCP) for energy and reserves. Under market clearing prices we mean prices which are compatible with the bid acceptance and balance constraints (see their formulation later). In other words, if the prices are equal to the market clearing prices, such an acceptance configuration of bids is possible (according to the bid acceptance rules), which ensures the balance of supply and reserve in every sub-market.

2.2.1. Supplementary Reserve Demand Bids

We assume that if uncertainty is present in the dispatch, in the spirit of the uncertain bidder pays principle, reserves must be allocated according to the measure of the uncertainty in question. We assume furthermore that these uncertain sources (being typically non-controllable units) are physically unable to provide reserves which could be used to handle the uncertainty implied by them. As we would like to make uncertain sources and consumers (bidders) pay for the implicated allocation of reserves, we assign obligatory reserve bids in the corresponding (positive, negative or both) reserve markets to each bid submitted in the energy sub-market. We call these compulsorily submitted reserve demand bids *supplementary reserve demand bids* (SRDBs). Both the bid price and bid quantity of these SRDBs are centrally regulated, they are not determined by the bidder. Uncertain energy bids together with the one or two connected SRDB(s) are called *orders*. As we will see later, the acceptance of the bids composing the order is dependent on the total income of the order, thus SRDBs and the related orders define interdependencies between the sub-markets.

Let us assume that y_j^{ESUb} is acceptance indicator of the energy supply bid of the bi-uncertain bidder k, the quantity of which is denoted by q_j^{ESUb}, while p_j^{ESUb} stands for price per unit (PPU) of the bid. In the proposed setup, implied by the bid corresponding to y_j^{ESUb}, bidder k also compulsorily submits a positive and a negative reserve demand bid, whose acceptance indicators are denoted by $y_j^{RD+\ ESUb}$ and $y_j^{RD-\ ESUb}$ respectively. The upper index in the notation refers to the set of positive/negative reserve demand bids implied by bi-uncertain energy supply bids.

As it is detailed in the following, the proposed concept of supplementary reserve demand bids may be introduced in the market gradually. In the beginning, it is the task of the ISO to allocate reserves and cover the connected costs. Furthermore, it is plausible that the ISO aims to ensure some of the required reserves in the day-ahead reserve markets. According to this consideration, we consider also reserve bids, which are not connected to uncertain energy bids. In the case when SRDBs cover all reserve needs, the model is completely functional without any non-SRDB reserve demand bid. The acceptance indicators of these (non-SRDB) bids are denoted by y_j^{RS+}, y_j^{RD+}, y_j^{RS-} and y_j^{RD-} in the case of positive reserve supply, positive reserve demand, negative reserve supply and negative reserve demand respectively.

Returning however to SRDBs, we need to consider the following. As positive deviations must be balanced by negative reserve and vice versa, positively uncertain ES bids (y_j^{ESU+}) imply negative reserve demand bids denoted by $y_j^{RD-\ ESU+}$ and negatively uncertain ES bids (y_j^{ESU-}) imply positive reserve demand bids denoted by $y_j^{RD+\ ESU-}$. In principle, the reserve amounts allocated for these demand bids cover the corresponding expected uncertainty, thus we may write

$$\begin{aligned} q_j^{RD+\ ESUb} &= -q_j^{ESUb} u_k^- \qquad & q_j^{RD-\ ESUb} &= -q_j^{ESUb} u_k^+ \\ q_j^{RD+\ ESU-} &= -q_j^{ESU-} u_k^- \qquad & q_j^{RD-\ ESU+} &= -q_j^{ESU+} u_k^+, \end{aligned} \tag{4}$$

We can see in Equation (4) that following the general convention, throughout the paper we use negative sign for the quantities of demand bids.

We also account for uncertainty in the case of energy demand bids—domestic retail electricity suppliers (who submit demand bids in the wholesale market, which is the subject of our study) may have for example, higher uncertainty compared to bidders corresponding to industrial demand. The notation is similar: the bi-uncertain energy demand bid y_j^{EDUb} implies the a positive and a negative reserve demand bids $y_j^{RD+\ EDUb}$ and $y_j^{RD-\ EDUb}$.

In our formalism, we consider demand with negative sign, so the row vectors in Equation (1) will be negative. Positive deviations in this case will mean less consumption, which must be balanced by negative reserves and mutatis mutandis. The SRDBs corresponding to demand bids are described by Equation (5).

$$\begin{aligned} q_j^{RD+\ EDUb} &= q_j^{EDUb} u_k^- \qquad & q_j^{RD-\ EDUb} &= q_j^{EDUb} u_k^+ \\ q_j^{RD-\ EDU+} &= q_j^{EDU+} u_k^+ \qquad & q_j^{RD+\ EDU-} &= q_j^{EDU-} u_k^-, \end{aligned} \tag{5}$$

We will suppose that the PPUs of these SRDBs are slightly higher compared to the highest PPU of the submitted reserve supply bids for the corresponding period. The difference is denoted by ε and corresponds to the unit of the market (e.g., 1 EUR/MW). The constant ε is introduced to avoid the possible overlap of supply and demand price curves in the case of the reserve sub-markets, which would potentially undermine the uniqueness of the optimal solution. The SRDB prices are formally defined as

$$p_j^{RD+\ SRDB} = \max_i (p_i^{RS+}) + \varepsilon \qquad p_j^{RD-\ SRDB} = \max_i (p_i^{RS-}) + \varepsilon \quad . \tag{6}$$

where $SRDB \in \{EDUb,\ EDU-\}$ in the case of positive reserve and $SRDB \in \{EDUb,\ EDU+\}$ in the case of negative reserve.

We assume that the amount of reserve supply is always enough to cover the total reserve demand. This assumption is usually valid in practice because regulators enforce power plants to offer reserve services. In this case the bid curves of the reserve spot markets (either positive or negative) will follow the qualitative scheme depicted in Figure 1.

As all SRDBs are accounted for on the price of the supply bid with the highest PPU $+\varepsilon$, the central line segment in the demand curve (labeled by SRDB in Figure 1) collects all the SRDBs. The line segments before and after it represent other reserve demand bids submitted to the reserve sub-market.

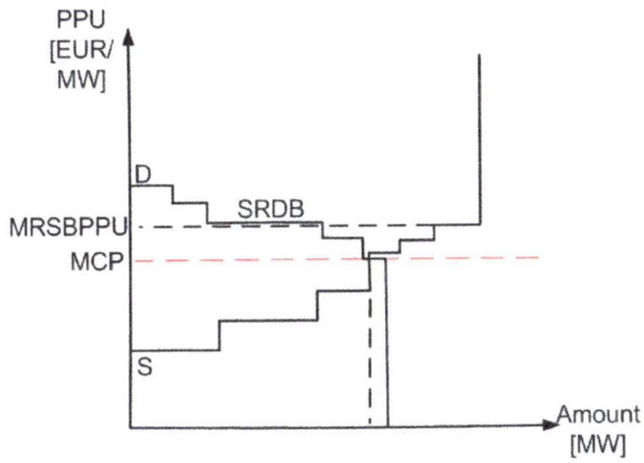

Figure 1. The scheme of the reserve spot market (both in the case of + and − reserve). *D*—demand, *S*—supply, *MCP*—market clearing price, *MRSBPPU*—Maximal reserve supply bid PPU, *SRDB*—Supplementary reserve demand bids. By definition, the PPU of SRDBs is equal to the PPU of the highest reserve supply bid + ε.

2.2.2. Minimum Surplus Conditions

Minimum Surplus Conditions for Uncertain Energy Supply Bids

In the proposed setup, without any additional considerations, it is possible that a submitted energy bid is rejected, while the connected SRDB(s) is/are accepted—this would naturally imply loss for the respective order, for the bidder is obliged to pay for the SRDB(s). Furthermore, even if the primary energy bid and the implied SRDB(s) is/are accepted, depending on the resulting MCPs, the surplus from the energy bid (originating from the energy sub-market) may not cover the cost of the SRDB(s) (originating from the reserve sub-markets) or the remaining surplus of the order (after extracting the costs of the SRDBs) may be very small. As the first step in the solution of this problem, we must calculate the incomes of the individual energy/reserve bids.

In order to formulate a linear computational framework, we take advantage of the dependence between MCPs and bid acceptance indicators (y) and use the description of income introduced in References [21,22], as follows. Let us denote the income of the bid corresponding to y_j^{ESUb} by I_j^{ESUb}. Intuitively I_j^{ESUb} may be calculated as

$$I_j^{ESUb} = MCP^E q_j^{ESUb} y_j^{ESUb} \tag{7}$$

where MCP^E stands for the market clearing price of energy.

Equation (7) holds however a quadratic expression of variables, namely the product of MCP^E and y_j^{ESUb}, which would result in a computationally demanding quadratically constrained problem (MIQCP). To overcome this issue we formulate the expressions for income as

$$y_j^{ESUb} > 0 \;\rightarrow\; I_j^{ESUb} = y_j^{ESUb} q_j^{ESUb} p_j^{ESUb} + q_j^{ESUb} MCP^E - q_j^{ESUb} p_j^{ESUb} \tag{8}$$

$$y_j^{ESUb} < 1 \;\rightarrow\; I_j^{ESUb} = y_j^{ESUb} q_j^{ESUb} p_j^{ESUb} \tag{9}$$

We implement the logical relations in the optimization framework based on the so called big-M method [23], using integer logical variables (denoted by z) as described in Appendix A.

To elucidate the Formulas (8) and (9), let us enumerate the following three possibilities:

- If the bid is entirely accepted ($y_j^{ESUb} = 1$), I_j^{ESUb} equals the product of q_j^{ESUb} and MCP^E according to (8).
- If the bid is partially accepted ($MCP^E = p_j^{ESUb}$), I_j^{ESUb} equals to $y_j^{ESUb} q_j^{ESUb} p_j^{ESUb}$. Both (8) and (9) are active in this case and they result in the same inequality.
- If the bid is entirely rejected ($y_j^{ESUb} = 0$), according to (9) $I_j^{ESUb} = 0$.

The above considerations may be naturally formulated also for income of bids corresponding to y_j^{ESU+} and y_j^{ESU-}—and also for bids corresponding to y_j^{ES} but their income wont be important in the proposed framework.

The incomes of uncertain energy demand bids I_j^{EDUb}, I_j^{EDU+}, I_j^{EDU-} and incomes of positive and negative SRDBs connected to energy bids, denoted by, $I_j^{RD+\ SRDB}$ and $I_j^{RD-\ SRDB}$ ($SRDB \in \{EDUb, EDU+, EDU-\}$) respectively, may be formulated similarly, taking into consideration in the latter two case that in the case of demand bids $q_j^{RD+\ SRDB}$ and $q_j^{RD-\ SRDB} < 0$, thus the income will mean practically expense because of the resulting negative sign.

According to these income calculations, now we may formulate constraints which exclude the scenario when the surplus of the primary energy bid does not meet the expense of the SRDB(s). In addition, we assume that for every uncertain order a surplus constant ($S > 0$) is defined, which describes how much the surplus of the primary bid must exceed the expenses (in other words it gives a lower bound for the total resulting surplus). In the proposed approach we assume that this constant may be determined by the bidder, thus it is diverse. However, under certain market conditions, it may be also plausible to assume that S is a parameter regulated by the central authority (system/market operator). Regarding the bid corresponding to y_j^{ESUb}, we denote this constant by S_j^{ESUb} (the notation is similar in the case of y_j^{ESU+} and y_j^{ESU-}). The constant $S_j^{ESUb} > 0$, will represent the minimum surplus value, which is required in the case of the acceptance of the order. According to this we may formulate the minimum surplus condition (MSC) for bi-uncertain energy supply bids as

$$S_j^{ESUb} - I_j^{RD+\ ESUb} - I_j^{RD-ESUb} \leq I_j^{ESUb} - p_j^{ESUb} q_j^{ESUb} y_j^{ESUb}, \tag{10}$$

where the right side is the surplus of the bid and the left side is the sum of the costs of the connected SRDBs (the incomes are negative because of demand) and the parameter S_j^{ESUb}. In the case of positively uncertain ES bids, we may write

$$S_j^{ESU+} - I_j^{RD-\ ESU+} \leq I_j^{ESU+} - p_j^{ESU+} q_j^{ESU+} y_j^{ESU+}, \tag{11}$$

while in the case of negatively uncertain ES bids, the formula becomes

$$S_j^{ESU-} - I_j^{RD+\ ESU-} \leq I_j^{ESU-} - p_j^{ESU-} q_j^{ESU-} y_j^{ESU-}. \tag{12}$$

Minimum Surplus Conditions for Uncertain Energy Demand Bids

In the case of bi-uncertain energy demand bids, the minimum surplus condition will state that total cost of the bid must be no more than the maximal potential cost of the bid, which would have been realized in the energy sub-market in the particular case if $MCP_i^E = p_j^{ED}$, minus a similar surplus constant (S_j^{EDUb}) as in the case of supply bids.

$$-I_j^{EDUb} - I_j^{RD+\ EDUb} - I_j^{RD-\ EDUb} \leq -p_j^{EDUb} q_j^{EDUb} y_j^{EDUb} - S_j^{EDUb}, \tag{13}$$

In the case of positively uncertain ED bids, we may write

$$-I_j^{EDU+} - I_j^{RD-\ EDU+} \leq -p_j^{EDU+} q_j^{EDU+} y_j^{EDU+} - S_j^{EDU+}, \tag{14}$$

while in the case of negatively uncertain ED bids, we may write

$$-I_j^{EDU-} - I_j^{RD+\ EDU-} \leq -p_j^{EDU-} q_j^{EDU-} y_j^{EDU-} - S_j^{EDU-}. \tag{15}$$

2.2.3. Bid Acceptance Constraints

For energy supply bids with no uncertainty

- $y_j^{ES} > 0 \rightarrow MCP^E \geq p_j^{ES}$.
- $y_j^{ES} < 1 \rightarrow MCP^E \leq p_j^{ES}$.

For bi-uncertain energy supply bids

- $y_j^{ESUb} > 0 \rightarrow$ Inequality (10) holds. This also implies $MCP^E \geq p_j^{ESUb}$.
- $y_j^{ESUb} < 1 \rightarrow MCP^E \leq p_j^{ESUb}$ or $y_j^{ESUb} = 0$, $y_j^{RD+\ ESUb} = 0$ and $y_j^{RD-\ ESUb} = 0$.

For positively uncertain energy supply bids

- $y_j^{ESU+} > 0 \rightarrow$ Inequality (11) holds. This also implies $MCP^E \geq p_j^{ESU+}$.
- $y_j^{ESU+} < 1 \rightarrow MCP^E \leq p_j^{ESU+}$ or $y_j^{ESU+} = 0$, and $y_j^{RD-\ ESU+} = 0$.

For negatively uncertain energy supply bids

- $y_j^{ESU-} > 0 \rightarrow$ Inequality (12) holds. This also implies $MCP^E \geq p_j^{ESU-}$.
- $y_j^{ESU-} < 1 \rightarrow MCP^E \leq p_j^{ESU-}$ or $y_j^{ESU-} = 0$, and $y_j^{RD+\ ESU-} = 0$.

For energy demand bids with no uncertainty

- $y_j^{ED} > 0 \rightarrow MCP^E \leq p_j^{ED}$.
- $y_j^{ED} < 1 \rightarrow MCP^E \geq p_j^{ED}$.

For bi-uncertain energy demand bids

- $y_j^{EDUb} > 0 \rightarrow$ Inequality (13) holds. This also implies $MCP^E \leq p_j^{EDUb}$.
- $y_j^{EDUb} < 1 \rightarrow MCP^E \geq p_j^{EDUb}$ or $y_j^{EDUb} = 0$, $y_j^{RD+\ EDUb} = 0$ and $y_j^{RD-\ EDUb} = 0$.

For positively uncertain energy demand bids

- $y_j^{EDU+} > 0 \rightarrow$ Inequality (14) holds. This also implies $MCP^E \leq p_j^{EDU+}$.
- $y_j^{EDU+} < 1 \rightarrow MCP^E \leq p_j^{EDU+}$ or $y_j^{EDU+} = 0$, and $y_j^{RD-\ EDU+} = 0$.

For negatively uncertain energy demand bids

- $y_j^{EDU-} > 0 \rightarrow$ Inequality (15) holds. This also implies $MCP^E \leq p_j^{ESU-}$.
- $y_j^{EDU-} < 1 \rightarrow MCP^E \geq p_j^{EDU-}$ or $y_j^{EDU-} = 0$, and $y_j^{RD+\ EDU-} = 0$.

For positive reserve supply bids

- $y_j^{RS+} > 0 \rightarrow MCP^{R+} > p_j^{RS+}$
- $y_j^{RS+} < 1 \rightarrow MCP^{R+} \leq p_j^{RS+}$

For negative reserve supply bids

- $y_j^{RS-} > 0 \rightarrow MCP^{R-} \geq p_j^{RS-}$
- $y_j^{RS-} < 1 \rightarrow MCP^{R-} \leq p_j^{RS-}$

For not SRDB positive reserve demand bids

- $y_j^{RD+} > 0 \rightarrow MCP^{R+} \leq p_j^{RD+}$
- $y_j^{RD+} < 1 \rightarrow MCP^{R+} \geq p_j^{RD+}$

For not SRDB negative reserve demand bids

- $y_j^{RD-} > 0 \rightarrow MCP^{R-} \leq p_j^{RD-}$
- $y_j^{RD-} < 1 \rightarrow MCP^{R-} \geq p_j^{RD-}$

For positive SRDBs

- $y_j^{RD+\ SRDB} > 0 \rightarrow$ Inequality (10), (12), (13) or (15) holds (depending on the type of the energy bid in the order of the SRDB) and $MCP^{R+} \leq p_j^{RD+\ SRDB}$, where $SRDB \in \{ESUb,\ ESU-,\ EDUb,\ EDU-\}$.
- $y_j^{RD+\ SRDB} < 1 \rightarrow MCP^{R+} \geq p_j^{RD+\ SRDB}$ or all acceptance indicators of the respective order are 0.

For negative SRDBs

- $y_j^{RD-\ SRDB} > 0 \rightarrow$ Inequality (10), (11), (13) or (14) holds (depending on the type of the energy bid in the order of the SRDB) and $MCP^{R-} \leq p_j^{RD-\ SRDB}$, where $SRDB \in \{ESUb,\ ESU+,\ EDUb,\ EDU+\}$.
- $y_j^{RD-\ SRDB} < 1 \rightarrow MCP^{R-} \geq p_j^{RD-\ SRDB}$ or all acceptance indicators of the respective order are 0.

The structure of the variable vector and the formulation of logical implications based thereon may be found in Appendix A.

2.2.4. Energy and Reserve Balances

The energy and reserve balances may be formulated as

$$\sum_{j=1}^{n_{ES}} y_j^{ES} q_j^{ES} + \sum_{j=1}^{n_{ESUb}} y_j^{ESUb} q_j^{ESUb} + \sum_{j=1}^{n_{ESU+}} y_j^{ESU+} q_j^{ESU+} + \sum_{j=1}^{n_{ESU-}} y_j^{ESU-} q_j^{ESU-} + \sum_{j=1}^{n_{ED}} y_j^{ED} q_j^{ED} + \sum_{j=1}^{n_{EDUb}} y_j^{EDUb} q_j^{EDUb} + \sum_{j=1}^{n_{EDU+}} y_j^{EDU+} q_j^{EDU+} + \sum_{j=1}^{n_{EDU-}} y_j^{EDU-} q_j^{EDU-} = 0, \tag{16}$$

$$\sum_{j=1}^{n_{RS+}} y_j^{RS+} q_j^{RS+} + \sum_{j=1}^{n_{RD+}} y_j^{RD+} q_j^{RD+} + \sum_{j=1}^{n_{ESUb}} y_j^{RD+\ ESUb} q_j^{RD+\ ESUb} + \sum_{j=1}^{n_{EDUb}} y_j^{RD+\ EDUb} q_j^{RD+\ EDUb} + \sum_{j=1}^{n_{ESU-}} y_j^{RD+\ ESU-} q_j^{RD+\ ESU-} + \sum_{j=1}^{n_{EDU-}} y_j^{RD+\ EDU-} q_j^{RD+\ EDU-} = 0, \tag{17}$$

$$\sum_{j=1}^{n_{RS-}} y_j^{RS-} q_j^{RS-} + \sum_{j=1}^{n_{RD-}} y_j^{RD-} q_j^{RD-} + \sum_{j=1}^{n_{ESUb}} y_j^{RD-\ ESUb} q_j^{RD-\ ESUb} + \sum_{j=1}^{n_{EDUb}} y_j^{RD-\ EDUb} q_j^{RD-\ EDUb} + \sum_{j=1}^{n_{ESU+}} y_j^{RD-\ ESU+} q_j^{RD-\ ESU+} + \sum_{j=1}^{n_{EDU+}} y_j^{RD-\ EDU+} q_j^{RD-\ EDU+} = 0. \tag{18}$$

2.2.5. The Objective Function

The objective function to maximize is the total social welfare (TSW). By definition the TSW is the total utility of consumption minus the total costs of production [24]. The TSW equals in this case the sum of the social welfare in the three sub-markets.

$$
\begin{aligned}
TSW &= TSW^{E} + TSW^{R+} + TSW^{R-} \\
TSW^{E} &= -\sum_{j=1}^{n_{ES}} y_j^{ES} q_j^{ES} p_j^{ES} - \sum_{j=1}^{n_{ED}} y_j^{ED} q_j^{ED} p_j^{ED} - \sum_{TEU} \sum_{j=1}^{n_{TEU}} y_j^{TEU} q_j^{TEU} p_j^{TEU} \\
TSW^{R+} &= -\sum_{j=1}^{n_{RS+}} y_j^{RS+} q_j^{RS+} p_j^{RS+} - \sum_{j=1}^{n_{RD+}} y_j^{RD+} q_j^{RD+} p_j^{RD+} \\
&\quad - \sum_{TEU} \sum_{j=1}^{n_{TEU}} y_j^{RD+\ TEU} q_j^{RD+\ TEU} p_j^{RD+\ TEU} \\
TSW^{R-} &= -\sum_{j=1}^{n_{RS-}} y_j^{RS-} q_j^{RS-} p_j^{RS-} - \sum_{j=1}^{n_{RD-}} y_j^{RD-} q_j^{RD-} p_j^{RD-} \\
&\quad - \sum_{TEU} \sum_{j=1}^{n_{TEU}} y_j^{RD-\ TEU} q_j^{RD-\ TEU} p_j^{RD-\ TEU}
\end{aligned} \tag{19}
$$

where $TEU \in \{ESUb, ESUp, ESUn, EDUb, EDUp, EDUn\}$ denotes set of possible types of uncertain energy bids. Negative signs are needed because of the quantity convention of bids: the amount of demand bids is negative (while supply is positive).

3. Simulation Results

We evaluate the proposed method in the case of a simple, single-period market clearing scenario, where supply and demand bids are submitted to energy and positive and negative reserve markets. SRDBs are created for uncertain energy bids and the market is cleared according to the rules described in Section 2.2. We assume that no network capacities or other limitations constrain the trading (in other words we assume a one-node market). We use the reference bid set described in Appendix B.

As the positive and negative uncertainty upper bounds $\overline{u}^{+}$ and $\overline{u}^{-}$ define the set of uncertain bids (as described in Section 2.1), decreasing these parameters from a sufficiently large value (which initially implies no uncertain bids) may be viewed also as gradual introduction of the uncertain bidder pays principle to the market. In this section, for the sake of simplicity, we assume that $\overline{u}^{+} = \overline{u}^{-} = \overline{u}$ and analyze the effect of decreasing $\overline{u}$.

3.1. Social Welfare of the Sub-Markets

Figure 2 shows how the TSW values of the energy and reserve sub-markets changes as the parameter $\overline{u}$ is decreased from 30% to 1% in 1% steps. As energy bids with uncertainty values over $\overline{u}$ are considered as uncertain bids, the decrease of this parameter implies an increasing number of uncertain bids. When a bid becomes uncertain, it is submitted with the respective SRDBs and the MSCs come into play. Even if the MCPs do not change, it can happen that, thanks to the newly occurring expenses of SRDBs, a formerly accepted energy bid does not meet the MSC conditions and will be rejected.

Regarding the energy sub-market it can be said that decreasing $\overline{u}$ adds additional constraints to the optimization problem (implied by the SRDBs and corresponding MSCs), thus the decreasing TSW in the first plot of Figure 2 is perfectly plausible.

Regarding the reserve sub-markets, the increase of the TSW depicted in the second plot of Figure 2 may be explained with the increasing number of demand bids. As more and more energy bids are classified as uncertain, more and more SRDBs appear in the markets.

As reserve bids do not correspond to physical production but to the allocation of potential production, usually the bid prices are lower compared to energy bids. This is also reflected in the bid set used for the example and described in Appendix B—the prices of reserve bids are lower. Considering that the volume traded on the reserve markets is also lower, this naturally results in a lesser value of social welfare compared to the energy sub-market.

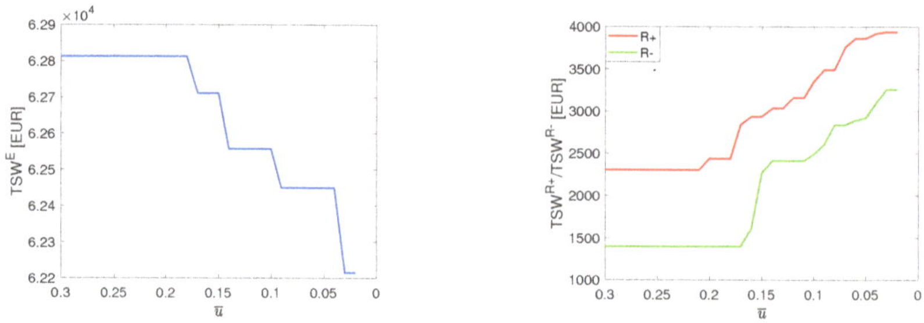

(a) (b)

Figure 2. (**a**) Total social welfare of the energy sub-market (TSW^E) as the parameter $\overline{u}$ is decreased. (**b**) Total social welfare of the reserve sub-markets as the parameter $\overline{u}$ is decreased. TSW^{R+} and TSW^{R-} denote the total social welfare of the positive and the negative reserve market respectively

3.2. Traded Volumes

The traded total volumes in the sub-markets (depicted in Figure 3) show a similar trend to TSWs: As more and more energy bids become uncertain and some of them do not meet the MSCs thus are disregarded. In contrast, with the increasing number of SRDBs, the demand ion the reserve sub-markets is increasing as the parameter $\overline{u}$ is decreased.

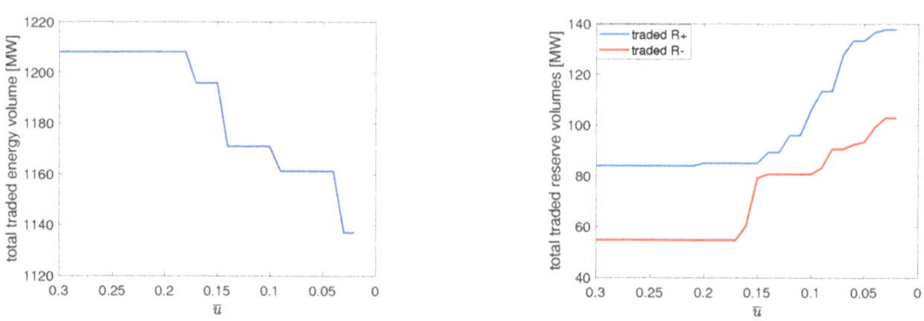

(a) (b)

Figure 3. (**a**) Total traded volume of the energy sub-market as the parameter $\overline{u}$ is decreased. (**b**) Total traded volumes of the reserve sub-markets as the parameter $\overline{u}$ is decreased.

3.3. Market Clearing Prices

Regarding the MCPs in different scenarios of the simulation, Figure 4 depicts the results.

The monotone rise in the reserve MCPs depicted in the second plot of Figure 4 may be explained by the increasing demand in the reserve sub-markets: The number of SRDBs, thus reserve demand increases with the number of uncertain energy bids. The effect of this phenomena on the MCP of energy (depicted in the first plot of Figure 4) is however twofold: Here bids become unacceptable due to increasingly occurring MSCs on both the demand and supply side. As the extent of 'lost' bids in supply/demand side potentially changes in every step, the change of MCP can not be predicted.

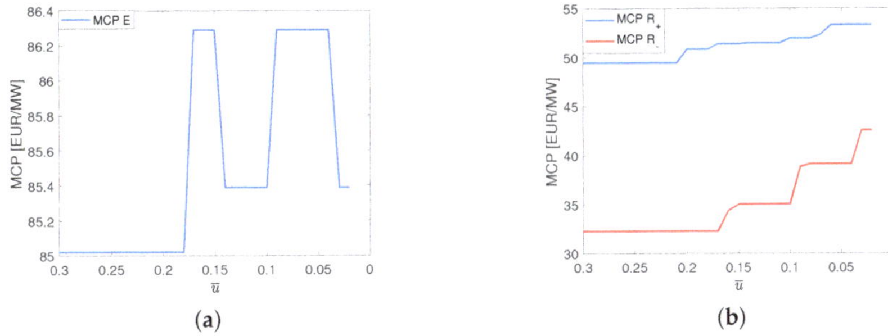

Figure 4. (**a**) Market clearing price (MCP^E) of the energy sub-market as the parameter $\overline{u}$ is decreased. (**b**) Market clearing prices of reserve sub-markets (MCP^{R+} / MCP^{R-}) as the parameter $\overline{u}$ is decreased.

3.4. Computational Properties

As discussed in References [25,26], some of the algorithms used or suggested for electricity market clearing (like EUPHEMIA [17]) contain heuristic elements. In contrast, the proposed model results in a standard MILP problem, which may be approached by any general solver.

To give an impression about the computational requirements and performance of the proposed framework, a small series of computational test were performed. The required computational time was measured as the function of the parameter $\overline{u}$ in the case of 3 different reference bid set containing various numbers of energy bids. The results regarding computational times and number of induced variables are depicted in Figure 5. It can be seen in the figure that as the uncertainty threshold $\overline{u}$ is lowered, with the increasing number of SRDBs, the computational demand shows an increasing trend. The computational demand is dominantly influenced by the number of integer variables, which also increases with the number of SRDBs.

The calculations were performed on a HP Z440 desktop computer, using the IBM CPLEX solver [27] called from MATLAB.

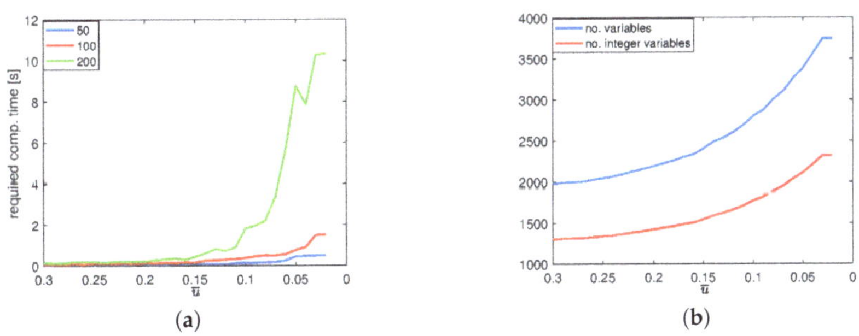

Figure 5. (**a**) Required computational time as $\overline{u}$ is decreased in the case of 50–50, 100–100 and 200–200 energy supply and demand bids (with similar uncertainty parameters as in the case of the previous example detailed in Appendix A). (**b**) Number of variables and integer variables in the 200 bids case.

4. Discussion

4.1. Additional Possible Phenomena in the Setup

Figure 6 depicts the TSW of reserve sub-markets in a similar experiment as detailed before and depicted in Figures 2 and 4: The parameter $\overline{u}$ is decreased.

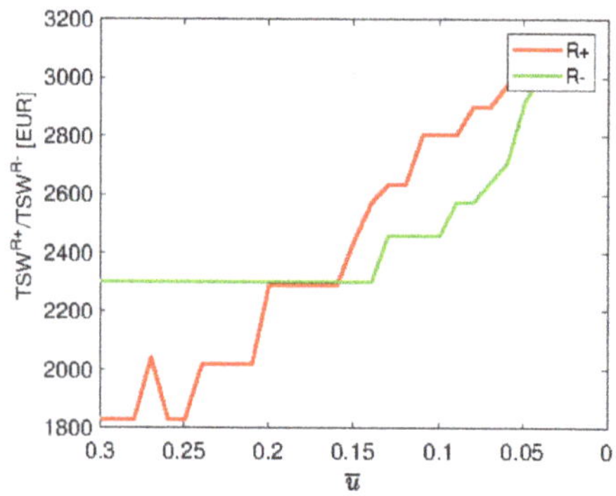

Figure 6. Total social welfare (TSW) of the reserve sub-markets as the parameter $\overline{u}$ decreases in example 2. TSW^{R+} and TSW^{R-} denote the total social welfare of the positive and the negative reserve market respectively.

In this figure however, the increase of TSW^{R+} is not monotone. The explanation for this is that if for example, an energy bid has a positive uncertainty of 27% and a negative uncertainty of 26%, the following happens. As $\overline{u}$ is decreased to 27%, the bid becomes positively uncertain. If the cost of the implied SRDB is acceptable and the MSC holds, the bid will be still accepted in the energy market and its SRDB will be also accepted, increasing the TSW of the reserve market (compared to the $\overline{u} = 28\%$ scenario). However, in the case of $\overline{u} = 26\%$, the bid becomes bi-uncertain and instead of one, two SRDBs must be paid for. In this case, it is plausible that the MSC does not hold anymore, resulting in the rejection of all three bids of the order (according to bid acceptance rules). This may be interpreted as a loss of a bid in the positive reserve sub-market, resulting in the decrease of the TSW.

4.2. Total Amount of Allocated Reserve

At a given value of $\overline{u}$, the amount of reserve resulting from the SRDBs is explicitly defined by Equation (4). On the other hand, as discussed before, several principles may be applied to determine the total amount of allocated reserves in the power system [6–9]. While the advantage of the proposed methodology is that it allocates reserves and the corresponding cost only in the case of accepted uncertain energy bids (due to MSCs) and it is flexible as the current power mix changes, there are no explicit guarantees for the amount of total allocated reserves for the whole system. The total reserve allocated by SRDBs may not meet the thumb rule that stating that the amount of allocated reserve must be at least equal the capacity of the largest unit in service (consider for example, a not-uncertain nuclear power plant and several smaller uncertain renewable sources). This approach is however is generally accepted in the context of event-driven reserves, while, as mentioned earlier, we are focussing on non event-driven resources. The main aim of the proposed approach is not to handle such large and conservative reserve needs (which most of the time may be handled by long term contracts by the system operator) but to provide a framework in which the (hour-level) actual reserve requirements and costs implied by the uncertainties of imminent power mix are automatically allocated.

On the other hand, Equation (4) may be modified by a normalization factor c as described in Equation (20) in order to tune the total amount of allocated reserves.

$$\begin{aligned} q_j^{RD+\ ESUb} &= -q_j^{ESUb} c u_k^- & q_j^{RD-\ ESUb} &= -q_j^{ESUb} c u_k^+ \\ q_j^{RD+\ ESU-} &= -q_j^{ESU-} c u_k^- & q_j^{RD-\ ESU+} &= -q_j^{ESU+} c u_k^+, \end{aligned} \quad (20)$$

If the computational capacity is sufficient, this tuning can be carried out via an outer control loop, calculating the dispatch and the total SRDB amount as a function of c. There are no guarantees that any amount of total SRDB-reserves may be allocated with this method (after increasing c above a certain level, none of the MSCs will hold, thus all uncertain bids will be rejected and no SRDB-reserve will be allocated) but this additional parameter may be a useful tool to achieve certain regulation aims.

4.3. Market Implementation

As proposed in Section 3, the values $\overline{u}-$ and $\overline{u}+$ may be decreased step-by-step from high values. During the simulation we analyzed how this decrease affects the various sub-markets assuming the same original bid set.

Regarding the possible market implementation of the method, this decrease of parameters may be carried out on for example, a monthly scale, thus progressively coupling the initially independent energy and reserve sub-markets via the SRDBs and meanwhile giving time for market participants to adapt to the changing regulation.

A further important question arises in connection with widely used special orders in electricity markets. Block orders [28] submitted to the energy sub-market and exhibiting the fill-or-kill property have not been discussed in the current paper for the aim of simplicity. There is no theoretical obstacle however to apply the fundamental principle of the approach for such orders. Considering a multi-period uncertain block order, the respective SRDBs may be defined for each affected period—in this case, the computations for each period must be carried out simultaneously and the income of the block order has to be formulated as the sum of incomes for the different periods. The MSC in this case can be included in the acceptance rules of the block order.

Minimum income condition orders [29], which are basically hourly step orders bound together by the minimum income condition, also define interdependencies between different periods. The novel income formulation of these orders proposed in Reference [21] and also used in this paper can be easily generalized to account also for the expenses of SRDBs, thus the acceptance rules of such bids may be also generalized for the proposed setup.

The proposed formulation is also compatible with general complex orders [22], orders including load gradient conditions (LGC orders) [30] and markets with so called 'PUN' orders, where the buyers pay uniform price despite multiple price zones [25].

One must keep in mind however that generalizing the proposed framework for interdependent multi-period clearing mechanisms may be computationally demanding because of the high number of integer variables originating from logical expressions. On the other hand, the MILP framework may be efficiently implemented using novel computational paradigms (e.g., Benders decomposition—see Reference [26]).

5. Conclusions and Future Work

In this article we proposed a method to implement the *uncertain bidder pays principle* in integrated portfolio-bidding electricity markets, in which the bidders of uncertain energy bids compulsorily submit predefined reserve demand bids (supplementary reserve demand bids or SRDBs) to the reserve markets to account for production/demand uncertainty. The profitability of orders corresponding to uncertain energy bids upon acceptance is ensured in the proposed framework by the minimum surplus conditions (MSC), which bounds the total surplus of the order composed by the uncertain energy bid and the connected SRDBs. As the set of uncertain bids is defined by an uncertainty threshold parameter $\overline{u}$, the proposed framework may be progressively applied, inducing increasing coupling between the initially independent energy and reserve sub-markets. The computational formulation results in a mixed integer linear programming problem (MILP).

Regarding the generalization perspectives of the framework, when we are discussing the implementation of the uncertain bidder pays principle in the context of portfolio-bidding markets, we think that the general question may be formulated as *'how the uncertainty of a unit/consumer becomes a reserve demand bid, which covers the cost of reserve allocation'*. In this paper we proposed an approach which works on the level of single bids. This approach has the benefit that the resulting SRDB (and thus the cost of the reserve) is straightforwardly assigned to a market participant—to the submitter of the uncertain energy bid. A different approach, where the uncertainty of multiple energy bids bids is represented by one reserve demand bid, would allow the more complex handling of uncertainties (e.g., with the possible application of risk measures [31]) but the problem of cost allocation would become more challenging as well. Nevertheless, the general problem of bid/order formulation from uncertain energy bids means a potential research direction for the future, which we plan to pursue based on the current results. A straightforward generalization of the concept is to formulate the approach for multi-period models with the inclusion of block orders and/or minimum income condition (MIC) orders.

Author Contributions: D.C., Á.S. and P.M.S. formulated the research problem and the proposed solution approach, D.C. and Á.S. developed the computational approach, D.C. implemented the codes, D.C. wrote the first draft of the paper, D.C., Á.S. and P.M.S. reviewed and wrote the paper.

Funding: This work has been supported by the Fund PD 123900 of the Hungarian National Research, Development and Innovation Office, by the János Bolyai Research Scholarship of the Hungarian Academy of Sciences and by the Fund KAP19-1.1-ITK of the Pázmány Péter Catholic University.

Conflicts of Interest: The authors declare no conflict of interest. The founding sponsors had no role in the design of the study; in the collection, analyses, or interpretation of data; in the writing of the manuscript, and in the decision to publish the results.

Abbreviations

The following abbreviations are used in this manuscript:

MIC	Minimum income condition
PPU	Price per unit
MCP	Market clearing price
SRDB	Supplementary reserve demand bid
MRSBPPU	PPU of the maximal reserve supply bid
MILP	Mixed integer linear problem
MIQCP	Mixed integer quadratically constrained problem
TSW	Total social welfare
MSC	Minimum surplus condition
EENS	Expected energy not supplied
LGC	Load gradient condition

The variable nomenclature is

H	Bid/schedule realization history
D	Deviation vector
u	Uncertainty indicator
$\overline{u}$	Uncertainty threshold
y	Bid acceptance indicator
q	Bid quantity
p	Bid price per unit (PPU)
I	Income
MCP	Market clearing price
S	Surplus constant
TSW	Total social welfare

The superscripts used in the variables stand for

ES	Energy supply
ED	Energy demand
ESU+	Positively uncertain energy supply bid
ESU−	Negatively uncertain energy supply bid
ESUb	Bi-uncertain energy supply bid
EDU+	Positively uncertain energy demand bid
EDU−	Negatively uncertain energy demand bid
EDUb	Bi-uncertain energy demand bid
RS+	Positive reserve supply bid
RS−	Negative reserve supply bid
RD+	Positive reserve demand bid
RD−	Negative reserve demand bid
RD+/RD− ESUb/ESU+/ESU−	Positive/negative reserve demand bid implied by bi-uncertain/positively uncertain/negatively uncertain energy supply bid
RD+/RD− EDUb/EDU+/EDU−	Positive/negative reserve demand bid implied by bi-uncertain/positively uncertain/negatively uncertain energy demand bid

Appendix A. Structure of the Variable Vector and Formulation of Logical Constraints

Let us assume that $\mathbb{R}_+$ denotes the nonnegative reals, $\mathbb{R}_{[0,1]}$ denotes the set of real numbers in $[0,1]$, while $\mathbb{B}$ denotes the set of binary numbers ($\mathbb{B} = \{0,1\}$).

We denote the numbers of the various bid types submitted to the market as summarized in Table A1.

Table A1. Number of various bid types submitted to the market.

n_{ES}	number of uncertainty-free energy supply bids
n_{ESUb}	number of bi-uncertain energy supply bids
n_{ESU+}	number of positively uncertain energy supply bids
n_{ESU-}	number of negatively uncertain energy supply bids
n_{ED}	number of uncertainty-free energy demand bids
n_{EDUb}	number of bi-uncertain energy demand bids
n_{EDU+}	number of positively uncertain energy demand bids
n_{EDU-}	number of negatively uncertain energy demand bids
n_{RS+}	number of non-SRDB positive reserve supply bids
n_{RD+}	number of non-SRDB positive reserve demand bids
n_{RS-}	number of non-SRDB negative reserve supply bids
n_{RD-}	number of non-SRDB negative reserve demand bids

The variable vector of the proposed formulation may be partitioned as

$$x = \begin{pmatrix} MCP \\ Y^{ES} \\ Y^{ED} \\ Y^{RS+} \\ Y^{RD+} \\ Y^{RS-} \\ Y^{RD-} \\ I^{ES} \\ I^{ED} \\ I^{RD+} \\ I^{RD-} \\ Z \end{pmatrix}, \tag{A1}$$

where MCP holds the market clearing prices,

$$MCP = \begin{pmatrix} MCP^{E} \\ MCP^{R+} \\ MCP^{R-} \end{pmatrix} \tag{A2}$$

$Y^{ES}, Y^{ED}, Y^{RS+}, Y^{RD+}, Y^{RS-}$ and Y^{RD-} hold the acceptance indicators,

$$Y^{ES} = \begin{pmatrix} y^{ES} \\ y^{ESUb} \\ y^{ESU+} \\ y^{ESU-} \end{pmatrix} \quad Y^{ED} = \begin{pmatrix} y^{ED} \\ y^{EDUb} \\ y^{EDU+} \\ y^{EDU-} \end{pmatrix}$$
$$Y^{RS+} = \left(y^{RS+} \right) \quad Y^{RD+} = \begin{pmatrix} y^{RD+} \\ y^{RD+\ ESUb} \\ y^{RD+\ EDUb} \\ y^{RD+\ ESU-} \\ y^{RD+\ EDU-} \end{pmatrix}$$
$$Y^{RS-} = \left(y^{RS-} \right) \quad Y^{RD-} = \begin{pmatrix} y^{RD-} \\ y^{RD-\ ESUb} \\ y^{RD-\ EDUb} \\ y^{RD-\ ESU+} \\ y^{RD-\ EDU+} \end{pmatrix} \tag{A3}$$

$Y^{ES} \in \mathbb{R}_{[0,1]}^{n_{ES}+n_{ESUb}+n_{ESU+}+n_{ESU-}},$
$Y^{ED} \in \mathbb{R}_{[0,1]}^{n_{ED}+n_{EDUb}+n_{EDU+}+n_{EDU-}},$
$Y^{RS+} \in \mathbb{R}_{[0,1]}^{n_{RS+}}, Y^{RD+} \in \mathbb{R}_{+}^{n_{RD+}+n_{ESUb}+n_{EDUb}+n_{ESU-}+n_{EDU-}},$
$Y^{RS-} \in \mathbb{R}_{[0,1]}^{n_{RS-}}, Y^{RD-} \in \mathbb{R}_{+}^{n_{RD-}+n_{ESUb}+n_{EDUb}+n_{ESU+}+n_{EDU+}}.$

The vectors I^{ES}, I^{ED}, I^{RD+} and I^{RD-} holding the incomes are composed as

$$I^{ES} = \begin{pmatrix} I^{ESUb} \\ I^{ESU+} \\ I^{ESU-} \end{pmatrix} \quad I^{ED} = \begin{pmatrix} I^{EDUb} \\ I^{EDU+} \\ I^{EDU-} \end{pmatrix}$$
$$I^{RD+} = \begin{pmatrix} I^{RD+\ ESUb} \\ I^{RD+\ EDUb} \\ I^{RD+\ ESU-} \\ I^{RD+\ EDU-} \end{pmatrix} \quad I^{RD-} = \begin{pmatrix} I^{RD-\ ESUb} \\ I^{RD-\ EDUb} \\ I^{RD-\ ESU+} \\ I^{RD-\ EDU+} \end{pmatrix} \tag{A4}$$

$I^{ESUb} \in \mathbb{R}_{+}^{n_{ESUb}}, I^{ESU+} \in \mathbb{R}_{+}^{n_{ESU+}}, I^{ESU-} \in \mathbb{R}_{+}^{n_{ESU-}},$
$I^{EDUb} \in \mathbb{R}_{+}^{n_{EDUb}}, I^{EDU+} \in \mathbb{R}_{+}^{n_{EDU+}}, I^{EDU-} \in \mathbb{R}_{+}^{n_{EDU-}},$
$I^{RD+\ ESUb} \in \mathbb{R}_{+}^{n_{ESUb}}, I^{RD+\ EDUb} \in \mathbb{R}_{+}^{n_{EDUb}},$
$I^{RD+\ ESU-} \in \mathbb{R}_{+}^{n_{ESU-}}, I^{RD+\ EDU-} \in \mathbb{R}_{+}^{n_{EDU-}},$
$I^{RD-\ ESUb} \in \mathbb{R}_{+}^{n_{ESUb}}, I^{RD-\ EDUb} \in \mathbb{R}_{+}^{n_{EDUb}},$
$I^{RD-\ ESU+} \in \mathbb{R}_{+}^{n_{ESU+}}, I^{RD-\ EDU+} \in \mathbb{R}_{+}^{n_{EDU+}}.$

The sub-vector z is a binary vector holding the auxiliary variables for logical implications, and as these binary variables are bound to acceptance variables, it is partitioned similarly to Y.

$$Z = \begin{pmatrix} Z^{ES} \\ Z^{ED} \\ Z^{RS+} \\ Z^{RD+} \\ Z^{RS-} \\ Z^{RD-} \end{pmatrix}, \tag{A5}$$

$$Z^{ES} = \begin{pmatrix} z^{ES} \\ z^{ESUb} \\ z^{ESU+} \\ z^{ESU-} \end{pmatrix} \quad Z^{ED} = \begin{pmatrix} z^{ED} \\ z^{EDUb} \\ z^{EDU+} \\ z^{EDU-} \end{pmatrix}$$
$$Z^{RS+} = \begin{pmatrix} z^{RS+} \end{pmatrix} \quad Z^{RD+} = \begin{pmatrix} z^{RD+} \\ z^{RD+\ ESUb} \\ z^{RD+\ EDUb} \\ z^{RD+\ ESU-} \\ z^{RD+\ EDU-} \end{pmatrix}$$
$$Z^{RS-} = \begin{pmatrix} z^{RS-} \end{pmatrix} \quad Z^{RD-} = \begin{pmatrix} z^{RD-} \\ z^{RD-\ ESUb} \\ z^{RD-\ EDUb} \\ z^{RD-\ ESU+} \\ z^{RD-\ EDU+} \end{pmatrix} \tag{A6}$$

To give an example how the logical implications corresponding to income formulations and bid acceptance constraints are implemented in the computational framework, let us consider the variable block $z^{ESUb} \in \mathbb{B}^{3n_{ESUb}}$ corresponds to the implications (8) and (9) describing the income of of bi-uncertain energy supply bids, and to the bid acceptance constraints of bi-uncertain energy supply bids. The logical implications using the z^{ESUb} variables are implemented as follows.

The constraints corresponding to income formulation described in Equations (8) and (9) may be written in the shorter form

$$\begin{aligned} y_j^{ESUb} > 0 \;&\rightarrow\; f_1^I \leq 0 \;\&\; f_1^I \geq 0 \\ y_j^{ESUb} < 1 \;&\rightarrow\; f_2^I \leq 0 \;\&\; f_2^I \geq 0 \end{aligned} \tag{A7}$$

where

$$f_1^I = y_j^{ESUb} q_j^{ESUb} p_j^{ESUb} + q_j^{ESUb} MCP^E - q_j^{ESUb} p_j^{ESUb} - I_j^{ESUb} \tag{A8}$$

and

$$f_2^I = y_j^{ESUb} q_j^{ESUb} p_j^{ESUb} - I_j^{ESUb} \tag{A9}$$

Bid acceptance constraints of ESUb bids may be written as

$$\begin{aligned} y_j^{ESUb} > 0 \;&\rightarrow\; f^{MSC} \leq b^{MSC} \\ y_j^{ESUb} < 1 \;&\rightarrow\; f_1^{BA} \leq b^{BA} \text{ or } f_2^{BA} \leq 0 \end{aligned} \tag{A10}$$

where, according to Equation (10) and the bid acceptance rules of ESUb bids

$$
\begin{aligned}
f^{MSC} &= p_j^{ESUb} q_j^{ESUb} y_j^{ESUb} - I_j^{ESUb} - I_j^{RD+\ ESUb} - I_j^{RD-\ ESUb} \\
b^{MSC} &= -S_j^{ESUb} \\
f_1^{BA} &= MCP^E \\
b^{BA} &= p_j^{ESUb} \\
f_2^{BA} &= y_j^{ESUb} + y_j^{RD+\ ESUb} + y_j^{RD-\ ESUb} \ .
\end{aligned}
\tag{A11}
$$

Let us note that (as every $y \in [0,1]$) $f_2^{BA} \leq 0 \Leftrightarrow y_j^{ESUb} = 0 \ \ y_l^{RD+\ ESUb} = 0 \ \ y_j^{RD-\ ESUb} = 0$. All together, the implications may be summarized and reformulated as

$$ y_j^{ESUb} \leq 0 \ \text{ and/or } \ \left(f_1^I \leq 0 \ \& - f_1^I \leq 0 \ \& \ f^{MSC} \leq b^{MSC}\right) \tag{A12} $$

$$ -y_j^{ESUb} \leq -1 \ \text{ and/or } \ \left(f_2^I \leq 0 \ \& - f_2^I \leq 0 \ \& \ (f_1^{BA} \leq b^{BA} \ \text{ or } \ f_2^{BA} \leq 0)\right) \tag{A13} $$

Formula (A12) may be implemented in the optimization framework as

$$
\begin{aligned}
& y_j^{ESUb} - z_{j1}^{ESUb} \leq 0 \\
& f_1^I - B_1^I(1 - z_{j1}^{ESUb}) \leq 0 \\
& - f_1^I - B_1^I(1 - z_{j1}^{ESUb}) \leq 0 \\
& f^{MSC} - B^{MSC}(1 - z_{j1}^{ESUb}) \leq b^{MSC}
\end{aligned}
\tag{A14}
$$

where the B-s are the so called 'big M'-s: $B_1^I = max(f_1^I), \ \ B^{MSC} = max(f^{MSC})$.

While, Formula (A13) is implemented as

$$
\begin{aligned}
& - y_j^{ESUb} - z_{j2}^{ESUb} \leq -1 \\
& f_2^I - B_2^I(1 - z_{j2}^{ESUb}) \leq 0 \\
& - f_2^I - B_2^I(1 - z_{j2}^{ESUb}) \leq 0 \\
& f_1^{BA} - z_{j3}^{ESUb} B_1^{BA} \leq b^{BA} \\
& f_2^{BA} - (1 - z_{j2}^{ESUb}) B_2^{BA} - (1 - z_{j3}^{ESUb}) B_2^{BA} \leq 0
\end{aligned}
\tag{A15}
$$

We can see that since we have an implication of the type $A \rightarrow B \ \text{ or } \ C$ a bi-uncertain energy supply bid requires 3 auxiliary binary variables. Bids, to which only simple acceptance constraints are connected like ES, ED, $RS+$ and so forth, require only 2 binary variables to formulate the two simple implications. Based on these considerations, the size of the z blocks may be easily determined. The other implications may be formulated analogously, using the appropriate variables.

In the case of SRDB-s, for example, $y_j^{RD+\ ESUb}$, (while the income-related constraints are totally analogous) we formulate the bid acceptance-related constraints as

$$
\begin{aligned}
& y_j^{RD+\ ESUb} > 0 \ \rightarrow \ f^{MSC} \leq b^{MSC} \ \& \ f_1^{BA} \leq b_1^{BA} \\
& y_j^{RD+\ ESUb} < 1 \ \rightarrow \ f_2^{BA} \leq b_2^{BA} \ \text{ or } \ f_3^{BA} \leq 0
\end{aligned}
\tag{A16}
$$

where f^{MSC} and b^{MSC} is the same as before (MSC condition for the order), and here $f_1^{BA} \leq b_1^{BA}$ corresponds to the appropriateness of MCP^{R+}: $f_1^{BA} = MCP^{R+}$, $b_1^{BA} = p_j^{RD+\ ESUb}$.

On the other hand, $f_2^{BA} = -MCP^{R+}$, $b_2^{BA} = -p_j^{RD+\ ESUb}$, and f_3^{BA} hold the acceptance indicators corresponding to the bids of the order.

Appendix B. Reference Bid Set

In this appendix, the reference bid set of the example detailed in Section 3 is described.

Table A2. Reference ES bid set: The columns correspond to the index of the bid (ID), quantity (q), bid price (p), positive and negative uncertainty (u^+, u^-), and S respectively.

ID	q [MW]	p [EUR/MW]	u^+	u^-	S [EUR]
1	32.08	54.04	0	0.02	0
2	35.76	67.03	0.16	0.14	27.06
3	72.78	109.6	0.07	0	0
4	43.2	83.5	0.01	0.17	0
5	74.77	82.05	0	0	22.3
6	75.63	92.84	0.07	0.02	0
7	76.18	91.3	0	0.01	78.34
8	28.99	109.2	0.01	0.12	17.48
9	56.92	67.69	0.03	0.09	0
10	21.34	56.35	0.05	0.5	0
11	24.86	59.68	0	0	0
12	36.13	52.4	0.01	0	16.5
13	30.41	61.69	0	0.07	6.92
14	33.66	101	0	0.18	25.98
15	46.86	71.59	0	0.17	27.26
16	45.29	82	0	0	19.06
17	55.52	71.47	0.02	0.12	0
18	75.26	103.2	0	0.42	70.78
19	51.23	55.33	0.03	0.03	34.66
20	24.94	72.62	0	0.09	22.4
21	45.05	100.7	0.02	0	19.72
22	26.92	55.15	0	0	0
23	66.69	51.58	0	0	37.82
24	21.87	102.6	0.07	0	10.32
25	62.03	108.9	0	0.04	133.3
26	43.21	65.7	0	0	0
27	67.12	68.3	0.04	0.07	0
28	57.16	86.29	0.03	0	0
29	42.01	98.65	0.11	0	0
30	70.98	106.4	0	0	98.5
31	78.39	104.2	0	0.03	0
32	22.96	82.1	0	0	0
33	48.56	89.24	0	0.32	43.88
34	54.63	102.8	0	0	0
35	69.34	94.12	0.18	0	77.9
36	37.74	96.92	0	0.16	61.82
37	47.68	90.75	0.14	0.01	76.42
38	50.2	50.35	0	0	28
39	63.17	63.71	0	0	1.22
40	76.42	110	0	0.05	81.08
41	23.56	77.62	0	0.01	9
42	34.04	82.91	0	0.09	0
43	40.81	106.1	0	0	0
44	51.88	62.6	0	0.07	0
45	74.69	93.82	0.02	0	0
46	24.02	67.05	0	0	29.88
47	55.86	85.02	0	0	20.36
48	30.22	70.81	0.08	0	0
49	74.31	90.97	0	0	0
50	75.13	72.05	0	0.17	0

Table A3. Reference ED bid set: The columns correspond to the index of the bid (ID), quantity (q), bid price (p), positive and negative uncertainty (u^+, u^-), and S respectively.

ID	q [MW]	p [EUR/MW]	u^+	u^-	S [EUR]
1	−36.32	104.4	0	0	0
2	−39.9	110.8	0.09	0.03	0
3	−41.39	75.68	0	0	0
4	−27.12	140.9	0	0	0
5	−29.76	101.8	0.14	0	0
6	−33.23	132.6	0.04	0	0
7	−41.42	96.16	0.31	0	0
8	−15.19	143.7	0	0	0
9	−27.33	113.2	0.34	0	0
10	−33.82	140	0.03	0.02	0
11	−-39.86	77.72	0	0	0
12	−10.38	134.4	0	0	0
13	−12.27	85.39	0	0	0
14	−27.64	94.7	0	0.1	0
15	−25.21	97.9	0	0.1	0
16	−44.12	79	0	0	0
17	−21.38	132.9	0.1	0	0
18	−20.38	144.6	0	0.07	0
19	−28.45	123.9	0.06	0	0
20	−24.55	76.77	0.09	0.06	0
21	−20.66	80.14	0	0.11	0
22	−14.17	114	0	0	0
23	−38.62	135.2	0.15	0	0
24	−22.05	81.56	0.13	0	0
25	−32.56	103.3	0	0.06	0
26	−25.72	88.34	0.29	0.32	0
27	−23.72	78.41	0.05	0.03	0
28	−35.95	138.4	0.03	0.04	0
29	−48.79	94.67	0	0.04	0
30	−29.24	140.4	0	0.16	0
31	−34.83	103.1	0.08	0.03	0
32	−34.81	118.1	0	0.07	0
33	−14.32	134.4	0.04	0	0
34	−13.19	98.94	0	0.06	0
35	−33.59	97.01	0	0.2	0
36	−44.54	121.2	0	0	0
37	−40.65	145	0	0.3	0
38	−39.18	107.4	0.15	0	0
39	−44.97	118.1	0	0	0
40	−49.31	142.9	0	0.06	0
41	−49.04	73.14	0	0	0
42	−24.23	136.6	0	0	0
43	−34.61	90.51	0.12	0.14	0
44	−15.27	140.7	0	0	0
45	−29.96	103.9	0	0.01	0
46	−28.88	129.7	0.04	0	0
47	−27.01	128.7	0.08	0	0
48	−47.28	114.3	0.15	0.1	0
49	−43.24	129.3	0	0	0
50	−20.66	124.2	0	0	0

Table A4. Reference (non-SRDB) RS+ bid set: The columns correspond to the index of the bid (ID), quantity (q) and bid price (p) respectively. The SRDB bids are generated according to $\overline{u}$, and the implied actual set of uncertain energy bids.

ID	q [MW]	p [EUR/MW]
1	7.11	28.34
2	4	28.38
3	19.94	51.98
4	16.2	63.76
5	16.89	67.17
6	2.64	31.28
7	15.06	64.25
8	8.73	34.73
9	12.37	26.15
10	2.78	67.01
11	9.23	60.32
12	6.36	52.39
13	18.08	53.33
14	18.64	69.82
15	4.7	26.11
16	18.41	31.66
17	3.61	57.21
18	5.15	32.18
19	10.62	67.7
20	15.34	55.26
21	3.26	38.77
22	18.87	63.57
23	18.04	51.49
24	4.92	32.92
25	15.46	67.54
26	13.85	49.47

Table A5. Reference (non-SRDB) RS- bid set: The columns correspond to the index of the bid (ID), quantity (q) and bid price (p) respectively. The SRDB bids are generated according to $\overline{u}$, and the implied actual set of uncertain energy bids.

ID	q [MW]	p [EUR/MW]
1	19.4	26.11
2	16	44.81
3	13.35	39.13
4	18.6	35.05
5	8.74	38.86
6	2.68	52.23
7	17.05	51.47
8	8.54	31.29
9	4.09	69.16
10	11.75	28.89
11	6.61	34.44
12	7.66	65.67
13	11.37	32.3
14	4.58	27.88
15	17.39	68.86
16	18.73	43.43

Table A6. Reference (non-SRDB) RD+ bid set: The columns correspond to the index of the bid (ID), quantity (q) and bid price (p) respectively. The SRDB bids are generated according to $\overline{u}$, and the implied actual set of uncertain energy bids.

ID	q [MW]	p [EUR/MW]
1	−12.76	45.55
2	−16.76	70.49
3	−12.6	54.63
4	−11.9	50.88
5	−19.93	51.42

Table A7. Reference (non-SRDB) RD- bid set: The columns correspond to the index of the bid (ID), quantity (q) and bid price (p) respectively. The SRDB bids are generated according to $\overline{u}$, and the implied actual set of uncertain energy bids.

ID	q [MW]	p [EUR/MW]
1	−18.23	56.8
2	−10.45	42.55
3	−3.49	63.49
4	−7.6	55.99
5	−5.8	35.06

References

1. Hernandez, L.; Baladron, C.; Aguiar, J.M.; Carro, B.; Sanchez-Esguevillas, A.J.; Lloret, J.; Massana, J. A survey on electric power demand forecasting: future trends in smart grids, microgrids and smart buildings. *IEEE Commun. Surv. Tutor.* **2014**, *16*, 1460–1495. [CrossRef]
2. Fallah, S.N.; Ganjkhani, M.; Shamshirband, S.; Chau, K.W. Computational Intelligence on Short-Term Load Forecasting: A Methodological Overview. *Energies* **2019**, *12*, 393. [CrossRef]
3. Raineri, R.; Rios, S.; Schiele, D. Technical and economic aspects of ancillary services markets in the electric power industry: An international comparison. *Energy Policy* **2006**, *34*, 1540–1555. [CrossRef]
4. Strbac, G.; Kirschen, D.S. Who should pay for reserve? *Electr. J.* **2000**, *13*, 32–37. [CrossRef]
5. Afshar, K.; Ehsan, M.; Fotuhi-Firuzabad, M.; Amjady, N. Cost-benefit analysis and MILP for optimal reserve capacity determination in power system. *Appl. Math. Comput.* **2008**, *196*, 752–761. [CrossRef]
6. Morales, J.M.; Conejo, A.J.; Pérez-Ruiz, J. Economic valuation of reserves in power systems with high penetration of wind power. *IEEE Trans. Power Syst.* **2009**, *24*, 900–910. [CrossRef]
7. Ortega-Vazquez, M.A.; Kirschen, D.S. Estimating the spinning reserve requirements in systems with significant wind power generation penetration. *IEEE Trans. Power Syst.* **2008**, *24*, 114–124. [CrossRef]
8. Margellos, K.; Haring, T.; Hokayem, P.; Schubiger, M.; Lygeros, J.; Andersson, G. A robust reserve scheduling technique for power systems with high wind penetration. In Proceedings of the International Conference on Probabilistic Methods Applied to Power Systems, Istanbul, Turkey, 10–14 June 2012.
9. Zugno, M.; Conejo, A.J. A robust optimization approach to energy and reserve dispatch in electricity markets. *Eur. J. Oper. Res.* **2015**, *247*, 659–671. [CrossRef]
10. Haring, T.W.; Kirschen, D.S.; Andersson, G. Efficient allocation of balancing and ramping costs. In Proceedings of the 2014 Power Systems Computation Conference, Wroclaw, Poland, 18–22 August 2014; pp. 1–9. [CrossRef]
11. Pinto, T.; Gazafroudi, A.S.; Prieto-Castrillo, F.; Santos, G.; Silva, F.; Corchado, J.M.; Vale, Z. Reserve costs allocation model for energy and reserve market simulation. In Proceedings of the 2017 19th International Conference on Intelligent System Application to Power Systems (ISAP), San Antonio, TX, USA, 17–20 September 2017; pp. 1–6.
12. Gazafroudi, A.S.; Shafie-Khah, M.; Abedi, M.; Hosseinian, S.H.; Dehkordi, G.H.R.; Goel, L.; Karimyan, P.; Prieto-Castrillo, F.; Corchado, J.M.; Catalão, J.P. A novel stochastic reserve cost allocation approach of electricity market agents in the restructured power systems. *Electr. Power Syst. Res.* **2017**, *152*, 223–236. [CrossRef]

13. Gazafroudi, A.S.; Afshar, K.; Bigdeli, N. Assessing the operating reserves and costs with considering customer choice and wind power uncertainty in pool-based power market. *Int. J. Electr. Power Energy Syst.* **2015**, *67*, 202–215. [CrossRef]
14. European Parliament. P8_TA-PROV (2019) 0227 Internal Market for Electricity I—European Parliament Legislative Resolution of 26 March 2019 on the Proposal for a Regulation of the European Parliament and of the Council on the Internal Market for Electricity (Recast) (COM(2016)0861-C8-0492/2016-2016/0379(COD)). *Off. J. Eur. Commun.* **2019**. Available online: http://www.europarl.europa.eu/doceo/document/TA-8-2019-0227_EN.pdf (accessed on 30 July 2019).
15. Hobbs, B.F.; Rothkopf, M.H.; O'Neill, R.P.; Chao, H.P. *The Next Generation of Electric Power Unit Commitment Models*; Springer Science & Business Media: Berlin, Germany, 2006; Volume 36.
16. Madani, M.; Van Vyve, M. Revisiting minimum profit conditions in uniform price day-ahead electricity auctions. *Eur. J. Oper. Res.* **2018**, *266*, 1072–1085. [CrossRef]
17. EUPHEMIA Public Description. Technical Report. 2015. Available online: http://www.belpex.be/wp-content/uploads/Euphemia-public-description-Nov-2013.pdf (accessed on 30 July 2019).
18. Arias, J.; Calle, M.; Turizo, D.; Guerrero, J.; Candelo-Becerra, J.E. Historical Load Balance in Distribution Systems Using the Branch and Bound Algorithm. *Energies* **2019**, *12*, 1219. [CrossRef]
19. Orosz, T.; Tamus, Z.Á. Impact of short-circuit impedance and tap changing method selection on the key-design parameters of core-form power transformers. *Electr. Eng.* **2018**, *100*, 1631–1637. [CrossRef]
20. Ji, Z.; Huang, X.; Xu, C.; Sun, H. Accelerated model predictive control for electric vehicle integrated microgrid energy management: A hybrid robust and stochastic approach. *Energies* **2016**, *9*, 973. [CrossRef]
21. Sleisz, Á.; Raisz, D. Efficient formulation of minimum income condition orders on the all-European power exchange. *Period. Polytech. Electr. Eng. Comput. Sci.* **2015**, *59*, 132–137. [CrossRef]
22. Sleisz, Á.; Divényi, D.; Raisz, D. New formulation of power plants' general complex orders on European electricity markets. *Electr. Power Syst. Res.* **2019**, *169*, 229–240. [CrossRef]
23. Raman, R.; Grossmann, I.E. Modelling and computational techniques for logic based integer programming. *Comput. Chem. Eng.* **1994**, *18*, 563–578. [CrossRef]
24. Madani, M. Revisiting European Day-Ahead Electricity Market Auctions: MIP Models and Algorithms. Ph.D. Thesis, Université Catholique de Louvain, Ottignies-Louvain-la-Neuve, Belgium, 2017.
25. Sleisz, Á.; Raisz, D. Integrated mathematical model for uniform purchase prices on multi-zonal power exchanges. *Electr. Power Syst. Res.* **2017**, *147*, 10–21. [CrossRef]
26. Madani, M.; Van Vyve, M. Computationally efficient MIP formulation and algorithms for European day-ahead electricity market auctions. *Eur. J. Oper. Res.* **2015**, *242*, 580–593. [CrossRef]
27. Manual, C.U. IBM ILOG CPLEX Optimization Studio. *Version* **1987**, *12*, 1987–2018.
28. Meeus, L.; Verhaegen, K.; Belmans, R. Block order restrictions in combinatorial electric energy auctions. *Eur. J. Oper. Res.* **2009**, *196*, 1202–1206. [CrossRef]
29. Sleisz, Á.; Divényi, D.; Polgári, B.; Sőrés, P.; Raisz, D. Challenges in the formulation of complex orders on European power exchanges. In Proceedings of the 2015 12th International Conference on the European Energy Market (EEM), Lisbon, Portugal, 19–22 May 2015; pp. 1–5.
30. Sleisz, Á.; Raisz, D. Complex supply orders with ramping limitations and shadow pricing on the all-European day-ahead electricity market. *Int. J. Electr. Power Energy Syst.* **2016**, *83*, 26–32. [CrossRef]
31. Artzner, P.; Delbaen, F.; Eber, J.M.; Heath, D. Coherent measures of risk. *Math. Financ.* **1999**, *9*, 203–228. [CrossRef]

© 2019 by the authors. Licensee MDPI, Basel, Switzerland. This article is an open access article distributed under the terms and conditions of the Creative Commons Attribution (CC BY) license (http://creativecommons.org/licenses/by/4.0/).

energies

MDPI

Article

Using Biofuels for Highly Renewable Electricity Systems: A Case Study of the *Jatropha curcas*

Petr Procházka [1,*], Luboš Smutka [2] and Vladimír Hönig [3,4]

1. Department of Economics, Faculty of Economics and Management, Czech University of Life Sciences Prague, Kamýcká 129, 165 00 Prague 6, Czech Republic
2. Department of Trade and Finance, Faculty of Economics and Management, Czech University of Life Sciences Prague, Kamýcká 129, 165 00 Prague 6, Czech Republic
3. Department of Chemistry, Faculty of Agrobiology, Food and Natural Resources, Czech University of Life Sciences Prague, Kamýcká 129, 165 00 Prague 6, Czech Republic
4. Department of Strategy, Faculty of Business Administration, University of Economics, Prague, W. Churchill Sq. 1938/4, 130 67 Prague 3, Czech Republic

* Correspondence: pprochazka@pef.czu.cz; Tel.: +420-224-382-131

Received: 27 June 2019; Accepted: 30 July 2019; Published: 6 August 2019

Abstract: Recent movements for the decarbonization of the electricity sector have become a priority for many countries around the world and will inevitably lead to the sharp decline of fossil-fuel-based energy. Energy from fossil fuels is to be replaced by renewable energy sources (RES), although the transition will neither be cheap nor smooth. One sustainable and environmentally friendly alternative to fossil fuels and which will take a considerable share in the increasing supply of renewable energy resources is biofuels. There are various types of biofuels used in practice; however, biodiesels represent one of the most popular and widespread ones. This paper focuses as a case study on the byproducts of *Jatropha curcas*, a crop and a plant that is already used for biofuel production and which is subsequently employed in electricity generation in *Jatropha curcas* producing regions. This paper identifies the limitations and prospects of *Jatropha curcas* utilization. Also, *Jatropha curcas* is compared to other materials suitable for biomass generation. An economic analysis for a 2 MW biofuel powerplant was conducted incorporating various market-related risks. The study shows that at current prices, net profitability can be achieved using *Jatropha curcas* byproducts for producing electricity.

Keywords: renewable energy; biofuels; electricity generation; power sector; biomass; *Jatropha curcas*

1. Introduction

On 1 January 2016, the 17 Sustainable Development Goals (SDGs) of the 2030 Agenda—adopted by world leaders in September 2015 at an historic United Nations (UN) summit—officially came into force. Over the next fifteen years, with these new Goals that universally apply to all, countries mobilized efforts to end all forms of poverty, fight inequalities, and tackle climate change, while ensuring that no one is left behind [1].

Electrification of rural areas in developing countries is considered fundamental for reducing energy poverty and meeting the SDGs. Provision of electricity to rural areas through national grids is costly per unit of electricity because rural consumers are more scattered and typically buy less electricity per consumer compared to urban consumers. Rural households are assumed to consume at least 250 kWh per year and urban households 500 kWh per year.

Instead of bringing the national grid to rural consumers, community scale electricity production units may be a more realistic solution for supplying electricity at a reasonable cost per kWh, and biomass-based electricity generation is deemed to have potential [2,3].

The SDGs build on strategies that create economic growth and addresses a range of social needs including education, health, social protection, and job opportunities, while tackling climate change and environmental protection [1]. This paper analyzed the use of biofuels in systems and aimed mainly on six goals of the SDGs, depicted in Figure 1. These goals are interconnected to each other, because one supports the rest.

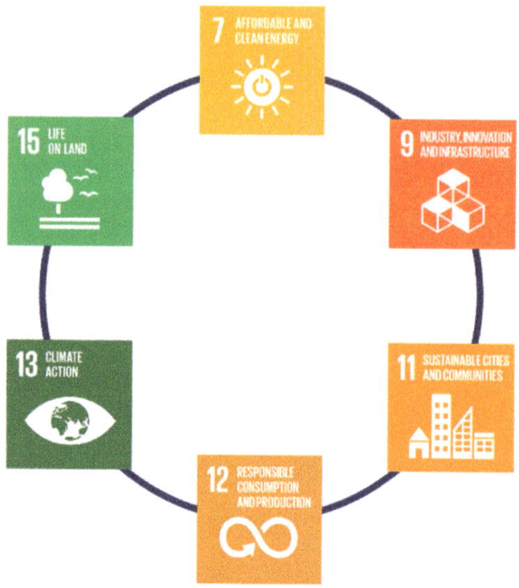

Figure 1. Contribution of this article to Sustainable Development Goals [1]. Source: Own results based on Reference [1].

Mitigating global climate change requires decarbonizing the electricity sector as it is a major source of global greenhouse gas (GHG) emissions. With roughly half GHG emissions coming from coal-fired power plants, electricity from natural gas presents another alternative—a lower carbon technology. The optimal strategy for picking ideal technology will depend on the ultimate costs of each technology as well as the social costs from GHG emissions. These analyses are increasingly discussed in many countries around the world [4]. Based on The National Climate Assessment, in just the US, due to the change in climate, hundreds of billions of dollars are lost, mainly because of heat-related deaths, coastal property losses, and lost wages in outdoor industries due to the presence of heat waves [5,6]. Transition to lower carbon emission technologies will neither be cheap nor smooth; however, the status quo is still fairly expensive, even though the cost is not directly visible.

The introduction of alternative fuels is a logical step, which is continuously being done around the world. The first generation of alternative fuels introduced in large numbers comprised oilseed, sugarcane, and other oil containing food and animal feed crops. First generation bioethanol is mainly produced from sugar containing plants or cereal (grain) crops. Vegetable oils are also used after a range of conversion to fatty acid methyl or ethyl esters. Even though, second, third, and fourth generation are currently under research both by commercial and scientific circles, first generation is still the main representative of alternative fuels [7].

This paper focused on the use of *Jatropha curcas* as a potentially useful source of renewable energy, which was discussed in October 2008 in the European Parliament. The proposal was that one-fifth of energy should come from non-food-related alternatives. Because this crop does not compete with food production, it can become a choice for assessment by international investors and biodiesel processors, energy producers and international institutions. In addition, it would be appropriate to support oilseeds which do not compete with food crops with certain subsidy incentives for areas, particularly in developing countries, because of employment policy.

The paper is organized as follows: Section 2 provides a short literature review. Section 3 focuses on the materials and methods used in our study. Section 4 presents the results and discussions. Finally, Section 5 concludes by summarizing the main findings and implications.

2. Literature Review

The genus *Jatropha curcas*, belonging to the family Euphorbiaceae, contains 322 genera and 8910 species ranging from large woody trees to simple weeds. The site of origin is tropical Central America, from where they spread to many tropical and subtropical areas, including India, Africa, and North America. The plants are monoecious and contain yellow to red latex. The lists are alternate, simple or palpated. The flowers are single-sex, the fruit is fleshy [8].

The most significant of the species is *Jatropha curcas* (hereto referred to as *Jatropha curcas*), which is widely grown in the tropics. It is a monoecious shrub or low tree, which grows up to 5 m. Its smooth shiny bark, which can appear greenish brown or yellow, has a paper-like look and tends to easily peel. The watery pink latex can be pulled out mechanically and, after use, the color turns brown. The branches are ascending, coarse, and glabrous [8].

In addition to the production of biofuels, oil from the fruits of this plant is used, including in the production of candles, soaps, hair conditioners, and lamp oils. *Jatropha curcas gossypifolia* and *Jatropha curcas multifida* also have the same area of origin. The seeds of these two species contain oil which can be used in a similar way. *Jatropha curcas multifida* is also popular as an ornamental and decorative plant, similar to *Jatropha curcas podarica* [9,10].

Of course, as a plant spread in the tropical and subtropical regions of most continents, *Jatropha curcas* has several distinct local names in each of these regions.

Jatropha curcas has high ecological adaptability. As a succulent plant that excretes water through its leaves during the dry season and also because it is deeply rooted, it is very well adapted to grow in semi-arid conditions. With increasing humidity, plant production also increases. Despite the fact that this plant survives the average annual rainfall of 250–300 mm very well, it needs an average annual rainfall of at least 600 mm to produce flowers and then fruit. Optimal precipitation for seed production is 1000–1500 mm. Higher precipitation can then cause fungal and fungal attack on the root system [11–16].

Jatropha curcas can withstand extremely high temperatures very well, which, however, affects yield, while even a small frost can permanently damage the plant. Enzymatic-catalytic chemical reactions of plants that affect biomass production are controlled by van't Hoff effect (increasing the temperature by 10 °C leads to an increase in the rate of reaction two to three times). The optimum temperature for its growth ranges from 20 °C to 28 °C. It can be grown at low altitudes. *Jatropha curcas* is not sensitive to daytime. It is very suitable for growing in areas of high light intensity, but it is not suitable for growing in shady places. Flower formation is not dependent on latitude, and the plant can bloom at any time of year.

The properties of the oil, especially its quality and density, are important for biodiesel production. Physical properties include density, viscosity, and low temperature properties. Density is important in relation to the calorific value of the fuel. Viscosity characterizes the degree of fluidity and, with a high viscosity, the fuel system is more stressed, or can cause that the fuel could not be pumped. Low temperature properties are important to ensure operation in winter. Of the chemical parameters, the acidity number in relation to the corrosive environment and especially the presence of unsaturated fatty acids is important. Multiple bonds in fatty acids are more susceptible to oxidation processes.

The physical and chemical properties of *Jatropha curcas* are quite variable. Characteristic properties are greatly influenced by the environment and genetic interaction, such as size, weight, and oilseed. The quality of the oil, especially the content of fatty acids, is further influenced by the maturity of the fruits during harvesting, the method of processing and storage. In general, it is necessary to ensure low oil contamination, low acid value, low phosphorus, water, and ash particles and to increase its oxidation stability. The crude oil is relatively viscous. It is characterized by low fatty acid content, which improves storage conditions. When stored, a higher content of linoleic acid may pose a problem, which may cause the oil to be more susceptible to oxidation. The high cetane number guarantees a low flash point. The oil contains a small amount of sulfur, resulting in lower SO_2 emissions during combustion [17–21].

It is possible to obtain an oil yield higher than 1500 kg per hectare of this plant. Yield starts from 18 months, but an economic yield is obtained from the third year after planting. Based on experimental field conditions, the average seed yield with existing varieties under irrigation conditions after 3 years is estimated to be 4 to 5 tons per hectare (4 to 6 kg/plant/year). However, farmers also recorded a yield of 6 tons of seed per hectare. The most limiting factor of *Jatropha curcas* besides water is the relatively high laboriousness. Therefore, large-scale *Jatropha curcas* plantations are not economically feasible today (under the current conditions) and it can be replaced only partially by mechanization. Three common planting densities can be identified: 1111, 1666 and 2500 trees per hectare equal to a squared spacing of 3.0 by 3.0, 2.4 by 2.4, and 2.0 by 2.0 meters per tree [22].

Jatropha curcas oil is adequate to be used as a raw material in biodiesel production which meets American and European standards [23,24]. Additionally, the press cake can be used as a fertilizer and the organic waste products can be digested to produce biogas (CH_4). Average oil content of dry seed on a mass basis is 34% [25]. Full composition is shown is in Table 1.

Table 1. Proximate and ultimate analysis, and higher heating values of the raw materials [21].

Sample	Moisture (%)	Proximate Composition (% Weight, Dry Basis)			Ultimate Composition (% Weight, Dry Basis)				
		Volatile matter	Fixed Carbon	Ash	C	H	N	S	O
Jatropha curcas seed cake	4.08	73.7	19.06	7.24	52.12	6.91	5.01	0.7	28
Seed cake biochar	9.28	40.02	46.62	13.36	61.31	3.55	3.77	0.38	17.6
Jatropha curcas Shell	10.57	71.52	17.64	10.84	40.8	5.9	1.53	0.43	40.5

Jatropha curcas is a promising plant for both bio-energy supply and socio-economic development in developing countries [26,27]. Full usage of *Jatropha curcas* is depicted in Figure 2.

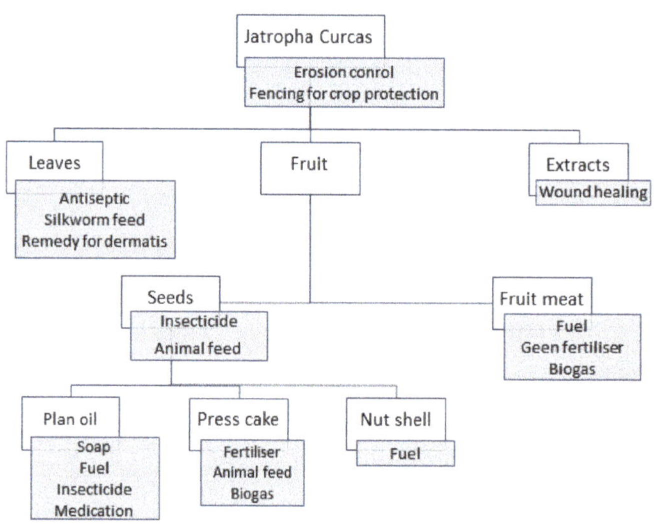

Figure 2. Usage of *Jatropha curcas*. Source: Own results.

For comparison of the biogas yield from *Jatropha curcas* cake with other raw materials which are used frequently for commercial biogas production see Table 2, which shows some examples of agricultural, industrial, and municipal residues as well as energy plants with properties of *Jatropha curcas*.

Table 2. Comparison of *Jatropha curcas* properties with other raw materials [28,29].

Raw Material	Yield [$m^3 \cdot t^{-1}$]
Jatropha curcas cake	360
Jatropha curcas mesocarp	200
Jatropha curcas cake and mesocarp mix	310
Fat (deep-fry)	960
Glycerin	930
Wheat	660
Rapeseed residues	640
Hay	450
Straw (barley)	345
Molasses	340
Fruit pomace	280
Grass cuttings	220
Maize silage	200
Organic municipal waste	100
Kitchen residues	80
Starch process water	65
Manure (poultry)	62
Potato mash	40
Manure (cattle)	22

Source: Own results.

The biogas yield of 360 m^3 of biogas per ton of *Jatropha curcas* is much higher than comparable energy plants, castor (*Ricinus communis*) and Camelina (*Camelina sativa*) [29].

The cake is excellently degradable. The cake residue has a high heating value between 16–17 MJ/kg; however, incineration would destroy the nutrients, which are needed on the ropa plantations to grow the trees. Therefore, *Jatropha curcas* is more suitable for biogas production than burning in steam powerplant [30–34]. Since the *Jatropha curcas* cake is excellently degradable, biogas production would be currently the best alternative. *Jatropha curcas* cake has a volume of 360 $m^3 \cdot t^{-1}$. Mesocarp, as opposed to the cake, is not well degradable. However, mesocarp could have significant value for biogas production and increase biogas yields in a powerplant Mesocarp yield is roughly 200 $m^3 \cdot t^{-1}$ [28]. When both *Jatropha curcas* cake and adequate amount of *Jatropha curcas* mesocarp (adequate portion from collected fresh fruit bunches) are processed into biogas, it would be approximately 310 $m^3 \cdot t^{-1}$ [35].

A significant number of articles about *Jatropha curcas* biodiesel are focused on fuel properties [36–41], different approaches and processes used to produce biodiesel from *Jatropha curcas* [29,42,43], comparative studies on fuel properties of *Jatropha curcas* with other biodiesels [44–47] or evaluation of its impact on countries [24,48–54]. As many studies were conducted that analyzed utilization of this plant for biodiesel, no comprehensive study is known to the authors that deals with the use of *Jatropha curcas* byproducts for a biogas power plant.

Jatropha curcas has the great advantage of being grown on agriculturally unsuitable soils, causing it to produce low or no carbon deficiencies in these soils offering immediate and lasting benefits. It is true that the yield of these plants is higher than that of *Jatropha curcas* On the other hand, the price of seeds can be four to five times higher than that of *Jatropha curcas*. Another disadvantage of these plants is the high demand for them in the food industry [55–57].

Biogas produced from *Jatropha curcas* can be further processed or used as is. Biogas as a mixture consisting mainly from methane and carbon dioxide can be purified in an absorption column, where the CO_2 is removed. The choice of a specific gas sweetening process depends mostly on the material itself, the required selectivity, costs, environmental requirements, and the final product. Usually amine compounds, physical solvents or hybrid (mixed-solvent systems that contain both amine and a physical solvent) are used [58]. This pure natural gas can be simply put into gas pipeline and distributed or liquified and either transported via LNG (liquid natural gas) tanker or used as high energy-density fuel for trucks or buses [59].

Another application, which will be also discussed in this paper, is burning biogas in a combustion chamber to produce stream and generate electricity by steam turbine. This application is an easier application of biogas than previous ones, because there is not any separation of gases. Detailed theoretical technology is discussed for example in Reference [60].

3. Materials and Methods

The process of making biofuels from *Jatropha curcas* was carried out as follows: after extraction of oil, *Jatropha curcas* cake and mesocarp were obtained. The first step of biogas production was pretreatment. Generally, all processes of this kind start with pretreatment step, where pH is regulated for optimal fermentation usually by adding acid or base. The bacteria of the individual process steps have different pH values at which they can grow optimally. The optimum pH for hydrolyzing and acid-forming bacteria is 4.5–6.3. Methane-forming bacteria need a pH in the neutral range of 6.8–7.5. If the fermentation process is in one tank—bioreactor—the pH must be properly maintained. The pH is usually adjusted spontaneously within the system via the alkaline and acidic metabolic products formed during anaerobic decomposition. Normally, the pH released by the carbon dioxide in the neutral range is balanced. If a drop in pH is observed, substrate delivery must be immediately reduced or stopped to give bacteria time to break down the acids present.

The fastest fermentation rate was observed when oil cake was incubated in 0.1 mol·L^{-1} NaCl at 20 °C. Biogas production rate went up to 0.0015 m^3/kg per day [61]. Temperature is important to monitor, because it affects all biochemical processes. As the temperature rises, the speed of all processes in the reaction increases, and another undesirable reaction may occur. Also, by changing the temperature and, hence, the speed of the processes, the dynamic equilibrium of the process is disrupted. It is, therefore, necessary to maintain a constant temperature for a stable course of anaerobic decomposition. Commonly, there are three typical temperature ranges that suit individual bacteria: psychrophilic (below 20 °C), mesophilic (25–40 °C) and thermophilic (above 45 °C).

After this pretreatment, the biogas production rate increased 5 times between the 1st and 5th days of fermentation when compared to fermentation without pretreatment. After this period, a sharp decrease in the biogas production rate was observed. The next step was fermentation. This process is usually done by species either from bacteria (*Bacillus licheniformis*, *Bacillus subtilis* or *Lactobacillus*) or microorganisms like *Filamentous fungi* [62,63]. The reactor was designed as a continuous stirred-tank reactor (CSTR), so the goal was to keep the highest production rate during the whole process. A CSTR was picked because it is easy to manage optimal temperature, it has low operational costs, and it always produces the same resulting product. On the other hand, the batch reactor had differences in product quality and the temperature could not be simply managed. Plug flow reactor (PFR) is by far the most expensive type of reactor and its strengths would not have been much used in this experiment.

A culture medium was continuously fed into the bioreactor to maintain the steady state. The bioreactor continuously leaves biogas and effluent with *Jatropha curcas* residue.

To ensure a stable and steady combustion of biogas, a compressor was put before the combustion chamber. The compressor kept the volume of biogas at a defined level so that process was safe.

A simplified process schema is depicted in Figure 2, below. After pre-treatment of the *Jatropha curcas* mix, the treated material enters the bioreactor where biogas is produced. Biogas is compressed and burnt in the combustion steam chamber to produce steam that enters a steam turbine which produces electricity.

An economic evaluation for biogas power plant fueled by the *Jatropha curcas* plant was done for a 2 MW biogas power plant. In economic evaluation, a mix of *Jatropha curcas* cake and mesocarp was considered, which means that 310 m^3 t^{-1} was taken as an average biogas yield. This and other main parameters for a 2 MW power plant are presented in the Table 3.

Table 3. Technical data for a *Jatropha curcas* power plant.

Raw Material	*Jatropha curcas* Cake/*Jatropha curcas* Mezocarp
Dry substance (%)	35
Biogas yield (m^3)	310
Energy yield (kWh t^{-1})	1695
Available raw material (t $year^{-1}$)	26,000
Operation period (d $year^{-1}$)	365
Net operation time (hr $year^{-1}$)	7884
Average daily operation time (hr.d^{-1})	21.6
Fertilizer production (t $year^{-1}$)	18,200
Annual biogas yield (m^3 $year^{-1}$)	8,060,000
Annual energy yield (MWh $year^{-1}$)	44.064
Total electrical capacity (MW)	2
Annual electricity production (MWh $year^{-1}$)	16,366
Plant electricity consumption (%)	5
Net electricity production (Q) (MWh $year^{-1}$)	15,519

Source: Own results.

Dry substance was considered 35%, which is the commonly required level. It was assumed that there would be technological breaks of 2.4 h every day in the process, so the operation period was 7884 h per annum. The analysis of the 2 MW biogas power plant determined it would produce fertilizer in a volume of approximately 18,200 tons if considering a 26,000 tons of input material and 30% decrease of volume after the biogas production process.

These technological parameters served as a basis for the economic evaluation of the proposed investment project. The profitability of the project depends on revenues and costs. Project revenues (TR) are calculated as a product of price (P) and quantity (Q).

$$TR = P \times Q \tag{1}$$

Quantity is deterministic and depends on the technological process. Price of electricity is considered to be fixed.

Costs for the biogas electric power plant consist of two parts: fixed cost (*FC*) and variable cost (*VC*).

$$TC = FC + VC \tag{2}$$

Fixed cost consists primarily of the cost for a power plant. The authors requested offers from various suppliers and calculated an average price for a power plant to be approximately $3 million. Another type of cost important for the analysis was annual operational costs that were mostly part of the variable costs. Operation can be broken down into the following items: personnel (operation, supervision), consumables (gasoline), maintenance (electro-mechanical, biological), insurance, and administration. Depreciation was set to 20 years [64,65]. Emission allowances for CO_2 were not considered in this analysis.

Another important part of the economic analysis is how much of the material input is needed. In order to fulfil needs for 2 MW a power plant, at least 26,000 tons of feedstock is needed. It is assumed that the material for biogas production is bought from a processor at 7% of the price of *Jatropha curcas* oil. *Jatropha curcas* oil is strongly correlated with palm oil price which is indexed at main commodity exchanges in Southeast Asia [43].

It is crucial to determine future movement in price of palm oil as it is a basis for *Jatropha curcas* oil price. For price prediction, ARIMA (AutoRegressive Integrated Moving Average) and Augmented Dickey Fuller test (ADF) were used [66,67].

ARIMA models are used for time-series forecasting. The aim of the ARIMA models is to describe the autocorrelations in the data. The ARIMA models, also called Box–Jenkins models, are models that may possibly include autoregressive terms (AR), moving average terms (MA), and differencing operations.

One of the most important models in econometrics is the random walk, which is basically an AR(1) process.

$$y_t = y_{t-1} + u_t \tag{3}$$

Equation (3) is the driftless random walk. If a constant is included, it becomes the random walk with drift. To determine if an AR(p) process is stationary involves examining the roots of its characteristic equation. Given the following AR(p) model, it can be said to be stationary if when written in the lag operator notation, the $\varphi(L)^{-1}$ converge to zero:

$$y_t = y_{t-1} + u_t \tag{4}$$

$$y_t = \varphi(L)^{-1}\, u_t \tag{5}$$

If this is the case, the autocorrelations decline to zero as the lag length is increased. For an AR(p) process to be stationary, the roots from the characteristic equation:

$$1 - \varphi_1 z - \varphi_2 z^2 - \ldots - \varphi_p z^p = 0 \tag{6}$$

all need to lie outside the unit circle, i.e., are greater than 1. The random walk is an example of a non-stationary process, as its roots lie on the unit circle not outside:

$$z = 1$$

$$\begin{aligned} y_t &= y_{t-1} + u_t \\ y_t &= L y_{t-1} + u_t \\ y_t(1 - L) &= + u_t \\ 1 - z &= 0, \end{aligned} \tag{7}$$

where $(1 - z)$ is the characteristic equation and the root (z) lies on the unit circle. The same principle applies to higher orders too:

$$z = 2$$

$$\begin{aligned} y_t &= y_{t-1} - 0.25\, y_{t-2} + u_t \\ y_t &= L y_t - 0.25 L^2 y_t + u_t \\ (1 - L + 0.25L^2) y_t &= u_t \\ 1 - L + 0.25L^2 &= 0 \\ (1 - 0.5z)(1 - 0.5z) &= 0 \end{aligned} \tag{8}$$

In the above example, both roots lie outside the unit circle, so the AR(2) process is stationary. The same applies for higher orders of lags too, although it becomes increasingly difficult to factorize these. Further characteristics of an AR(p) process are that the mean and variance of an AR(1) process are:

$$E(y_t) = \frac{\mu}{1 - \varphi_1}, var(y_t) = \frac{\sigma^2}{\left(1 - \varphi_1^2\right)} \tag{9}$$

The Augmented Dickey Fuller test (ADF) is test for a unit root in a time-series sample. This test is for a larger and more complicated set of time-series models and a statistic value of test is a negative number. The more negative it is, the stronger the rejection of the hypothesis that there is a unit root at some level of confidence. The null hypothesis is that the variable contains a unit root, and the alternative is that the variable was generated by a stationary process 28.

The extended regression used in the ADF test can be expressed in its most general form as:

$$\Delta Y_t = \mu + \gamma Y_{t-1} + \sum_{j=1}^{p} \alpha_j \Delta Y_{t-j} + \beta_t + \omega_t \quad (10)$$

where μ is the drift term, t denotes the time trend, and p is the largest lag length used.

One of the most important economic factors that affect profitability and therefore feasibility of using *Jatropha curcas* byproducts is the price of this crop on the market. As *Jatropha curcas* is not a commodity, it is necessary to derive its price based upon other oil producing crop in this region, such as palm oil. Palm oil price evolution given by index Mundi [68] is depicted in the Figure 3.

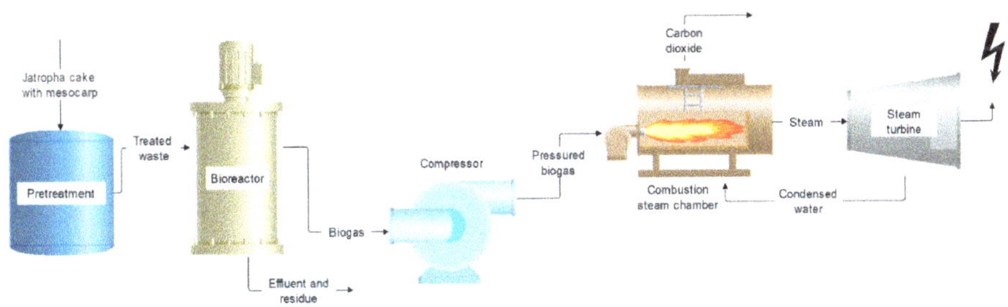

Figure 3. *Jatropha curcas* biogas power plant. Source: Own results.

Data in Figure 4 show that while in between 2006 and 2008, there has been a growing trend, in 2008 the price fell to a minimum level of approximately $450 per ton. After 2008, prices recovered to an all-time high of approximately $1250 in 2011. After that the price has been falling. In January 2017, the price reached a local maximum at $825 per ton and since then, the prices are going down to a level of $600 and below per ton.

Based on the numbers in the analysis, the following indicators of profitability are calculated: payback period, net present value, and break-even point.

Payback period (PP) shows how long it takes to recover the initial investment (C_0) through annualized cash flow (CF). It is calculated as follows.

$$PP = \frac{C_0}{CF} \quad (11)$$

Net present value is the difference between the present value of cash inflows and the present value of cash outflows. Net present value is used in capital budgeting to analyze the profitability of an investment or project, where r stands for a discount rate.

$$NPV = -C_0 + \frac{CF_1}{(1+r)^1} + \frac{CF_2}{(1+r)^2} + \ldots + \frac{CF_t}{(1+r)^t} = \sum_{t=0}^{n} \frac{CF_t}{(1+r)^t} \quad (12)$$

Net present value is calculated for the period of t = 20 years as it represents best the physical depreciation of the power plant.

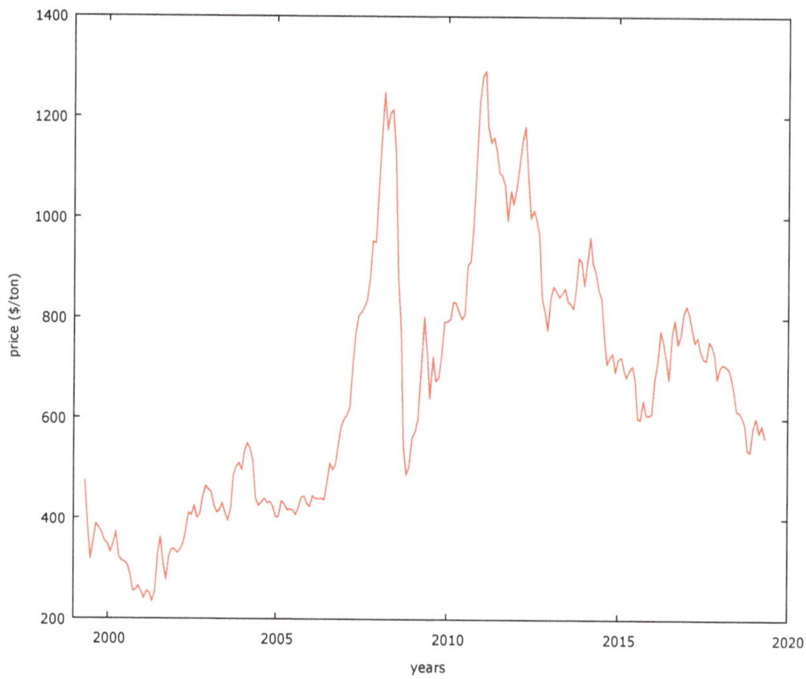

Figure 4. Evolution of *Jatropha curcas* price since 1999. Source: Own results.

Break-even point (*BEP*) is the point at which cost and revenue are equal. It means that there is no net loss or gain, and one has "broken even".

$$BEP = \frac{FC}{P - VC} \tag{13}$$

where *P* stands for output price (electricity), *FC* are fixed costs, *VC* are variable costs.

In order to conduct the profitability calculations, it is necessary to define the economic variables. They are given in the Table 4 below.

Table 4. Economic data for *Jatropha curcas* power plant.

Items	Value
Fixed cost ($)	3 million
Operational cost ($p.a.)	152,000
Input price ($/ton)	39.76
Output price ($/kWh)	0.11
Discount rate (%)	0.5

Source: Own results.

Fixed cost was simplified to only the purchasing price of the biogas powerplant. Input price was calculated as 7% fraction of the predicted *Jatropha curcas* price. Output price was given as fixed, based on the currently prevailing prices in *Jatropha curcas* producing regions. Discount rate was set to 8% on average, based on risks given for similar projects in the area [69,70]. In the calculations of NPV, the discount rate was simulated in the range of 1–16%. Cost structure is represented in Table 5 below.

Table 5. Cost structure for a *Jatropha curcas* power plant.

Items	Value
Jatropha curcas cake and mesocarp (%)	87
Operational cost (%)	13

Source: Own results.

The cost structure shows that the material costs for *Jatropha curcas* cake and mesocarp are the most important ones with nearly nine-tenth of total costs.

4. Results and Discussion

In order to proceed to an estimation of an ARIMA model, it is essential to test the variable for the existence of a unit root. The most common test is the Augmented Dickey Fuller test as introduced in the methodological section. For data $N = 232$, it includes 2 lags of $(1 - L)$ price (max was 12, criterion modified AIC) with unit–root null hypothesis: a = 1, two tests were conducted: test with constant and no trend and test with constant and trend.

The first test takes the form of the model $(1 - L)y = \mathrm{b}_0 + (\mathrm{a} - 1)y(-1) + \ldots + \mathrm{e}$.

1st order autocorrelation coefficient for e = 0.011, lagged differences: F(2, 113) = 14.796 (0.0000), estimated value of (a – 1) = 0.0518326, tau_c(1) = −2.39326 and asymptotic *p*-value = 0.1436.

The second test takes the form of model $(1 - L)y = \mathrm{b}_0 + \mathrm{b}_1 t + (\mathrm{a} - 1)y(-1) + \ldots + \mathrm{e}$.

1st order autocorrelation coefficient for e = −0.001, lagged differences: F(8, 100) = 4.779 (0.0001), estimated value of (a − 1): −0.0492756, tau_ct(1) = −1.64713 and asymptotic *p*-value = 0.7743.

The results showed that a null hypothesis of the unit–root was rejected. Therefore, it was not necessary to difference the time-series for the ARIMA model. Using the autocorrelation function and partial autocorrelation function, the authors identified the ARIMA model to be ARIMA(1,0,1). Definitions of the roots lead to estimation of an ARMA(1,1) model.

The results in Table 6 show that both roots were significant. Hence, this test can be used for the prediction of the future price of palm oil, which will provide guidance for *Jatropha curcas* oil prices. This prediction is graphically given in Figure 5 for the next two years.

Table 6. Results of ARMA model.

	Coefficient	SD	*z*	*p*-Value
const	627.468	115.799	5.419	6.01×10^{-8}
phi_1	0.971330	0.0138850	69.96	0.0000
theta_1	0.360203	0.0516843	6.969	3.19×10^{-12}

Source: Own results.

The price ranges from $552 per ton to $584 per ton. The results of the prediction for the next 24 months show that the palm oil price has probably reached a bottom at $552 per ton and a slight growth can be expected in the coming months. Given the uncertainty of the prediction (green shaded 95% interval), a middle value of $568 was then used for the economic analysis of *Jatropha curcas* biogas powerplant profitability. When using the predicted price one must be cautious, as it is important to take into account other fundamental factors which may influence the price, such as prices of substitutes, etc.

Using all the aforementioned analytical results, the authors calculated the payback period, net present value, and break-even point.

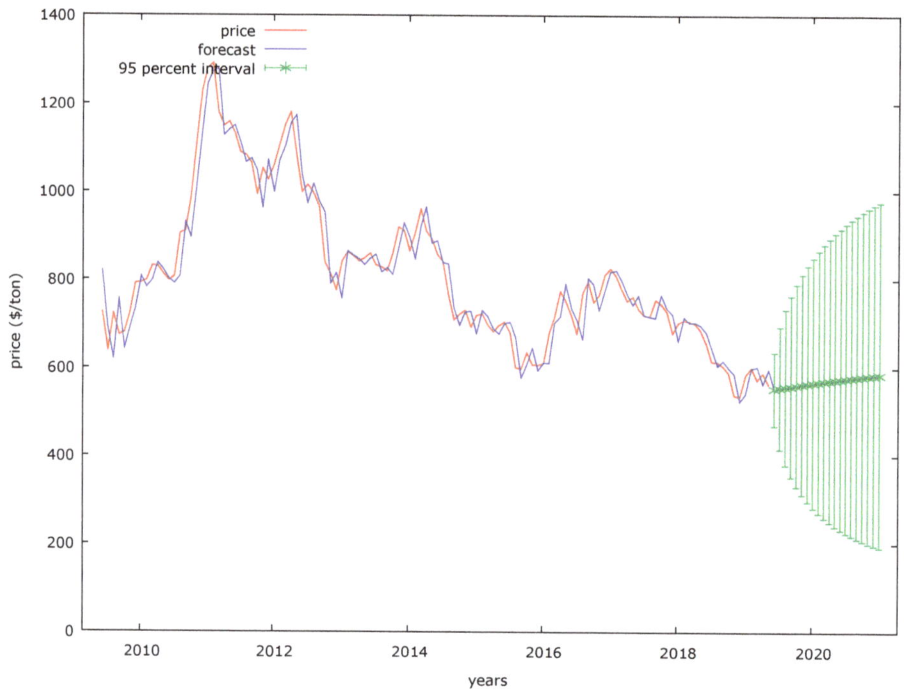

Figure 5. Prediction of future price for palm oil. Source: Own results.

Payback period in Table 7 shows a relatively good outcome. Results could be even better if we assumed rising price of electricity. It is, however, important to be cautious as some of the assumptions are relatively strong. For example, it is assumed that CO_2 allowances will not be paid for by the power plant.

Table 7. Results of the economic analysis of biogas generation.

Model Variable	Result
Payback period (years)	7.91
Break event point, volume (MWh)	123,358

Source: Own results.

Finally, net present value was also calculated for different discount rates. The result of this analysis is presented in Figure 6. The shaded area in the diagram contains 80% confidence interval.

The results show that net present value is mostly positive for discount rates ranging from 1 to 11%. If the prevailing discount rate 8% is taken in account, NPV is positive in the entire interval of prediction. It is obvious that the price of *Jatropha curcas* or palm oil has a large impact on NPV.

The results presented in this paper are not discussed by many authors, as many studies focus on the use of *Jatropha curcas* seed and oil primarily for biodiesel production [51,71,72]. For example, Verma and Gaur [73] estimate the costs of production in the similar range.

A study by Bouffaron et al. [74] described the chemical process and economic model, but with uncertain conclusions for the viability of the project.

Other authors discuss energy investments in Indonesia. For example, according to [75], investment into diesel replacement with palm oil showed positive net present values and quicker payback period thanthe project analyzed in this paper. For China, Deng et al. [76] calculated the cost efficiency of

Jatropha curcas biodiesel with positive results albeit with a smaller discount rate. Use of biofuels was also discussed for Germany and the United States [77–79].

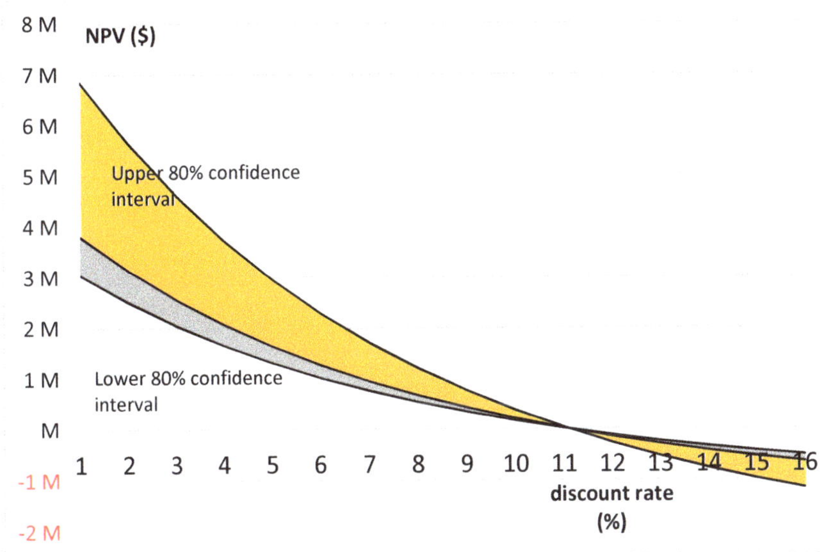

Figure 6. Net present value calculations. Source: Own results.

5. Conclusions

Our results provide a unique analysis of how to utilize byproducts from *Jatropha curcas* production in the form of cake and mesocarp. We propose utilization of these byproducts for electric power plants making it an excellent source of cheap electricity and an essential element in high renewables electricity systems. We quantified all the cost and benefit components of the project and set important parameters for their calculation including a floating discount rate. Viability of this venue was calculated. All profitability indicators used in the analysis showed very promising results. *Jatropha curcas*, based on this analysis, may be a very successful crop for biogas production. It would be, however, important to also evaluate other impacts of this project in terms of its social and environmental impact by incorporation of, for example, emission allowances.

The analysis presented in the paper shows that using byproducts of *Jatropha curcas* may be profitable. For some of the processors this can be a method of diversifying their businesses to use the byproducts of *Jatropha curcas* seed production. In terms of circular economy, the proposed use of *Jatropha curcas* mesocarp and cake in electricity production is also beneficial as it provides great benefits and contributes to a lower waste burden. All of these features make *Jatropha curcas* a novel and interesting element in a highly renewable electricity system.

Author Contributions: P.P., L.S. and V.H. designed methodology; P.P., L.S. and V.H. performed the calculation and analyzed the data; P.P., L.S. and V.H. wrote the paper.

Funding: This research was funded by Internal Grant Agency (IGA) of Faculty of Economics and Management, CULS Prague, grant number 2019B001.

Conflicts of Interest: The authors declare no conflict of interest.

References

1. United Nations. The Sustainable Development Agenda. In Proceedings of the United Nations Sustainable Development Summit, New York, NY, USA, 25–27 September 2015.
2. Grimsby, L.K.; Aune, J.B.; Johnsen, F.H. Human energy requirements in Jatropha oil production for rural electrification in Tanzania. *Energy Sustain. Dev.* **2012**, *16*, 297–302. [CrossRef]
3. OECD/IEA. *Energy Poverty: How to Make Modern Energy Access Universal?* OECD/IEA: Paris, France, 2010.
4. Wendling, Z.A. Bridges beyond renewable energy: Decarbonizing the global electricity sector under uncertainty. *Energy Res. Soc. Sci.* **2019**, *48*, 235–245. [CrossRef]
5. Wuebbles, D.J.; Fahey, D.W.; Hibbard, K.A.; De Angelo, B.; Doherty, S.; Hayhoe, K.; Horton, R.; Kossin, J.P.; Taylor, P.C.; Waple, A.M.; et al. *Executive summary. Climate Science Special Report: Fourth National Climate Assessment, Volume I*; U.S. Global Change Research Program: Washington, DC, USA, 2017. [CrossRef]
6. Reidmiller, D.R.; Avery, C.W.; Easterling, D.R.; Kunkel, K.E.; Lewis, K.L.M.; Maycock, T.K.; Stewart, B.C. *Impacts, Risks, and Adaptation in the United States: The Fourth National Climate Assessment, Volume II*; U.S. Global Change Research Program: Washington, DC, USA, 2018.
7. Dutta, K.; Daverey, A.; Lin, J.G. Evolution retrospective for alternative fuels: First to fourth generation. *Renew. Energy* **2014**, *69*, 114–122. [CrossRef]
8. Maghuly, F.; Vollmann, J.; Laimer, M. Biotechnology of *Euphorbiaceae* (*Jatropha curcas, Manihot esculenta, Ricinus communis*). In *Applied Plant Genomics and Biotechnology*; Elsevier: Amsterdam, The Netherlands, 2015; pp. 87–114. ISBN 978-0-08-100068-7.
9. Wu, Q.; Patocka, J.; Nepovimova, E.; Kuca, K. *Jatropha gossypiifolia* L. and its biologically active metabolites: A mini review. *J. Ethnopharmacol.* **2019**, *234*, 197–203. [CrossRef] [PubMed]
10. Laviola, B.G.; Alves, A.A.; Rosado, T.B.; Bhering, L.L.; Formighieri, E.F.; de Azvedo Peixoto, L. Establishment of new strategies to quantify and increase the variability in the Brazilian *Jatropha genotypes*. *Ind. Crop. Prod.* **2018**, *117*, 216–223. [CrossRef]
11. Garg, K.K.; Karlberg, L.; Wani, S.P.; Berndes, G. *Jatropha* production on wastelands in India: Opportunities and trade-offs for soil and water management at the watershed scale. *Biofuels Bioprod. Bioref.* **2011**, *5*, 410–430. [CrossRef]
12. Keneni, Y.G.; Hvoslef-Eide, A.K.; Marchetti, J.M. Mathematical modelling of the drying kinetics of *Jatropha curcas* L. seeds. *Ind. Crop. Prod.* **2019**, *132*, 12–20. [CrossRef]
13. Chaudhary, A.; Gupta, A.; Kumar, S.; Kumar, R. Pool fires of jatropha biodiesel and their blends with petroleum diesel. *Exp. Therm. Fluid Sci.* **2019**, *101*, 175–185. [CrossRef]
14. Baumert, S.; Khamzina, A.; Vlek, P.L.G. Greenhouse gas and energy balance of *Jatropha* biofuel production systems of Burkina Faso. *Energy Sustain. Dev.* **2018**, *42*, 14–23. [CrossRef]
15. EL-Seesy, A.I.; Hassan, H.; Ookawara, S. Performance, combustion, and emission characteristics of a diesel engine fueled with *Jatropha* methyl ester and graphene oxide additives. *Energy Convers. Manag.* **2018**, *166*, 674–686. [CrossRef]
16. Resende, O.; Ullmann, R.; Siqueira, V.C.; Chaves, T.H.; Ferreira, L.U. Modelagem matemática e difusividade efetiva das sementes de pinhão-manso (*Jatropha curcas* L.) durante a secagem. *Eng. Agríc.* **2011**, *31*, 1123–1135. [CrossRef]
17. Ntaribi, T.; Paul, D.I. Status of *Jatropha* plants farming for biodiesel production in Rwanda. *Energy Sustain. Dev.* **2018**, *47*, 133–142. [CrossRef]
18. Ntaribi, T.; Paul, D.I. The economic feasibility of *Jatropha* cultivation for biodiesel production in Rwanda: A case study of Kirehe district. *Energy Sustain. Dev.* **2019**, *50*, 27–37. [CrossRef]
19. Ramesh, A.; Palanichamy, K.; Tamizhdurai, P.; Umasankar, S.; Sureshkumar, K.; Shanthi, K. Sulphated Zr–Al_2O_3 catalysts through jatropha oil to green-diesel production. *Mater. Lett.* **2019**, *238*, 62–65. [CrossRef]
20. Manurung, R.; Wever, D.A.Z.; Wildschut, J.; Venderbosch, R.H.; Hidayat, H.; van Dam, J.E.G.; Leijenhorst, E.J.; Broekhuis, A.A.; Heeres, H.J. Valorisation of *Jatropha curcas* L. plant parts: Nut shell conversion to fast pyrolysis oil. *Food Bioprod. Process.* **2009**, *87*, 187–196. [CrossRef]
21. Ramírez, V.; Martí-Herrero, J.; Romero, M.; Rivadeneira, D. Energy use of *Jatropha* oil extraction wastes: Pellets from biochar and *Jatropha* shell blends. *J. Clean. Prod.* **2019**, *215*, 1095–1102. [CrossRef]
22. Nepomuk, W.; Tina, H.; Christine, M.; Florian, L.F. Insights into *jatropha* projects worldwide. *SSRN Electron. J.* **2012**. [CrossRef]

23. Huerga, I.R.; Zanuttini, M.S.; Gross, M.S.; Querini, C.A. Biodiesel production from *Jatropha curcas*: Integrated process optimization. *Energy Convers. Manag.* **2014**, *80*, 1–9. [CrossRef]
24. Kamel, D.A.; Farag, H.A.; Amin, N.K.; Zatout, A.A.; Ali, R.M. Smart utilization of jatropha (*Jatropha curcas* Linnaeus) seeds for biodiesel production: Optimization and mechanism. *Ind. Crop. Prod.* **2018**, *111*, 407–413. [CrossRef]
25. Achten, W.M.J.; Verchot, L.; Franken, Y.J.; Mathijs, E.; Singh, V.P.; Aerts, R.; Muys, B. *Jatropha* bio-diesel production and use. *Biomass Bioenergy* **2008**, *32*, 1063–1084. [CrossRef]
26. Axelsson, L.; Franzen, M.; Ostwald, M.; Berndes, G.; Ravindranath, N.H. Performance of *Jatropha* biodiesel production and its environmental and socio-economic impacts—A case study in Southern India. In Proceedings of the World Renewable Energy Congress, Linköping, Sweden, 8–13 May 2011; pp. 2470–2477.
27. Kalinda, C.; Moses, Z.; Lackson, C.; Chisala, L.; Donald, Z.; Darius, P.; Exildah, C.K. Economic impact and challenges of *Jatropha curcas* L. Projects in north-western province, Zambia: A case of Solwezi district. *Sustainability* **2015**, *7*, 9907–9923. [CrossRef]
28. Sinbuathong, N.; Marr, J.M.; Sillapacharoenkul, B.; Chulalaksananukul, S. Effect of the solid content on biogas production from *Jatropha curcas* seed cake. *Int. J. Glob. Warm.* **2011**, *3*, 403–416. [CrossRef]
29. Patil, P.D.; Gude, V.G.; Deng, S. Biodiesel production from *Jatropha curcas*, waste cooking, and *Camelina sativa* oils. *Ind. Eng. Chem. Res.* **2009**, *48*, 10850–10856. [CrossRef]
30. Zhang, F.; Tian, X.F.; Fang, Z.; Shah, M.; Wang, Y.T.; Jiang, W.; Yao, M. Catalytic production of *Jatropha* biodiesel and hydrogen with magnetic carbonaceous acid and base synthesized from *Jatropha* hulls. *Energy Convers. Manag.* **2017**, *142*, 107–116. [CrossRef]
31. Laviola, B.G.; Rodrigues, E.V.; Teodoro, P.E.; de Azvedo Peixoto, L.; Bhering, L.L. Biometric and biotechnology strategies in *Jatropha* genetic breeding for biodiesel production. *Renew. Sustain. Energy Rev.* **2017**, *76*, 894–904. [CrossRef]
32. Deeba, F.; Kumar, V.; Gautam, K.; Saxena, R.K.; Sharma, D.K. Bioprocessing of *Jatropha curcas* seed oil and deoiled seed hulls for the production of biodiesel and biogas. *Biomass Bioenergy* **2012**, *40*, 13–18. [CrossRef]
33. Silva Alves, R.; Teodoro, P.E.; de Azevedo Peixoto, L.; do Amaral Santos de Carvalho Rocha, J.R.; Silva, L.A.; Galveas Laviola, B.; de Resende, M.D.V.; Lopes Bhering, L. Multiple-trait BLUP in longitudinal data analysis on *Jatropha curcas* breeding for bioenergy. *Ind. Crop. Prod.* **2019**, *130*, 558–561. [CrossRef]
34. Haas, W.; Mittelbach, M. Detoxification experiments with the seed oil from *Jatropha curcas* L. *Ind. Crop. Prod.* **2000**, *12*, 111–118. [CrossRef]
35. Raheman, H.; Mondal, S. Biogas production potential of *jatropha* seed cake. *Biomass Bioenergy* **2012**, *37*, 25–30. [CrossRef]
36. Choudhury, S.; Bose, P.K. *Jatropha Derived Biodiesel—Its Suitability as CI Engine Fuel*; Technical report No. 2008–28–0040; Dr B C Roy Engineering College: Durgapur, India, 2008.
37. Basha, S.A.; Gopal, K.R.; Jebaraj, S. A review on biodiesel production, combustion, emissions and performance. *Renew. Sustain. Energy Rev.* **2009**, *13*, 1628–1634. [CrossRef]
38. Parawira, W. Biodiesel production from *Jatropha curcas*: A review. *Sci. Res. Essays* **2010**, *5*, 1796–1808.
39. Ganapathy, T.; Gakkhar, R.P.; Murugesan, K. Influence of injection timing on performance, combustion and emission characteristics of *Jatropha* biodiesel engine. *Appl. Energy* **2011**, *88*, 4376–4386. [CrossRef]
40. Chauhan, B.S.; Kumar, N.; Cho, H.M. A study on the performance and emission of a diesel engine fueled with *Jatropha* biodiesel oil and its blends. *Energy* **2012**, *37*, 616–622. [CrossRef]
41. Ashraful, A.M.; Masjuki, H.H.; Kalam, M.A.; Rizwanul Fattah, I.M.; Imtenan, S.; Shahir, S.A.; Mobarak, H.M. Production and comparison of fuel properties, engine performance, and emission characteristics of biodiesel from various non-edible vegetable oils: A review. *Energy Convers. Manag.* **2014**, *80*, 202–228. [CrossRef]
42. Koh, M.Y.; Ghazi, T.I.M. A review of biodiesel production from *Jatropha curcas* L. oil. *Renew. Sustain. Energy Rev.* **2011**, *15*, 2240–2251. [CrossRef]
43. Patil, V.K.; Bhandare, P.; Kulkarni, P.B.; Naik, G.R. Progeny evaluation of *Jatropha curcas* and *Pongamia pinnata* with comparison to bioproductivity and biodiesel parameters. *J. Res.* **2015**, *26*, 137–142. [CrossRef]
44. Gui, M.M.; Lee, K.T.; Bhatia, S. Feasibility of edible oil vs. non-edible oil vs. waste edible oil as biodiesel feedstock. *Energy* **2008**, *33*, 1646–1653. [CrossRef]
45. Karaj, S.; Huaitalla, R.M.; Müller, J. Physical, mechanical and chemical properties of *Jatropha curcas* L. seeds and kernels. In Proceedings of the Conference on International Agricultural Research for Development, Stuttgart-Hohenheim, Germany, 7–9 October 2008.

46. Patil, P.D.; Deng, S. Optimization of biodiesel production from edible and non-edible vegetable oils. *Fuel* **2009**, *88*, 1302–1306. [CrossRef]
47. Sahoo, P.K.; Das, L.M. Process optimization for biodiesel production from *Jatropha*, Karanja and Polanga oils. *Fuel* **2009**, *88*, 1588–1594. [CrossRef]
48. Jain, S.; Sharma, M.P. Prospects of biodiesel from *Jatropha* in India: A review. *Renew. Sustain. Energy Rev.* **2010**, *14*, 763–771. [CrossRef]
49. Castro Gonzáles, N.F. International experiences with the cultivation of *Jatropha curcas* for biodiesel production. *Energy* **2016**, *112*, 1245–1258. [CrossRef]
50. Yang, C.Y.; Fang, Z.; Li, B.; Long, Y. Review and prospects of Jatropha biodiesel industry in China. *Renew. Sustain. Energy Rev.* **2012**, *16*, 2178–2190. [CrossRef]
51. Silitonga, A.S.; Atabani, A.E.; Mahlia, T.M.I.; Masjuki, H.H.; Badruddin, I.A.; Mekhilef, S. A review on prospect of *Jatropha curcas* for biodiesel in Indonesia. *Renew. Sustain. Energy Rev.* **2011**, *15*, 3733–3756. [CrossRef]
52. Indrawan, N.; Thapa, S.; Rahman, S.F.; Park, J.H.; Park, S.H.; Wijaya, M.E.; Gobikrishnan, S.; Purwanto, W.W.; Park, D.H. Palm biodiesel prospect in the Indonesian power sector. *Environ. Technol. Innov.* **2017**, *7*, 110–127. [CrossRef]
53. Akbar, E.; Yaakob, Z.; Kamarudin, S.K.; Ismail, M.; Salimon, J. Characteristic and composition of *Jatropha curcas* oil seed from Malaysia and its potential as biodiesel feedstock feedstock. *Eur. J. Sci. Res.* **2009**, *29*, 396–403.
54. Thapa, S.; Indrawan, N.; Bhoi, P.R. An overview on fuel properties and prospects of *Jatropha* biodiesel as fuel for engines. *Environ. Technol. Innov.* **2018**, *9*, 210–219. [CrossRef]
55. Nygaard, I.; Bolwig, S. The rise and fall of foreign private investment in the jatropha biofuel value chain in Ghana. *Environ. Sci. Policy* **2018**, *84*, 224–234. [CrossRef]
56. Kalam, M.A.; Ahamed, J.U.; Masjuki, H.H. Land availability of *Jatropha* production in Malaysia. *Renew. Sustain. Energy Rev.* **2012**, *16*, 3999–4007. [CrossRef]
57. Aung, M.M.; Yaakob, Z.; Abdullah, L.C.; Rayung, M.; Li, W.J. A comparative study of acrylate oligomer on *Jatropha* and Palm oil-based UV-curable surface coating. *Ind. Crop. Prod.* **2015**, *77*, 1047–1052. [CrossRef]
58. Kidnay, A.J.; Parrish, W.R. *Fundamentals of Natural Gas. Processing*, 1st ed.; CRC Press: Boca Raton, FL, USA, 2006; ISBN 978-0-8493-3406-1.
59. Hönig, V.; Prochazka, P.; Obergruber, M.; Smutka, L.; Kučerová, V. Economic and technological analysis of commercial LNG production in the EU. *Energies* **2019**, *12*, 1565. [CrossRef]
60. Sher, F.; Pans, M.A.; Afilaka, D.T.; Sun, C.; Liu, H. Experimental investigation of woody and non-woody biomass combustion in a bubbling fluidised bed combustor focusing on gaseous emissions and temperature profiles. *Energy* **2017**, *141*, 2069–2080. [CrossRef]
61. Jabłoński, S.J.; Kułażyński, M.; Sikora, I.; Łukaszewicz, M. The influence of different pretreatment methods on biogas production from *Jatropha curcas* oil cake. *J. Environ. Manag.* **2017**, *203*, 714–719. [CrossRef] [PubMed]
62. Soccol, C.R.; da Costa, E.S.F.; Letti, L.A.J.; Karp, S.G.; Woiciechowski, A.L.; de Vandenberghe, L.P.S. Recent developments and innovations in solid state fermentation. *Biotechnol. Res. Innov.* **2017**, *1*, 52–71. [CrossRef]
63. Phengnuam, T.; Suntornsuk, W. Detoxification and anti-nutrients reduction of *Jatropha curcas* seed cake by *Bacillus* fermentation. *J. Biosci. Bioeng.* **2013**, *115*, 168–172. [CrossRef]
64. Cucchiella, F.; D'Adamo, I.; Gastaldi, M. An economic analysis of biogas-biomethane chain from animal residues in Italy. *J. Clean. Prod.* **2019**, *230*, 888–897. [CrossRef]
65. Govender, I.; Thopil, G.A.; Inglesi-Lotz, R. Financial and economic appraisal of a biogas to electricity project. *J. Clean. Prod.* **2019**, *214*, 154–165. [CrossRef]
66. Adhikari, R.; Agrawal, R.K. An introductory study on time series modeling and forecasting. *arXiv* **2009**, arXiv:1302.6613.
67. Mushtaq, R. Augmented dickey fuller test. *SSRN J.* **2011**. [CrossRef]
68. IndexMundi Palm Oil—Monthly Price—Commodity Prices—Price Charts, Data, and News—IndexMundi. Available online: https://www.indexmundi.com/commodities/?commodity=palm-oil (accessed on 23 June 2019).
69. Kgathi, D.L.; Mmopelwa, G.; Chanda, R.; Kashe, K.; Murray-Hudson, M. A review of the sustainability of *Jatropha* cultivation projects for biodiesel production in southern Africa: Implications for energy policy in Botswana. *Agric. Ecosyst. Environ.* **2017**, *246*, 314–324. [CrossRef]
70. Wang, Z.; Calderon, M.M.; Lu, Y. Lifecycle assessment of the economic, environmental and energy performance of *Jatropha curcas* L. biodiesel in China. *Biomass Bioenergy* **2011**, *35*, 2893–2902. [CrossRef]

71. Foidl, N.; Foidl, G.; Sanchez, M.; Mittelbach, M.; Hackel, S. *Jatropha curcas* L. as a source for the production of biofuel in Nicaragua. *Bioresour. Technol.* **1996**, *58*, 77–82. [CrossRef]
72. Mofijur, M.; Masjuki, H.H.; Kalam, M.A.; Hazrat, M.A.; Liaquat, A.M.; Shahabuddin, M.; Varman, M. Prospects of biodiesel from *Jatropha* in Malaysia. *Renew. Sustain. Energy Rev.* **2012**, *16*, 5007–5020. [CrossRef]
73. Verma, K.C.; Gaur, A.K. *Jatropha curcas* L.: Substitute for conventional energy. *World J. Agric. Sci.* **2009**, *5*, 552–556.
74. Bouffaron, P.; Castagno, F.; Herold, S. Straight vegetable oil from *Jatropha curcas* L. for rural electrification in Mali—A techno-economic assessment. *Biomass Bioenergy* **2012**, *37*, 298–308. [CrossRef]
75. Procházka, P.; Hönig, V. Economic analysis of diesel-fuel replacement by crude palm oil in Indonesian power plants. *Energies* **2018**, *11*, 504. [CrossRef]
76. Deng, X.; Han, J.; Yin, F. Net energy, CO_2 emission and land-based cost-benefit analyses of *Jatropha* biodiesel: A case study of the Panzhihua region of Sichuan province in China. *Energies* **2012**, *5*, 2150–2164. [CrossRef]
77. Kapustová, Z.; Kapusta, J.; Bielik, P. Food-biofuels interactions: The case of the U.S. biofuels market. *AOL* **2018**, *10*, 27–38. [CrossRef]
78. Lajdová, Z.; Kapusta, J.; Bielik, P. Assessing interdependencies between food and energy prices: The case of biodiesel in Germany. *AOL* **2017**, *9*, 51–59. [CrossRef]
79. Svoboda, R.; Severová, L. Regional cooperation of farmers and producers of organic food in the Czech Republic and EU (Federal Republic of Germany). In Proceedings of the International Scientific Conference on Opportunities and Threats to Current Business Management in Cross-Border Comparison, Pilsen, Czech Republic, 23 May 2015; pp. 93–99.

© 2019 by the authors. Licensee MDPI, Basel, Switzerland. This article is an open access article distributed under the terms and conditions of the Creative Commons Attribution (CC BY) license (http://creativecommons.org/licenses/by/4.0/).

energies

Article

Efficient Methods of Market Pricing in Power Industry within the Context of System Integration of Renewable Energy Sources

Evgeny Lisin [1,*], Galina Kurdiukova [1], Pavel Okley [2] and Veronika Chernova [3,4]

1 Department of Economics in Power Engineering and Industry, National Research University "Moscow Power Engineering Institute", Krasnokazarmennaya st. 14, Moscow 111250, Russia
2 OAC INTER RAO UES, Bolshaya Pirogovskaya st. 27, Moscow 119435, Russia
3 Faculty of Economics, Peoples' Friendship University of Russia, Miklukho-Maklaya Str. 6, Moscow 117198, Russia
4 Department of Marketing, State University of Management, 99 Ryazan Avenue, Moscow 109542, Russia
* Correspondence: lisinym@mpei.ru

Received: 24 July 2019; Accepted: 19 August 2019; Published: 23 August 2019

Abstract: Currently, the majority of world economies (even those located in the sunbelt (+/− 35 degrees of latitude with good sunshine with low seasonality) uses various types of fossil fuels as the main source of energy for their economies. However, this represents a very volatile and unsustainable strategy, since according to various estimates, the fossil fuel era will inevitably end as all carbon fuels are going to be spent in the next few centuries. Unlike traditional energy, renewable energy sources (RES) are not based on energy resources, but rather rely upon natural energy flows. With regard to its unique property, there has been an active construction of power plants of renewable energy and their gradual integration into national energy supply systems in recent decades. At the same time, the existing models of electricity markets were unprepared for their wide distribution. Hence, determination of the market value of energy generated by power plants using renewable energy sources becomes a particularly significant issue. This market value has to take into account the prevention of costs from the use of fossil fuels, as well as the resulting environmental benefits. Our paper proposes methods for solving this problem, contributing to the increase of economic efficiency of investment projects for the construction of renewable energy facilities and the formation of economic incentives for their propagation in energy supply systems. The proposed methods are based on the dynamic differentiation of tariffs for consumers with renewable energy sources depending on their structure of electricity consumption. Its effectiveness is demonstrated by calculating the cost of electricity for households located in the Krasnodar region using renewable energy sources. It is shown that this approach to the formation of tariffs for consumers allows the household to receive additional savings from the efficient use of energy installations on RES and energy storage devices in terms of alignment of the energy consumption schedule. This creates a significant incentive for households to use them and contributes to increasing the effectiveness of government renewable energy support programs, including by solving the acute problem of raising electricity tariffs from the grid.

Keywords: renewable energy; system integration; power industry; state policy; smart metering; tariff rates

---

1. Introduction

Over the past few decades, the views on the role of traditional hydrocarbon energy changed dramatically. An increasing stress in concepts of development of a human society of the future is made on power engineering on the basis of using the renewable energy sources, as Melas et al., Augutis et al., Bhattacharya et al., and Umbach et al. have consistently shown through examples of various countries across the world [1–4]. The answer to the question of why this is happening lies not only in the consideration of actions of economic mechanisms and incentives in the energy sector. One can see that the answer also partly lies in a political subtext of energy development based on renewable energy, its scientific, environmental, and social aspects and so on. All of them form the outer visible part of the energy economic and policy iceberg.

Within the core of the hidden processes of close attention of governments of many countries to the development of renewable energy generation one can find careful attempts to find the scale and depth of the next energy crisis. Its occurrence is only a matter of time, due to many factors, the main of which are the following two: (i) physically limited amount of extracted energy sources, and (ii) the constant growth of energy consumption by human civilization, as Qureshi et al., Uz, or Ang et al. point out in their papers in the relevant energy-focused journals [5–7]. Particularly noteworthy is the fact that electricity is no longer just an additional benefit. As a result of the serious growth of the world's population and the concomitant restructuring of all vital industries nowadays, energy has become the key to the survival of billions of people all around the world.

In an attempt to avoid the consequences of a social and economic shock as a result of the energy crisis, various countries took steps to break the vicious circle and provide the basis for ensuring the minimum necessary level of power generation within the national energy systems [8,9]. This explains the emergence of keen interest in renewable energy based on the use of natural energy flows. Their integration into power systems is one of the priority tasks of the energy policy of a number of countries aimed at ensuring energy security, despite the fact that renewable energy sources in many cases are characterized by low economic efficiency and form many problems associated with the optimization and management of operating modes of power systems. For instance, some researchers [10–13] explain this by the uneven production of renewable energy and its inconsistency with the electricity consumption schedule. Moreover, the introduction of renewable energy often leads to an increase in the cost of energy for the end consumer. Some other researchers [14–16] note that the increase in the cost of electricity is primarily due to an increase in the grid tariff due to a decrease in electricity consumption from the grid and an increase in the demand for peak power capacities.

Most recently, academic research and engineering projects conducted in the field of development of renewable energy have started to gain serious attention.

Serious state support is provided for scientific and technological developments of renewable energy power plants, the emphasis of which is shifted towards increasing their installed capacity. As noted in the papers [17–19], an increase in the unit capacity of power plants leads to a reduction in the cost of electricity production.

Another direction in the field of the use of renewable energy sources which is developing no less intensively is the development, creation and operation of small power plants. According to many authors [20–23], they focus on the needs of private households or small settlements. It is worth noting that the military departments of most countries are strongly interested in creating such autonomous systems, which ensures a constant flow of funds to finance such projects.

To achieve the goals of national strategies for the development of renewable energy and increase their investment attractiveness at the level of governments and energy departments of countries, it is necessary to develop state tariff policy tools aimed at introducing and using renewable energy sources in energy supply systems. They should include economic incentives to optimize energy consumption through the introduction of autonomous energy sources and be based on differentiated pricing by hour area of the day. At the same time, as practice shows, differentiated tariffs are often less profitable for the consumer than a fixed tariff during the billing period [14,15]. This is due to the discrepancy between the fixed values of the tariff rates established for the time zones and the actual daily energy consumption schedule. Taking into account the development of automated systems for commercial metering of electricity, it is necessary to develop a pricing method that takes into account real-time changes in energy consumption and stimulates consumers and manufacturers to introduce and effectively use renewable energy sources.

This paper aims to solve the following important problems and issues:

- Inconsistency between the existing systems for differentiating electricity tariffs by time zones and real daily energy consumption schedules which leads to inefficiency of the tariff policy in the field of energy management of consumers and their irrational use of renewable energy sources.
- The need to reduce the growth in the cost of electricity for consumers from the power supply system caused by suboptimal loading of the production capacity of the energy system when consumers use autonomous energy sources based on renewable energy sources.
- Stimulating consumers to align the schedule of electricity consumption during the day and increase the efficiency of using autonomous energy sources based on renewable energy sources (RES), including the use of electric energy storage technology.

2. Overview of the Global Renewable Energy Market

According to the data from 2016, the renewable energy market is represented by power plants with a total capacity of more than 2000 GW (Figure 1) [24–26]. The main suppliers of renewable energy are hydropower and wind as well as solar energy and bioenergy.

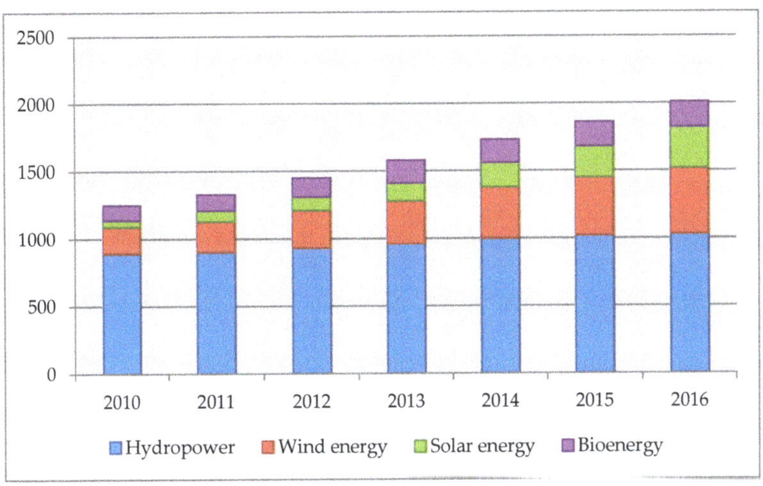

Figure 1. World production capacity of renewable energy sources, GW.

At the same time, solar and wind energy markets are the fastest growing in the world. The world wind power capacity over the past 7 years increased 2.5 times, reaching a total capacity of 487 GW, while the global solar photovoltaic power capacities increased more than 10 times and amounted to 303 GW (Figure 2) [24–26].

Additionally, there is a significant reduction in the cost of individual technologies of renewable energy. This contributes to the innovative development of the industry, in particular, reducing the cost of installation and production of solar photovoltaic cells, as well as the modernization of wind turbine structures and the emergence of new materials. In terms of the cost, the renewable energy sources in many countries are now close to competing with new fossil-fired power plants (Figure 3) [27–29].

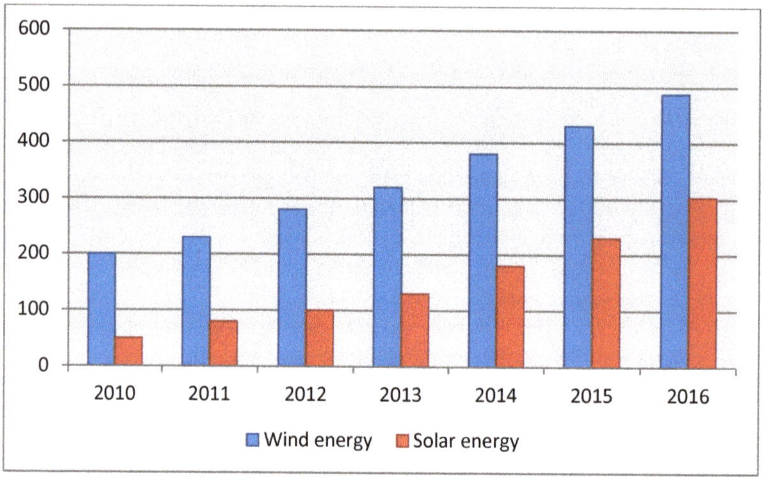

Figure 2. World wind and solar production capacity, GW.

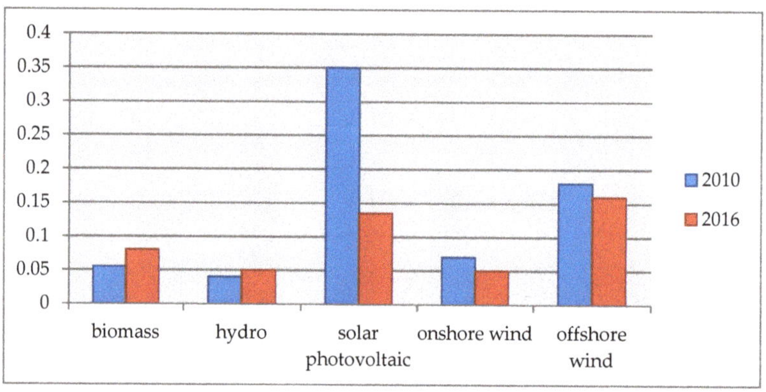

Figure 3. Estimated cost of electricity on renewable energy installations, US dollars per kWh.

One of the most competitive renewable energy sources is coastal wind installations, primarily for countries where energy resources are expensive. Where fossil fuels are fairly cheap (for example, fairly low gas prices in the United States and Russia), wind turbines can compete with traditional energy sources only under favorable weather conditions and state support programs for the development of the industry. At the same time, the increased interest of many countries in environmental protection and energy security leads to the active development of renewable energy [30–34].

Nowadays, more than 160 countries of the world (82%) pursue a policy of supporting the sector, of which in 20 countries the share of renewable energy sources in the total energy balance exceeds 20%. According to the EU energy strategy, by 2030 countries should ensure a 40% reduction in greenhouse gas emissions and an increase to 32% of the share of renewable energy [35]. In the longer term, many countries will likely go significantly further. In particular, Germany plans to reach 60% of the share of renewable energy in the country's total energy balance by 2050. Figure 4 presents a generalized model of factors influencing the development of renewable energy in the world.

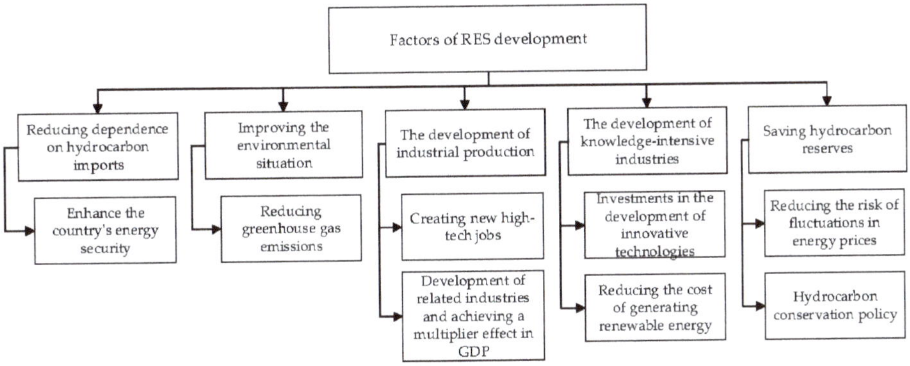

Figure 4. Factors renewable energy development in the world.

Overall, it becomes apparent that high rates of growth in production of renewable energy are largely due to significant state support. It allows in a number of countries to make renewable energy attractive, even in cases where its initial economic indicators are more than 50% worse than when using fossil fuel resources. At the same time, renewable energy is successfully developing in the world in spite of the crisis phenomena in the global economy, and, in fact, is one of the most effective ways to overcome the energy crisis.

3. Analysis of the Methods and Consequences of State Support for the Development of RES

Over the past few decades, the development of renewable energy sources in the world has achieved great success in increasing production efficiency and reducing equipment cost. Due to the development of new technologies, renewable energy has become capable of the cost of production to compete with traditional power generation in countries with favorable for the use of natural energy. However, in spite of all their achievements, today renewable energy technologies face great challenges in ensuring the required level of competitiveness due to the peculiarities of energy production and the high cost of building an installed capacity unit, which makes it impossible to develop them without state support [36,37].

Government programs to support renewable energy sources are usually aimed at solving several problems at once [38–42]:

- decarbonization of power systems and an increase in domestic electricity production,
- improving the technological competitiveness of renewable energy by reducing production costs and creating new jobs.
- In some countries, state programs of support for renewable energy have existed for more than 20 years, which made it possible to analyze the experience of the best practices in implementing such programs. There are four categories of government policy tools to support renewables:
- tax incentives;
- state subsidies;
- regulatory tools;
- privileged access policy.

Many countries use several RES support tools at once. At the same time, it is possible to trace the relationship between the number of tools used to support renewable energy sources and the country's income. Low-income countries use an average of 2.2 different mechanisms, while high-income countries use 4.8 [43]. Table 1 presents a summary of the use of renewable energy support mechanisms in the world.

Table 1. Application of state support mechanisms for renewable energy sources (RES) in the world.

Instruments of State Support for RES	2005	2016
Countries with approved state programs for the development of RES	48	160
Countries applying special tariffs for generating renewable energy	34	108
Countries using emission quotas for greenhouse gases	11	99

In recent years, more and more political tools have been created for the development of RES. For 2018, 164 countries are already implementing state programs for the development of RES. At the same time, half of them are countries with developing or transitional economies [43,44]. Tax relief and earmarked loans are commonly used to support renewable energy. In high-income countries investment subsidies and grants are mainly used.

The dominant mechanisms for the support of renewable energy are schemes for the introduction of a special fixed (green) tariff (feed-in tariff) and netting of electricity (net metering), applied in both developing and developed countries. Today, about 66% of countries use these tools.

Net metering assumes that in windy or sundial hours the owner of an installation of renewable energy sources delivers surplus electricity to the grid, which he cannot consume himself, and during windless or night hours he consumes electricity from the grid to offset the volumes supplied earlier. If according to the results of the billing period, such a participant in energy supply consumed more electricity than he supplied, then he compensates for the difference at the usual rate. If, on the contrary, he consumes less energy than he supplied, he may use the surplus in subsequent periods or even receive payments from the power utility. When using net metering, only one meter is required, which will show the difference between the volumes of electricity taken and delivered to the grid. A feed-in tariff allows owners of a power plant for renewable energy to receive compensation for electricity supplied to the network that exceeds the retail tariff; however, as the number of owners of such installations grows, the feed-in tariff usually decreases and ultimately compares to the retail tariff. This scheme requires the installation of two separate meters.

Despite the advantages, the use of these tools to support renewable energy often leads to the emergence of cross-subsidization of consumers and generation, as well as an increase in the tariff for electricity transmission throughout the network. So for the period 2008–2016, Australia, Spain, Portugal and Germany were characterized by an exceptional increase in electricity prices (a 112% price increase) [14,15,43]. This is due to the fact that in these countries, generation and sale of electricity are competitive activities, while transmission and distribution networks are structurally separated and regulated by the monopolies. Reducing the demand for electricity from the grid and reducing the uniformity of the electricity consumption schedule due to the spread of generation to renewable energy sources led to an increase in the network tariff, which was to provide the necessary regulated revenue to grid companies and compensate them for the investments made over the past five years to develop the network infrastructure. In turn, consumers responded to the growth of tariffs by reducing consumption, which led to an increase in network tariff rates for the next year in order to provide the necessary gross revenue to grid companies and the second wave of changes in electricity demand from the grid.

The described negative processes in the energy supply system aggravated the state programs for subsidizing consumers who commissioned installations for renewable energy sources, in particular, solar photovoltaic panels, which were compensated by raising prices for power grid supply. The sharp increase in fixed-rate tariff obligations for RES (feed-in tariff) led to the third wave of network tariff increases.

Another consequence of the wrong state policy in support of renewable energy is the emergence of indirect cross-subsidization, when consumers who have not installed or cannot afford installations on renewable energy, in fact, pay the grid company for consumers who switch to renewable energy, forced to raise the network tariff, and generation company, which increases the uneven loading of power equipment and the number of hours of use of expensive peak power while reducing the amount of electricity generation [44–46].

The way out of the situation described is the development of a tariff that stimulates the consumer to equalize the load during the day period of electricity consumption [47–49]. This suggests the use of a differentiated tariff for the zones of the day, the key disadvantage of which in this situation is its static nature, when the selected zones may not correspond to a dynamically changing load schedule and, thus, instead of reducing the price of electricity for the consumer, lead to its increase. It is necessary to develop a methodology for dynamic differentiation of the electricity tariff, which with the development of automated systems for commercial metering of electricity and their intellectualization becomes applicable in practice.

4. Proposal for the Method of Dynamic Differentiation of Electricity Tariffs

To solve the problem of reducing power supply costs, reducing the need for expensive peak capacity of the power system and increasing the efficiency of using autonomous power plants for renewable energy, it is necessary to manage power consumption by encouraging consumers to use measures to level the daily consumption schedule. These measures should encourage consumers to introduce power storage equipment into the autonomous power supply system (for example, based on solar photovoltaic panels), allowing them to distribute the produced energy over a wide range of hours during the day, as well as to change their institutional behavior, which will be expressed in adapting the daily pattern of electricity consumption to a production schedule of available autonomous energy sources.

Such a measure to stimulate the optimization of power consumption and the effective use of autonomous sources is the formation of an economically sound tariff according to the zonal principle. The zones of base, semi-peak and peak load on the power system are distinguished. For each zone, depending on the amount of electricity consumption, its own tariff rate is set.

At the same time, as shown in the next part of the paper, the zonal differentiation of the tariff does not always contribute to improving the efficiency of electricity supply and may even be less beneficial for the consumer than the single-value undifferentiated tariff. This is due to the fixed values of the peak and half-peak tariff rates and transition time zones, which do not correspond to the actual daily load curve of consumers. Also, a contribution is made by the existing tariff differentiation in many countries in terms of amount of energy consumption, which is often poorly founded.

The development of automated systems for commercial metering of electricity along the path of using advanced (smart) electricity meters, equipped with communication tools for monitoring and data transmission on energy consumption in real time, allows us to develop pricing methods for electricity with much greater differentiation of time zones consumption and tariff rates than the existing ones [50–52]. Thus, it is possible to solve the problem of inconsistency of the tariff differentiation system and real daily energy consumption schedules.

The following algorithm is proposed for dynamic pricing of electricity, taking into account the fact that consumer load charts on weekends and working days differ significantly:

1. Based on the statistical analysis of data for the previous period, daily graphs of consumer load are compiled on a typical work day and a day off: $P_w(t)$ и$P_h(t)$
2. According to consumer load schedules, values are calculated:

    a. Minimum power consumption per unit of time:

$$P_w^{\min} = \min P_w(t), P_h^{\min} = \min P_h(t)$$

b. Average power consumption per unit of time:

$$\overline{P}_w = \frac{\sum_{t=0}^{n-1} P_w(t)}{n}, \overline{P}_h = \frac{\sum_{t=0}^{n-1} P_h(t)}{n}$$

c. Half-peak power consumption per unit of time:

$$P_w^{pp} = \overline{P}_w + 2\sqrt{\frac{\sum_{t=0}^{n-1} \left(P_w(t) - \overline{P}_w\right)^2}{n}}, P_h^{pp} = \overline{P}_h + 2\sqrt{\frac{\sum_{t=0}^{n-1} \left(P_h(t) - \overline{P}_h\right)^2}{n}}$$

d. Maximum power consumption per unit of time:

$$P_w^{\max} = \max P_w(t), P_h^{\max} = \max P_h(t)$$

3. Energy consumption zones are determined:

a. Basic power consumption (B):

$$P_w^B(t) \in [0, P_w^{\min}], P_h^B(t) \in [0, P_h^{\min}], t = \overline{0, 23}$$

b. Half-peak power consumption (PP):

$$P_w^{PP}(t_w) \in (P_w^{\min}, P_w^{pp}], P_h^{PP}(t_h) \in (P_h^{\min}, P_h^{pp}], t_w = \overline{t_{P_w^{\min}}, t_{P_w^{PP}}}, t_h = \overline{t_{P_h^{\min}}, t_{P_h^{PP}}}, P_w(t_{P_w^{\min}}) = P_w^{\min},$$
$$P_h(t_{P_h^{\min}}) = P_h^{\min}, P_w(t_{P_w^{PP}}) = P_w^{PP}, P_h(t_{P_h^{PP}}) = P_h^{PP}$$

c. Peak power consumption (P):

$$P_w^P(t_w) \in (P_w^{pp}, P_w^{\max}], P_h^P(t_h) \in (P_h^{PP}, P_h^{\max}], t_w = \overline{t_{P_w^{PP}}, t_{P_w^{\max}}}, t_h = \overline{t_{P_h^{PP}}, t_{P_h^{\max}}}, P_w(t_{P_w^{PP}}) = P_w^{PP},$$
$$P_h(t_{P_h^{PP}}) = P_h^{PP}, P_w(t_{P_w^{\max}}) = P_w^{\max}, P_h(t_{P_h^{\max}}) = P_h^{\max}$$

4. A fixed tariff rate is established for the base consumption zone, and an interval of tariff rates is established for the half-peak and peak power consumption zones:

a. Basic power consumption (B):

$$C^B(t) = C^B, t = \overline{0, 23}$$

b. Half-peak power consumption (PP):

$$C^{PP}(t) \in [C_{\min}^{PP}, C_{\max}^{PP}], C_{\min}^{PP} = C^{PP}(P(t) = P_{\min}), C_{\min}^{PP} > C^B, C_{\max}^{PP} = C^{PP}(P(t) = P^{PP}), t = \overline{t_{P\min}, t_{PPP}}$$

c. Peak power consumption (P):

$$C^P(t) \in [C_{\min}^P, C_{\max}^P], C_{\min}^P = C^P(P(t) = P^{PP}), C_{\min}^P > C_{\max}^{PP}\ C_{\max}^P = C^{PP}(P(t) = P^{\max}), t = \overline{t_{PPP}, t_{P\max}}$$

5. We believe that the increase or decrease in energy consumption proportionally reduces or increases the cost of electricity in the corresponding zone with the coefficient K:

a. Half-peak power consumption (PP):

$$K^{PP} = \frac{C^{PP}_{\max} - C^{PP}_{\min}}{t_{pPP} - t_{p\min}}$$

b. Peak power consumption (P):

$$K^{P} = \frac{C^{P}_{\max} - C^{P}_{\min}}{t_{p\max} - t_{pPP}}$$

6. The chain growth rates are calculated, which characterize the increments of energy consumption in the areas of half-peak and peak consumption:

a. Half-peak power consumption (PP):

$$G^{PP}(t) = \frac{P^{PP}(t)}{P^{PP}(t-1)}$$

b. Peak power consumption (P):

$$G^{P}(t) = \frac{P^{P}(t)}{P^{P}(t-1)}$$

7. A dynamic calculation of the cost of electricity for the consumers at each time point for working days and days off are produced. In case of increase in energy consumption in the half-peak and peak zones, an additional penalty is imposed $[G(t) - 1]K$. Accordingly, when energy consumption decreases in comparison with the previously registered value, the tariff rate is decreased:

$$TC_w(t) = \begin{cases} C^B P_w(t), P_w(t) = P_w^{\min} \\ C^{PP}(t)P_w(t) + [G^{PP}(t) - 1]K^{pp}, P_w(t) \in (P_w^{\min}, P_w^{PP}] \\ C^{P}(t)P_w(t) + [G^{P}(t) - 1]K^{p}, P_w(t) \in (P_w^{PP}, P_w^{\max}] \end{cases},$$

$$TC_h(t) = \begin{cases} C^B P_h(t), P_h(t) = P_h^{\min} \\ C^{PP}(t)P_h(t) + [G^{PP}(t) - 1]K^{pp}, P_h(t) \in (P_h^{\min}, P_h^{PP}] \\ C^{P}(t)P_h(t) + [G^{P}(t) - 1]K^{p}, P_h(t) \in (P_h^{PP}, P_h^{\max}] \end{cases}$$

8. The total cost of electricity per day is calculated, taking into account the discrete values of the indications of smart meters:

$$TC_w^D = \sum_{i=1}^{n} TC_w(t_i)(t_i - t_{i-1}), TC_h^D = \sum_{i=1}^{n} TC_h(t_i)(t_i - t_{i-1})$$

The above algorithm implements the principle of dynamic pricing of electricity for the consumer, when at each moment in time it is determined in which energy consumption zone (base, half-peak, peak) the consumer load curve is found. Based on the calculation of the chain load growth rates for each energy consumption zone, a system of tariff rates is formed. The growth of energy consumption in the semi-peak zone and its coverage due to the use of energy system resources, rather than renewable energy, is punished by higher tariff rates, reduction of energy consumption is encouraged by a dynamic reduction in rates.

5. Empirical Model: A Case Study of Krasnodar Region of Russia

Our empirical model focuses on the Krasnodar region which is one of the largest regions-consumers of electricity in Russia. At the same time, it is characterized by a high level of energy deficiency. Hence, the development of territorial energy system in the region is given great importance [53,54].

In 2016, electricity production amounted to about 7 billion kWh with a consumption volume of 22 billion kWh. From 2006 to 2016 electricity consumption in the region increased by more than two times. The increase in consumption in the territorial energy system was mainly due to the growth of electricity consumption by industrial enterprises, the intensive development of the resort and recreation complex and the buildup of electrical loads during the construction of Olympic facilities in Sochi.

The shortage of generating capacity of the territorial energy system of the Krasnodar region was covered by the flow of electricity and power through the intersystem power lines from adjacent power systems. In January 2017, the total flow of electricity into the power system of the Krasnodar region amounted to more than 1 billion kWh, including 130 million kWh via interstate lines. In total, the Krasnodar region satisfies its electricity needs by 35% due to preferential energy production at thermal power plants [55].

Krasnodar region is one of the most attractive in Russia for the development of generation based on the use of renewable energy sources in terms of its natural and climatic characteristics. According to existing estimates, by realizing the potential of renewable energy in the Krasnodar region, it is possible to produce up to 1300 MW instead of burning hydrocarbon fuels. The most important types of renewable resources for the Krasnodar region are solar and wind energy, which account for 80% of the energy potential of renewable energy sources in the region [52,56,57].

Special attention in the region is paid to the development of distributed solar energy, which should make it possible to reduce capital expenditures on the construction of new sources of traditional energy, as well as on the construction of new kilometers of transmission lines. The potential of solar energy in the region is about 1400 kWh/m^2, which corresponds to the level of countries that actively use solar energy today, such as the United States and Australia [14,28,52].

Solar energy is used primarily for the needs of households using electricity generators based on photovoltaic cells and panels [58]. Figures 5 and 6 show the averaged graphs of the daily load of a household on a typical working day and on a day off with a graph of energy generated by solar installations.

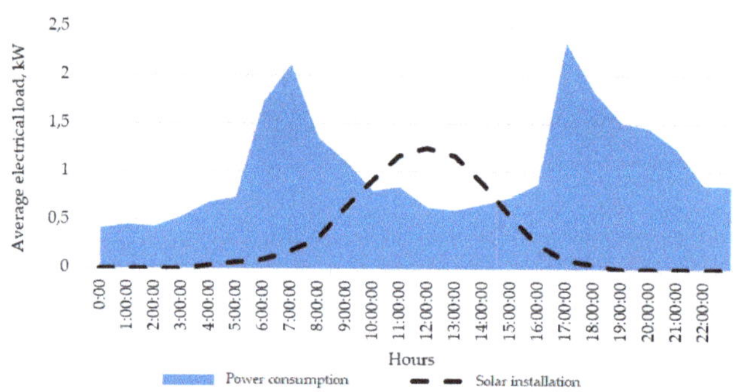

Figure 5. Household electrical load graph on a working day with a superimposed graph of electricity generated by solar installations.

As can be seen from the graphs presented, the peak of the electricity produced by solar power plants does not correlate with the time of day with the peaks of household electricity consumption, especially on working days. The significant difference between daily power consumption and autonomous power generation of households with solar panels leads to inefficient use of solar generation. Thus, the economic potential of individual solar installations is not fully disclosed, which leads to low efficiency of state programs to subsidize the development of solar energy.

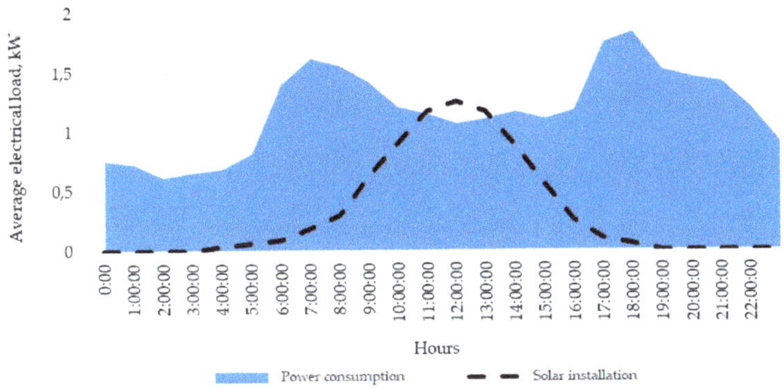

Figure 6. Household electrical load graph on a day off with a superimposed graph of electricity generated by solar installations.

One of the ways to increase the efficiency of using solar installations is to use energy storage technologies, which make it possible to distribute electricity generated from solar power plants during the day (Figure 7) [38,59–61].

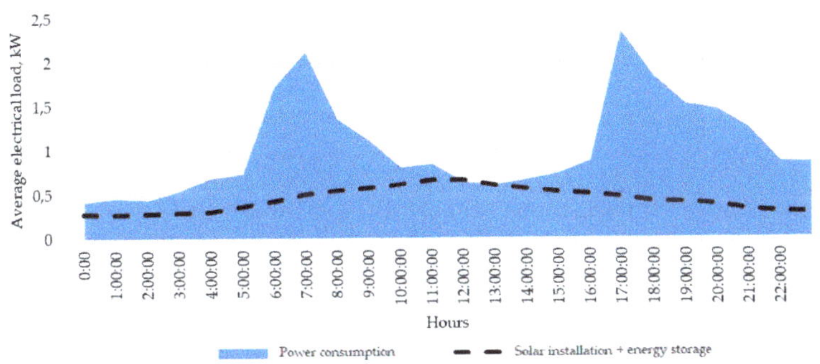

Figure 7. Household electrical load graph on a working day with a superimposed graph of electricity produced by solar installations with energy storage.

As can be seen from the graph, as a result of the use of energy storage, it becomes possible to use electricity generated from solar installations more evenly throughout the day. Consumption during peak hours is reduced by about 30% and household energy consumption at night is reduced by 80%.

The use of solar power and energy storage by households is largely related to the current tariff menu for electricity from the network in the region. Today, one-rate tariffs are applied in Kranodar region, differentiated by two and three zones of the day, as well as the volume of consumption (Table 2) [59].

According to the current tariff menu and the "dynamic" differentiated tariff proposed in the previous part of the paper, the cost of electricity per month was estimated for households consuming electricity only from the grid, households with solar panels and households with solar panels and storage using given electrical load graphs. The calculation results are shown in Table 3.

Table 2. Electricity tariffs in the Krasnodar region.

Category	One-Rate Tariff, Rubles/kWh	One-Rate Tariff, Differentiated by Two Zones of the Day, Rubles/kWh		One-Rate Tariff, Differentiated by Three Zones of the Day, Rubles/kWh		
		Day Zone (Peak and Half-Peak)	Night Zone	Peak Zone	Half-Peak Zone	Night Zone
for the amount of electricity consumed up to 250 kWh per month	1.10	1.10	0.77	1.65	1.10	0.44
for the amount of electricity consumed from 250 to 600 kWh per month	1.44	1.44	1.01	2.16	1.44	0.58
for the amount of electricity consumed in excess of 600 kWh per month	4.95	4.95	3.47	7.43	4.95	1.98

Table 3. Comparison of the results of calculating the cost of electricity for households according to the current tariff menu and the proposed dynamic tariff differentiation.

Types of Power Grid Tariffs	The Cost of Electricity for Households, Rubles/month		
	with Consumption only from the Power Grid	with Solar Installations	with Solar Installations and Storage
One-rate tariff	204,496	106,756	88,515
One-rate tariff, differentiated by two zones of the day	235,367	126,757	96,784
One-rate tariff, differentiated by three zones of the day	221,299	118,334	90,327
"Dynamic" differentiated tariff	220,157	98,245	73,346

As can be seen from the data, the cost of electricity when applying the existing tariff differentiation scheme for two and three times zones of the day is the highest for households that do not use solar power. These results are also true for households with solar panels and storage, which suggests that the existing tariff menu does not encourage consumers to use solar energy efficiently. This is due to the discrepancy between the time zones used in the existing differentiated tariffs and the actual peak and half-peak areas of household electricity consumption, as well as the high values of household electricity consumption in the morning and evening hours. Thus, it is beneficial for the consumer to use the one-rate undifferentiated tariff regardless of the availability of renewable energy sources and the use of energy storage devices.

This problem can be solved using the proposed method of dynamic tariff differentiation. For households that do not use solar installations, the simplest one-rate tariff still remains the cheapest. However, for households that use solar installations, and the solar systems with stored energy, there are obvious advantages for a "dynamic" differential tariff. Reducing the cost of electricity for households will stimulate the population to more active and, most importantly, rational use of solar installations, will reduce the load on the energy system, as well as reduce the need for building new generating capacity of traditional energy with high capital costs.

6. Conclusions

Overall, it can be stated that the need and prospects for the development of renewable energy across the world are confirmed today by various stimulating legal and economic acts and laws, including those located in the area of creating differentiated power grid tariffs for consumers.

At the same time, as the conducted study showed, the applied system of state tariff differentiation often not only does not lead to the formation of economic incentives for optimizing electricity consumption when introducing power plants for renewable energy, but also turns out to be less beneficial for the consumer than simple one-rate tariffs. This is due to the fixed values of the transition to tariff zones at half-peak and peak rates, which do not correspond to the real daily consumption graph of the region's households. The situation is also aggravated by poorly justified differentiation of the tariff by the volume of energy consumed during the period.

Analysis of the developed daily energy consumption graphs of households in the Krasnodar region on workdays and weekends showed that the rational use of solar power plants leads to a significant decrease in the amount of energy consumed from the power system in the half-peak zone of the load curve. At the same time, it is not possible to achieve a uniform distribution of the load on the power system during the day due to the use of autonomous solar installations. Energy consumption at peak hours is reduced, but the unevenness of the daily load graph remains. Consequently, there remains a need for the availability of maneuverable peak equipment in the power system, which is rapidly gaining power, even if the accumulative power equipment is installed to reduce the irregularity of the power consumption graph. At the same time, solar energy sources will displace half-peak power plants from the load graph, reducing the need of this generation for power system to ensure continuous power supply to consumers.

The development of automated systems for commercial metering of electricity along the path to the use of smart electricity meters allows developing electricity pricing methods with a much greater differentiation of consumption zones and tariff rates than the existing ones. Thus, it is possible to solve the problem of inconsistency of the tariff differentiation system by time zones and real daily energy consumption graphs.

Using the example of the Krasnodar region, it is shown that the proposed method of dynamic differentiation of the electricity tariff allows a household to receive additional savings from the effective use of solar power and energy storage in terms of equalizing the energy consumption schedule. This creates a significant incentive for their use by households and contributes to the effectiveness of government programs to support renewable energy sources, including solving the acute problem of increasing power grid tariffs.

Author Contributions: Conceptualization, E.L. and G.K.; methodology, E.L., V.C., and G.K.; formal analysis, E.L., P.O., and V.C.; investigation, E.L.; resources, P.O. and V.C.; data curation, E.L., P.O., G.K., and V.C.; writing—original draft preparation, E.L., V.C., and P.O.; writing—review and editing, E.L.; visualization, E.L.

Funding: This research was funded by the Ministry of Education and Science of the Russian Federation, grant number 26.9593.2017/8.9.

Acknowledgments: This paper was supported by the Ministry of Education and Science of the Russian Federation, research project No. 26.9593.2017/8.9.

Conflicts of Interest: The authors declare no conflict of interest.

Nomenclature

A List of Symbols

$P_w(t), P_h(t)$	daily graphs of consumer load on a typical work day and a day off
P_w^{min}, P_h^{min}	minimum power consumption on a typical work day and a day off
$\overline{P}_w, \overline{P}_h$	average power consumption on a typical work day and a day off
P_w^{pp}, P_h^{pp}	half-peak power consumption on a typical work day and a day off
P_w^{max}, P_h^{max}	maximum power consumption on a typical work day and a day off
C^B	fixed tariff rate for the basic power consumption zone
$[C_{min}^{PP}, C_{max}^{PP}]$	interval of tariff rates for the half-peak power consumption zone
$[C_{min}^{P}, C_{max}^{P}]$	interval of tariff rates for the peak power consumption zone
K^{PP}	cast coefficient for the half-peak power consumption zone
K^P	cast coefficient for the peak power consumption zone
$G^{PP}(t)$	chain growth rates for the half-peak power consumption zone
$G^P(t)$	chain growth rates for the peak power consumption zone
$TC_w(t), TC_h(t)$	tariff rates at each time point for working days and days off
TC_w^D, TC_h^D	total cost of electricity per day for working days and days off

Abbreviations

RES	renewable energy sources
B	basic power consumption zone
P	peak power consumption zone
PP	half-peak power consumption zone

References

1. Melas, V.; Lisin, E.; Tvaronavičienė, M.; Peresadko, G.; Radwański, R. Energy security and economic development: Renewables and the integration of energy systems. *J. Secur. Sustain. Issues* **2017**, *7*, 133–139. [CrossRef]
2. Augutis, J.; Martišauskas, L.; Krikštolaitis, R.; Augutienė, E. Impact of the Renewable Energy Sources on the Energy Security. *Energy Procedia* **2014**, *61*, 945–948. [CrossRef]
3. Bhattacharya, M.; Paramati, S.; Ozturk, I.; Bhattacharya, S. The effect of renewable energy consumption on economic growth: Evidence from top 38 countries. *Appl. Energy* **2016**, *162*, 733–741. [CrossRef]
4. Umbach, F. Global energy security and the implications for the EU. *Energy Policy* **2010**, *38*, 1229–1240. [CrossRef]
5. Qureshi, M.; Rasli, A.; Zaman, K. Energy crisis, greenhouse gas emissions and sectoral growth reforms: Repairing the fabricated mosaic. *J. Clean. Prod.* **2016**, *112*, 3657–3666. [CrossRef]
6. Uz, D. Energy efficiency investments in small and medium sized manufacturing firms: The case of California energy crisis. *Energy Econ.* **2018**, *70*, 421–428. [CrossRef]
7. Ang, B.; Choong, W.; Ng, T. Energy security: Definitions, dimensions and indexes. *Renew. Sustain. Energy Rev.* **2015**, *42*, 1077–1093. [CrossRef]
8. Filipović, S.; Radovanović, M.; Golušin, V. Macroeconomic and political aspects of energy security—Exploratory data analysis. *Renew. Sustain. Energy Rev.* **2018**, *97*, 428–435. [CrossRef]
9. Sovacool, B. Differing cultures of energy security: An international comparison of public perceptions. *Renew. Sustain. Energy Rev.* **2016**, *55*, 811–822. [CrossRef]
10. Abdmouleh, Z.; Alammari, R.; Gastli, A. Review of policies encouraging renewable energy integration & best practices. *Renew. Sustain. Energy Rev.* **2015**, *45*, 249–262. [CrossRef]
11. Oree, V.; Hassen, S.; Fleming, P. Generation expansion planning optimisation with renewable energy integration: A review. *Renew. Sustain. Energy Rev.* **2017**, *69*, 790–803. [CrossRef]
12. Radovanović, M.; Filipović, S.; Pavlović, D. Energy security measurement-A sustainable approach. *Renew. Sustain. Energy Rev.* **2017**, *68*, 1020–1032. [CrossRef]
13. Maheshwari, Z.; Ramakumar, R. Smart integrated renewable energy systems (SIRES): A novel approach for sustainable development. *Energies* **2017**, *10*, 1145. [CrossRef]
14. Simshauser, P. Distribution network prices and solar PV: Resolving rate instability and wealth transfers through demand tariffs. *Energy Econ.* **2016**, *54*, 108–122. [CrossRef]
15. Strielkowski, W.; Štreimikienė, D.; Bilan, Y. Network charging and residential tariffs: A case of household photovoltaics in the United Kingdom. *Renew. Sustain. Energy Rev.* **2017**, *77*, 461–473. [CrossRef]
16. Ntanos, S.; Kyriakopoulos, G.; Chalikias, M.; Arabatzis, G.; Skordoulis, M. Public perceptions and willingness to pay for renewable energy: A case study from Greece. *Sustainability* **2018**, *10*, 687. [CrossRef]
17. Caldés, N.; del Río, P.; Lechón, Y.; Gerbeti, A. Renewable Energy Cooperation in Europe: What Next? Drivers and Barriers to the Use of Cooperation Mechanisms. *Energies* **2018**, *12*, 70. [CrossRef]
18. Ntanos, S.; Skordoulis, M.; Kyriakopoulos, G.; Arabatzis, G.; Chalikias, M.; Galatsidas, S.; Katsarou, A. Renewable energy and economic growth: Evidence from European countries. *Sustainability* **2018**, *10*, 2626. [CrossRef]
19. Noailly, J.; Smeets, R. Directing technical change from fossil-fuel to renewable energy innovation: An application using firm-level patent data. *J. Environ. Econ. Manag.* **2015**, *72*, 15–37. [CrossRef]
20. Cucchiella, F.; D'Adamo, I.; Gastaldi, M. Future trajectories of renewable energy consumption in the European Union. *Resources* **2018**, *7*, 10. [CrossRef]
21. Kim, K.; Kim, Y. Role of policy in innovation and international trade of renewable energy technology: Empirical study of solar PV and wind power technology. *Renew. Sustain. Energy Rev.* **2015**, *44*, 717–727. [CrossRef]

22. Brecl, K.; Topič, M. Photovoltaics (PV) System Energy Forecast on the Basis of the Local Weather Forecast: Problems, Uncertainties and Solutions. *Energies* **2018**, *11*, 1143. [CrossRef]
23. Kammen, D.; Sunter, D. City-integrated renewable energy for urban sustainability. *Science* **2016**, *352*, 922–928. [CrossRef] [PubMed]
24. International Energy Agency. Statistics Data Browser. Available online: https://www.iea.org/statistics (accessed on 20 June 2019).
25. BP Statistical Review of World Energy. 2018. Available online: https://www.bp.com/content/dam/bp/business-sites/en/global/corporate/pdfs/energy-economics/statistical-review/bp-stats-review-2018-full-report.pdf (accessed on 20 June 2019).
26. Global Energy Statistical Yearbook 2019. Available online: https://yearbook.enerdata.net/renewables/renewable-in-electricity-production-share.html (accessed on 20 June 2019).
27. International Energy Agency. Renewables 2017. Available online: https://www.iea.org/renewables2017 (accessed on 20 June 2019).
28. International Renewable Energy Agency. Available online: https://www.irena.org/-/media/Files/IRENA/Agency/Publication/2018/Jan/IRENA_2017_Power_Costs_2018.pdf (accessed on 20 June 2019).
29. University of Oxford. Our World in Data. Renewable Energy. Available online: https://ourworldindata.org/renewable-energy (accessed on 20 June 2019).
30. Sander, M. Conceptual proposals for measuring the impact of international regimes on energy security. *Energy Policy* **2013**, *63*, 449–457. [CrossRef]
31. Li, Y.; Shi, X.; Yao, L. Evaluating energy security of resource-poor economies: A modified principle component analysis approach. *Energy Econ.* **2016**, *58*, 211–221. [CrossRef]
32. Löschel, A.; Moslener, U.; Rübbelke, D.T. Indicators of energy security in industrialised countries. *Energy Policy* **2010**, *38*, 1665–1671. [CrossRef]
33. Proskuryakova, L. Energy technology foresight in emerging economies. *Technol. Forecast. Soc. Chang.* **2017**, *119*, 205–210. [CrossRef]
34. Hoffmann, C. Beyond the resource curse and pipeline conspiracies: Energy as a social relation in the middle east. *Energy Res. Soc. Sci.* **2018**, *41*, 39–47. [CrossRef]
35. European Commission. 2030 Climate & Energy Framework. Available online: https://ec.europa.eu/clima/policies/strategies/2030_en (accessed on 20 June 2019).
36. Lisin, E.; Rogalev, A.; Strielkowski, W.; Komarov, I. Sustainable modernization of the Russian power utilities industry. *Sustainability* **2015**, *7*, 11378–11400. [CrossRef]
37. Lisin, E.; Strielkowski, W. Modelling new economic approaches for the wholesale energy markets in Russia and the EU. *Transform. Bus. Econ.* **2014**, *13*, 566–580.
38. Strielkowski, W.; Lisin, E.; Tvaronavičienė, M. Towards energy security: Sustainable development of electrical energy storage. *J. Secur. Sustain. Issues* **2016**, *6*, 235–244. [CrossRef]
39. Ćetković, S.; Buzogány, A. Varieties of capitalism and clean energy transitions in the European Union: When renewable energy hits different economic logics. *Clim. Policy* **2016**, *16*, 642–657. [CrossRef]
40. Ringel, M. Energy efficiency policy governance in a multi-level administration structure - Evidence from Germany. *Energy Effic.* **2017**, *10*, 753–776. [CrossRef]
41. Kasmaei, M.P.; Lehtonen, M.; Sanz, J.C.; Mantovani, J.R.S. Carbon Footprint Management: A Pathway toward Smart Emission Abatement. *IEEE Trans. Ind. Inform.* **2019**. [CrossRef]
42. Cerna, F.V.; Pourakbari-Kasmaei, M.; Contreras, J.; Gallego, L.A. Optimal Selection of Navigation Modes of HEVs Considering CO_2 Emissions Reduction. *IEEE Trans. Veh. Technol.* **2019**, *68*, 2196–2206. [CrossRef]
43. The Renewable Energy Policy Network of the 21st Century (REN21). Renewables 2016. Available online: https://www.ren21.net/wp-content/uploads/2019/05/REN21_GSR2016_FullReport_en_11.pdf (accessed on 20 June 2019).
44. Salas-Fumás, V.; Rosell-Martínez, J.; Delgado-Gómez, M. Capacity, investment and market power in the economic value of energy firms. *Energy Econ.* **2016**, *53*, 28–39. [CrossRef]
45. Amrouche, S.; Rekioua, D.; Rekioua, T.; Bacha, S. Overview of energy storage in renewable energy systems. *Int. J. Hydrog. Energy* **2016**, *41*, 20914–20927. [CrossRef]
46. Lisin, E.; Kindra, V.; Horvathova, Z. Sustainable development of regional heat supply systems in the context of the Eurasian economic union energy markets association. *J. Secur. Sustain. Issues* **2017**, *6*, 215–245. [CrossRef]

47. McPherson, M.; Tahseen, S. Deploying storage assets to facilitate variable renewable energy integration: The impacts of grid flexibility, renewable penetration, and market structure. *Energy* **2018**, *145*, 856–870. [CrossRef]
48. Lisin, E.; Shuvalova, D.; Volkova, I.; Strielkowski, W. Sustainable Development of Regional Power Systems and the Consumption of Electric Energy. *Sustainability* **2018**, *10*, 1111. [CrossRef]
49. Lisin, E.; Sobolev, A.; Strielkowski, W.; Garanin, I. Thermal efficiency of cogeneration units with multi-stage reheating for Russian municipal heating systems. *Energies* **2016**, *9*, 269. [CrossRef]
50. Jokar, P.; Arianpoo, N.; Leung, V.C. A survey on security issues in smart grids. *Secur. Commun. Netw.* **2016**, *9*, 262–273. [CrossRef]
51. Dincer, I.; Acar, C. Smart energy systems for a sustainable future. *Appl. Energy* **2017**, *194*, 225–235. [CrossRef]
52. Cucchiella, F.; D'Adamo, I.; Gastaldi, M. Economic analysis of a photovoltaic system: A resource for residential households. *Energies* **2017**, *10*, 814. [CrossRef]
53. Daus, Y.; Kharchenko, V.; Yudaev, I. Evaluation of solar radiation intensity for the territory of the Southern Federal District of Russia when designing microgrids based on renewable energy sources. *Appl. Sol. Energy* **2016**, *52*, 151–156. [CrossRef]
54. Smeets, N. Similar goals, divergent motives. The enabling and constraining factors of Russia's capacity-based renewable energy support scheme. *Energy Policy* **2017**, *101*, 138–149. [CrossRef]
55. Federal State Statistics Service. Main Statistics. 2018. Available online: http://www.gks.ru/wps/wcm/connect/rosstat_main/rosstat/ru/statistics (accessed on 20 June 2019).
56. System Operator UES. Report on the functioning of the UES of Russia in 2017. Available online: https://so-ups.ru/fileadmin/files/company/reports/disclosure/2018/ups_rep2017.pdf (accessed on 20 June 2019).
57. Kozlova, M.; Collan, M. Modeling the effects of the new Russian capacity mechanism on renewable energy investments. *Energy Policy* **2016**, *95*, 350–360. [CrossRef]
58. Vasileva, E.; Viljainen, S.; Sulamaa, P.; Kuleshov, D. RES support in Russia: Impact on capacity and electricity market prices. *Renew. Energy* **2015**, *76*, 82–90. [CrossRef]
59. PJSC "Kubanenergo". Annual Report. 2017. Available online: http://kubanenergo.ru/media/media/kuban_web_oneside%2005-06.pdf (accessed on 20 June 2019).
60. Faessler, B.; Schuler, M.; Preißinger, M.; Kepplinger, P. Battery storage systems as grid-balancing measure in low-voltage distribution grids with distributed generation. *Energies* **2017**, *10*, 2161. [CrossRef]
61. Subramani, G.; Ramachandaramurthy, V.K.; Padmanaban, S.; Mihet-Popa, L.; Blaabjerg, F.; Guerrero, J.M. Grid-tied photovoltaic and battery storage systems with Malaysian electricity tariff—A review on maximum demand shaving. *Energies* **2017**, *10*, 1884. [CrossRef]

© 2019 by the authors. Licensee MDPI, Basel, Switzerland. This article is an open access article distributed under the terms and conditions of the Creative Commons Attribution (CC BY) license (http://creativecommons.org/licenses/by/4.0/).

energies

MDPI

Article

Merit Order Effect Modeling: The Case of the Hellenic Electricity Market

Stelios Loumakis *, Eugenia Giannini and Zacharias Maroulis

Laboratory of Process Analysis and Design, National Technical University of Athens, 15780 Athens, Greece; evgenia@gianninilaw.eu (E.G.); maroulis@mail.ntua.gr (Z.M.)

* Correspondence: steliosloumakis@gmail.com; Tel.: +30-694-477-7331

Received: 10 September 2019; Accepted: 10 October 2019; Published: 12 October 2019

Abstract: A simple effective model is proposed for the day-ahead electricity market. The model considers the main factors which govern the process, predicts the seasonal and daily variation of electricity demand, renewable production, system marginal price, and merit order effect. The accuracy of the model is increased by fitting to historic data of the Hellenic electricity market. During the period between October 2016 and December 2018, the Hellenic electricity market calculated explicitly the merit order effect using an innovative mechanism to directly charge the electricity suppliers (retailers). On the basis of the proposed model and the market recorded data, the effect of the renewable penetration on the wholesale Hellenic electricity prices is revealed. The model is further used to analyze the market future behavior when basic factors (electricity demand, conventional power, and renewable penetration) are known or estimated. The effect of merit order effect on the Hellenic legislation is discussed and the appropriate measures adopted by the Hellenic authorities are analyzed and evaluated.

Keywords: system marginal price; renewable energy sources; photovoltaics; day ahead market; merit order curve; electricity demand; seasonal and daily variation; RES (renewable energy sources) surcharge

---

1. Introduction

The integration of renewable energy sources (RES) in electricity markets has been significantly reinforced by policy support measures in all European countries. The effect of their participation on the determination of electricity wholesale marginal prices has become, however, a topic of political debate. Considering that renewables enter with priority into the day-ahead market (DAM) (zero pricing bids), the merit order curve (MOC) is shifted to the right or equivalent, the demand curve is shifted to the left, most expensive plants are driven out, and subsequently, the clearing wholesale marginal electricity price is diminished. This phenomenon is the so-called merit order effect (MOE). Since intense discussions have taken place about the economic surcharge that passes to the final consumers due to renewable supporting mechanisms, a broad spectrum of literature exists on the analysis, quantification, and evaluation of the MOE phenomenon.

The methods for the examination of MOE are based on two main approaches [1,2], which are as follows: (a) the development of electricity market models, which simulate the operation of DAM and calculate the resulted spot electricity price for various scenarios [3–9] and (b) the regression analysis approach, which uses historical price and generation data in order to quantify the actual achieved reduction in spot price for a given period of time [2,3,10–14]. Studies that combine both approaches are also met [15]. Although each country has its unique characteristics in terms of energy mix, economic growth, incentives policy, etc., useful information and knowledge can be retrieved by relevant research.

The impact of the RES generation in Germany was analyzed by Sensfuß et al. [8], by developing a detailed electricity market simulation model. The results showed the high impact of the diffusion of renewables on spot market prices, reaching a reduction of 7.8€/MWh (2006). For the same country, Weigt [6] investigated the impact of wind penetration, and indicated a significant decline in market price, especially during peak periods (10€/MWh). In Spain, two relevant studies were found in the literature. De Miera [7] assessed the effect of support schemes on power prices for the case of wind-generation electricity, and concluded that the reduction of spot price was much higher than the increase in cost charged to the consumers for the financial support of the technology. Ciarreta et al. [9] analyzed the Spanish electricity market for four years (2008–2012) and calculated the cost of "green" energy as the difference between the savings gained due to MOE and the incentive amounts. According to the results, after the wide deployment of RES, a positive net cost was computed, but a difference in the cost among technologies was observed.

The significant impact of wind generation on spot prices was also proven by Jónsson et al. [11], who used a non-parametric regression model to analyze the Danish electricity market. Similarly, based on the regression approach, Luňáčková et al. [14] quantified the MOE in the Czech market by using six years of hourly, daily, and weekly data (2010–2015), which led to the outcome of a 10% increase in RES deployment, except from solar which had a negative impact in general, and caused a 2.5% decrease in electricity price. Gelabert et al. [10] used a multivariate regression model to estimate the MOE of the RES and cogeneration in Spain between 2005 and 2010 and showed that a marginal increase of 1 GWh of electricity production from the previous technologies leads to a decrease of 2€/MWh in electricity prices. In the case of Italy, Clò et al. [3] used empirical data from the DAM and proved that wholesale electricity prices were reduced by 2.3€/MWh and 4.2€/MWh due to an increase of 1 GWh in the hourly average of daily produced energy from solar and wind systems, respectively.

In Greece, the share of RES in electricity production mixture increased from to 6.9% in 2004 to 16.3% in 2017, while the national target for 2020 is 18% [16]. The main incentive for the RES wide penetration in the country has been the feed-in tariff (FIT) mechanism [17,18]. The impact of this deployment on the Hellenic electricity market was evaluated by Simoglou et al. [19,20], by developing a simulation model for the market operation under various scenarios of RES capacity. The results indicated that RES integration caused a significant reduction in the SMP (system marginal price) and CO_2 emissions, but the payment of consumers was increased for the examined years (2009 and 2011).

This paper aims to extend the current literature and quantify, as well as analyze the MOE in the Hellenic wholesale electricity DAM by: (1) analyzing historic data, (2) proposing and validating a simple model describing the phenomenon, (3) analyzing future market behavior when the crucial factors are known or predicted, and (4) analyzing and evaluating the related surcharge mechanism used by the Hellenic authorities. The MOE analysis refers to seasonal and daily variation of the MOE along with the effect of renewable technology on the MOE. The political significance of MOE in electricity pricing along with the Hellenic authorities curing experiment (2016–2018) are analyzed in Section 2, and the corresponding recording data are statistically analyzed in Section 4.1. A robust mathematical model is proposed in Section 3, which is fitted to recording data in Section 4.2 and used to analyze future market behavior in Section 4.3. Section 5 summarizes the conclusions of the analysis.

2. Merit Order Effect and the Hellenic Electricity Market

Greece enjoys a renewable energy sources (RES) installed capacity of approximately 5 GW operating under the feed-in tariff (FIT) model, initially introduced by law 3468/2006 [21]. Priority to the grid applies also to net-metering, virtual net-metering, and feed in premium scheme lastly introduced through law 4414/2016 [22] leading, as well, to a further decline of the system marginal price (SMP).

The RES account introduced through law 2773/1999 [23], is responsible for payments of renewable electricity production under feed-in supporting schemes. Its revenue architecture faced significant challenges and deficits in Greece during the last decade, leading the RES investments all over the country to payment delays, financial problems, and unfortunately retroactive cuts in contracted tariffs even for operating projects through laws 4093/2012 [24], 4152/2013 [25], and finally 4254/2014 [26]. Although at first sight one could claim that the RES account deficits were solely driven by rapid expansion of renewables during 2011 to 2013 without a simultaneous and adequate increase in the RES levy (called ETMEAR in Greece) in consumer electricity bills, the continuation of financial problems of the RES account even after retroactive cuts in tariffs of 2012–2014, indicated the problematic revenue architecture of the account.

Focusing closer on the RES account structure [27], a major revenue component has to do with its income from the wholesale market, corresponding to the renewable electricity produced and infused into the grid multiplied by the hourly SMP. In order to provide priority for renewables to the grid over other conventional sources such as lignite or gas-fired plants or imports, feed-in operating models (feed-in tariff or feed in premium) provide that RES are not truly participating in the wholesale market with price bids that compete with fossil fuel bids but instead their bidding price remains zero every hour in order to be always preferable and selected to feed-in [28].

Therefore, regardless of the specific operating model, such as FIT, FIP, or net-metering, it is the priority in entering the market (grid) that renewable energy enjoys through zero bidding prices that diminishes SMP in DAM and, consequently, RES account revenues for renewable overall electricity production infused. This naturally means that a further increase in renewable capacity under priority, even at comparable to SMP price levels, is expected to expand MOE further, diminishing SMP even more and consequently further triggering the need for a RES levy increase, in order for the RES account to keep its financial ability to support contracted prices for renewables. This systematic decline of SMP due to RES priority is benefiting suppliers not only for the renewable electricity they buy at this lower marginal wholesale market price but for all their electricity needs drawn from the pool, because for each hour the wholesale price is one and unique [29].

On the one hand, the RES levy not only works to finance the true cost difference between renewable and fossil fuel electricity, but as a tool to cover the continuously increasing distortion because of MOE and the missing money problem it causes for the RES account [29]. It is obvious that a general withdrawal of RES as a priority aiming to minimize MOE and hence the RES account distortion in revenues through SMP, could not be part of a viable solution. The stochastic nature of wind and solar energy production could cause them to not survive direct competition against fossil fuels, unless storage solutions become reliable and economically effective for them.

On the other hand, one could argue that for the consumer the distortion of MOE finally plays no role regarding total electricity bill cost, since the increased RES levy is counteracted by the reduced wholesale electricity cost based on the diminished SMP. In other words, the added amount of money that the regulated part of the bill (through RES levy) needs to carry because of the distortion of MOE, is balanced by the decrease in the nonregulated or so called "competitive" part of the bill. Furthermore, if new renewables entering the system are remunerated with tariffs close to the SMP levels, as it now happens, then electricity bills for the consumer will no longer increase. However, even new renewables, because of their priority entrance under zero pricing bidding, will trigger a further SMP decline and need for a RES levy increase not only for them but for all other previously existing and operating RES penetration, and hence the vicious circle continues even more aggressively.

Real life in Greece during the recession years showed that regulated electricity costs bare political costs, meaning that authorities are, in general, reluctant or negative towards increasing a part of the electricity bill, especially when they cannot easily explain to consumers why this is happening (MOE is difficult to describe) and at the same time when they cannot directly control the nonregulated part of the bill that should be equally decreased.

During 2015, the RES account in Greece started showing deficits again due to a SMP drop. Suppliers did not proceed to electricity price reductions in the competitive part but instead kept high profitability margins benefiting from the MOE in order to face the growing phenomenon of unpaid consumer bills. At the same time, the RES levy was officially frozen against increases by the State. Less than a year later, during 2016, delays in payments for renewables, due to the RES account increasing cumulative deficit, started to rapidly rise jeopardizing national policies against climate change and of course Greece's credibility as an investment destination. Therefore, public dialogue and the challenge during that time was to determine how to take out of the RES levy its distorted part due to the MOE and charge it to suppliers that were systematically benefitting from high margins out of it.

Towards this aim, the idea was to introduce a parallel virtual wholesale market being resolved every hour exactly as the official one but excluding renewables that operate under priority. Out of this virtual market, a virtual SMP could be calculated and its difference from the official SMP would give the decline of the SMP because of MOE. This difference between the virtual SMP (VSMP) and the official one multiplied by the amount of MWh each supplier drew from the pool in order to cover the demand of its customers, would represent a new charge for suppliers. This new charge, domestically called "PXEFEL", aimed to make suppliers return to the RES account their systematic economic benefit and increased margin because of the RES and the MOE they caused.

In this respect, by having the RES account as an additional regular revenue through the MOE charge counteracting the MOE, the RES levy could be kept to the necessary, fair (not distorted) lower levels. The aim of the RES levy was to make it correspond to whatever difference in the production costs truly exists between fossil and renewable sources. On the other hand, for suppliers, the MOE charge consisted of a new cost component incorporated into the "competitive" part of their consumer bills, whereby suppliers had the ability to disperse it in a variety of manners among different customer groups in correlation to their hourly consuming profile. The MOE differs significantly among different hours in the DAM, since it mainly has to do with electricity demand and RES active penetration.

A radical alternative solution proposed at that time was to fully integrate the RES levy into the wholesale cost of suppliers, and hence to the "competitive" part of their bills. To do so, the market operator would have to charge them a "below the line" RES levy on the basis of electricity drawn from the DAM pool. Suppliers would then fully incorporate the cost of the RES levy in their pricing policies, mixed of course with their benefit from the MOE that the RES at the same time cause. Suppliers' bills would then become more market oriented, since there would be no regulated part for the RES electricity cost that suppliers do not "own". This model would be somehow equivalent to suppliers being contracted directly by the RES producers through bilateral contracts to purchase their production at their fixed tariffs.

3. Day-Ahead Electricity Market Model

The proposed mathematical model aims to predict the MOE versus the crucial factors affecting the DAM. It is basically a deterministic model which considers the "predictable" part of the seasonal and daily variation of the demand and renewable production, while the "random" part is considered as just a noise which is eliminated by fitting to historical data. Random variation could be considered by adding random functions describing the noise, but this is out of the scope of the present paper. A stochastic model for probabilistic forecasting has been proposed recently by Bello et al. [30].

3.1. *Definition of the Merit Order Effect*

Mathematically, the MOE is defined as the difference in system marginal price between (a) the value calculated when all the demand is considered into market clearance procedure, the virtual system marginal price (VSMP) and (b) the value calculated when the renewable production is excluded from the demand, the system marginal price (SMP). That is, MOE = VSMP − SMP.

This definition is graphically presented in Figure 1.

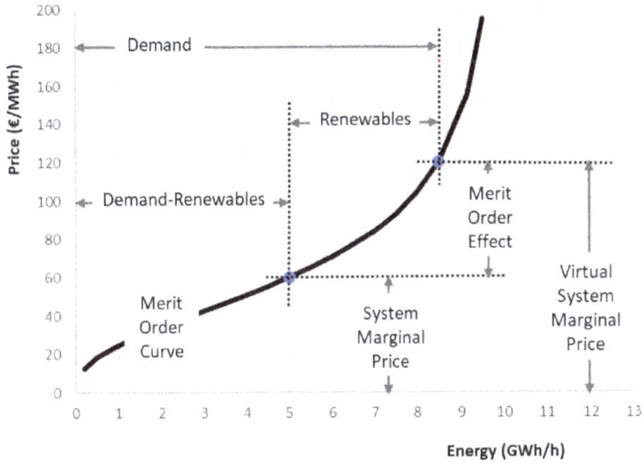

Figure 1. Definition of the merit order effect (MOE) using the demand and supply curve (merit order curve) in the day-ahead electricity market.

3.2. The Merit Order Curve

The supply curve in the DAM is called the merit order curve (MOC). The following equation is proposed and used in this paper to describe it:

$$P = P_{med}\left(\frac{S}{S_{max} - S}\right)^n \tag{1}$$

where the supply price P (€/MWh) is calculated versus the dispatchable electricity supply S (GWh/h), when the following three parameters are known:

- S_{max} GWh/h the maximum feasible electricity supply;
- P_{med} €/MWh the electricity price at half of maximum supply;
- n - empirical shape constant.

Obviously, the maximum feasible electricity supply is analogous to the conventional installed power (lignite-fired, gas-fired, and large hydro) that are dispatchable, meaning that they have the ability to determine the SMP through their hourly price bidding in the DAM.

The above analysis assumes that other factors affecting the MOC remain constant during the analysis period. Other factors such as the suppliers bidding strategy, the CO_2 emissions price, and the fuel price can be introduced in the parameters P_{med} and n through appropriate functions, but this is out of the scope of this paper.

3.3. The Electricity Demand

The electricity demand appears to have both seasonal and daily variations. The seasonal variation of the electricity demand appears as two peaks, one during the winter and one during the summer. Loumakis et al. [31] considered the following three additive kinds of demand with different variation each: (a) A constant demand independently of the season, (b) a winter activities demand, and (c) a summer activities demand. A normal distribution was proposed to describe both the winter and summer activities with different characteristics. The resulting equation is [31]:

$$D_i = \frac{D_T(1 - d_w - d_s)}{365} + d_w D_T{\cdot}N(t_{wo}, \Delta t_w, i) + d_s D_T{\cdot}N(t_{so}, \Delta t_s, i) \tag{2}$$

where the electricity demand D_i (GWh/day) during the day i (1, 2, ... , 365) is calculated, when the following seven parameters are known:

- D_T GWh/year the total electricity demand during the year;
- d_w - the portion of the total annual demand for winter activities;
- d_s - the portion of the total annual demand for summer activities;
- t_{wo} days the time of the peak of winter activities;
- t_{so} days the time of the peak of summer activities;
- Δt_w days the typical duration of the winter activities;
- Δt_s days the typical duration of the summer activities.

The normal distribution is expressed by the following equation: $N(\mu, \sigma, t) = \frac{1}{\sqrt{2\pi\sigma^2}} exp\left(-\frac{(t-\mu)^2}{2\sigma^2}\right)$.

The daily variation of the demand also appears to have two peaks, one at noon and another in the evening. The above idea concerning the seasonal variation is also used to describe the daily variation:

$$D_{ij} = \frac{D_i(1 - d_n - d_e)}{24} + d_n D_i \cdot N(t_{no}, \Delta t_n, j) + d_e D_i \cdot N(\mathrm{t_{eo}}, \Delta t_e, j) \quad (3)$$

where the electricity demand D_{ij} (GWh/hour) during the hour j (= 1, 2, ... , 24) of the day i (1, 2, ... , 365) is calculated, when the total demand D_i (GWh/day) of the day i is calculated from Equation (2) and the following six parameters are additionally known:

- d_n - the portion of the total daily demand for noon activities;
- d_e - the portion of the total daily demand for evening activities;
- t_{no} hours the time of the peak of noon activities;
- t_{eo} hours the time of the peak of evening activities;
- Δt_n hours the typical duration of the noon activities;
- Δt_e hours the typical duration of the evening activities.

3.4. Renewable Electricity Production

The renewable electricity production appears also to have both seasonal and daily variations. Loumakis et al. [31] proved that all types of renewables follow a cosine seasonal variation with different amplitude and peak time. Thus, the resulting total renewable also appears cosine variation with characteristics depended on renewable mixture. Instead, concerning daily variation, all renewables, except photovoltaics, are random or almost constant without any deterministic variation. Photovoltaics follows a well-known deterministic variation during daytime sunshine hours. On the basis of these remarks the renewables can be divided into two categories, concerning the daily variation profile, the photovoltaics (PV), and the other (W). Thus:

$$R_i = W_i + PV_i \quad (4)$$

where:

- i days the day of the year (1, 2, ... , 365);
- R_i GWh/day the electricity generated from renewables during the day i;
- W_i GWh/day the electricity generated from renewables except photovoltaics during the day i;
- PV_i GWh/day the electricity generated from photovoltaics during the day i.

The seasonal variation of both kinds of renewables is described by the cosine equation:

$$W_i = \frac{W_T}{365} - \Delta W\, cos\left(2\pi \frac{i - i_{Wo}}{365}\right) \quad (5)$$

$$PV_i = \frac{PV_T}{365} - \Delta PV\, cos\left(2\pi \frac{i - i_{PVo}}{365}\right) \quad (6)$$

Equation (5) calculates the electricity W_i (GWh/day) generated from renewables except photovoltaics during the day of the year i (1, 2, … , 365), when the following three parameters are known:

- W_T GWh/year the total annual electricity generated by renewables except photovoltaics;
- ΔW GWh/day the seasonal variation amplitude of renewables except photovoltaics;
- i_{Wo} days the day of minimum production for renewables except photovoltaics.

Equation (6) calculates the electricity PV_i (GWh/day) generated from photovoltaics during the day of the year i (1, 2, … , 365), when the following three parameters are known:

- PV_T GWh/year the total annual electricity generated by photovoltaics;
- ΔPV GWh/day the seasonal variation amplitude of photovoltaics;
- i_{PVo} days the day with the minimum production for photovoltaics.

Concerning the daily variation, the following equations are proposed:

$$W_{ij} = \frac{W_i}{24} \tag{7}$$

$$PV_{ij} = cf\frac{PV_i}{24}max\left(cos\left(2\pi\frac{j - j_{PVo}}{24}\right), 0\right) \tag{8}$$

Equation (7) calculates the electricity W_{ij} (GWh/hour) generated from renewables except photovoltaics during the hour j (1, 2, … , 24) of the day of the year i (1, 2, … , 365), when the total daily electricity is calculated by Equation (5) (uniform distribution).

Equation (8) calculates the electricity PV_{ij} (GWh/hour) generated from photovoltaics during the hour j (1, 2, … , 24) of the day i (1, 2, … , 365), when the total daily electricity is calculated by Equation (6) and the following parameter is additionally known:

- j_{PVo} hours the hour with maximum production from photovoltaics.

The correction factor *cf* is given by the equation: cf $= PV_i \ / \sum_{j=1}^{24} PV_{ij}$

3.5. Regression Analysis

The resulting mathematical model of Equations (1)–(8) calculates all crucial quantities of the electricity market (electricity demand, renewable generation, merit order curve, system marginal price, and merit order effect) when twenty-three parameters are known. The information flow diagram of the proposed model is presented in Figure 2.

The parameters can be estimated using three separate regressions:

- Equations (2) and (3) are simultaneous fitted to real data of demand;
- Equations (5)–(8) are simultaneous fitted to real data of renewable electricity generation;
- Equation (1) is fitted to real data of system marginal price and virtual marginal price.

The appropriate data were retrieved from the Hellenic electricity market operator [32] and the Hellenic energy exchange group [33].

During the period from October 2016 to December 2018 (27 months) the auction clearing process of the DAM was calculated twice (a) using the demand and supply including renewables in order to obtain the system marginal price (SMP) and (b) using the demand and supply excluding renewables in order to obtain the corresponding SMP without the renewables, denoted as virtual system marginal price (VSMP).

The following data were retrieved for the whole year 2017:

- D_{ij} GWh/h Hourly electricity demand;

- R_{ij} GWh/h Hourly renewables electricity;
- SMP_{ij} €/MWh Hourly system marginal price;
- $VSMP_{ij}$ €/MWh Hourly virtual system marginal price.

where i is the day identification number (1, 2, ... , 365) and j the hour identification number (1, 2, ... , 24) per day.

The resulting fitted model can be used to forecast by changing the following parameters (factors):

- S_{max} GWh/h the maximum feasible electricity supply;
- W_T GWh/year the total annual electricity generated by renewables except photovoltaics;
- PV_T GWh/year the total annual electricity generated by photovoltaics;
- D_T GWh/year the total electricity demand during the year.

All other model parameters cannot be considered as factors since they remain long-term constants. They are not analogous to the market size, but they are dependent on the country activities and weather characteristics.

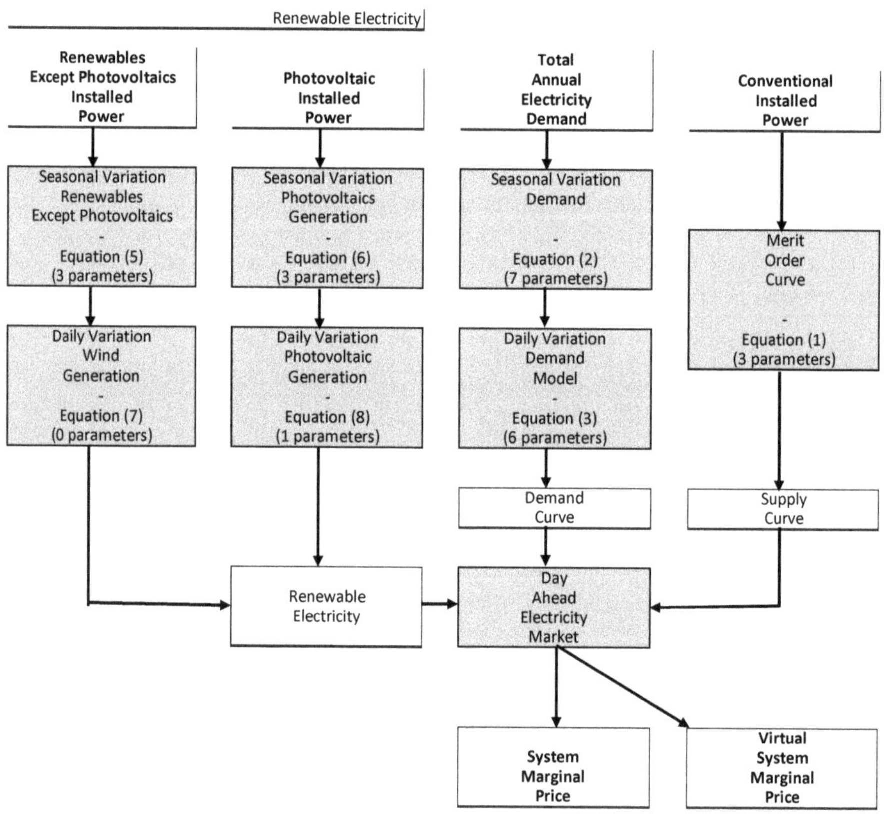

Figure 2. Model information flow diagram.

4. Results and Discussion

4.1. Statistical Analysis

The recording quantities (D, R, VSMP, and SMP) along with the calculated ones (DR = D − R and MOE = VSMP − SMP) are presented in Figure 3. Hourly data are plotted, that is 8760 points for each quantity. The time duration curves are also presented in Figure 3. Furthermore, the basic statistics are presented in Table 1 and the corresponding fitted normal distributions in Figure 4.

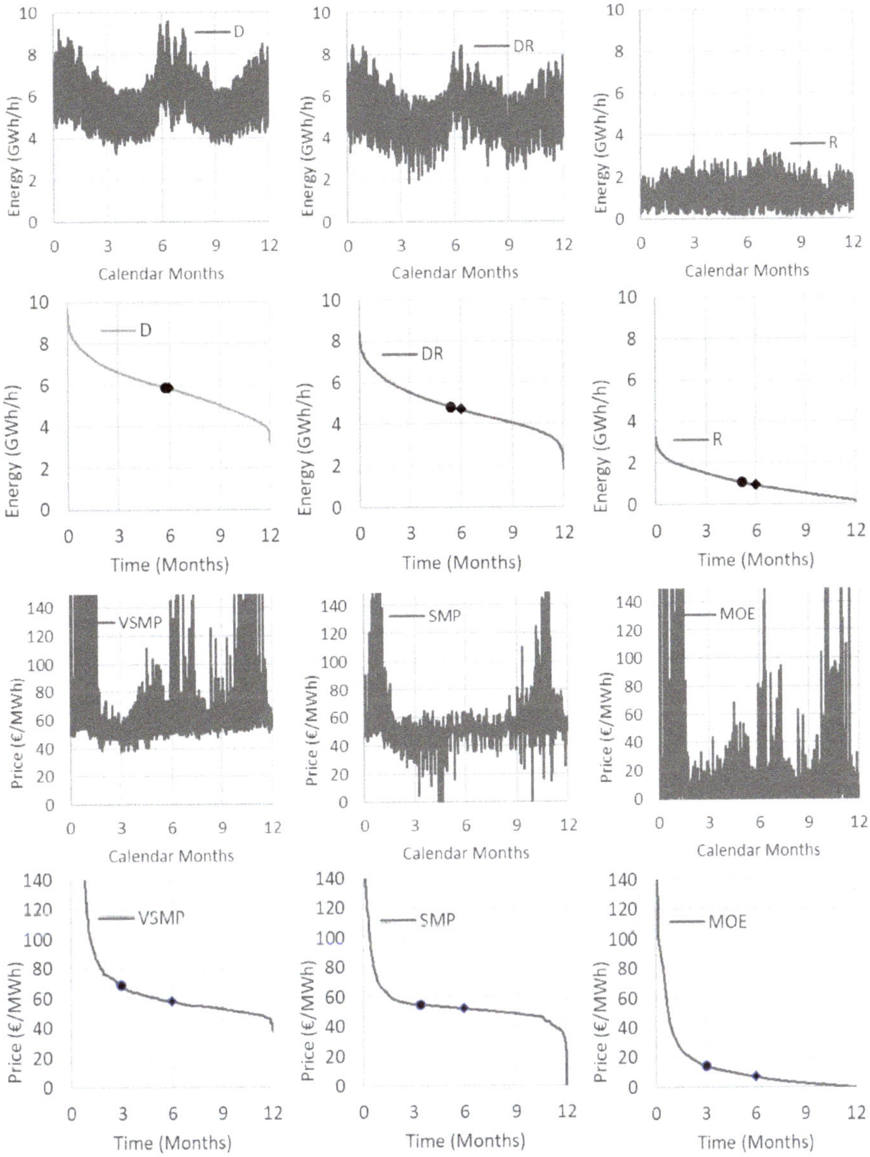

Figure 3. Real data versus time along with the corresponding duration curves: D = demand, DR = demand renewables, R = renewables, VSMP = virtual system marginal price, SMP = system marginal price, MOE – merit order effect. Data from Hellenic electricity market operator [32] and the Hellenic energy exchange group [33].

Table 1. Basic statistics.

Variable	Symbol	St Dev	Mean	Median	Min	Max	Units
Demand	D	1.13	5.89	5.84	3.24	9.61	GWh/h
Demand renewables	DR	1.09	4.84	4.70	1.81	8.48	GWh/h
Renewables	R	0.64	1.05	0.91	0.12	3.29	GWh/h
Virtual system marginal price	VSMP	33.9	68.8	57.6	37.8	>150	€/MWh
System marginal price	SMP	16.8	54.7	51.7	0.0	>150	€/MWh
Merit order effect	MOE	24.5	14.2	6.9	0.0	>150	€/MWh

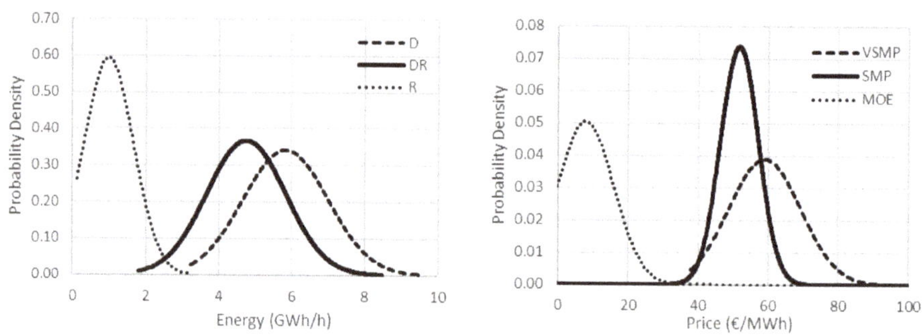

Figure 4. Fitted normal distribution: D = demand, DR = demand renewables, R = renewables, VSMP = virtual system marginal price, SMP = system marginal price, MOE = merit order effect.

The examined quantities, as analyzed in the previous sections, appear as seasonal and daily variation. Seasonal variation is revealed using 24 h moving averages which eliminate the daily variation. The results are presented in Figure 5.

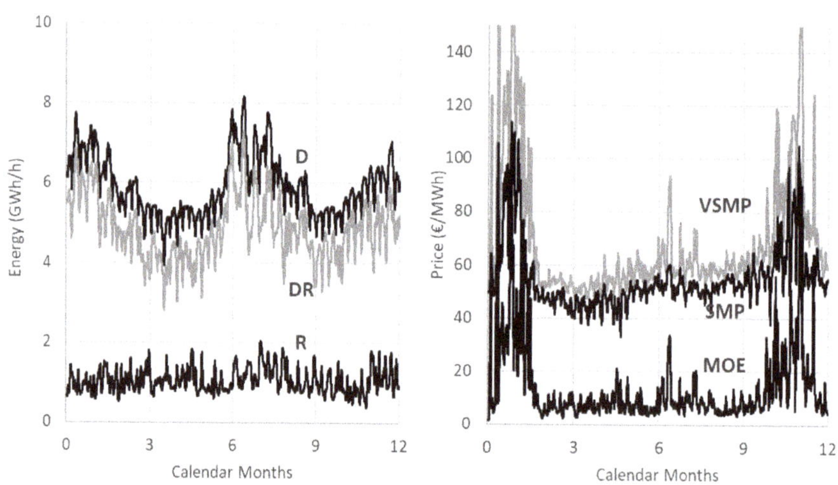

Figure 5. Seasonal variation using 24 hour moving averages: D = demand, DR = demand renewables, R = renewables, VSMP = virtual system marginal price, SMP = system marginal price, MOE = merit order effect.

Daily variation is revealed by averaging data separately for each hour of the day. Averaging is performing for every month since due to seasonal variation every month different daily characteristics appear. The results are presented in Figures 6 and 7.

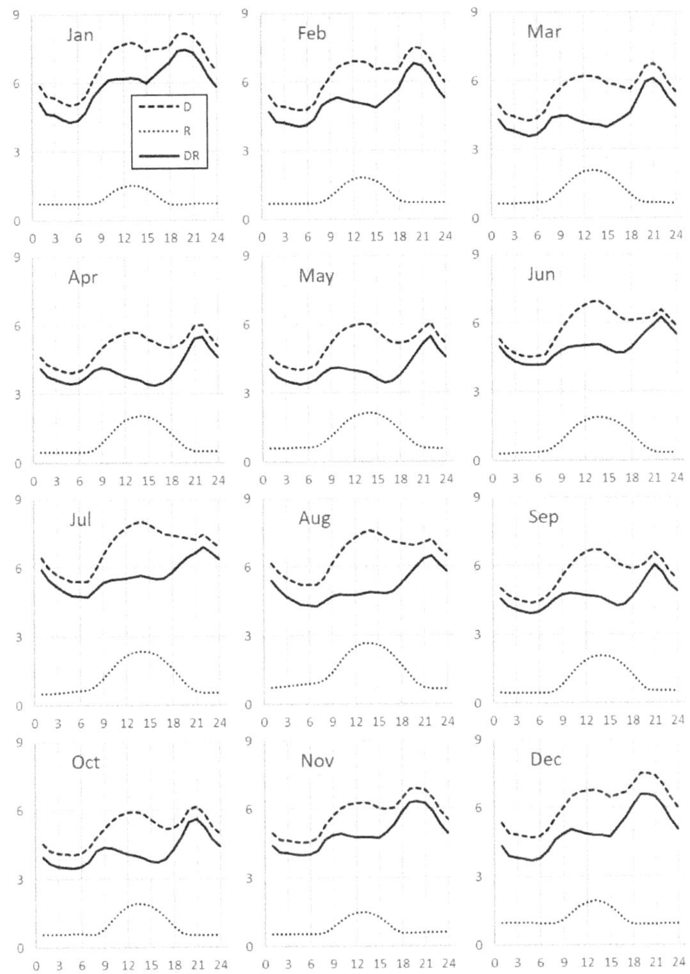

Figure 6. Daily variation using monthly-averaged hourly data: D = demand, DR = demand renewables, R = renewables. Horizontal axis = hour of the day, vertical axis = energy in GWh/h.

It must be noticed that, during January and July 2017, peak demand conditions were met resulting, as expected, in higher SMP, VSMP prices, and MOE.

Due to the pan-European electricity crisis triggered by the withdrawal of several nucleal plants in France for service and repair reasons in late Autumn 2016, wholesale electricity prices all over Europe climbed to very high levels during December 2016 to January 2017. To cover the gap, France turned to imports of electricity from neighbour countries dispersing crisis all over Europe. The extra needs for electricity were covered mostly from natural gas-fired plants leading to a gas market crisis as well. This climb also happened with the SMP in Greece and of course the VSMP. Winter peaking demand reinforced, as expected, market thirst and anxiety for electricity. In February of 2017, the Regulatory Authority of Energy (RAE) imposed a ceiling of 15€/MWh per hour for MOE in order to avoid its extreme values leading to excessive burden for suppliers. The same electricity demand anxiety phenomenon occurred also at the end of 2017 mostly due to bad cold weather conditions reinforced by the memory and fear of the previous year's crisis as well.

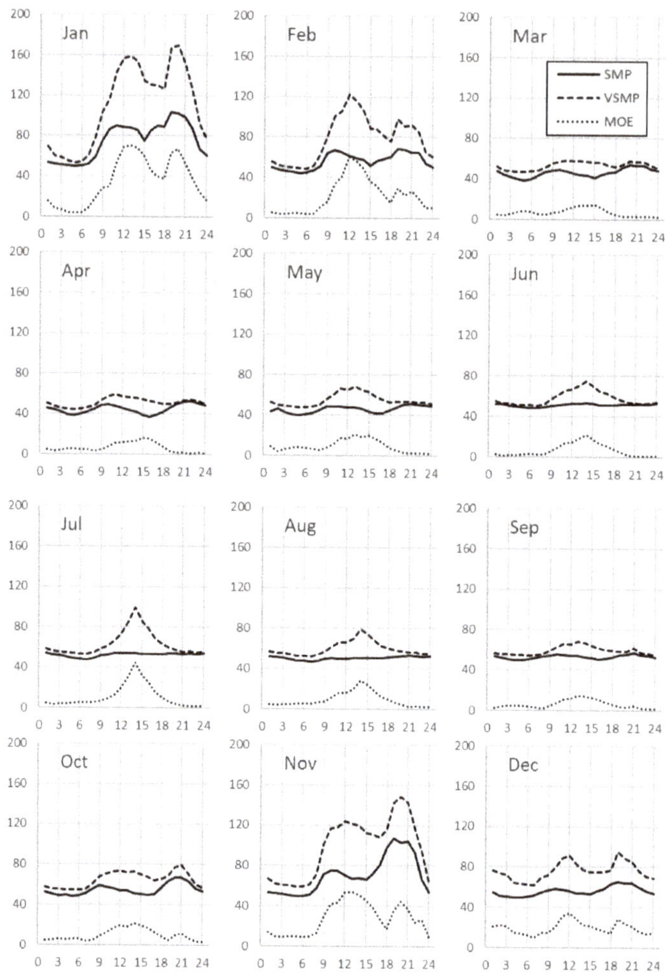

Figure 7. Daily variation using monthly-averaged hourly data: VSMP = virtual system marginal price, SMP = system marginal price, MOE = merit order effect. Horizontal axis = hour of the day, vertical axis = price in €/MWh.

Two electricity demand peaks per day are observed in Figure 6, one during noon and the second in the evening. Noon peak demand is almost completely covered by the PV production all over the year. Noon peak is higher compared to evening peak during summer months while the contrary happens in winter.

Figure 7 shows how the SMP prices escalate when electricity demand is peaking and what happens to wholesale prices (VSMP) if the RES were not present. Furthermore, the PV penetration leads to time concentrated electricity production during noon hours and reduces the SMP during these hours at the most, resulting in the highest MOE values.

4.2. Regression Analysis

Table 2 presents the parameter estimates from the regression analysis as described in the previous sections. Three different regressions were applied and the comparison between real and calculated values are presented in Figures 8–11.

Table 2. Model parameter estimation results.

Merit Order Curve Parameters	Equation (1)		Units
Price at half of maximum supply	P_{med}	55.2	€/MWh
Maximum feasible electricity supply	S_{max}	9.88	GWh/h
Empirical shape constant	n	0.412	-
Demand Seasonal Variation Parameters	**Equation (2)**		
Annual demand	D_T	51.6	TWh/y
Winter activities fraction	d_w	0.099	-
Winter activities peak day	t_{wo}	13.1	days
Winter activities standard period	Δt_w	44.5	days
Summer activities fraction	d_s	0.085	-
Summer activities peak day	t_{so}	205	days
Summer activities standard period	Δt_s	33.3	days
Demand Daily Variation Parameters	**Equation (3)**		
Noon activities fraction	d_n	0.147	-
Noon activities peak hour	t_{no}	12.5	h
Noon activities standard period	Δt_n	3.38	h
Evening activities fraction	d_e	0.127	-
Evening activities peak hour	t_{eo}	21.1	h
Evening activities standard period	Δt_e	3.03	h
Renewable Electricity Variation Parameters	**Equations (5)–(8)**		
Annual production of other renewables	W_T	5.48	TWh/y
Seasonal relative variation of other renewables	ΔW	0.00	-
Other renewables generation peak day	i_{wo}	0.00	days
Annual production of photovoltaics	PV_T	3.71	TWh/y
Seasonal relative variation of photovoltaics	ΔPV	0.294	-
Photovoltaics production minimum day	i_{PVo}	3.58	days
Photovoltaic production peak hour	j_{PVo}	13.7	h

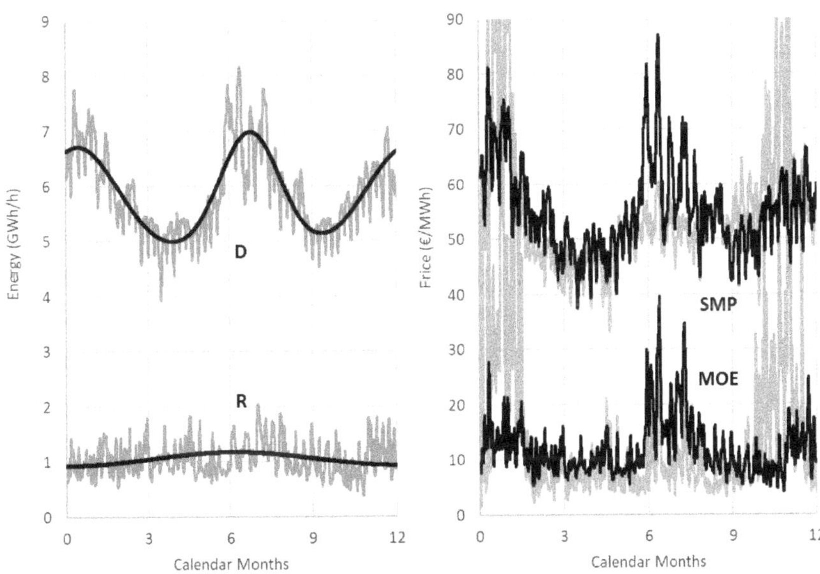

Figure 8. Seasonal variation. Comparison between real (grey lines) and model calculated (black lines) values. D = demand, R = renewables, SMP = system marginal price, MOE = merit order effect.

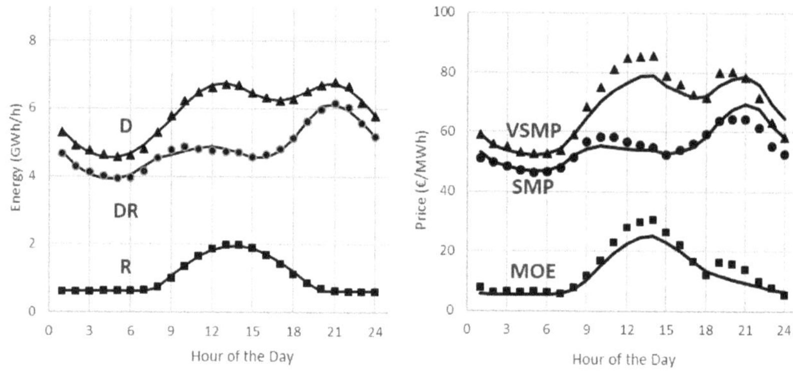

Figure 9. Daily variation: Comparison between real (points) and model calculated (lines) values. D = demand, DR = demand renewables, R = renewables, VSMP = virtual system marginal price, SMP = system marginal price, MOE = merit order effect.

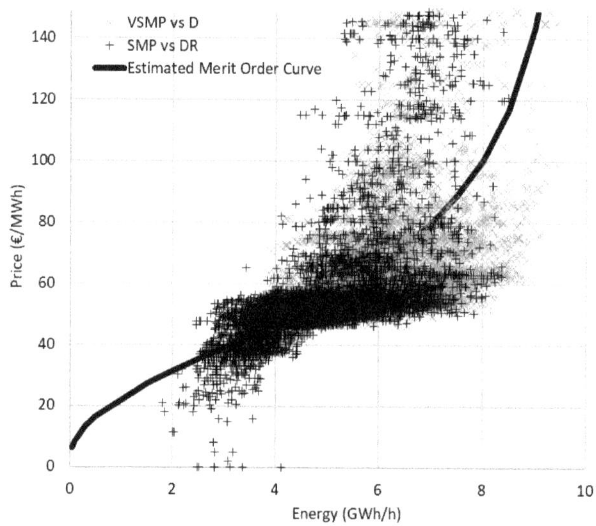

Figure 10. Merit order curve: Comparison between real and model calculated values.

Parameters in Table 2 express either (a) the market characteristics (weather, demand, generation technologies, etc) which remain constant long term or (b) the market size (demand, conventional and renewable installed power, etc) which follow the growth of economy.

The four parameters which express the market size are:

- S_{max} GWh/h the maximum feasible electricity supply;
- W_T GWh/year the total annual electricity generated by renewables except photovoltaics;
- PV_T GWh/year the total annual electricity generated by photovoltaics;
- D_T GWh/year the total electricity demand during the year.

S_{max} is analogous to the installed power of conventional electricity generated systems that are dispatchable, meaning that they determine SMP through hourly price bidding in the day-ahead market. Similarly, W_T and PV_T are analogous to the renewable installed power, non-photovoltaics and photovoltaics, respectively. Finally, D_T expresses the society activities and is analogous to the economic growth.

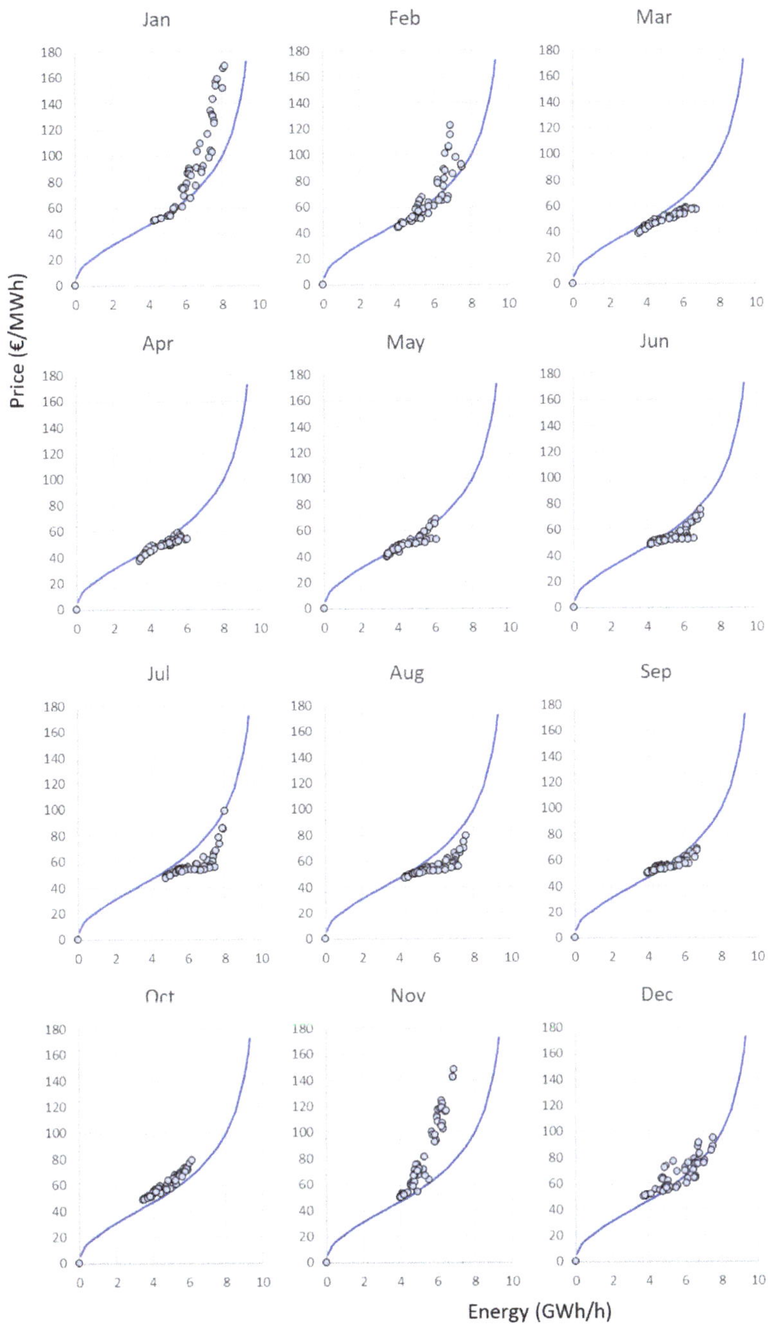

Figure 11. Merit order curve: Comparison between real and model calculated values. Monthly averaged values.

Thus, it is useful to compare the estimated values of these parameters with the recorded values:

- Estimated values of D_T and PV_T are exactly as recorded 51.9 TWh/y and 3.72 TWh/y, respectively, while estimated value of W_T is a little higher than recorded 5.84 TWh/y;
- S_{max} of 9.88 GWh/h is a decline from the expected installed conventional power of 12 GW, because the total capacity does not participate continuously into the market.

4.3. Sensitivity Analysis

The proposed mathematical model can be used to predict electricity market behavior in the future when the parameter estimates in the Table 2 are adjusted to future values. The four parameters which express the size of the market are considered as factors. All other parameters express market characteristics which remain constant long term. Thus, the factors which affect market future behavior are the following four parameters:

- S_{max} GWh/h the maximum feasible electricity supply;
- W_T GWh/year the total annual electricity generated by renewables except photovoltaics;
- PV_T GWh/year the total annual electricity generated by photovoltaics;
- D_T GWh/year the total electricity demand during the year.

S_{max} is analogous to the installed power of conventional electricity generated systems that are dispatchable, meaning that they determine SMP through hourly price bidding in the day-ahead market. Similarly, W_T and PV_T are analogous to the renewable installed power, non-photovoltaics and photovoltaics, respectively. Finally, D_T expresses the society activities and is analogous to the economic growth.

In the sensitivity analysis of Figure 12 the effect of the above four factors on the merit order effect (MOE), separately, is presented. Demand is the crucial factor; it can double the MOE when it increases by about 25%. Renewables appear to have a smaller effect, while conventional installed power appears to have a negative effect. Obviously, all these effects are interpreted from Figure 1 by changing the interception of the demand and supply curves.

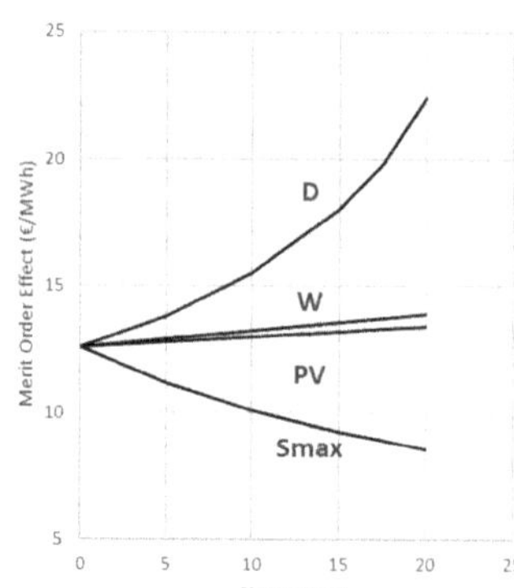

Figure 12. Sensitivity analysis (one factor at a time): Effect of crucial market factors on the merit order effect. D = demand, W = renewables except photovoltaics, PV = photovoltaics, and S_{max} = conventional.

The effect of the factors when they are changed simultaneously is presented in the scenario analysis of Figure 13. The following scenarios are examined:

- D Demand is increased, all other factors are kept constant;
- D + R Demand and renewables are increased by the same rate and conventionals are kept constant;
- D + C Demand and conventionals are increased with the same rate and renewables are kept constant;
- A All factors are increased by the same rate;

- A − D All factors except Demand are increased by the same rate.

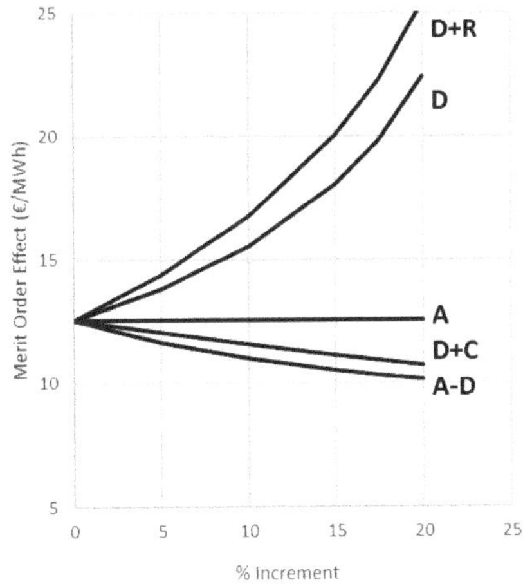

Figure 13. Scenario analysis (simultaneous factor variation): D + R = demand and renewables are increased by the same rate, conventionals are kept constant; D = demand is increased, all other factors are kept constant; A = all factors are increased by the same rate; D + C = demand and conventionals are increased with the same rate, renewables are kept constant; and A − D = all factors, except demand, are increased by the same rate.

Obviously, the higher positive effect is obtained when demand and renewables are increased simultaneously, and the higher negative effect is obtained when demand remains constant and all other factors are increased. Similarly, any other combination of changes can be examined.

5. Conclusions

During the period between October 2016 and December 2018 the Hellenic DAM calculated explicitly the MOE using an innovative mechanism to directly charge the electricity suppliers. Through the MOE charge (called PXEFEL in Greeks), suppliers were returning to the RES account the financial benefit they were enjoying because of MOE, namely the lower SMP values in the DAM due to RES penetration.

The above mechanism needs an appropriate model to evaluate and analyze daily, seasonal, and long-term variations of the MOE towards an optimum MOE charge strategy. Thus, a simple model for the DAM is proposed and validated to real data. The model, considering the main factors which govern the process, predicts the seasonal and daily variation of electricity demand, renewable production, SMP, and MOE. The model was fitted adequately to historic data of the Hellenic DAM during the year 2017.

The model innovation is based on the separate direct simulation of the supply and demand curves and the seasonal and daily variation of the electricity demand and renewable generation.

On the basis of the proposed model and the market recorded data, the effect of the renewable penetration on the wholesale electricity prices is analyzed.

The model can further be used to predict future market behavior when the basic factors (electricity demand, conventional power, conventional costs, and renewable penetration) are known or estimated. Thus, the effect of the evolution of the RES penetration on the MOE can be estimated and analyzed towards an optimum RES penetration supporting strategy.

The proposed model was applied successfully to the Hellenic electricity market, but it can be used to any other similar market.

Author Contributions: All authors contributed equally.

Funding: This research received no external funding.

Conflicts of Interest: The authors declare no conflict of interest.

Nomenclature

cf	the correction factor (-)
d_e	the portion of the total daily demand for evening activities (-)
D_i	the electricity demand during the day i (GWh/day)
D_{ij}	the electricity demand during the hour j of the day i (GWh/hour)
d_n	the portion of the total daily demand for noon activities (-)
d_s	the portion of the total annual demand for summer activities (-)
D_T	the total electricity demand during the year (GWh/year)
d_w	the portion of the total annual demand for winter activities (-)
i	the day of the year (1, 2, ... , 365) (days)
i_{PVo}	the day with the minimum production for photovoltaics (days)
i_{Wo}	the day of minimum production for renewables except photovoltaics (days)
j	the hour with maximum production from photovoltaics (hours)
j_{PVo}	the hour with maximum production from photovoltaics (hours)
n	empirical shape constant (-)
P	the supply price (€/MWh)
PV_i	the electricity generated from photovoltaics during the day i (GWh/day)
PV_{ij}	the electricity generated from photovoltaics during the hour j of the day i (GWh/hour)
P_{med}	the electricity price at half of maximum supply (€/MWh)
PV_T	the total annual electricity generated by photovoltaics (GWh/year)
R_i	the electricity generated from renewables during the day i (GWh/day)
R_{ij}	the electricity generated from renewables during the hour j of the day i (GWh/hour)
S	the dispatchable electricity supply (GWh/h)
S_{max}	the maximum feasible electricity supply (GWh/h)
SMP	the System Marginal Price (€/MWh)
SMP_{ij}	the System Marginal Price during the hour j of the day i (€/MWh)
t_{eo}	the time of the peak of evening activities (hours)
t_{no}	the time of the peak of noon activities (hours)
t_{so}	the time of the peak of summer activities (days)
t_{wo}	the time of the peak of winter activities (days)
$VSMP$	the Virtual System Marginal Price (€/MWh)
$VSMP_{ij}$	the Virtual System Marginal Price during the hour j of the day i (€/MWh)
W_i	the electricity generated from renewables except photovoltaics during the day i (GWh/day)
W_{ij}	the electricity generated from renewables except photovoltaics during the hour j of the day i (GWh/hour)
W_T	the total annual electricity generated by renewables except photovoltaics (GWh/year)
ΔPV	the seasonal variation amplitude of photovoltaics (GWh/day)
Δt_e	the typical duration of the evening activities (hours)
Δt_n	the typical duration of the noon activities (hours)
Δt_s	the typical duration of the summer activities (days)
Δt_w	the typical duration of the winter activities (days)
ΔW	the seasonal variation amplitude of renewables except photovoltaics (GWh/day)

Abbreviations

A	all
C	conventionals
CO_2	carbon dioxide
D	demand
DAM	day-ahead market
DR	demand minus renewables
ETMEAR	surcharge on electricity price (Greek abbreviation)
FIP	feed-in premium
FIT	feed-in Tariff
MOC	merit order curve
MOE	merit order effect
PVs	photovoltaics

PXEFEL	merit order effect charge (Greek abbreviation)
R	renewables
RAE	regulatory authority of energy
RES	renewable energy sources
SMP	system marginal price
VSMP	virtual system marginal price
W	renewables except photovoltaics

References

1. Cludius, J.; Forrest, S.; MacGill, I. Distributional effects of the Australian Renewable Energy Target (RET) through wholesale and retail electricity price impacts. *Energy Policy* **2014**, *71*, 40–51. [CrossRef]
2. Cludius, J.; Hermann, H.; Matthes, F.C.; Graichen, V. The merit order effect of wind and photovoltaic electricity generation in Germany 2008–2016 estimation and distributional implications. *Energy Econ.* **2014**, *44*, 302–313. [CrossRef]
3. Clò, S.; Cataldi, A.; Zoppoli, P. The merit-order effect in the Italian power market: The impact of solar and wind generation on national wholesale electricity prices. *Energy Policy* **2015**, *77*, 79–88. [CrossRef]
4. Traber, T.; Kemfert, C. Gone with the wind? Electricity market prices and incentives to invest in thermal power plants under increasing wind energy supply. *Energy Econ.* **2011**, *33*, 249–256. [CrossRef]
5. McConnell, D.; Hearps, P.; Eales, D.; Sandiford, M.; Dunn, R.; Wright, M.; Bateman, L. Retrospective modeling of the merit-order effect on wholesale electricity prices from distributed photovoltaic generation in the Australian National Electricity Market. *Energy Policy* **2013**, *58*, 17–27. [CrossRef]
6. Weigt, H. Germany's wind energy: The potential for fossil capacity replacement and cost saving. *Appl. Energy* **2009**, *86*, 1857–1863. [CrossRef]
7. De Miera, G.S.; del Río González, P.; Vizcaíno, I. Analysing the impact of renewable electricity support schemes on power prices: The case of wind electricity in Spain. *Energy Policy* **2008**, *36*, 3345–3359. [CrossRef]
8. Sensfuß, F.; Ragwitz, M.; Genoese, M. The merit-order effect: A detailed analysis of the price effect of renewable electricity generation on spot market prices in Germany. *Energy Policy* **2008**, *36*, 3086–3094. [CrossRef]
9. Ciarreta, A.; Espinosa, M.P.; Pizarro-Irizar, C. Is green energy expensive? Empirical evidence from the Spanish electricity market. *Energy Policy* **2014**, *69*, 205–215. [CrossRef]
10. Gelabert, L.; Labandeira, X.; Linares, P. An ex-post analysis of the effect of renewables and cogeneration on Spanish electricity prices. *Energy Econ.* **2011**, *33*, S59–S65. [CrossRef]
11. Jónsson, T.; Pinson, P.; Madsen, H. On the market impact of wind energy forecasts. *Energy Econ.* **2010**, *32*, 313–320. [CrossRef]
12. Forrest, S.; MacGill, I. Assessing the impact of wind generation on wholesale prices and generator dispatch in the Australian National Electricity Market. *Energy Policy* **2013**, *59*, 120–132. [CrossRef]
13. Woo, C.K.; Horowitz, I.; Moore, J.; Pacheco, A. The impact of wind generation on the electricity spot-market price level and variance: The Texas experience. *Energy Policy* **2011**, *39*, 3939–3944. [CrossRef]
14. Lu, P.; Pr, J.; Janda, K. The merit order effect of Czech photovoltaic plants. *Energy Policy* **2017**, *106*, 138–147. [CrossRef]
15. Aineto, D.; Iranzo-Sánchez, J.; Lemus-Zúñiga, L.G.; Onaindia, E.; Urchueguía, J.F. On the influence of renewable energy sources in electricity price forecasting in the Iberian market. *Energies* **2019**, *12*, 2082. [CrossRef]
16. Eurostat. Europe 2020 targets: Statistics and indicators for Greece. In *Share of Renewable Energy*; European Commission: Brussels, Belgium, 2019. Available online: https://ec.europa.eu (accessed on 4 September 2019).
17. Giannini, E.; Moropoulou, A.; Maroulis, Z.; Siouti, G. Penetration of Photovoltaics in Greece. *Energies* **2015**, *8*, 6497–6508. [CrossRef]
18. Karteris, M.; Papadopoulos, A.M. Legislative framework for photovoltaics in Greece: A review of the sector's development. *Energy Policy* **2013**, *55*, 296–304. [CrossRef]
19. Simoglou, C.K.; Member, S.; Biskas, P.N.; Zoumas, C.E.; Bakirtzis, A.G.; Member, S. Evaluation of the Impact of RES Integration on the Greek Electricity Market by Mid-term Simulation. In Proceedings of the 2011 IEEE Trondheim PowerTech, Trondheim, Norway, 19–23 June 2011; pp. 1–8.

20. Simoglou, C.K.; Biskas, P.N.; Bakirtzis, A.G. Impact of Increased RES Penetration on the Operation of the Greek Electricity Market. In Proceedings of the 7th Mediterranean Conference and Exhibition on Power Generation, Transmission, Distribution and Energy Conversion, Agia Napa, Cyprus, 7–10 November 2010; Available online: https://ieeexplore.ieee.org/document/5716014 (accessed on 4 September 2019).
21. Electricity Production from Renewable Energy Sources and High Efficiency Cogeneration of Heat and Power, and Other Provisions. Law 3468/2006, Government Gazette A129, Greece. 2006. Available online: http://www.et.gr/index.php/anazitisi-fek (accessed on 29 October 2019).
22. New Support Scheme for Power Plants from Renewable Energy Sources and High Efficiency Cogeneration of Heat and Power. Law 4414/2016, Government Gazette A149, Greece. 2016. Available online: http://www.et.gr/index.php/anazitisi-fek (accessed on 29 October 2019).
23. Liberalization of Electricity Market. Regulation of Energy Policy Issues and Other Provissions. Law 2773/1999, Government Gazette A286, Greece. 1999. Available online: http://www.et.gr/index.php/anazitisi-fek (accessed on 29 October 2019).
24. Ratification of Mid-term Fiscal Strategy 2013–2016—Urgent Regulations Relating to the Implementation of Law 4046/2012 and the Midterm Fiscal Strategy 2013–2016. Law 4093/2012, Government Gazette A222, Greece. 2012. Available online: http://www.et.gr/index.php/anazitisi-fek (accessed on 29 October 2019).
25. Emergency Measures for the Application of Laws 3919/2011, 4093/2012 and 4127/2013. Law 4152/2013, Government Gazette A107, Greece. 2013. Available online: http://www.et.gr/index.php/anazitisi-fek (accessed on 29 October 2019).
26. Support Measures for the Stabilization and Development of Greek Economy and Other Provisions. Law 4254/2014, Government Gazette A85, Greece. 2014. Available online: http://www.et.gr/index.php/anazitisi-fek (accessed on 29 October 2019).
27. *Approval of RES Administrator and Guarantees of Origin Code according to paragraph 3 of art 117E of Law 4001/2011*; Decision N°509/2018; Government Gazette B2307: Greece, 2018.
28. *Electricity Exchange Code*, 5.3 ed.; Energy Exchange (EnEx) Group: Athens, Greece, 2019. Available online: http://www.enexgroup.gr/fileadmin/groups/EDRETH/Manuals/20190910_KSIE_Ekdosi_5.3.pdf (accessed on 29 October 2019).
29. *Modification of Electricity Exchange Code and Electricity Exchange Code Manual, According to the Provisions of Paragraph 3.a, (bb) of Article 143 and Law 4001/2011*; RAE Decision No 334/2016; Government Gazette B3169: Greece, 2016.
30. Bello, A.; Bunn, D.W.; Reneses, J.; Munoz, A. Medium-Term Probabilistic Forecasting of Electricity Prices: A Hybrid Approach. *IEEE Trans. Power Syst.* **2017**, *32*, 334–343. [CrossRef]
31. Loumakis, S.; Giannini, E.; Maroulis, Z. Renewable Energy Sources Penetration in Greece: Characteristics and Seasonal Variation of the Electricity Demand Share Covering. *Energies* **2019**, *12*, 2441. [CrossRef]
32. Hellenic Electricity Market Operator. Houlry Day Ahead Market Data. Available online: http://www.lagie.gr (accessed on 7 January 2018).
33. Hellenic Energy Exchange Group. Houlry Day Ahead Market Data. Available online: http://www.enexgroup.gr (accessed on 8 September 2019).

© 2019 by the authors. Licensee MDPI, Basel, Switzerland. This article is an open access article distributed under the terms and conditions of the Creative Commons Attribution (CC BY) license (http://creativecommons.org/licenses/by/4.0/).

energies

Article

Electricity Market Empowered by Artificial Intelligence: A Platform Approach

Yueqiang Xu [1,*], Petri Ahokangas [1], Jean-Nicolas Louis [2] and Eva Pongrácz [2]

1 Martti Ahtisaari Institute, University of Oulu, 90014 Oulu, Finland; petri.ahokangas@oulu.fi
2 Water, Energy and Environmental Engineering Research Unit, Faculty of Technology, University of Oulu, 90014 Oulu, Finland; jean-nicolas.louis@oulu.fi (J.-N.L.); eva.pongracz@oulu.fi (E.P.)
* Correspondence: yueqiang.xu@oulu.fi; Tel.: +358-503-411-827

Received: 28 August 2019; Accepted: 24 October 2019; Published: 30 October 2019

Abstract: Artificial intelligence (AI) techniques and algorithms are increasingly being utilized in energy and renewable research to tackle various engineering problems. However, a majority of the AI studies in the energy domain have been focusing on solving specific technical issues. There is limited discussion on how AI can be utilized to enhance the energy system operations, particularly the electricity market, with a holistic view. The purpose of the study is to introduce the platform architectural logic that encompasses both technical and economic perspectives to the development of AI-enabled energy platforms for the future electricity market with massive and distributed renewables. A constructive and inductive approach for theory building is employed for the concept proposition of the AI energy platform by using the aggregated data from a European Union (EU) Horizon 2020 project and a Finnish national innovation project. Our results are presented as a systemic framework and high-level representation of the AI-enabled energy platform design with four integrative layers that could enable not only value provisioning but also value utilization for a distributed energy system and electricity market as the new knowledge and contribution to the extant research. Finally, the study discusses the potential use cases of the AI-enabled energy platform.

Keywords: electricity market; energy market; artificial intelligence; digital platform; peer-to-peer

---

1. Introduction

The energy transformation, so sought after world-wide, would only be effective via the increased integration of renewable energies that are supported and enabled by intelligent grids, or smart grids. Smart-grid technologies will make it possible to use the available renewable energy sources efficiently and sustainably to create added value to the energy service as well as reducing costs for energy consumers and prosumers while supporting a decentralized and open architecture and design for the energy system [1]. However, energy scholars focusing on power systems acknowledge the challenges with increasingly complex energy systems. As such, the electric grid is a massive infrastructure with e.g., physical limitations [2].

The European Commission [3] envisions that tomorrow's power grids will be made up of interconnected and diverse systems, with a growing number of distributed energy generation and consumption equipment and appliances that generate a large volume of data [4]. Considering only smart meters, if the average packet size is about 200 bytes [5], with a reading interval of 15 min as suggested in the European Union (EU) regulations and the 200 million smart meters that are deployed in 2020 [6], the total amount of memory in Europe is 5 606 TB of information. Reduce the sampling size to every second for near-real-time network measurement, and this is around 5 exabytes of data to be collected within a year only from smart meters. In particular, the electricity industry demands big data capabilities and novel architectures that can enhance power system management of the more complex and decentralized grids [3]. Thus, over the last few decades, research efforts and scholarly discussion around the world have been focused on ways to effectively predict and optimize energy use.

Artificial intelligence (AI) as an emerging technological field has been altering numerous sectors of the world's industries and economies. AI is seen as an innovation front and enabling technology [7,8]. AI includes a broad collection of computational techniques to extract insights from a variety of data sources—including the so-called "small data" that is generated by the algorithm itself) that assist in decision-making [9]—and create actionable information [10]. Generally, AI is deemed a general-purpose technology that can have significant technological, social, economic and political implications [9,11,12]. Some describe AI as parallel to the steam engines or electricity technologies that have changed many aspects of human life.

Evidently, AI will radically transform the energy sector. General Electric (GE) estimates that AI can enhance the production of a wind farm by as much as 20% [13]. At the same time, as claimed by Nagy et al. [1], the AI transformation in the energy industry will directly influence the international energy stability and economic prosperity. For instance, the use of artificial neural networks is of great importance for electricity companies to improve their productivity, the quality, and safety of the productions, and the stability of the electricity. As a case in point, the Google Date Center utilized the DeepMind AI to effectively achieve 40% of the energy-saving for the data center operation [14].

However, as AI technology adoption and penetration rates increase in the energy sector, the existing AI and energy-related research remain under-explored without sufficient scholarly effort on AI applications in the energy and electricity market design on top of the physical and technical energy infrastructure.

The study of Quan and Sanderson [15] introduces a holistic framework to analyze AI systems and platforms, including AI core technologies, AI platforms, and AI applications. Based on this framework, the extant AI and energy research has been largely focused on AI core technologies. For example, AI technology has penetrated the technical applications rapidly in the industrial systems. Technically, in the power and electrical engineering domain, AI techniques, such as expert systems, neural networks, and fuzzy logic have been utilized to solve various technical challenges [16], including but not limited to (1) energy forecasting [17,18], (2) energy market price prediction [19], (3) smart grid fault detection [20], (4) demand-side management [21], (5) building energy management [22], (6) smart home demand response management [23], and (7) smart grid data security with AI and blockchain [24]. In terms of driving the energy transition process, the greatest potential in the use of AI is forecasting renewable energy potential, big data management and optimization of hybrid renewable energy systems, e.g., [25,26].

Research Gaps Relevant to the Study

Several gaps can be identified in the extant research. **The first gap** in the current research is the lack of AI platform discussion dedicated to energy systems and market design. As argued by Reich [27] and Singer et al. [28], to the present day, AI systems (including AI-empowered energy systems) have hardly reached broad use in practice for several reasons. Research has often focused on narrow and specialized technical subtasks, not on larger and more integrated problems. Specifically, for AI's application in renewable systems, it is suggested by [29] that most studies discuss the utilization of AI approaches in wind- and solar-based systems. AI approaches have great potential, but their effective utilization in future research is needed to provide more novel approaches to managing renewable energy sources, such as in the case of AI for hybrid renewables.

Although AI is identified as a promising optimization method [25], there is little discussion on it at the systemic level. A "platform" as a system model or architecture can tackle this problem. As stated by Landahl et al. [30], when developing a variety of products for a large number of customers/end users, the producers strive for commonalities and distinguishing features of the developed product variants. A common approach for this is the adoption of a platform model [31], which is typically achieved by modularizing the product's architecture [32].

As suggested by Ramos and Porto [2], some of the important problems in energy supply technology and in the energy market require capabilities such as (1) heuristic search, (2) logical thinking, (3) perception as well as (4) uncertainty tolerance. All these needs give rise to the opportunity for cross-investigation between energy research and AI technologies. At a broader systemic level (e.g., a system platform perspective), AI does not only enhance a certain operation but can be potentially an effective enabler of the new solutions in the energy ecosystem in the areas of enhanced artificial neural networking and decision making, operation, maintenance, market monitoring, as well as optimized trading in the electricity markets [2].

For another emerging paradigm of energy internet thinking, the energy internet is considered to be a new and advanced paradigm for smart grids with massive inter-connected devices and bi-directional power and information flows. This emerging paradigm also demands more advanced techniques to optimize the entire energy system rather than individual applications [33], including big data, AI-based control, and optimization as well as cloud computing. Thus, the AI platforms for energy system and market design at a systemic level is a research front that requires more attention and research effort.

The second gap relates to the synergy between AI applications and electricity market design. The research and development of AI focus primarily on developing new algorithms and software that tackle a focal problem [29]. While most of the previous studies cover the applications of AI techniques, such as fuzzy logic or artificial neural networks in renewable systems [29], in theoretical research there is relatively limited discussion on AI's potential in the electricity market such as facilitating market trading and renewable matchmaking in the electricity marketplace, especially in the emerging context of massive renewable integration into the smart grids. In contrast, empirically, companies in the energy industry have started experimenting with different use cases. For example, ABB focuses on advanced testing methods of 3D-simulation of power plants with a holistic view of the system which is based on smart human-computer interaction systems [34].

Fundamentally, markets have played a key role in giving market participants (or market actors) the opportunity to profit from the trade. Market design literature (e.g., [35]) suggests that the design of a market needs structure and a diverse range of market operations and transaction mechanisms to operate efficiently [36]. Market design studies often delve into fundamental problems: how the market agents can learn the patterns of the market to determine the optimal transaction prices? Or, how the market agents can have an enhanced understanding of the market environment where they operate?

This research suggests that in the energy/electricity market, it is possible to harness the improved predictions provided by data and AI algorithms to enable the marketplaces to better predict consumer and producer demand and supply. This proposition is supported by the previous literature of Ramo and Porto [2], which calls for further study on the economic perspective of the energy industry. However, we need to develop a deeper understanding of how such a platform is constructed than an ontological description. This is addressed in this study.

The third gap pertains to the value utilization of the electricity market. As mentioned, existing AI technologies have been focused on how to optimize energy production and distribution for the legacy energy systems. The issue arises when the platform thinking is adopted for the existing energy systems that overemphasize on the energy production and energy supply chain. In platform literature, a platform-like marketplace is formed because of the participation from both supply and demand, or value provisioning and utilization that help create the network effect for a platform to succeed [37]. By adopting platform thinking, AI can impact the utilization side of the marketplace or platform is limited from the existing energy research and literature.

By developing a high-level systemic architecture framework for an AI-enabled energy platform, this paper contributes to the need for a better understanding and articulation of the potential AI platform in the energy industry as well as AI's roles and applications in electricity market design and operation. More specifically, the research addresses two parallel research questions: firstly, what can be the AI-enabled energy platform for the electricity market? and secondly what are the potential AI applications for the value provisioning and utilization of the energy platform in order to provide better services for energy consumers?

Overall, by addressing these two questions, the research contributes to the existing literature in three ways: (1) the study enriches the existing AI and energy research discussion with a more holistic view: expanding from the technical focus of narrow AI technical applications to the utilization of AI at a systemic level; (2) the research introduces platform thinking and a platform architecture/model for the potential integration of AI into the energy systems. An AI-empowered energy system can have the capability to manage massive data and complex grid and market operations when the shares of the renewables increase significantly; (3) the research explores and proposes a number of AI applications in managing and optimizing the market operation and design. Previous literature [2] only gives a proposition that AI technology can transform the energy market operation without providing details and in-depth discussion. This research gap is filled in this research, including discussing how AI and blockchain can transform the further electricity market based on the extant studies of [38–40].

As addressed by [1], the direction of AI development in the energy industry is still an open path. To facilitate more effective and efficient human–AI collaboration, more research efforts are required to progress towards the development of new synergies AI and other parts of the energy system. Thus, this study is inspired and built on this call for action. The research question of this study is as follows: *"How can we integrate AI technologies into the future energy system architecture, in order to enhance transparency and participation of users in different markets?"* The focus is on AI's potential at the utilization side of the electricity market, in order to enable smart energy services and increase value utilization. The paper focuses on AI-enabled energy market architecture at the system level. Therefore, we are open to a variety of AI technologies that demonstrate potential in the energy industry rather than a particular AI technique.

This paper contains the following sections: Section 2 encompasses a number of definitions on AI, AI's application in smart grid and renewable researches. This section further introduces and explains the platform thinking and elaborates on the research method. Section 3 shows the outcome of the study by constructing and articulating the architectural model of an AI platform in the context of smart grids by expanding the 4C (connection, content, context, and commerce) framework in ICT (information and communication technology) and digitalization research. Section 4 discusses the potential applications of AI in the electricity market through the framework created in the previous section. Finally, Section 5 provides the conclusions of this study and heads to further research direction.

2. Literature Review and Methods

Nilsson [41] defines that "AI encompasses the intelligent behavior in artifacts, which involves perception, reasoning, learning, communicating and acting in complex environments". The main purpose of AI is basically considered as the creation of intelligent machines that can perform human-like capabilities and beyond. Furthermore, it is recommended by the scholars (e.g., [42]) that the capabilities of AI can be divided into subgroups or categories such as reasoning, problem-solving, natural language processing, perception as well as general intelligence [42] to solve complex problems [43].

From the computer science perspective, AI is deemed as an approach for the software programs to analyze focal problems and devise appropriate solutions. AI core technologies are classified as machine learning, artificial neural network, and heuristic search [44]. To this date, numerous studies have been conducted to explore, experiment and investigate the capability and use of AI in a variety of fields such as big data analytics, information system, production engineering, and medicine, just to name a few [29].

2.1. *Artificial Intelligence (AI) Application in Smart Grids and Renewables*

Based on the framework of Quan and Sanderson [15], AI in energy research is mostly directed in different techniques and solutions for designing, optimizing, and managing the operations of different domains. In energy and renewables research, the literature review shows that AI has been studied in areas such as solar, wind, geothermal and hydro [29].

A substantial number of studies have focused on energy demand prediction [45,46]. For instance, prediction models can be categorized into the following types [47]: physical-oriented, data-oriented, hybrid, and large-scale prediction methods. In the area of energy prediction, the artificial neural network technique is well-known for the accurate forecasting of energy usage [29]. However, there is a challenging issue related to accuracy when the scale is reduced (e.g., neighborhood or household level) although precise load forecasting is possible at the aggregated level (e.g., national level), In this vein, the study of Ahmadi [48] suggests that the artificial neural network is developed in as many as 40% of the energy artificial neural networking algorithms [49].

The use of AI in solar energy: AI applications in solar energy are reviewed in the literature [50–52]. Applications often involve the use of artificial neural network methods for solar modeling [51] in both single and hybrid approaches [53]. For instance, the use of machine learning can improve the solar forecasting accuracy with a range of 30% to 50% increase, e.g., [54,55] compared to conventional forecasting models.

The use of AI in wind energy: empirical research and trial by industry incumbents like GE show that through the use of IoT (Internet of Things) sensors, data networks, and advanced analytics can optimize the wind turbines to have as high as 20% peak efficiency in artificial neural network energy production [13]. By reviewing the academic research, Jha et al. [29] identify a number of studies related to the physical model, statistical model as well as artificial neural network methods for wind modeling. Basically, the applications of AI techniques in wind energy cover several categories, such as neural, statistical and evolutionary learning [29].

Overall, the artificial neural network, reinforcement learning, genetic algorithms, and multiagent systems are common techniques for AI to solve the problems of classification, forecasting, artificial neural networking, optimization, and control [2]. The current literature and research also show the disintegrated gap and low visibility of studies that integrate ICT systems, power/energy systems as well as energy market research [2]. However, this research suggests that AI and machine learning is an emerging application in the power and energy field, quickly gaining a footing within the intelligent energy systems domain.

2.2. Platform Thinking and Approach

The literature offers numerous platform types with different definitions and properties. The platform concept and thinking can find its presence in both economic and technology/engineering literature [56]. The work of Gawer [56] seeks to integrate technological platforms.

Digital platform architectures can generally be classified into two types which can have multiple hybrid combinations [57]: the economic-oriented transactional platform and the engineering-oriented technology platform. First, an economic-focused transaction platform facilitates exchanges by fragmented groups of consumers and/or suppliers. The platform provides a matchmaking mechanism to connect supply and demand that are normally fragmented [58]. This definition is in line with an economic point of view, and digital platforms are seen as markets where the platform facilitates exchanges between actors who otherwise would not be able to do business with each other. An example of this type of platform is Uber, as the platform connects drivers with travelers and facilitates the transaction between the two groups as a two-sided market or platform [59]. The economic platform thinking can also be extended to multi-sided markets [60,61].

The engineering/technological platform is seen as a digital platform with technological architectures that enable innovation [56,62,63]. Such a platform architecture builds a foundational technology and distribution system that enables the other technologies to be integrated into the platform, enhancing value for the whole system. For instance, the mobile app ecosystem of Apple (App Store) is a well-known case of such a digital platform [63]. The core idea is that the platform includes modules that can be used to decompose complex systems into manageable components connected by interfaces.

The modules provide information-sharing because each module does not need information about the entire system, which in turn allows access control over specific data [59,64].

The research of Gandia and Parmentier [65] shows that it is possible to combine and integrate both the economic and engineering/technological conceptualization of the platform. For example, economic thinking would consider the declining costs of information gathering and market mediation for the empirical multi-faceted platforms on the Internet, such as eBay, Amazon, and Airbnb. In parallel, the technological perspective of a platform is rooted in presenting the platform as an integral system of software and hardware that offers standards, interfaces, and mechanisms that enable supplement technology vendors or providers to be matched with the end users as well as interacting with other technology providers. In this vein, the platform innovators' and the complementary partners' success depend on the platform owner's continuous innovation [65].

2.3. Research Method

In an effort to address the research gaps, this research adopts two approaches: First, system architecture thinking, e.g., [66], where a "*systems architecture is the conceptual model that defines the structure, behavior, and more views of a system*", and action design research (ADR) [67]. In the paper, the 4C systemic framework is used: connection, content, context, and commerce [68] and the 4C layered typologies [69] that can incorporate both the value provisioning and value utilization of a digital system, which bridges the gap in existing energy research. The 4C framework can be reconciled with the market layer of the smart grid architecture model (SGAM) that is the formal system framework established by the standardization organizations in the European Union [70]. Furthermore, the paper follows the system architecture logic and align the discussion based on system architecture's three key constructs: the system structure, the system (and the component) behavior, and the view of a system.

Second, this study utilizes the constructive and action-oriented research approach for theory building and concept proposition for the AI energy platform architecture as an IT (information technology) artifact, aiming at bringing novelty value to the energy system and market design studies. Information system research has rich literature on the relevant approaches for dealing with the development and design of system architecture. For instance, action research is one stream of studies that incorporate theory generation with researcher intervention to tackle organization-related research problems and connecting the theory with practices [67,71,72]. In another stream of information system research, design science is used to create IT artifacts with a constructive approach. The IT artifacts are "shaped by the interests, values, and assumptions of a wide variety of communities of developers, investors, users" [73].

Seins et al. [67] have proposed a new IT artifact (e.g., software architecture) development approach that combines action research and design science, namely, ADR. The ADR approach stresses the importance of the relevance cycle through explicit guidance for integrating the process of building, intervention, and evaluation in a concerted effort [67]. In the context of this study, ADR addresses two challenge areas for creating the system architecture of the AI-enabled energy platform; (1) addressing a problem situation in a specific organizational setting which is the ecosystem as an organizational form in the energy sector; (2) constructing and validating an IT artifact, that is the 4C platform architecture that addresses the issues/needs in the energy ecosystem. Thus, this paper follows the four processes of the ADR to formulate, build, reflect and formalize the AI-enabled energy platform as an artifact. This means that the aim of the research is to generate theory and a framework with the help of empirical data, rather than the deductive approach that uses hypothesis and theory testing based on stringent empirical data [74].

This study is built on two large=scale research projects, including the EU P2P-SmartTest European Horizon 2020 project [75] and the VirpaD Finnish national innovation project [76]. The EU-level P2P-SmartTest project focuses on the investigation of a more intelligent electricity distribution system through the incorporation of 5G technology, local energy markets and novel business models for future power grids. Peer-to-peer (P2P)-based distributed smart energy grids have the capability to integrate multiple sources of distributed renewable generations at the local level, such as the microgrids. Such an approach can provide qualitative and quantifiable value from tremendously enhanced interaction and integration of the ICT technologies and energy networks at the systemic level. Furthermore, regulatory, standardization and commercial frameworks are proposed to enable distributed energy trading and market design (e.g., through the P2P platform).

The VirpaD project focuses on the digitalization of building and energy services. This project is under the umbrella of the Finnish national research and innovation project that promotes the utilization of IoT, the sharing economy ideology and AI to develop digital platforms in the built environment (e.g., smart buildings, energy efficiency). The project provides and supports the creation and operation of digital platforms that are enabled by big data and AI algorithms. Furthermore, the VirpaD project provides the inputs of how an AI platform can be designed and what are its capabilities at different levels/layers of the platform.

Following the ADR logic, this research involves two main stages and four specific processes. The first stage was to review the existing digital architectural design and models of platforms in the two research projects. The data was collected from the platform development and engagement workshops with a stakeholder engagement approach [77]. The workshops involved multiple stakeholders that participated in the projects, representing energy utilities, energy service companies (ESCO), ICT and telecommunication operators, consumers/prosumers, regulators and so on. Moreover, the project technical reports and relevant project deliverables were collected and utilized. The P2P-SmartTest project provided the inputs and models of the distributed energy market design, while the VirpaD project provided the digital platform architecture with AI integration. The second stage involved the use of the ADR approach [78–80] with four specific processes that enable the mapping and reconciliation of the platform design into one model, the 4C systemic model that is presented in detail in the next section. This action design-oriented approach is an established approach both in the information system and organization research domain.

Process (1) Problem formulation: in this process, the research focused on conceptualizing the formulated problem, which is the focal research question of this study: what is the potential of AI-enabled system architecture for the future energy markets? In particular, this question was posed to address the two challenges stemming from (1) incorporation of massive renewables that are integrated into the grids and create new operational, technological and economic challenges for the future energy market and; (2) an emerging shift from the utility-centric model to a consumer-centric model. As the World Energy Council [81] addresses, the dynamic consumer-centric energy utilities are 85% more likely to outperform their peers with respect to the rate of profitability growth, yet only 14% of energy companies today are in this category.

Process (2) Building and intervention: in this process, the research involved multiple stakeholders in the energy industry. Ecosystem actor workshops were organized that focused on the creation of the energy and AI platform architecture. The same 4C framework was utilized in both P2P-SmarTest and VirpaD projects, where the P2P-SmarTest project focused on a decentralized energy market as the context and VirpaD project focus was on the AI-enabled platform in smart building context (including building energy consumption). As suggested by [67], this process involved reciprocal shaping and concurrent evaluation.

Process (3) Reflection and learning: this is a continuous process and parallels the first two stages as suggested by Sein et al. [67]. This process involved iterative cycles of conscious reflection on the problem framing and the theories chosen to reach the target model or architecture as the conceptual outcome. In this process, the researchers had in-depth queries and discussion on the platform architecture from process 2 with energy ecosystem actors involved in the project and workshops.

Process (4) Formalization of the learning: in this process, the research formalized the results of the study to cast the artifact instance into the artifact class or a more generalized model. The architecture design outcomes (instances) of the P2P-SmarTest and VirpaD projects were cast into a more generalized 4C architecture. Furthermore, the study also used secondary empirical cases to validate the architectural model developed. The empirical case information was collected from the industry reports, industry news media, case company websites and public reports as secondary data. It is of importance to note that the cases included in the invalidation step are only to demonstrate the explainability and usability of the proposed model. The cases are not from an exhaustive list of empirical companies.

3. Results

This study adopts the system architecture thinking and ADR approach for the design of the AI energy platform. Thus, the results are presented and discussed by focusing on the three key constructs: the structure, the behavior and components, and the view of the system [82].

3.1. System Structure: The Architectural Framework of the AI Energy Platform

The platform architecture may also define the business/market layer from the Smart Grid Architecture Model (SGAM) [83]. SGAM has three dimensions: domains on the x-axis (Generation, Transmission, Distribution, Distributed Energy Resources, Customer premises), Zones of ICT-based control systems on the y-axis (Process, Field, Station, Operation, Enterprise, Market), and the Interoperability layer on the z-axis (Components, Communication Information, Function, Business). Therefore, the AI platform architecture contributes to enhancing the smart grid platform as it exists at the moment to integrate the high penetration of renewable energy sources. From this point on, the business layer [84] is, therefore, decomposed into four layers, those from the 4C framework in Table 1 below.

Table 1. The 4C framework for artificial intelligence (AI) energy platform (adapted from [69,85]).

Layer	Description
Commerce	The information and communication technology (ICT) solutions that provide all stakeholders with an application or marketplace for trading alternative connectivity solutions, content or context data.
Context	The ICT solutions that provide data and information-related contextual-based services.
Content	The ICT solutions that provide any data, information, and content that the users would want or need.
Connection	The connectivity-related solutions to connect one or several networks.

The 4C systemic framework is applied in various empirical-inspired digitalization studies [68,69,85], which is used in this study as the foundation for theory and framework building. The framework has been used in value-related research in smart energy and the smart grid. The framework encompasses four types of value typologies in a digital system (e.g., smart energy and renewable system): connectivity, content, context, and commerce (as shown in Table 1). The 4C framework resonates with the SGAM architecture as the formal system model in the EU. The framework connects the service logic and value-based provisioning and utilization [85] within the focal ecosystem.

Each of the 4C layers is formed from the smart grid architecture and shall integrate the different actors within the value chain of the smart grid infrastructure. Therefore, the AI platform is transversal between the energy actors but also throughout the domains and hierarchical zones of the smart grid. In the 4C framework, the content and context layers are particularly relevant to the field of AI as data and contextual-aware services from these two layers are empowered by the technological infrastructure. For instance, AI and more specifically deep-learning technologies have been used for price prediction for intra-day trade in Nordpool (the Nordic regional energy-trading platform) based on the research of [86]. The simplified conceptual framework developed and utilized in this study is shown in Figure 1 and the detailed interconnection and integration between the 4C framework as the business/market layer of the SGAM are presented in Figure 2.

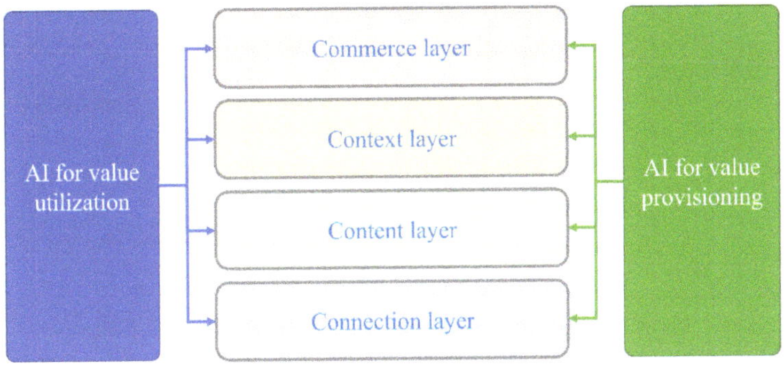

Figure 1. The simplified overall conceptual framework of the study.

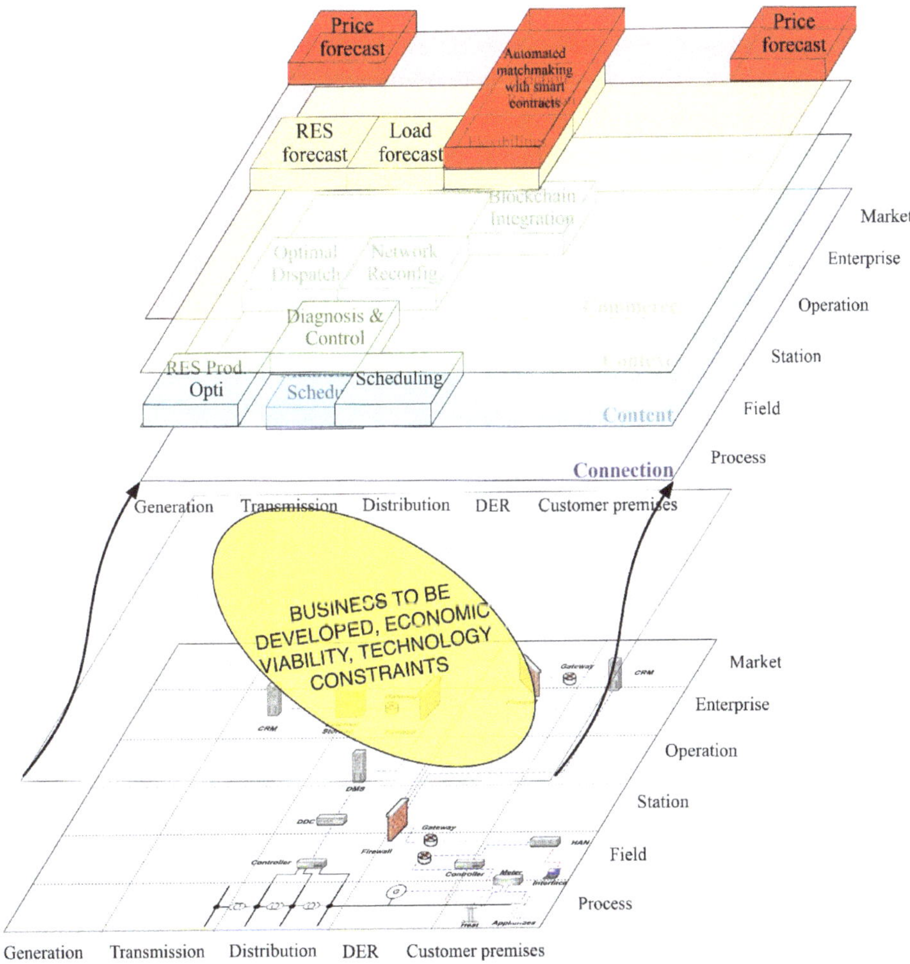

Figure 2. Integrating the 4C framework within the smart grid architecture model (SGAM).

From a technology perspective, the computation value of a digitalized ecosystem is realized through the integration of AI technology and algorithms into the existing ICT systems. In particular, the AI algorithms are developed and trained through various frameworks (e.g., TensorFlow, PyTorch) and hardware and software computation resources. Building on top of the establishment of network connectivity and the extraction, collection, and storage of data, AI algorithms are trained to build meaningful models to enable the execution of automated tasks at scale.

3.2. System Behavior: AI Energy Platform from Market Design and Value Utilization Perspective

Conventionally, a "simple" market model suggests that the business model is to facilitate the value created in the market to flow from providers to users through a central entity as the aggregator. The concept paves the way to the simple platform model proposed by [58]. The platform models can be two- or multi-sided platforms that connect the heterogeneous groups of value providers and value users through market mechanisms such as matchmaking and bridging [87].

The work of Gandia and Parmentier [65] describes the design of a digital platform from the value utilization point of view. The platform operators normally need to subsidize one side of the platform such as the platform service users [65]. In a multi-sided platform, the end-user may consume the product, service or other value that is created and provided by the other actors on the platform while producing value to other actors. Furthermore, a multi-sided platform can bridge a variety of user groups to interact with each other and create positive network effects [88]. Expanding this stream of design for the AI platform market led to the incorporation of the work from Milgrom and Tadelis [37] that focuses on how AI can impact generic market design in two major ways.

First, AI can promote trust in digital marketplaces or platforms (hereafter, digital platform). According to Milgrom and Tadelis [37], trust is a critical issue on digital platforms, mainly due to the threat from anonymous trading and asymmetric distribution of information. Confidence and trust on different sides of the platform are essential for the platform actors to conduct transactions and form a market. The existing digital platforms use online review, feedback and reputation systems to establish digital trust among platform actors. However, Milgrom and Tadelis [37] argue that the current online reputation mechanisms potentially inflate the seller rating and do not provide accurate information for buyers. For instance, the average positive feedback for eBay merchants is about 99.4%. This can mislead the buyers to make the right choice on the platform today. AI techniques such as natural language processing can be implemented to access the online communication and provide more accurate evaluations of the quality and trustworthiness of the buyers and sellers [89].

Second, AI can reduce matchmaking friction. Existing literature (e.g., [37,90]) shows that a key deployment of AI and more specifically machine learning, for digital platform is to enhance and optimize the way in which potential actors engage with the platform through the search function. Today, search engines like Google or Bing also employ AI algorithms as their core technology provides better search result quality, enhances the matchmaking capability of the platform and provides a great user experience. The sharing economy platforms like Airbnb also utilizes AI to deliver better search results for the users from its platform to make customers happy. Therefore, on the utilization side of the platform, when the energy system and market become highly distributed, AI can impact, enhance and optimize for better user experience such as better trust mechanisms and matchmaking friction reduction.

3.3. System Behavior: AI Energy Platform from Technology Innovation and Value-Provisioning Perspective

The value-provisioning side of an AI platform is focused on the creation and development of AI functionality and capability to create value for the platform users. Here, the 4C framework is used for the AI platform as the model goes beyond the boundaries of a company and enables collaborative value creation and capture with other ecosystem actors [91] and for the joint and system-wide development of innovations [92].

The 4C concept is manifested as the stacks and layers of the platform model where lower layers serve as the foundations for the higher layers enable the commerce or electricity market trading [85].

The layered approach is similar to the EU's SGAM model as the formal and technical-oriented model or architecture of smart grids, where there are five layers from bottom to top: physical component layer, component layer, data layer, functional layer, and business layer). Compared to SGAM, the layers in the 4C framework can also be dynamically combined and stacked to provide more versatile value to the energy system [85].

Overall, the value provisioning of the AI platform is more concerned with the creation of value by AI, for example, optimized grid operation and management (connection layer), the enhanced prediction of energy supply (content layer), load balancing (context layer) or market pricing for whole electricity market (commerce).

3.4. System View: the Proposition of an AI-Empowered Energy Platform

This study sees that an AI platform model/framework can be established as a market model of the energy industry. In such a platform, energy users and producers (e.g., consumers and prosumers) can contribute to the electricity market by providing distributed renewable energy supplies and network balancing capability in exchange for other types of value, such as a tangible monetary reward for active market participation or intangible value like environmental and social value. In this case, a multi-sided platform can be created not only for the exchange of monetary value but also to enable the market participants to capture other types of value [85]. AI can be utilized as an orchestrating entity to facilitate, optimize, manage and automate the market transactions. The increased number of market transactions and activity logs can serve as data and training input to improve the AI and algorithms (e.g., how Google uses neural networks to improve its data center operation's energy saving). This type of AI platform builds on the existing electricity market transactions and operations that are dependent on human decision making and control.

4. Discussion

Referring to the framework in the previous section, the mapping of the AI platform model and validation of such an architectural model are discussed in this section. The initial mapping of the AI platform architecture is shown in Figure 3.

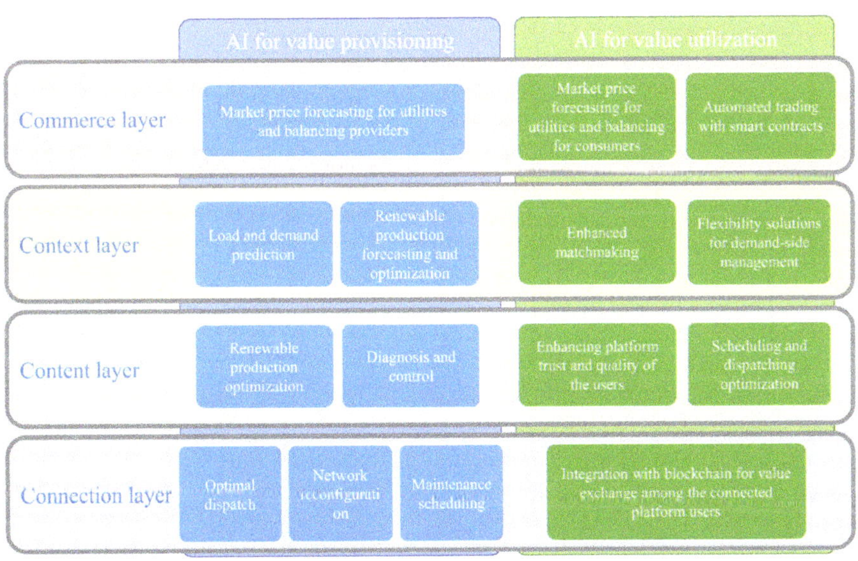

Figure 3. How AI can impact the value provisioning and utilization on the energy platform.

4.1. Using AI to Enhance Value Provisioning on the Energy Platform

AI has demonstrated its effectiveness in the smart grids in renewable integration. At the connection layer, AI can support artificial neural networking and scheduling for the electric grid and network operations such as optimal dispatch, network reconfiguration, and maintenance schedule. According to Ramos and Porto [2], a number of AI techniques such as artificial neural networks or fuzzy systems are frequently used to solve the issues related to these operation areas.

At the content layer, AI can provide enhanced energy supply and production. For example, GE utilizes intelligent algorithms to improve and optimize the operation of wind turbines. Furthermore, when massive nodes of renewable supplies are connected to the grids, AI can perform well in the areas of diagnosis and control. For example, intelligent tutoring systems are trained and experimented with by utilities to improve network control [29].

At the context layer, contextual information and data can provide input to train AI models for load and demand prediction [93] as load and supply predictions are not only used by the traditional energy utilities.

At the commerce layer that relates to market trading, fuzzy logic has been implemented to provide price forecasting for electricity markets. Moreover, deep learning as a recently popular technique is utilized in cross-border electricity trading and price prediction [86]. The study of [94] shows the progress in using AI algorithms to assist distribution system operators (DSO) in managing high levels of renewables on a local flexibility trading market.

4.2. Using AI to Enhance Value Utilization on the Energy Platform

Based on the framework of the research, a number of areas that AI can impact on the utilization side of the smart grids and renewables are identified.

First, similar to the market price prediction on the commerce layer that is mainly used by the energy utilities, aggregators and balancing service providers in the electricity market trading today, AI can provide similar data, information, and prediction for the consumers who are on the utilization side of the platform at the commerce layer. Moreover, the energy-intensive industry branches could use AI to schedule processes based on their energy intensity to maximize profits and minimize costs.

Second, as proposed by Milgrom and Tadelis [37], a potential use case of AI in the context layer is to enhance the way potential consumers interact with the digital user interface to search for products or services provided by the platform. In our case, it can be the renewable trading products and other energy services provided on a digital energy platform in the context layer. Moreover, there is also a market for storage capacity. Apart from providing emergency reserves in the intra-day market, there is also a vision that electricity consumers could not only buy battery power from public facilities but also sell.

Third, at the content layer of the value utilization, AI can improve and transform the trust mechanisms of today's digital platform (e.g., at the content layer). As mentioned previously, natural language processing can be used to extract semantic information and meaning from the platform actors' communication messages in order to provide better insight into the trustworthiness and quality of the platform actors so that it enables better use of the energy platform at the commerce layer. From a technical perspective, the study of Chui et al. [95] introduces a novel approach, the genetic algorithm support vector machine multiple kernel learning (GA-SVM-MKL) algorithm to detect 20 types of home appliances with improved accuracy of 7%. Such an application can support better optimization of home energy use at a more granular level.

At the connection layer, AI can be connected with other technologies such as blockchain to promote trusted and automated connections that enable advanced market matchmaking and transaction on other layers. By integrating AI, smart contracts and distributed-ledger technologies, there is potential to bridge the exchange of excess renewable generation, storages and electric vehicles (EV) through automated matchmaking with little or no human intervention, which can be a game-changer for tomorrow's electricity market [38].

4.3. Validating the Platform Model with Empirical Cases

Based on the action design approach, this study includes a validation process on the proposed system architecture with the empirical cases collected with desk research. The findings from analyzing the empirical cases are two-fold (Figure 4):

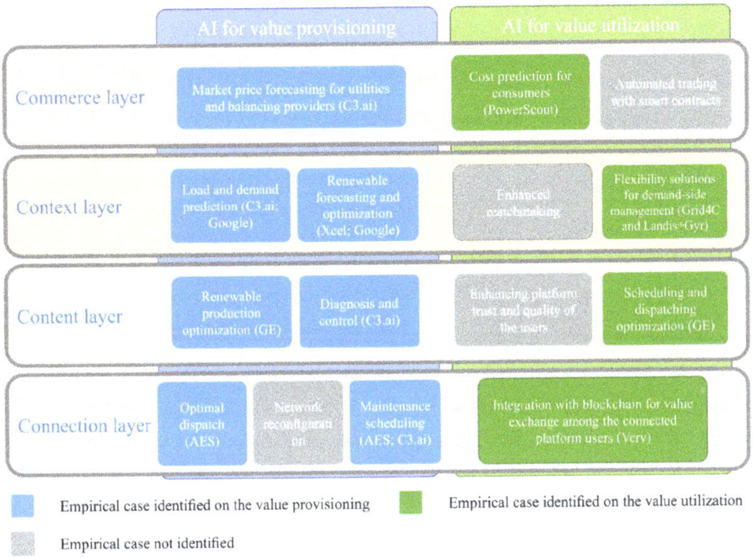

Figure 4. Validation results of the proposed platform architectural model.

First, the existing empirical cases resonate with the initial argument of the paper in that the AI applications in energy are rather narrowly focused on a specific use-case domain. However, these cases can be mapped with the 4C framework on both value-provisioning and utilization sides. Second, there are emerging companies that have adopted the platform model to develop an AI-enabled energy platform, a platform that spans two or three layers within the 4C framework. Although these cases do not fully cover the whole 4C framework, the empirical results demonstrate that the 4C framework can be used to analyze and explain the architectural design of an AI-enabled energy platform in practice.

On the value-provisioning side of the framework, a number of examples show that AI has been applied at different layers. The context layer can be associated with supply prediction. Xcel, as one of the largest energy suppliers in the US, is utilizing AI technology in Colorado to address weather forecasting challenges. AI-based data mining approach is used to create in-depth weather reports with high accuracy. Xcel's system collects data from local satellites, weather stations, and wind farms to identify data patterns and make predictions to better inform the company's planning decision in terms of energy supply [96]. Google as an ICT and AI giant has used machine learning to make predictions for wind farms. For instance, Google's London-based subsidiary DeepMind has utilized its own AI system to forecast wind-farm production. DeepMind can predict wind power output 36 h ahead by using DeepMind's neural networks [97].

At the content layer, GE has used its Predix platform to optimize GE's wind-farm production. The platform can integrate data from different sensors from the wind turbine and performs detailed analysis to predict production and operation failure in the machine before it happens. Furthermore, the Predix platform can also optimize a collection of wind turbine assets to maximize the electricity product output [96].

At the connection layer, AES (an energy company that has 36 gigawatts of energy capacity in 17 countries) has been developing advanced neural networks, natural language processing, and machine intelligence. The key application area of the AES' AI is to improve the awareness, efficiency, and maintenance of the electric grids that connect the company's solar and gas generation assets, as the so-called "preventative maintenance" for grid operations [98].

On the value utilization side of the framework, several new AI companies have emerged in the energy market. PowerScout [99] is a California–based startup that can be considered as providing the solution for the consumers on the commerce layer of the 4C framework. The company uses AI and machine learning to improve consumer awareness and participation in the energy market. PowerScout utilizes industry data and AI to demonstrate potential savings on power costs for the consumers. PowerScout's AI platform collects data from over 100 billion data points that are connected to 45 million households. The data is collected from several sources and predicts whether or not a given household should be investing in solar energy and helps the solar installers to match with the potential buyers.

At the content and context layers, GE has been focusing on utilizing AI to optimize how electricity flows out of batteries and points of consumption [100]. Grid4C and Landis+Gyr (one of the leading smart meter manufacturers) have formed a partnership to provide utilities with granular real-time predictions and actionable insights for operations and customer-facing applications. The so-called "AI grid edge" solution is the core technology that enables the applications to achieve granular load forecasting as well as optimization for the distributed energy resources for home energy management at the appliance level. The new application can predict and detect faults in both grid assets and home appliances and can be used to reduce peak demand at consumer home premises [101]. These cases demonstrate that AI technologies are actually used for demand-side management at consumer locations.

In addition to the empirical AI applications that are focused on a single layer within the 4C framework, the research has also identified a breed of emerging AI platforms that span multiple layers of the framework and resemble the platform architecture proposed in the paper.

C3.ai [102] is an AI-as-a-Service (AIaaS) platform that enables the utility companies to utilize a variety of data sources (from the grid operations) that underpin AI and machine learning algorithms to optimize grid asset management and forecasting systems, enhance the energy efficiency, and enrich customer service engagement with real-time predictive insights. C3.ai's AI energy platform offers solutions that can cover several 4C layers: (1) at the commerce layer, C3.ai has the software solutions for digital customer experience and customer segmentation and targeting; (2) at the context layer, the platform offers energy-management analytics to reduce utility operation costs and enhance operation via real-time tracking, analytics, and optimization. AI techniques are used to provide more accurate forecasting and enable more effective demand response; (3) at the content layer, the C3.ai optimizes DER management by integrating real-time energy data across systems and sensor networks. The platform can enable secure API (application programming interface) to access various data for a utility's grid edge control and operational platforms. With the AI algorithms, the C3.ai platform enables the management of distribution asset capacity constraints and mitigation of active and reactive voltage issues; (4) at the connection layer, the C3.ai provides predictive maintenance that can estimate asset failures in advance for the generation, transmission, and distribution systems. Both supervised and unsupervised learning algorithms are used to process the data streams from sensors, SCADA (supervisory control and data acquisition) systems, and asset management systems to identify anomalies and predict malfunction probability of the assets.

Another empirical case is Verv [103], which is an AI energy startup that integrates AI and blockchain: (1) at the connection layer, the company's device is connected directly to the smart meter in a home. Through the hardware device, Verv's platform then monitors the whole electricity usage of the consumer premises. Verv's device has a "sample rate" that is up to 5 million times faster than a typical smart meter [103]. This is known as high-frequency data which allows the Verv device to gain more detail from the electricity mains.; (2) the content and context are interconnected in Verv's case, where the collected high-frequency electricity data allow the AI algorithms to recognize the household appliances via their unique energy signatures, eventually identifying new appliances in consumer homes and providing smarter insights into usage patterns. (3) At the commerce layer, the company builds a blockchain-enabled electricity trading platform that enables P2P energy trading. The households can sell the excess solar generation directly to other consumers/prosumers through the matchmaking mechanism provided by Verv Trading. This approach helps provide low-cost electricity for the consumers without solar panels and more return on investment for the prosumers who have solar panels installed.

Overall, based on the outcome of the validation, the proposed platform model demonstrates the expandability of the empirical cases. On one hand, the proposed model shows that the stand-alone and hybrid empirical AI applications can be mapped on the architecture model at all four layers and two sides of value provisioning and utilization, except for the connection layer on the value utilization side. On the other hand, the results show that there is a gap between theory and practice. The AI energy concepts are developed ahead of the empirical cases that can actually implement. This is particularly true for the value utilization side of the platform. For example, there is no empirical case on automated trading at the commerce layer for consumers. However, Fortum as a Nordic energy supplier is providing the Nordpool spot market price for the electricity consumers today. There is potential to incorporate AI-enabled pricing prediction for the consumers and prosumers in the future.

The enhanced matchmaking (context) and platform trust (content) enabled by AI are missing. However, the ICT and eCommerce literature suggests that AI algorithms can be used to optimize and generate highly relevant results and reduce search friction and cost to the user [37] in other empirical cases. Currently, there is no massive energy platform like Amazon for eCommerce platforms or Airbnb for home-sharing platforms, which means enhanced matchmaking and search friction reduction is not critical at the time of this paper as the energy platform has not reached a critical mass to demand these functions. However, technology is mature enough to provide such capability. For instance, Google receives more than 63,000 searches per second on any given day, which is equivalent to 5.6 billion searches per day [104].

5. Conclusions

In the ever-increasing pace of renewable integration and the adoption of smart grids as the next generation of the energy system, the integration of massive distributed energy supply and resources is the key. Evidently, AI technology can tackle many of today's energy-system challenges that have numerous non-linear and high-uncertainty issues.

The literature used in the paper represents two streams of scientific research, the energy market design literature and the information system literature. Our key contribution to energy market design is to use introduce platform thinking for the AI-enabled energy market. The platform is a well-established concept and theory in energy economics. For the information system literature, we bring in the business model perspective as the information system comprises the operation process and business process. Even though platforms and data have been related to both research streams, existing research has not incorporated both streams to investigate AI as a general-purpose technology [9] that can enable and influence the technical and market architecture of the energy platforms. It is key to distinguish that this paper focuses on the system architecture of the energy market, while deriving mathematical modeling is not within the scope.

Above all, a consumer- and prosumer-oriented electricity system architecture enabled by AI technology is becoming a crucial area for the energy sector and energy market worldwide [105]. A more autonomous, optimized and flexible design of an energy system can be enabled by AI technology that is supported by the advancement in big data, IoT technology as well as computing technology. Multiple studies show that AI can improve the operating efficiency, reliability and intelligent ability of the energy system. Overall, AI is expected to be one of the means to develop security, economy, and reliability of the power industry.

The key contributions of the study are as follows. First, the study proposes a new energy system/market design architecture that is enabled by AI and big data. AI techniques have been tested, experimented upon and proven effective in numerous technical areas of smart energy systems and renewable productions. However, the lack of a holistic view of how these AI techniques can be integrated from the energy system perspective has been missing. This study tackles this issue by utilizing the 4C framework that has earlier been used in the ICT and energy ecosystem studies [69,85]. The end results show how AI technologies can be integrated into various parts of the energy system architecture.

Second, through the proposed platform model, the study identifies the research gap that the current AI and energy studies have been focused on narrow AI applications. This paper discusses the possibility of an AI platform that can incorporate, coordinate and manage different AI applications so that to create extended value for a complex system of the energy industry and market. The study introduces platform thinking to AI and energy research, suggesting that an AI-empowered energy platform or marketplace can be a potential solution for the next-generation energy systems for the incorporation of massive distributed renewable resources. Companies like Google, Amazon, Airbnb have proven that AI has the capability to manage and automate a digital system and platform that can go beyond human limitations e.g., by handling tens of thousands of research queries per second without compromising the resulting quality.

Third, the advantage or benefit of using the platform approach is to see how extant energy research has focused on value provisioning for the energy systems. This is largely due to the fact that the legacy energy systems are built on the paradigm of centralization and where the energy supply chain thinking is dominant. Platform thinking brings forward a perspective shift to help us see not only value provisioning but also value utilization in the electricity market. Thus, coordination and optimization can and should take place on both provisioning and utilization sides of the market. This inspires us to see AI's capability on the utilization sides of the electricity market in cases such as enhancing market trust, reducing research and matchmaking friction and cost to the market participants. It is important to note that AI's capabilities on the value utilization side are not only limited to the areas discussed in this paper. In fact, this research encourages future research to explore and discover more on the utilization side of the energy platform to enrich energy research and literature.

The limitation of the study is that it is built on secondary data and is inductive research. AI platform thinking is a novel concept that steers us towards the future energy industry and new AI technologies, and applications are still emerging. The paper focuses on concept development and proposition rather than validation. It is recommended for future research to utilize the deductive approach to survey and collect more empirical case studies to further test and improve the framework developed in this study.

This paper contributes to how AI can be combined with platform thinking to develop a holistic view on the AI energy platform. We argue that this is a potential model and design for the future and for more distributed energy and electricity markets. In addition, theoretical research and the practical application of AI in energy markets are further encouraged, For instance, a more holistic energy system comprises more than just renewable electricity exchange, it also includes services targeted at smart cities, industries, and transportation among others [95]. From the geopolitical perspective, with the increased share of renewables entering the scene of electricity and the energy market, tough competition will be faced by traditional powerplants and utilities to stay profitable, although baseload power will still be required. Therefore, there is massive potential for the use of AI technologies today and in the future for a regional market beyond country borders, such as the pan-European energy and electricity market.

Author Contributions: Conceptualization, Y.X., P.A. and E.P.; Data curation, Y.X. and P.A.; Formal analysis, Y.X., E.P. and J.-N.L.; Funding acquisition, P.A.; Investigation, Y.X., P.A., E.P. and J.-N.L.; Methodology, Y.X. and P.A.; Project administration, Y.X. and P.A.; Resources, P.A. and E.P.; Supervision, P.A. and E.P.; Validation, Y.X., P.A., E.P. and J.-N.L.; Visualization, Y.X.; Writing—original draft, Y.X., P.A., E.P. and J.-N.L.; Writing—review and editing, Y.X., P.A., E.P. and J.-N.L.

Funding: This research is funded by Business Finland (the Finnish public funding agency for research and technology development), the VirpaD project and Academy of Finland, SEN2050 project (Smart energy networks 2050, funding number: 287748).

Acknowledgments: The authors would like to thank the P2P-SmarTest project, the VirpaD project, and the SEN2050 project consortia for their support.

Conflicts of Interest: The authors declare no conflict of interest.

References

1. Nagy, K.; Hajrizi, E. Beyond the age of oil and gas—How artificial intelligence is transforming the energy portfolio of the societies. *IFAC Pap.* **2018**, *51*, 308–310. [CrossRef]
2. Ramos, C.; Porto, P. AI in power systems and energy markets. *IEEE Intell. Syst.* **2011**, *26*, 5–8. [CrossRef]
3. Grawitz, C. *The European Commission Horizon 2020 Work Programme for Research & Innovation 2018–2020*; European Commission: Brussels, Belgian, 2018.
4. Lytras, M.D.; Raghavan, V.; Damiani, E. Big data and data analytics research: From metaphors to value space for collective wisdom in human decision making and smart machines. *Int. J. Semant. Web Inf. Syst.* **2017**, *13*, 1–10. [CrossRef]
5. Andreadou, N.; Kotsakis, E.; Masera, M. Smart meter traffic in a real LV distribution network. *Energies* **2018**, *11*, 1156. [CrossRef]
6. The European Commission Smart Grids and Meters. Available online: http://ec.europa.eu/energy/en/topics/markets-and-consumers/smart-grids-and-meters (accessed on 11 August 2019).
7. Campbell, J.A. On artificial intelligence. *Artif. Intell. Rev.* **1986**, *1*, 3–9. [CrossRef]
8. Hernández, J.; Sanz, R.; Corredera, Á.; Palomar, R.; Lacave, I. A fuzzy-based building energy management system for energy efficiency. *Buildings* **2018**, *8*, 14. [CrossRef]
9. Teece, D.J. Profiting from innovation in the digital economy: Enabling technologies, standards, and licensing models in the wireless world. *Res. Policy* **2018**, *47*, 1367–1387. [CrossRef]
10. Xu, Y.; Ahokangas, P.; Pirttikangas, S.; Turunen, M.; Mäntymäki, M. From PFI to PFAI: Platform business models for smart buildings. In Proceedings of the 25th Nordic Academy of Management Conference, Vaasa, Finland, 22–24 August 2019.
11. Maglogiannis, I.G. *Emerging Artificial Intelligence Applications in Computer Engineering: Real Word AI Systems with Applications in Ehealth, HCI, Information Retrieval and Pervasive Technologies*; IOS Press: Amsterdam, The Netherlands, 2007; Volume 160.
12. Trajtenberg, M. AI as the Next GPT: A Political-Economy Perspective. Available online: https://www.nber.org/papers/w24245.pdf (accessed on 15 August 2019).
13. General Electric GE Launches the Next Evolution of Wind Energy Making Renewables More Efficient, Economic: The Digital Wind Farm. Available online: https://www.genewsroom.com/press-releases/ge-launches-next-evolution-wind-energy-making-renewables-more-efficient-economic (accessed on 11 August 2019).
14. Yao, W. Analysis on the application of the artificial intelligence neural network on the new energy micro grid. In Proceedings of the 2017 4th International Conference on Machinery, Materials and Computer (MACMC 2017), Xi'an, China, 27–29 Nobember 2017.
15. Quan, X.I.; Sanderson, J. Understanding the artificial intelligence business ecosystem. *IEEE Eng. Manag. Rev.* **2018**, *46*, 22–25. [CrossRef]
16. Bose, B.K. Artificial intelligence techniques in smart grid and renewable energy systems—Some example applications. *Proc. IEEE* **2017**, *105*, 2262–2273. [CrossRef]
17. Asare-Bediako, B.; Kling, W.L.; Ribeiro, P.F. Day-ahead residential load forecasting with artificial neural networks using smart meter data. In Proceedings of the 2013 IEEE Grenoble Conference, Grenoble, France, 16–20 June 2013; pp. 1–6.
18. Johannesen, N.J.; Kolhe, M.; Goodwin, M. Relative evaluation of regression tools for urban area electrical energy demand forecasting. *J. Clean. Prod.* **2019**, *218*, 555–564. [CrossRef]
19. Ghoddusi, H.; Creamer, G.G.; Rafizadeh, N. Machine learning in energy economics and finance: A review. *Energy Econ.* **2019**, *81*, 709–727. [CrossRef]

20. Ford, V.; Siraj, A.; Eberle, W. Smart grid energy fraud detection using artificial neural networks. In Proceedings of the 2014 IEEE Symposium on Computational Intelligence Applications in Smart Grid (CIASG), Orlando, FL, USA, 9–12 December 2014; pp. 1–6.
21. Macedo, M.N.Q.; Galo, J.J.M.; De Almeida, L.A.L.; de C. Lima, A.C. Demand side management using artificial neural networks in a smart grid environment. *Renew. Sustain. Energy Rev.* **2015**, *41*, 128–133. [CrossRef]
22. Mocanu, E.; Nguyen, P.H.; Gibescu, M.; Kling, W.L. Deep learning for estimating building energy consumption. *Sustain. Energy Grids Netw.* **2016**, *6*, 91–99. [CrossRef]
23. Lu, R.; Hong, S.H.; Yu, M. Demand response for home energy management using reinforcement learning and artificial neural network. *IEEE Trans. Smart Grid* **2019**, *6*, 6629–6639. [CrossRef]
24. Salah, K.; Rehman, M.H.U.; Nizamuddin, N.; Al-Fuqaha, A. Blockchain for AI: Review and open research challenges. *IEEE Access* **2019**, *7*, 10127–10149. [CrossRef]
25. Zahraee, S.M.; Khalaji Assadi, M.; Saidur, R. Application of Artificial Intelligence Methods for Hybrid Energy System Optimization. *Renew. Sustain. Energy Rev.* **2016**, *66*, 617–630. [CrossRef]
26. Allam, Z.; Dhunny, Z.A. On big data, artificial intelligence and smart cities. *Cities* **2019**, *89*, 80–91. [CrossRef]
27. Reich, Y. Artiicial intelligence in bridge engineering: Towards matching practical needs with technology. *Microcomput. Civ. Eng. Comput. Bridg. Eng.* **1995**, 1–25. Available online: http://citeseerx.ist.psu.edu/viewdoc/summary?doi=10.1.1.56.2150 (accessed on 15 August 2019).
28. Singer, D.; Bügler, M.; Borrmann, A.; Program, K. Knowledge based bridge engineering—Artificial intelligence meets building information modeling. In Proceedings of the 23rd EG-ICE International Workshop, Kraków, Poland, 29 June–1 July 2016; Available online: https://www.semanticscholar.org/paper/Knowledge-based-Bridge-Engineering-Artificial-meets-Singer-B%C3%BCgler/ffb32ffe51205cb95783a035d8c1e2698409be05 (accessed on 15 August 2019).
29. Jha, S.K.; Bilalovic, J.; Jha, A.; Patel, N.; Zhang, H. Renewable energy: Present research and future scope of Artificial Intelligence. *Renew. Sustain. Energy Rev.* **2017**, *77*, 297–317. [CrossRef]
30. Landahl, J.; Panarotto, M.; Johannesson, H.; Isaksson, O.; Lööf, J. Towards Adopting Digital Twins to Support Design Reuse during Platform Concept Development. In Proceedings of the NordDesign 2018, Linköping, Sweden, 14–17 August 2018.
31. Meyer, M.H.; Lehnerd, A.P. *The Power of Product Platforms*; Free Press: New York, NY, USA, 1997.
32. Erixona, G.; Anders, Y.; Arnström, A. Modularity—The basis for product and factory reengineering. *Cirp Ann. Manuf. Technol.* **1996**, *45*, 1–6. [CrossRef]
33. Yue, D.; Han, Q.-L. Guest editorial special issue on new trends in energy internet: Artificial intelligence-based control, network security, and management. *IEEE Trans. Syst. Man Cybern.* **2019**, *49*, 1551–1553. [CrossRef]
34. Gebhardt, C. The strategic relevance of artificial intelligence for corporate success in the energy market. *Int. J. Technol. Policy Manag.* **2002**, *2*, 194–217. [CrossRef]
35. Vickrey, W. Counterspeculation, auctions, and competitive sealed tenders. *J. Financ.* **1961**, *16*, 8–37. [CrossRef]
36. Milgrom, P.R. *Discovering Prices, Auction Design in Markets with Complex Constraints*; Columbia University Press: New York, NY, USA, 2017; ISBN 978-0-231-54457-3.
37. Milgrom, P.R.; Tadelis, S. How artificial intelligence and machine learning can impact market design. *Econ. Artif. Intell.* **2018**, 567–586. [CrossRef]
38. Mylrea, M. AI enabled blockchain smart contracts: Cyber resilient energy infrastructure and IoT. In Proceedings of the 2018 AAAI Spring Symposium Series, Paolo Alto, CA, USA, 26–28 March 2018.
39. Xu, Y.; Ahokangas, P.; Yrjölä, S.; Koivumäki, T. The fifth archetype of electricity market: The blockchain marketplace. *Wirel. Netw.* **2019**, 1–17. [CrossRef]
40. Andoni, M.; Robu, V.; Flynn, D.; Abram, S.; Geach, D.; Jenkins, D.; McCallum, P.; Peacock, A. Blockchain technology in the energy sector: A systematic review of challenges and opportunities. *Renew. Sustain. Energy Rev.* **2019**, *100*, 143–174. [CrossRef]
41. Nilsson, N. *Artificial Intelligence: A New Synthesis*; Morgan Kaufmann: Burlington, MA, USA, 1998; ISBN 9780080948348.
42. Ilter, D.; Dikbas, A. A review of the artificial intelligence applications in construction dispute resolution. In Proceedings of the CIBW78 Managing IT in Construction 26th International Conference, Istanbul, Turkey, 1–3 October 2009.
43. Coppin, B. *Artificial Intelligence Illuminated*; Solomon, S., Ed.; Jones and Bartlett Publishers, Inc.: Burlington, MA, USA, 2004; ISBN 0-7637-3230-3.

44. Liu, L.; Chen, S. The application of artificial intelligence technology in energy internet. In Proceedings of the 2018 2nd IEEE Conference on Energy Internet and Energy System Integration (EI2), Beijing, China, 20–22 October 2018; pp. 1–5.
45. Ahmad, T.; Chen, H. Utility companies strategy for short-term energy demand forecasting using machine learning based models. *Sustain. Cities Soc.* **2018**, *39*, 401–417. [CrossRef]
46. Ahmad, T.; Chen, H.; Shah, W.A. Effective bulk energy consumption control and management for power utilities using artificial intelligence techniques under conventional and renewable energy resources. *Int. J. Electr. Power Energy Syst.* **2019**, *109*, 242–258. [CrossRef]
47. Ahmad, T.; Chen, H.; Guo, Y.; Wang, J. A comprehensive overview on the data driven and large scale based approaches for forecasting of building energy demand: A review. *Energy Build.* **2018**, *165*, 301–320. [CrossRef]
48. Ahmadi, A.A. An empirical analysis on the adoption of electronic banking in the financial institutes using structural, behavioral and contextual factors. *Manag. Sci. Lett.* **2012**, *2*, 1669–1682. [CrossRef]
49. Kumar, B.R.S.; Varalakshmi, N.; Lokeshwari, S.S.; Rohit, K.; Manjunath; Sahana, D.N. Eco-friendly IOT based waste segregation and management. In Proceedings of the 2017 International Conference on Electrical, Electronics, Communication, Computer, and Optimization Techniques (ICEECCOT), Mysuru, India, 15–16 December 2017; pp. 297–299.
50. Zhao, H.X.; Magoulès, F. A review on the prediction of building energy consumption. *Renew. Sustain. Energy Rev.* **2012**, *16*, 3586–3592. [CrossRef]
51. Kalogiro, S.A. Artificial neural networks in renewable energy systems applications: A review. *Renew. Sustain. Energy Rev.* **2001**, *5*, 373–401. [CrossRef]
52. Dounis, A.I. Artificial intelligence for energy conservation in buildings. *Adv. Build. Energy Res.* **2010**, *4*, 267–299. [CrossRef]
53. Mellit, A.; Pavan, A.M. A 24-h forecast of solar irradiance using artificial neural network: Application for performance prediction of a grid-connected PV plant at Trieste, Italy. *Sol. Energy* **2010**, *84*, 807–821. [CrossRef]
54. Lu, S.; Youngdeok, H.; Khabibrakhmanov, I.; Marianno, F.J.; Shao, X.; Zhang, J.; Hodge, B.M.; Hamann, H.F. Machine learning based multi-physical-model blending for enhancing renewable energy forecast—Improvement via situation dependent error correction. In Proceedings of the 2015 European Control Conference (ECC), Linz, Austria, 15–17 July 2015; pp. 283–290.
55. Brancucci Martinez-Anido, C.; Botor, B.; Florita, A.R.; Draxl, C.; Lu, S.; Hamann, H.F.; Hodge, B.M. The value of day-ahead solar power forecasting improvement. *Sol. Energy* **2016**, *129*, 192–203. [CrossRef]
56. Gawer, A. Bridging differing perspectives on technological platforms: Toward an integrative framework. *Res. Policy* **2014**, *43*, 1239–1249. [CrossRef]
57. Evans, P.C.; Gawer, A. *The Rise of the Platform Enterprise A Global Survey*; The Center for Global Enterprise: New York, NY, USA, 2016.
58. Rochet, J.-C.; Tirole, J. Platform Competition in Two-Sided Markets. *J. Eur. Econ. Assoc.* **2003**, *1*, 990–1029. [CrossRef]
59. Andersson, E.; Eckerwall, K. Enabling Successful Collaboration in the Manufacturing Industry: A Study of Digital Twins. Master's Thesis, Lulea University of Techology, Luleå, Sweden, 2019.
60. Evans, D.; Schmalensee, R. The Industrial Organization of Markets with Two-Sided Platforms. *Cpi J.* **2007**, *3*, 151–179.
61. Rochet, J.C.; Tirole, J. Two-sided markets: A progress report. *Rand J. Econ.* **2006**, *37*, 645–667. [CrossRef]
62. Henfridsson, O.; Bygstad, B. The Generative Mechanisms of Digital Infrastructure Evolution. *Mis Q. Manag. Inf. Syst.* **2013**, *37*, 907–931. [CrossRef]
63. Teece, D.J.; Linden, G. Business models, value capture, and the digital enterprise. *J. Organ. Des.* **2017**, *6*, 1–14. [CrossRef]
64. Uhlemann, T.H.J.; Lehmann, C.; Steinhilper, R. The Digital Twin: Realizing the Cyber-Physical Production System for Industry 4.0. *Procedia CIRP* **2017**, *61*, 335–340. [CrossRef]
65. Gandia, R.; Parmentier, G. Optimizing value creation and value capture with a digital multi-sided business model. *Strat. Chang.* **2017**, *26*, 323–331. [CrossRef]
66. Jaakola, H. Service Design with Third Agers—Introducing the Possibilities of Mobile Devices. Master's Thesis, Laurea University of Applied Sciences, Vantaa, Finland, 2015.

67. Sein, M.; Henfridsson, O.; Purao, S.; Rossi, M.; Lindgren, R. Action design research. *Manag. Inf. Syst. Q.* **2011**, *35*, 37–56. [CrossRef]
68. Wirtz, B.W.; Schilke, O.; Ullrich, S. Strategic development of business models: Implications of the web 2.0 for creating value on the internet. *Long Range Plann.* **2010**, *43*, 272–290. [CrossRef]
69. Yrjola, S.; Ahokangas, P.; Matinmikko, M. Evaluation of recent spectrum sharing concepts from business model scalability point of view. In Proceedings of the 2015 IEEE International Symposium on Dynamic Spectrum Access Networks, DySPAN 2015, Stockholm, Sweden, 29 September–2 October 2015; pp. 241–250.
70. CEN/CENELEC/ETSI. *CEN-CENELEC-ETSI Smart Grid Coordination Group: Smart Grid Information Security*; CEN/CENELEC/ETSI: Brussels, Belgium, 2012.
71. Babüroglu, O.N.; Ravn, I. Normative Action Research. *Organ. Stud.* **1992**, *13*, 19–34. [CrossRef]
72. Baskerville, R.; Wood-Harper, A.T. Diversity in information systems action research methods. *Eur. J. Inf. Syst.* **1998**, *7*, 90–107. [CrossRef]
73. Orlikowski, W.J.; Iacono, C.S. Research commentary: Desperately seeking the "IT" in IT research—A call to theorizing the IT artifact. *Inf. Syst. Res.* **2001**, *12*, 121–134. [CrossRef]
74. Eisenhardt, K.M.; Graebner, M.E. Theory Building from Cases: Opportunities and Challenges. *Acad. Manag.* **2007**, *50*, 25–32. [CrossRef]
75. Pouttu, A.; Haapola, J.; Ahokangas, P.; Xu, Y.; Kopsakangas-Savolainen, M.; Porras, E.; Matamoros, J.; Kalalas, C.; Alonso-Zarate, J.; Gallego, F.D.; et al. P2P model for distributed energy trading, grid control and ICT for local smart grids. In Proceedings of the 2017 European Conference on Networks and Communications (EuCNC), Paris, France, 29 June–2 July 2017; pp. 1–6.
76. Xu, Y.; Turunen, M.; Ahokangas, P.; Mäntymäki, M.; Heikkilä, J. Contextualized business model: The case of experiential environment and AI. In Proceedings of the The 2nd Business Model Conference, Florence, Italy, 6–7 June 2018.
77. Grunewald, P.H.; Cockerill, T.T.; Contestabile, M.; Pearson, P.J.G. The socio-technical transition of distributed electricity storage into future networks-System value and stakeholder views. *Energy Policy* **2012**, *50*, 449–457. [CrossRef]
78. Peffers, K.; Tuunanen, T.; Rothenberger, M.; Chatterjee, S. A Design Science Research Methodology for Information Systems Research. *J. Manag. Inf. Syst.* **2007**, *24*, 45–77. [CrossRef]
79. Coghlan, D. Action research: Exploring perspectives on a philosophy of practical knowing. *Acad. Manag. Ann.* **2011**, *5*, 53–87. [CrossRef]
80. Eden, C.; Huxham, C. Researching organizations using action research. In *The SAGE Handbook of Organization Studies*; Clegg, S.R., Hardy, C., Lawrence, T.B., Nord, W.R., Eds.; Sage Publications: London, UK, 2006; pp. 389–408. ISBN 9781848608030.
81. Word Energy Council. *World Energy Scenarios 2019*; Word Energy Council: London, UK, 2019; Volume 1.
82. Jaakkola, H.; Thalheim, B. Architecture-driven modelling methodologies. *Front. Artif. Intell. Appl.* **2011**, *225*, 97–116.
83. Uslar, M.; Rohjans, S.; Neureiter, C.; Andrén, F.P.; Velasquez, J.; Steinbrink, C.; Efthymiou, V.; Migliavacca, G.; Horsmanheimo, S.; Brunner, H.; et al. Applying the smart grid architecture model for designing and validating system-of-systems in the power and energy domain: A European perspective. *Energies* **2019**, *12*, 258. [CrossRef]
84. Pavlovic, M.; Gawron-Deutsch, T.; Neureiter, C.; Diwold, K. SGAM business layer for a local flexibility market. *IET Conf. Publ.* **2016**, *1–4*. [CrossRef]
85. Xu, Y.; Ahokangas, P.; Reuter, E. EaaS: Electricity as a service? *J. Bus. Model.* **2018**, *6*, 1–23.
86. Kolberg, J.; Waage, K. *Artificial Intelligence and Nord Pool's Intraday Electricity Market Elbas: A Demonstration and Pragmatic Evaluation of Employing Deep Learning for price Prediction*; Norwegian School of Economics: Bergen, Norway, 2018.
87. Amit, R.; Han, X. Value creation through novel resource configurations in a digitally enabled world. *Strat. Entrep. J.* **2017**, *11*, 228–242. [CrossRef]
88. Eisenmann, T.; Parker, G.; Alstyne, M.W. Van Strategies for Two-Sided Markets. *Harv. Bus. Rev.* **2006**, *84*, 12.
89. Masterov, D.V.; Mayer, U.F.; Tadelis, S. Canary in the e-Commerce Coal Mine: Detecting and Predicting Poor Experiences Using Buyer-to-Seller Messages. In Proceedings of the EC '15 Sixteenth ACM Conference on Economics and Computation, Portland, OR, USA, 15–19 June 2015; pp. 81–93.

90. Blake, T.; Nosko, C.; Tadelis, S. Returns to Consumer Search. In Proceedings of the EC '16 2016 ACM Conference on Economics and Computation, Maastricht, The Netherlands, 24–28 July 2016; Volume V, pp. 531–545.
91. Chesbrough, H.; Vanhaverbeke, W. A Classification of Open Innovation and Open Business Models. In *New Frontiers in Open Innovation.*; Oxford University Press: Oxford, UK, 2014; pp. 50–68. ISBN 9780199682676.
92. Saebi, T.; Foss, N.J. Business models for open innovation: Matching heterogeneous open innovation strategies with business model dimensions. *Eur. Manag. J.* **2015**, *33*, 201–213. [CrossRef]
93. Sangpetch, T.; Lo, K.L. Stochastic modelling and AI techniques for power system reinforcements in a competitive energy market. In *Proceedings of the International Conference on Power System Management and Control*; IET: London, UK, 2002; pp. 377–383.
94. Eck, B.; Fusco, F.; Gormally, R.; Purcell, M.; Tirupathi, S. AI Modelling and Time-series Forecasting Systems for Trading Energy Flexibility in Distribution Grids. In Proceedings of the Tenth ACM International Conference on Future Energy Systems, e-Energy 2019, Phoenix, AZ, USA, 25–28 June 2019; pp. 381–382.
95. Chui, K.T.; Lytras, M.; Visvizi, A. Energy sustainability in smart cities: Artificial intelligence, smart monitoring, and optimization of energy consumption. *Energies* **2018**, *11*, 2869. [CrossRef]
96. Georgiou, M. The Role of AI Technology in Improving the Renewable Energy Sector. Available online: https://www.imaginovation.net/blog/artificial-intelligence-in-renewable-energy/ (accessed on 5 October 2019).
97. Statt, N. Google and DeepMind Are Using AI to Predict the Energy Output of Wind Farms. Available online: https://www.theverge.com/2019/2/26/18241632/google-deepmind-wind-farm-ai-machine-learning-green-energy-efficiency (accessed on 19 August 2019).
98. Goldstein, P. What Is the Potential for AI in the Energy Industry? Available online: https://biztechmagazine.com/article/2017/10/what-potential-ai-energy-industry (accessed on 7 October 2019).
99. PowerScout PowerScout. Available online: https://powerscout.com/ (accessed on 7 October 2019).
100. Hirtenstein, A. GE Working on Robot That It Says Can Save $200 Billion of Power. Available online: https://www.bloomberg.com/news/articles/2017-09-18/ge-working-on-robot-that-it-says-can-save-200-billion-of-power (accessed on 5 October 2019).
101. T&D World AI-Powered Analytics in Grid-Edge Smart Meters. Available online: https://www.tdworld.com/metering/ai-powered-analytics-grid-edge-smart-meters (accessed on 2 October 2019).
102. C3.AI Transforming Utilities with Enterprise AI. Available online: https://c3.ai/industries/utilities/ (accessed on 6 October 2019).
103. Verv Verv. Available online: https://verv.energy/ (accessed on 6 October 2019).
104. Aleksandra 63 Fascinating Google Search Statistics Fascinating Google Stats. Available online: https://seotribunal.com/blog/google-stats-and-facts/ (accessed on 11 August 2019).
105. Sun, Q.; Yang, L. From independence to interconnection—A review of AI technology applied in energy systems. *CSEE J. Power Energy Syst.* **2019**, *5*, 21–34.

© 2019 by the authors. Licensee MDPI, Basel, Switzerland. This article is an open access article distributed under the terms and conditions of the Creative Commons Attribution (CC BY) license (http://creativecommons.org/licenses/by/4.0/).

energies

MDPI

Article

Research and Application of Continuous Bidirectional Trading Mechanism in Yunnan Electricity Market

Xuguang Yu [1,2], Gang Li [1,2,*], Chuntian Cheng [1,2], Yongjun Sun [3] and Ran Chen [3]

1 Institute of Hydropower and Hydroinformatics, Dalian University of Technology, Dalian 116024, China; yuxuguang@mail.dlut.edu.cn (X.Y.); ctcheng@dlut.edu.cn (C.C.)
2 Key Laboratory of Ocean Energy Utilization and Energy Conservation of Ministry of Education, Dalian 116024, China
3 Kunming Power Exchange Center, Kunming 650011, China; sunyongjunlmy@126.com (Y.S.); crandy@foxmail.com (R.C.)
* Correspondence: glee@dlut.edu.cn; Tel.: +86-0411-8470-8468

Received: 17 November 2019; Accepted: 5 December 2019; Published: 8 December 2019

Abstract: To further promote market competition, enrich trading varieties, alleviate information asymmetry, and improve trading efficiency during electricity market reform in China, the continuous bidirectional transaction (CBT) was designed and applied in the Yunnan electricity market (YNEM), which is dominated by medium- and long-term power energy trading. The clearing model for the CBT with the goal of maximum social welfare is proposed in two bidding stages, including call auction (CA) and continuous double auction (CDA). Correspondingly, the integrated two-stage market clearing algorithm is also introduced to ensure the data consistency and business continuity. Finally, the analysis of the practical application shows that the proposed model, algorithm, and various key implementation strategies of the trading platform support the bidding and clearing of the CBT well. In addition, the research and application of CBT may also provide valuable insights for other electricity market construction.

Keywords: electricity market; trading mechanism; clearing model; clearing algorithm; trading platform

1. Introduction

Over the last few decades, the installed power-generation capacity of China has exploded, especially for hydropower capacity, increasing from 20 GW in 1980 to 352 GW in 2018 [1,2]. However, wasted hydropower is often associated with the rapid deployment of hydropower because of a lack of energy planning, delayed construction of power grid, and multiple interest game [3]. According to official statistics, wasted hydropower exceeded 28.7 terawatt hours (TWh) and 31 TWh, respectively, in Sichuan and Yunnan in 2016 [4], which are the two hydropower-dominated power grids. This is a very serious problem and also partly motivated the electricity market reform. Therefore, the "Opinions on Further Deepening the Reform of the Electric Power Systems: Document No. 9" was promulgated in March 2015 by the State Council and the Central Committee of the Communist Party [5].

Since the new round of electricity market reform in 2015, various regions and provinces in China have created medium- and long-term electricity markets, and a great deal of progress has been made. Two regional trading centers in Beijing and Guangzhou, as well as provincial trading centers in each province, have been set up to handle cross-provincial, cross-regional, and intra-provincial transactions. From the perspective of the whole country, the market reform has been deepened year by year due to positive factors, such as diversified market entities, rich trading varieties, and diversified market services. In 2018, the total amount of electricity traded (including the generation rights traded) in the national electricity market was 2065.4 billion kilowatt hours, up 26.5% from the previous year. The proportion of electricity traded in the market was 30.2% of the total electricity consumed by the whole society, 4.3 percentage points higher than in the previous year. The proportion of electricity traded in the market was 37.1% of the electricity sold by power grid enterprises [6].

Yunnan Province, as one of the first batches of comprehensive pilot provinces of electricity market reform, has always been in the forefront of domestic electricity market reform. After several years implementing successful operations, much progress has been made in, for instance, the market scale, the market supervision and the trading mechanism. By the end of December 2018, a total of 7662 market entities had completed registration and gained access. This scale is three times that of the initial stage of the Yunnan electricity market (YNEM). In 2018, more than 85 billion kilowatt-hours of electricity were traded in the provincial market, an increase of more than 21% from the previous year. Among this electricity, the proportion of clean energy trading electricity accounted for more than 97%, ranking highest in the country. Trading electricity accounted for more than 62% of the electricity sold in the Yunnan power grid (YNPG) [7]. This increase in electricity trading volume indicates an increase in market demand, which promotes the reduction of hydropower curtailment to a certain extent. In addition, medium- and long-term trading mechanisms are becoming increasingly mature, and trading varieties are abundant, providing diversified choices for market players [8]. As a pioneer in the new round of electricity market reform in China, the newly reformed YNEM has achieved initial results, where an increasingly mature, efficient, and fair electric power trading market has been gradually established, which provides valuable experience for the construction of electricity markets in other provinces.

The market operators of the YNEM have been committed to enriching the construction of the trading mechanism. To further promote market competition, enrich trading varieties, alleviate information asymmetry, and improve trading efficiency, the continuous bidirectional transaction (CBT) was designed and applied in the YNEM in 2018 [9], which is characterized by a sequential two-stage bidding process with real-time market information disclosure that integrates call auction (CA) and continuous double auction (CDA) [10,11]. The CBT is quite different from the original trading methods in the YNEM. How to design and implement CBT business, such as bidding and market clearing, is a problem worth studying.

Research on electricity trading mechanisms attracts much attention. Many experts and scholars have done a great deal of work on this topic [12]. On the one hand, game theory, auction theory, and other mechanism design theories are widely used in power market design. Using game theory, Carlos Silva et al. designed a new mechanism. In the deregulated environment, efficiency (economic dispatch) can be achieved despite the information barrier of effective distribution of power production [13]. Based on the theory of signal game, Liu Z et al. took the degree of information interference as an important factor to evaluate the bidding mechanism and put forward a semi-random matching (GSM) incentive bidding mechanism [14]. In order to alleviate the market power of participants and improve the social welfare of the power market, Xiaoyan Zou took the social welfare contribution of each participant as the key factor of market clearing, payment and settlement, and transaction matching rules, and proposed a new double-sided auction mechanism [15]. Under the condition of satisfying the characteristics of mechanism design theory of individual rationality, incentive compatibility and payment cost minimization, Zou P et al. analyzed, derived and designed a customized pool-based market mechanism by using the economic mechanism design theory [16]. In order to institutionally stabilize the market, Chen Zhang and Wei Yan restricted the abuse of market power by restricting the arbitrage opportunities generated by multiple markets and developed a mechanism [17]. In order to realize the long-term stability of electricity price in the day-ahead electricity market [18–20], Ying Yu et al. designed an incentive contract menu. In the framework of a one-leader and multi-follower game, a bilevel Stackelberg game model was also proposed to find the optimal incentive mechanism [21]. On the other hand, with the development of power system, the design of power market in the environment of rapid penetration of renewable energy and distributed energy is also concerned [22]. Agalgaonkar et al. examined placement and penetration level of the distributed generations under the standard market design (SMD) framework [23]. Based on the indirect mechanism design framework, Khazaei and Zhao aggregated renewable energy power generation companies in a two-settlement electricity market and designed the revenue distribution mechanism by solving the competitive equilibrium of the special market with transferable revenue [24]. Liu et al. discussed P2P power trading mechanisms based on auction and bilateral contract and analyzed their effectiveness in the future power distribution system, to reasonably manage power trading between users [25]. To develop strategies for the integration of prosumers into competitive electricity markets, Parag and Sovacool identified and discussed three potential retail markets related to the integration of retail power grid, peer-to-peer models, and retail community groups [26]. Relying on blockchain and CDA mechanism, Wang et al. proposed a decentralized electricity transaction mode for microgrids [27]. Mengelkamp et al. derived seven market components as a framework to build an efficient microgrid energy market [28].

Although there are a lot of studies on market mechanism, its literature and real application on integrated CA and CDA in the electricity market, which is dominated by medium- and long-term power energy trading, are relatively less. This paper aims to research and design the CBT and put it into operation in YNEM. To this end, the clearing model with the goal of maximum social welfare for the CBT is proposed, which is the theoretical basis of market clearing based on CBT trading rules. Correspondingly, based on the difference and connection analysis of the two stages, the design of a two-stage integrated market clearing algorithm to ensure the data consistency and business continuity is also introduced. Moreover, the practical experiences in implementing the CBT-based electricity trading platform (CETP) are described. Finally, the practical application is analyzed, and the CBT mechanism, clearing model, algorithm, and platform construction experience mentioned in this paper may provide valuable insight for other power market constructions.

The rest of the paper is structured as follows. In Section 2, the continuous bidirectional transaction of the YNEM is briefly introduced. The two-stage integrated clearing model of the CBT is proposed in Section 3. Section 4 illustrates the design of the market-clearing algorithm of the CBT. Section 5 presents the architecture and some implementation strategies adopted in the CETP. The practical application of the CBT based on the CETP in the YNEM is shown in Section 6. Finally, conclusions are drawn in Section 7.

2. Organization of Continuous Bidirectional Trading Mechanism

Unlike the major electricity markets operated by Pennsylvania-New Jersey-Maryland(PJM), California Independent System Operator (CAISO), and Electric Reliability Council of Texas (ERCOT) in North America [29–31], British Electricity Trading and Transmission Arrangements (BETTA) in Great Britain [32] and Nord Pool Spot in Nordic Europe [33], the YNEM is dominated by medium- and long-term electrical energy transactions briefly described in Figure 1. Moreover, all trading contracts require physical delivery, which means that the final transaction results need to be checked by YNPG [34]. The continuous bidirectional transaction (CBT) is new to the YNEM; it was introduced in the beginning of 2018, and currently, this bidding scheme is applied to the monthly market, where electricity consumers and producers submit a set of price-volume pairs, indicating their willingness to buy and sell electricity for the next operating month [9]. The CBT, together with the original bilateral transaction and matching transaction, constitute the main transaction varieties in the monthly market.

The CBT is a multistage double auction with numerous buyers and sellers, where the following rules are applied:

1. It is a two-stage double auction online.
2. The first stage includes a call auction (CA), i.e., a one-shot game, where buyers and sellers submit sealed bids with discrete price–quantity combinations, which state the amount of energy that the company is pleased to sell or buy at specified prices.
3. During the first stage, the sellers cannot obtain the buyer's demand curves. In the same way, sellers' supply curves are not available to the buyers. All market participants can bid or cancel a bid until the time is up. After the closure of this stage, the submitted seller and buyer bids are aggregated, resulting in the first-stage unconstrained results by the CETP of Kunming Power

Exchange (KMPEX). The results include the market clearing prices (MCPs) of this stage and bid matchings according to the submitted bids [35]. Orders with remaining unmatched volume will enter the second stage first with the same timestamp.

4. The second stage consists of a continuous double auction (CDA) [36]. Participants use the CETP to advertise offers to sell or bids to buy energy. In such a computerized marketplace, all participants can observe the quantity and price submitted by other parties, but they cannot learn the identity of any party who submits an order. When a party enters a new bid, the CETP checks whether there is a matching offer within the bid-delivery period. If it finds that the differences of prices (DOP) between the buyer's and the seller's bids is not negative, the transaction is automatically concluded, and the price and quantity are displayed for all participants to view. Otherwise, the unmatched new bid is added to the list of uncompleted bids and will remain until a transaction can be concluded or until the bid is withdrawn. A similar procedure occurs each time a new bid is entered into the system [37].
5. Since the auction is calculated in a sequential machine, the bidding information is registered in time-sharing. Therefore, even if two bids are submitted at exactly same time, they are registered in order, which is determined by chance [38].
6. If no time is left in the second period, the auction is finished. Here, the total unconstrained results are submitted to YNPG for security checks, which take into account the technical requirements, such as spatial–temporal coupling among reservoirs, line congestion on both the buyer and seller side, and the risk of water spillage, rather than economic factors, such as price or cost. For instance, when the transmission line is blocked, the power generation needs to be curtailed according to the capacity, which means that power generation is substantially controlled by the grid [39].
7. The results of the security check are returned to KMPEX in the form of the caps of the power generation, which KMPEX regards as constraints to the final bid clearing. Then, the final constraint results will be part of the monthly transaction contracts [40].

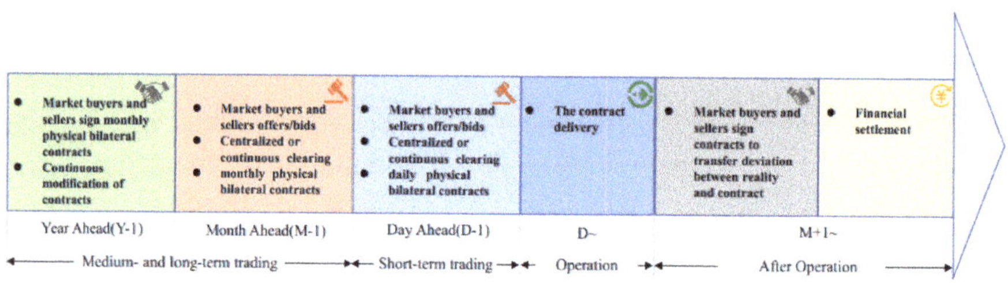

Figure 1. Process of the medium- and long-term electrical energy transactions in the Yunnan electricity market (YNEM).

The organizational process of the CBT is shown in Figure 2. The regular schedule of each stage listed in the figure may be adjusted and disclosed to market participants before the transaction is actually carried out.

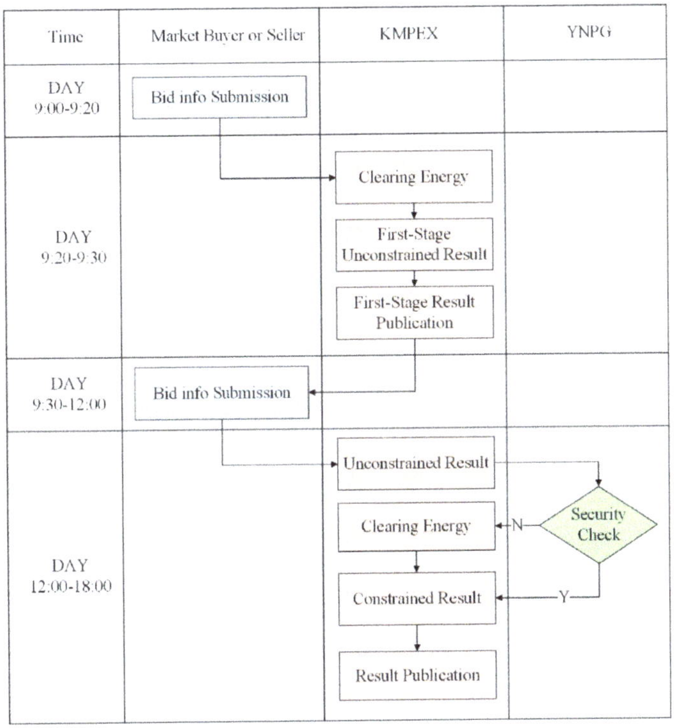

Figure 2. Flowchart of the continuous bidirectional transaction (CBT).

3. Two-Stage Clearing Model of the CBT

In this paper, the expected total revenue of buyers and sellers is regarded as the whole social welfare (also denoted as 'surplus') from the auction [15,35,41]. The CBT is also an auction mechanism maximizing the whole social welfare, which is equivalent to maximizing the expected total revenue of sellers and buyers under the following restrictions. The market-clearing model for the CBT can be described as follows.

3.1. Call Auction Stage

Objective function—the social welfare of the call auction stage can be expressed as follows:

$$\pi_1 = \max \sum_{i \in I} \sum_{j \in J} ((P_{bb,i} - P_{bd,ij}) \times Q_{bd,ij} + (P_{sd,ij} - P_{sb,j}) \times Q_{sd,ij}) \quad (1)$$

where i and I represent the index and set of market buyers, respectively; j and J represent the index and set of market sellers; $P_{bb,i}$ (yuan/kWh) represents the bid price of market buyer i; $P_{bd,ij}$ (yuan/kWh) represents the deal price of market buyer i with market seller j; $Q_{bd,ij}$ (kWh) represents the energy cleared of market buyer i with market seller j; $P_{sb,j}$ (yuan/kWh) represents the bid price of market seller j; $P_{sd,ij}$ (yuan/kWh) represents the deal price of market seller j with market buyer i; $Q_{sd,ij}$ (kWh) represents the energy cleared of market seller j with market buyer i; and π_1 (yuan) represents the social welfare of the call auction stage.

All the constraints are given by the KMPEX before the bidding. The constraints faced by the call auction stage include bidding constraints and clearing constraints.

Market bidding constraints are defined as follows:

$$\begin{cases} 0 \le Q_{bb,i} \le \overline{Q}_{bb,i} \\ 0 \le Q_{sb,j} \le \overline{Q}_{sb,j} \end{cases} \tag{2}$$

$$\begin{cases} P_\alpha \le P_{bb,i} \le P_\beta \\ P_\alpha \le P_{sb,j} \le P_\beta \end{cases} \tag{3}$$

where $\overline{Q}_{bb,i}$ (kWh) represents the quantity caps of buyer bid; $\overline{Q}_{sb,j}$ (kWh) represents the quantity caps of seller bid; and P_α (yuan/kWh) and P_β (yuan/kWh) represent the lower and upper bounds of the bid price of buyer i and seller j, respectively.

Market clearing constraints are defined as follows:

$$\begin{cases} \sum\limits_{j \in J} Q_{bd,ij} \le Q_{bb,i} \\ \sum\limits_{i \in I} Q_{sd,ij} \le Q_{sb,j} \end{cases} \tag{4}$$

$$\sum_{i \in I} \sum_{j \in J} Q_{bd,ij} = \sum_{i \in I} \sum_{j \in J} Q_{sd,ij} \tag{5}$$

$$\begin{cases} Q_{bd,ij} \ge 0 \\ Q_{sd,ij} \ge 0 \end{cases} \tag{6}$$

$$\begin{cases} P_{\Delta,ij} = P_{bb,i} - P_{sb,j} \\ P_{bd,ij} = P_{bb,i} - P_{\Delta,ij}/2 \\ P_{sd,ij} = P_{sb,j} + P_{\Delta,ij}/2 \end{cases} \tag{7}$$

Equation (4) indicates that the accumulated transaction quantity does not surpass the bid quantity. Equation (5) is the power balance constraint, which represents that the total amount of electricity traded between buyers and sellers is balanced. In Equation (6), the lower bound takes nonnegative values for $Q_{bd,ij}$ and $Q_{sd,ij}$, which means that transfers of electricity between buyers or between sellers are not allowed in this transaction.

All sellers ($i \in I$) and buyers ($j \in J$) are paired into $I \times J$ pairs, and the DOP of $I \times J$ pairs are calculated by Equation (7), where $P_{\Delta,ij}$ represents the DOP between marker buyer i and marker seller j, which is not negative to meet the transaction conditions. $P_{bd,ij}$ and $P_{sd,ij}$ represents the buying and selling prices for the financial settlement of each pair, respectively, which are determined according to a modified pay-as-bid (PAB) principle that adjusts the effective prices through their DOP [4].

3.2. Continuous Auction Stage

In the continuous stage, every pair matched successfully is recorded as one round. That is, there is only one matching pair of trades per round.

Objective function—the social welfare of the continuous auction stage can be expressed as follows:

$$\pi_2 = \sum_{r \in R} \max((P'_{bb,i,r} - P'_{bd,ij,r}) \times Q'_{bd,ij,r} + (P'_{sd,ij,r} - P'_{sb,j,r}) \times Q'_{sd,ij,r}) \tag{8}$$

where r, R represents the index and set of trade rounds during the second stage (continuous double auction); $P'_{bb,i,r}$ (yuan/kWh) represents the bid price of market buyer i in the round r; $P'_{bd,ij,r}$ (yuan/kWh) represents the deal price of market buyer i with market seller j in the round r; $Q'_{bd,ij,r}$ (kWh) represents the energy cleared of market buyer i with market seller j in the round r; $P'_{sb,j,r}$ (yuan/kWh) represents the bid price of market seller j in the round r; $P'_{sd,ij,r}$ (yuan/kWh) represents the deal price of market seller j with market buyer i in the round r; $Q'_{sd,ij,r}$ (kWh) represents the energy cleared of market seller j with market buyer i in the round r; and π_2 (yuan) represents the social welfare of the second stage.

Market bidding constraints are defined as follows:

$$\begin{cases} Q'_{bb,i,r} \leq \overline{Q}'_{bb,i,r} \\ Q'_{sb,j,r} \leq \overline{Q}'_{sb,j,r} \end{cases} \tag{9}$$

$$\begin{cases} \overline{Q}'_{bb,i,r} = \overline{Q}_{bb,i} - \sum\limits_{j \in J} Q_{bd,ij} - \sum\limits_{r'=1}^{r-1} \sum\limits_{j \in J} Q'_{bd,ij,r'} \\ \overline{Q}'_{sb,j,r} = \overline{Q}_{sb,j} - \sum\limits_{i \in I} Q_{sd,ij} - \sum\limits_{r'=1}^{r-1} \sum\limits_{i \in I} Q'_{sd,ij,r'} \end{cases} \tag{10}$$

$$\begin{cases} P_\alpha \leq P'_{bb,i,r} \leq P_\beta \\ P_\alpha \leq P'_{sb,i,r} \leq P_\beta \end{cases} \tag{11}$$

where $\overline{Q}'_{bb,i,r}$ represents the bid quantity caps of buyer i in the round r; $\overline{Q}'_{sb,j,r}$ represents the bid quantity caps of seller j in the round r; Equation (10) explains that how to calculate the bid quantity constraint value of the round r.

Market clearing constraints are defined as follows:

$$\begin{cases} \sum\limits_{j \in J} Q'_{bd,ij,r} \leq Q'_{bb,i,r} \\ \sum\limits_{i \in I} Q'_{sd,ij,r} \leq Q'_{sb,j,r} \end{cases} \tag{12}$$

$$\sum_{i \in I} \sum_{j \in J} Q'_{bd,ij,r} = \sum_{i \in I} \sum_{j \in J} Q'_{sd,ij,r} \tag{13}$$

$$\begin{cases} Q'_{bb,i,r} \geq 0 \\ Q'_{sb,j,r} \geq 0 \end{cases} \tag{14}$$

$$\begin{cases} P'_{\Delta,ij,r} = P'_{bb,i,r} - P'_{sb,j,r} \\ P'_{bd,ij,r} = P'_{bb,i,r} - P'_{\Delta,ij,r}/2 \\ P'_{sd,ij,r} = P'_{sb,j,r} + P'_{\Delta,ij,r}/2 \end{cases} \tag{15}$$

Equation (12) indicates that the accumulated transaction quantity must be less than their bid quantity in the per round. Equation (13) represents that the total amount of electricity traded between buyers and sellers is balanced. In Equation (14), the lower bound takes nonnegative values for $Q'_{bb,i,r}$ and $Q'_{sb,j,r}$, which means that transfers of electricity between buyers or between sellers are not allowed in this transaction. The clearing price mechanism is illustrated in Equation (15). $P'_{\Delta,ij,r}$ represents the DOP between marker buyer i and marker seller j in the round r, and $P'_{bd,ij,r}$ $P'_{sd,ij,r}$ are the buying and selling prices for the financial settlement of each pair, respectively.

4. Two-Stage Integrated Clearing Algorithm

As mentioned above, the CBT includes a bidding process that integrates CA and CDA. In order to better design a two-stage integrated market clearing algorithm, it is necessary to analyze its main differences and connections.

Connections: After the market clearing of the first stage, the orders with remaining unmatched volume will be directly included in the second stage, at the same time as the earliest bidding information in the second stage.

Differences: The market clearing takes place only once after all the biddings are collected in the first stage, and the bidding and clearing are obviously sequential; the market clearing is real-time in the second stage, and once the transaction conditions are met, the matching will be carried out. The real-time market-clearing couples the continuous bidding and clearing together.

To ensure the data consistency and business continuity in an efficient way, based on the above analysis, a two-stage integrated market-clearing algorithm is designed. The algorithm used to solve the whole problem is represented by the flowchart in Figure 3. The main steps involved in the algorithm implementation are described as follows.

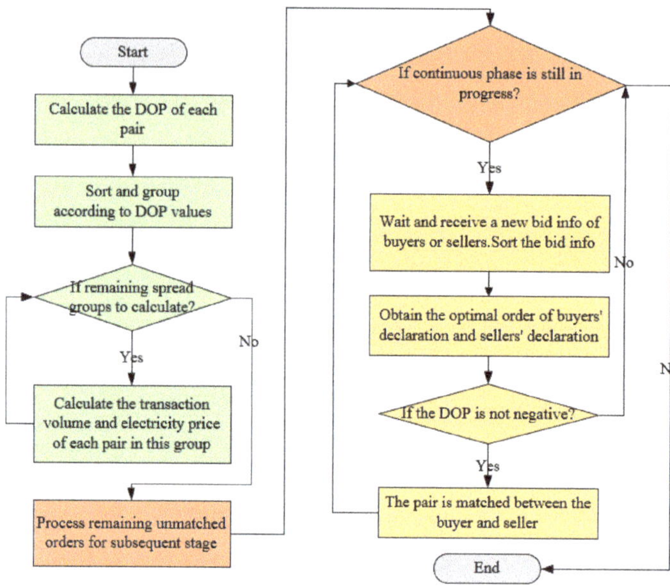

Figure 3. Flowchart of the algorithm.

Step 1: pair each buyer and seller, and the DOPs are calculated for all matching pairs.
Step 2: these pairs are then sorted in descending order based on DOP values. Obviously, a pair with a negative DOP value will not be contracted. The same DOP pairs are divided into a group.
Step 3: start with the group with the largest DOP. If there are any remaining spread groups to calculate, proceed to the next step; otherwise, the market clearing of the first phase is over, so move on to Step 5.
Step 4: the total transaction amount of this group is the smaller value of the quantity declared by all the buyers and the quantity declared by all the sellers in this group. Then, the transaction quantity of each pair is allocated according to the bid quantity in this group. The prices are given by Equation (7), which is mentioned above. After the contract is formed, move on to step 3.
Step 5: orders with remaining unmatched volume are marked with the same timestamp and grouped by price. After entering the second stage, each group is listed as a new order at the earliest time.
Step 6: if the continuous phase is still in progress, then enter the next step; otherwise, the market clearing in the second phase is over.
Step 7: wait and receive new bid info of buyers or sellers. All buyer bids are sorted according to bid-price values in descending order, while seller bids are sorted in ascending order. Once the bid price between buyers or between sellers is the same, it shall be sorted according to the principle of early declaration priority.
Step 8: obtain the optimal order of buyers' declaration and sellers' declaration, and calculate the DOP between buyers' and sellers' bids.
Step 9: if the DOP is not negative, go to the next step; otherwise, move on to step 6.
Step 10: the pair is matched between the buyer and seller, whose transaction amount is the smaller value of the quantity declared by the buyer and the quantity declared by the seller. The buyer price

and seller price are given by Equation (15), which is mentioned above. After the contract is formed, move on to step 6.

5. CBT-Based Electricity Trading Platform

As the software application platform for the CBT in the YNEM, the CETP mainly serves market participants and the market operators. The former need to submit a set of price-volume bids through the CETP to obtain the electricity contracts. The latter is responsible for matching the electricity bids submitted by buyers and sellers and for determining the prices and quantities of the energy traded in the market, which needs to be completed through the CETP.

The framework of the CETP and some key implementation strategies adopted in the system are described below.

5.1. The Overall Architecture of CETP

The conceptual framework of the CETP is demonstrated in Figure 4, employing a multitier architecture: client-side application (presentation tier); application server cluster (application tier); and database cluster (data tier) [42]. In the overall deployment, the responsibilities of each tier are different, and each tier can have one or more components. It should be noted that these tiers may not correspond to the physical distribution and are purely abstract. The components that handle the user interface and user interaction are at the presentation tier. Components that work together to solve business logic problems, such as user bidding, order validation, and market clearing, belong to the application layer. The business logic tier uses the data tier as the repository of the whole system. The CETP selects both relational and nonrelational databases [43], according to the business characteristics of the CBT. For the sake of high performance, clustering technology and dual hot backup machine technology are adopted for each application server and database server in the system.

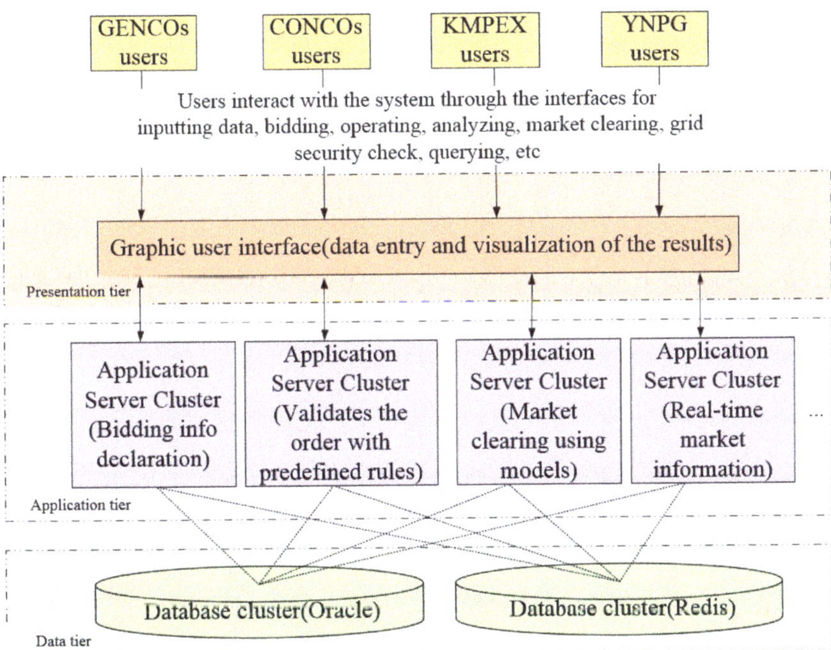

Figure 4. Conceptual framework of the trading system with multitiered architecture.

Generating company (GENCO) users, consumption company (CONCO) users, market operators of the KMPEX, etc. can independently and synchronously access the user interfaces for data entry, bidding, operation, analysis, market clearing, grid security checks, querying information, and so on. The requests received in the presentation tier are proposed in the application tier. Components consisting in the application tier constitute four main modules: message queue-based order processing, multilevel order logic verification with predefined rules, market clearing using models (i.e., market clearing models), and real-time push and disclosure of market information. The data generated by the application layer will be stored in the database for archiving. The results are displayed to the user in the presentation tier.

5.2. Process Flow of Bidding Order

The CBT orders run through the kernel business of the CETP. Figure 5 graphically represents the order data flows in the CETP and analyzes the data flows during order processing. During processing, the orders in the CETP are stored in a message queue. All the orders submitted or canceled are first placed in the original queue. If two bids are made at exactly the same time, the order of entry into the original queue is determined by chance. Next, the orders in the original queue will be marked with a timestamp, as a key indicator of order sorting for the clearing algorithm, and then enter the queue to be detected. Furthermore, based on the trading rules, all orders in the order queue to be detected are checked according to the rules, and those that do not meet the conditions are put into the order queue without pass validation; otherwise, they are put into the order queue with pass validation. Only orders that pass validation can be used for market clearing, and the orders that meet the transaction conditions will be matched to form the final transaction results stored in the result queue. In addition, the market information needs to be calculated logically based on the market process information and pushed to the market participants through the real-time push service. All process and result information for processing an order is stored in a persistent manner.

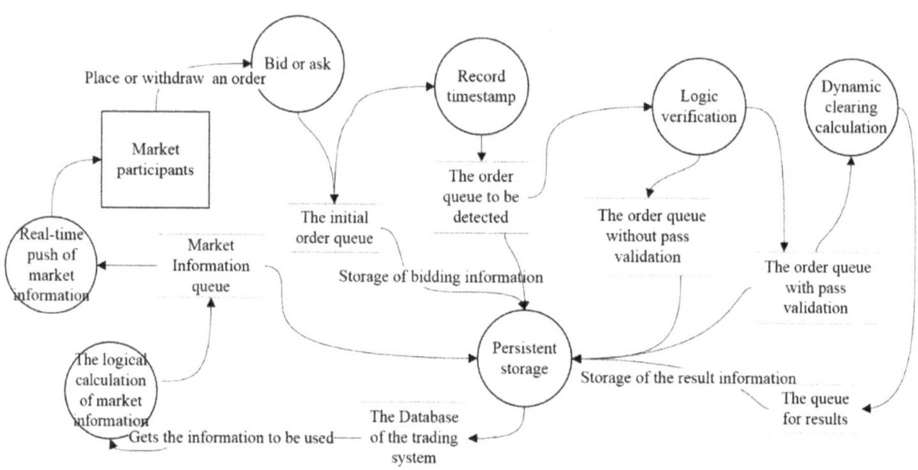

Figure 5. Data flow diagram of CBT.

5.3. Logic Verification for Trading Rules

Reasonable and effective order verification can ensure the standardization and effectiveness of bidding orders, to ensure the normal operation of transactions. A multilevel order logic verification method with predefined rules is put into use in the CETP.

The verification content of each level is as follows.

Level One: verify the information integrity of the bidding data, such as the unique identification code of the bidding subject and the bidding timestamp, which is regarded as the information integrity validation.

Level Two: verify the basic specifications of bidding power and electricity price in each stage, such as data type and accuracy, which is regarded as business standardization verification.

Level Three: verify the number of bidding segments in each stage, the logic of placing and withdrawing orders, and other types of business constraints, which is regarded as the business logic validation.

As shown in Figure 6, the verification of the three layers is carried out in sequence. Once the order cannot pass the verification of one layer, the verification failure will be directly determined. It should be noted that the check items in each layer can be adjusted, invalidated, or added when the business rules change. After the verification process, the verification module forwards the order to the verification result queue, and participants get notified of a successful or failed order submission.

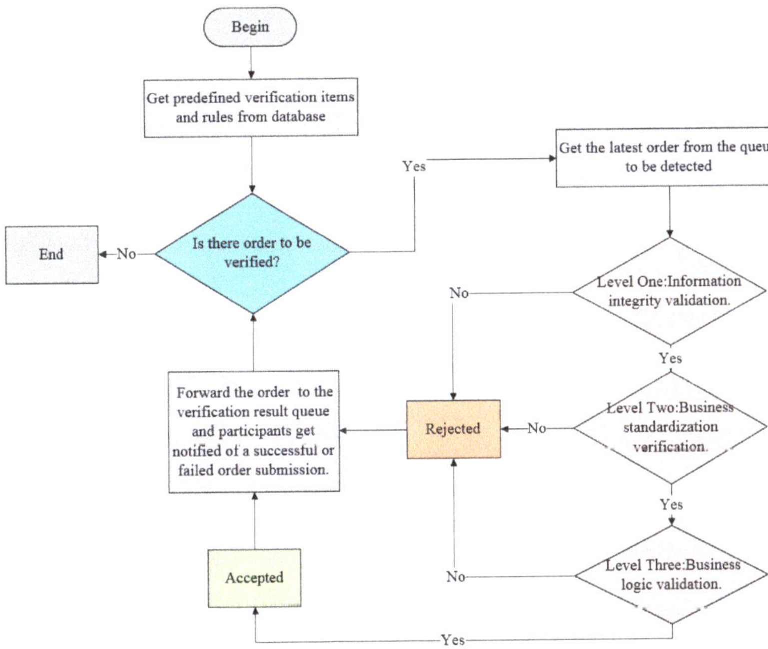

Figure 6. Schematic diagram of the verification process.

5.4. Market Quotation Information

The information-disclosure mechanism is very important to increase market transparency, reduce transaction costs, improve market fairness, and enhance power-system security [44]. In the bidding process of the CBT, market participants can obtain not only their own private information, such as the submitted quantity price, but also the public information of the market, i.e., market information. The real-time display of market information is one of the major features of the CBT in the continuous bidding stage. Normative and transparent information disclosure can weaken information asymmetry. The market information of the CBT mainly includes top of book, market summary, last trades, and candlestick charts, which are effective guides for market participants to make bidding decisions. The market information is explained below, as shown in Figure 7.

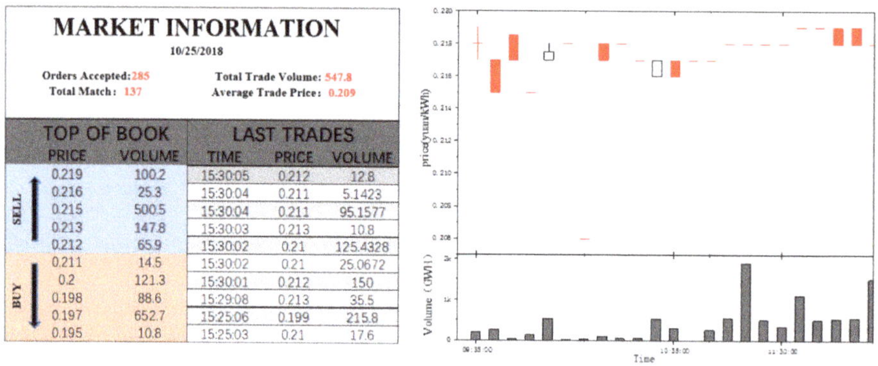

Figure 7. Schematic diagram of market information.

Top of book: all the bids are queued in an order book. The five best bids or asking prices submitted to the market are listed on the publicly observable board. The buyers' bids in the order book are arranged in descending order of price, while the sellers' bids are sorted in ascending order (the higher the buyer's price is, the more likely the transaction will be caused by the lower the seller's price is).

Market summary: during the bidding process, the market summary can be disclosed based on the latest real-time market data, which include the number of orders received, the total trade volume, the average hammer price, etc.

Last trades: the last trades are listed in a contract table. The disclosure information is restricted to the paired transaction time, the clearing price, and the volume for each contract. The trades in the table are sorted in decreasing order of matching time.

Candlestick charts: as trade progresses, candlestick charts for different timescales, such as 5 min and 10 min, will be drawn based on cumulative data. In practice, the relationship among opening, high, low, and closing prices in a continuous trading process is utilized by proponents of candlestick charts, to predict price movements.

All the abovementioned information is updated when there is a new order or trade.

6. Case Study and Application

6.1. Results and Analysis for CBT

To illustrate the effect and characteristics of CBT, take the power energy trading market in November 2018 as an example. Its process and schedule are shown in Figure 8. The bidding period of the first stage was 09:00–09:20, and that of the second stage was 09:30–12:00. A total of 45 CONCOs and 34 GENCOs participated in this transaction. The bidding and clearing of this transaction are described below.

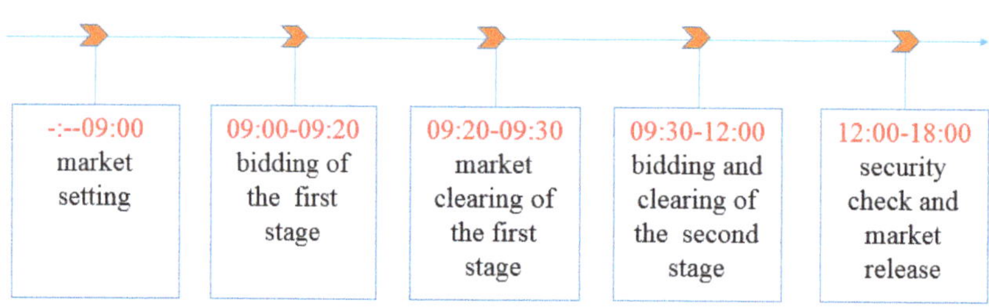

Figure 8. Process and schedule of the CBT in November 2018.

A total of 81 orders (including six orders withdrawn) were declared during the first stage, among which the CONCOs declared 101.929 GWh of electricity, with an average price of 0.19867 yuan/kWh. The declared electric quantity of GENCOs was 2774.818 GWh, with an average price of 0.22544 yuan/kWh. The market bidding data also reflected the characteristics of the YNEM, where supply far exceeds demand. However, during the process of the first stage, the market information was not transparent, which was precious for the bidding-strategy adjustment. The one-round auction made it difficult to realize full competition. As Figure 9 illustrates, buyers and sellers had great differences in price intention. On the demand side, the bid price changed greatly and dispersedly, the bid volume was relatively small, and the demand curve was steep. On the supply side, hydropower plants stored water to high reservoir levels after the wet season, and bid price was usually higher at this point. However, for such hydropower stations with poor regulating performance, they were willing to declare a lower price, to improve the probability of winning the bid, in order to increase its power generation and reduce the risk of water abandonment. The transaction volume of this stage was only 14 GWh, and the average transaction price was 0.21380 yuan/kWh. A total of 24 contracts were concluded, and the rate of trade for the users was only 13.74%.

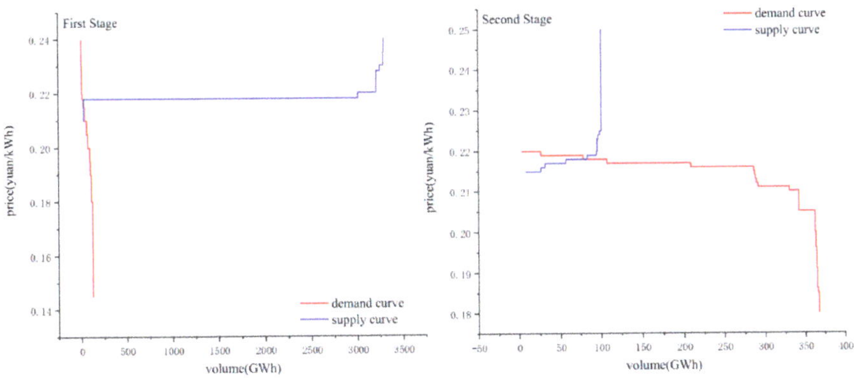

Figure 9. Supply-and-demand curves of two stages of CBT.

Before the beginning of the second stage, the clearing price and contract volume of the first stage were announced, which played a certain price-guiding role in the subsequent bidding. During the second stage, a total of 248 orders (including 100 orders withdrawn) were declared, among which the cumulative effective order quantity of 124.05771 GWh was placed by CONCOs, and the average price was 0.21562 yuan/kWh. The effective order quantity of GENCOs was 83.444 GWh, with an average price of 0.21839 yuan/kWh. In contrast to the first stage, the market information was disclosed in real time in the continuous bidding process, and the market participants could directly respond to the price signals through their trading decisions. Figure 9 shows that the difference in price intention between buyers and sellers was much smaller than previously. As opposed to the one-shot nature of the first stage's auction, the transaction intentions were reached gradually through the strategy adjustment of both sides in the process of constant gaming until market closure. As the deal closed, the accumulated transaction power reached 93.99562 GWh, the average transaction price was 0.21770 yuan/kWh, 83 contracts were reached, and the deal rate for the user was up to 75.77%.

Finally, all the two-stage unconstrained transaction results passed the security check by YNPG.

According to the two-stage clearing model of the CBT, the social welfare of the first stage and the second stage were 106,519 yuan and 14,880.08 yuan, respectively. Although the volume of matched deals in the second stage was far more than that in the first stage, the social welfare was less than that in the first stage. The reasons accounting for this phenomenon mainly included the following two aspects: (1) due to the high transparency of price in the second stages, the price willingness of both parties was gradually close to each other's, and the DOP of the two matched sides was small or even zero. (2) The second stage was a continuous transaction based on the principle with first come, first serve, which existed the loss of allocative efficiency. In other words, the previously matched result maybe lost the greater social welfare that was possibly brought by the subsequent bidding [45].

To further illustrate the dynamic bidding process in the continuous bidding stage, i.e., the second stage, a large user (user A) was taken as an example to describe the dynamic bidding and game process based on market information. Figure 10 shows the average transaction price during the period of the entire market used to indicate the market price signal. The bidding price and trading volume of user A, which are drawn in Figure 10, represent its bidding strategy and strategic results, respectively.

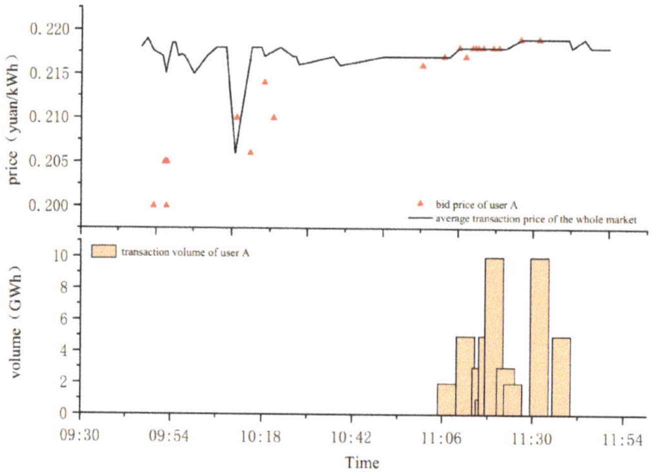

Figure 10. The dynamic bidding and game process of user A.

After the market was opened, the first hammer price of the whole market was approximately 0.218 yuan/kWh, and user A failed to win the bid for the first order, with a price of 0.2 yuan/kWh. In the early period of the market (approximately 09:30–10:00), the average transaction price fluctuated in the price range of 0.215–0.218 yuan/kWh; however, the highest bid of user A was only 0.205 yuan/kWh, so the contract was not reached. Then, the average price showed a large drop (approximately 10:10); at this time, user A had raised the price to 0.209 yuan/kWh but did not take the initiative and still did not win the bid. Subsequently, user A continued to adjust the bidding strategy but failed to satisfy the conditions for the conclusion of the business. The average transaction price was relatively stable at 0.217 yuan/kWh in the middle term of the market (approximately 10:25–11:00), but user A did not bid. As the market gradually entered the later period, the user significantly increased the order price to 0.217 yuan/kWh and won the bid for the first time, reaching a contract volume of nearly 2 GWh. Later, the quotation price of user A was steadily above 0.218 yuan/kWh, winning the bid many times, and the transaction contracts were concluded. With the deadline of the continuous bidding stage, the final cumulative transaction volume of user A was fixed at 46 GWh.

It can be seen from the above description that, in the whole continuous bidding stage, market participants can flexibly adjust their bidding strategies and conduct a dynamic game with the market itself by considering their own resource endowment, risk preference, and other factors to reach trading intention and maximize their own interests.

According to the above two aspects of the analysis, the combination of the two bidding stages accelerates the discovery of market prices. The real-time disclosure of market information weakens information asymmetry. Market players adjust their strategies based on market information, and dynamic games promote full competition in the market. Continuous clearing increases transaction efficiency.

6.2. Overall Application and Prospect

The CETP has been in operation since the end of 2017, so it supports the effective implementation of the CBT. By December 2018, 12 CBTs were successfully launched in the YNEM. A total of 232 CONCOs participated in the bidding, with a total of 2178 Gigawatt hours (GWh) declared. A total of 167 GENCOs participated in the declaration, with a total of 32196 GWh declared [7]. In the whole year of 2018, a total of 5278 orders were processed, 5136 CBT contracts were concluded, and the total volume of the energy cleared was up to 1018 GWh.

Table 1 shows aggregated data from the CETP in 2018, including the total volume and average price of each transaction. Data analysis shows that the transaction size of the CBT varies from month to month, which is affected by the primary energy supply, market demand, and other trading varieties within a month. As a hydro-dominated electricity market, the YNEM has abundant incoming water in the wet season (from June to October in Yunnan), and the supply far exceeds the market demand. The typical buyer's market causes trading prices to be lowered to "the floor price", defined by trading rules (0.13 yuan/kWh) in many months of the wet season. In addition, compared with the dry season, most of the market demand has been met by other varieties (for example, the bilateral negotiated transaction) before the CBT is carried out in the same month during the wet season, leaving less demand space, resulting in a small amount of CBT contracts. In addition, from the point of the two-phase contrast of each month, the contract volume in the continuous bidding stage is more than that in the call auction stage. The longer trading period in the continuous bidding stage has a positive impact on the contract volume, but the more important reason is that the real-time disclosure of market information is conducive to market players tracking market trends, capturing price signals, and forming more contracts through more sufficient competition. Although the transaction amount in the call auction stage is small, it still has many functions: First, in bilateral transactions, many market players take the average transaction price of the first stage as the "benchmark" and determine the final transaction price of bilateral transaction electricity [9]; second, in the continuous bidding stag, market players usually bid or offer by reference to the average transaction price of the call auction stage [7]. More public information about the CBT can be found at www.kmpex.com.

Table 1. The market-clearing overview of the CBT in 2018 (data source: CETP).

Month	The Two-Stage Summation		Call Auction Stage		Continuous Auction Stage	
	Volume (GWh)	Price (Yuan/kWh)	Volume (GWh)	Price (Yuan/kWh)	Volume (GWh)	Price (Yuan/kWh)
2018-01	249.276568	0.22452	58.20	0.22416	191.076568	0.22463
2018-02	84.792	0.1799	46.034	0.18473	38.758	0.17416
2018-03	72.707	0.22859	5.45	0.22307	67.257	0.22904
2018-04	141.33481	0.22834	13	0.22933	128.33481	0.22824
2018-05	105.6456	0.21783	5.4806	0.21691	100.165	0.21788
2018-06	134.588	0.16188	13.513	0.16644	121.075	0.16137
2018-07	4.88	0.13	0.31	0.13	4.57	0.13
2018-08	1.46	0.13103	0.60	0.1325	0.86	0.13
2018-09	0.165	0.13	0.12	0.13	0.045	0.13
2018-10	10.11	0.13	0.11	0.13	10	0.13
2018-11	107.99562	0.21719	14	0.2138	93.99562	0.2177
2018-12	104.742	0.22258	0.752	0.22597	103.99	0.22256

Through the analysis of the application in the whole year, it shows that any mechanism can give full play to its advantages only under suitable conditions. The CBT needs to be further improved and adjusted based on the above practical operation analysis. Examples of improvements include the following:

1. In the medium- and long-term market, the proportion of bilateral negotiation transaction and CBT can be coordinated to release more bidding space for the CBT, which may further play the advantages of the CBT.
2. In the stage of the call auction, the degree of information disclosure can be improved, such as real-time disclosure of virtual transaction price, virtual transaction volume, and virtual non-transaction volume in the bidding stage, but the order information is not disclosed, which may help to further play the function of price discovery.

7. Conclusions

Since the beginning of 2015, China has been in the process of electricity market reform, and all provinces or regions are actively exploring the suitable market mechanism and construction. In this paper, a new trading mechanism is proposed, which is characterized by a sequential bidding and clearing process that integrates CA and CDA. The clearing model of the CBT, with the maximum goal of social welfare for the two bidding stages, is introduced. A two-stage integrated market clearing algorithm is designed to ensure data consistency and business continuity. Finally, the trading platform based on the CBT mechanism is established and has been applied in the YNEM. The analysis results and practical application throughout one year illustrate that the proposed model, algorithm, and various key measures in the trading platform support the business development of the CBT well. It shows that sellers and buyers actively take part in the electricity market and can take the initiative to adjust their bidding wills in two stages, according to the price signals of the market, in order to win the desired electric volume. From a practical point of view, the CBT can promote market competition and price discovery, alleviate information asymmetry, and improve transaction efficiency.

Certainly, the CETP also has some weakness, including the following two points: (1) market entity could not timely receive market quotation information because of poor network condition; (2) the logical verification is less efficient for many bidding orders in a few paralleled electricity markets because the program is running in the same server machine. We will look at solving these problems as our next step.

Data Statement

We report on aggregated data from the CBT-based electricity trading platform of the YNEM. These data are reported in the figure in which they are shown, but original datasets are subject to commercial confidentiality requirements and are hence not publishable.

Author Contributions: Conceptualization, X.Y. and G.L.; methodology, X.Y. and C.C.; software, G.L. and X.Y.; validation, G.L., X.Y., Y.S., and R.C.; investigation, C.C. and Y.S.; writing—original draft preparation, X.Y.; writing—review and editing, C.C., G.L., and R.C.

Funding: This research was supported by the National Natural Science Foundation of China (No. 51879030).

Acknowledgments: The authors are very grateful to the anonymous reviewers and editors for their constructive comments.

Conflicts of Interest: The authors declare no conflicts of interest.

References

1. China Electric Power Press. *China Electric Power Yearbook 2016*; China Electric Power Press: Beijing, China, 2017.
2. A list of Statistical Express of the National Power Industry in 2018. ec.org.cn 2019. Available online: http://www.cec.org.cn/d/file/guihuayutongji/tongjxinxi/niandushuju/2019-01-22/4fedb4c956f6059c5998913b10a6233a.pdf (accessed on 19 January 2019).
3. Li, X.; Chen, Z.; Fan, X.; Cheng, Z.J. Hydropower development situation and prospects in China. *Renew. Sustain. Energy Rev.* **2018**, *82*, 232–239. [CrossRef]

4. Cheng, C.; Chen, F.; Li, G.; Ristić, B.; Mirchi, A.; Qiyu, T.; Madani, K. Reform and renewables in China: The architecture of Yunnan's hydropower dominated electricity market. *Renew. Sustain. Energy Rev.* **2018**, *94*, 682–693. [CrossRef]
5. State Council. Opinions on Further Deepening the Reform of the Electric Power Systems: Document No. 9. Beijing. Available online: http://tgs.ndrc.gov.cn/zywj/201601/t20160129_773852.html (accessed on 15 March 2015).
6. Trading in the National Electricity Market in 2018. Sohu.com 2019. Available online: http://www.sohu.com/a/300817741_722664 (accessed on 11 March 2019).
7. Kunming Power Exchange. Operational Analysis of Yunnan Electricity Market of 2018 and Market Forecast of 2019. Available online: https://www.kmpex.com:8443/file/download/a5dd58fa5699445dab6d880d9bd986a6 (accessed on 2 February 2019).
8. Yang, Q.; Cai, H.; Yan, M.; Zhang, M.; Liu, S.; Xing, Y. Design and application of electricity market mechanisms for Yunnan based on theory of Incentive compatibility and games in depth. *Autom. Electr. Power Syst.* **2017**, *41*, 27–34. [CrossRef]
9. Yunnan Provincial Industry and Information Technology Commission. Implementation Plan of the Yunnan Electricity Market in 2018. Available online: http://www.ynetc.gov.cn/Item/17434.aspx (accessed on 21 November 2017).
10. Fabra, N.; Von der Fehr, N.H.; Harbord, D. Modeling electricity auctions. *Electr. J.* **2002**, *15*, 72–81. [CrossRef]
11. Haoyong, C.; Xiuli, W.; Xifan, W.; Jian, G. Auction Theory and Its Application in Electricity Market Auction Design Part Two: Application of Auction Theory in Electricity Market Auction Design. *Autom. Electr. Power Syst.* **2003**, *4*, 17–23.
12. Stoft, S. Power System Economics: Designing Markets for Electricity. *J. Energy Lit.* **2002**, *99*, 58–63.
13. Silva, C.; Wollenberg, B.F.; Zheng, C.Z. Application of mechanism design to electric power markets (republished). *IEEE Trans. Power Syst.* **2001**, *16*, 862–869. [CrossRef]
14. Liu, Z.; Zhang, X.; Lieu, J. Design of the incentive mechanism in electricity auction market based on the signaling game theory. *Energy* **2010**, *35*, 1813–1819. [CrossRef]
15. Zou, X. Double-sided auction mechanism design in electricity based on maximizing social welfare. *Energy Policy* **2009**, *37*, 4231–4239. [CrossRef]
16. Zou, P.; Chen, Q.; Xia, Q.; He, C.; Kang, C. Incentive compatible pool-based electricity market design and implementation: A Bayesian mechanism design approach. *Appl. Energy* **2015**, *158*, 508–518. [CrossRef]
17. Zhang, C.; Yan, W. Spot Market Mechanism Design for the Electricity Market in China Considering the Impact of a Contract Market. *Energies* **2019**, *12*, 1064. [CrossRef]
18. Pezzutto, S.; Grilli, G.; Zambotti, S.; Dunjic, S. Forecasting Electricity Market Price for End Users in EU28 until 2020—Main Factors of Influence. *Energies* **2018**, *11*, 1460. [CrossRef]
19. Bobinaitė, V.; Konstantinavičiūtė, I.; Lekavičius, V. Theoretical model for electricity market price forecasting. *Econ. Manag.* **2012**, *17*, 944–951. [CrossRef]
20. Nowotarski, J.; Tomczyk, J.; Weron, R. Robust estimation and forecasting of the long-term seasonal component of electricity spot prices. *Energy Econ.* **2013**, *39*, 13–27. [CrossRef]
21. Yu, Y.; Jin, T.; Zhong, C. Designing an incentive contract menu for sustaining the electricity market. *Energies* **2015**, *8*, 14197–14218. [CrossRef]
22. Hobbs, B.F.; Oren, S.S. Three waves of US reforms: Following the path of wholesale electricity market restructuring. *IEEE Power Energy Mag.* **2019**, *17*, 73–81. [CrossRef]
23. Agalgaonkar, A.P.; Kulkarni, S.V.; Khaparde, S.A.; Soman, S.A. Placement and penetration of distributed generation under standard market design. *Int. J. Emerg. Electr. Power Syst.* **2004**, *1*. [CrossRef]
24. Khazaei, H.; Zhao, Y. Indirect mechanism design for efficient and stable renewable energy aggregation. *IEEE Trans. Power Syst.* **2019**, *34*, 1033–1042. [CrossRef]
25. Liu, Y.; Wu, L.; Li, J. Peer-to-peer (P2P) electricity trading in distribution systems of the future. *Electr. J.* **2019**, *32*, 2–6. [CrossRef]
26. Parag, Y.; Sovacool, B.K. Electricity market design for the prosumer era. *Nat. Energy* **2016**, *1*, 16032. [CrossRef]
27. Wang, J.; Wang, Q.; Zhou, N.; Chi, Y. A novel electricity transaction mode of microgrids based on blockchain and continuous double auction. *Energies* **2017**, *10*, 1971. [CrossRef]
28. Mengelkamp, E.; Gärttner, J.; Rock, K.; Kessler, S.; Orsini, L.; Weinhardt, C. Designing microgrid energy markets: A case study: The Brooklyn Microgrid. *Appl. Energy* **2018**, *210*, 870–880. [CrossRef]

29. Ott A, L. Experience with PJM market operation, system design, and implementation. *IEEE Trans. Power Syst.* **2003**, *18*, 528–534. [CrossRef]
30. Huang, J.; Yalla, P.; Yong, T. New real time market applications at the California Independent System Operator (CAISO). In Proceedings of the IEEE PES Power Systems Conference and Exposition, New York, NY, USA, 10–13 October 2004; pp. 1228–1233.
31. Amin, M. North America's electricity infrastructure: Are we ready for more perfect storms? *IEEE Secur. Priv.* **2003**, *99*, 19–25. [CrossRef]
32. Roques, F.A.; Newbery, D.M.; Nuttall, W.J. Investment incentives and electricity market design: The British experience. *Rev. Netw. Econ.* **2005**, *4*. [CrossRef]
33. Flatabo, N.; Doorman, G.; Grande, O.S.; Randen, H.; Wangensteen, I. Experience with the Nord Pool design and implementation. *IEEE Trans. Power Syst.* **2003**, *18*, 541–547. [CrossRef]
34. Chen, F.; Cao, R.; Cheng, C.; Li, G.; Li, X. Contracted generation verification in hydro-dominated electricity markets. In Proceedings of the 2017 IEEE International Conference on Environment and Electrical Engineering and 2017 IEEE Industrial and Commercial Power Systems Europe (EEEIC/I&CPS Europe), Milan, Italy, 6–9 June 2017; pp. 1–6.
35. Fang, D.; Ren, Q.; Yu, Q. How Elastic Demand Affects Bidding Strategy in Electricity Market: An Auction Approach. *Energies* **2019**, *12*, 9. [CrossRef]
36. Chen, K.; Lin, J.; Song, Y. Trading strategy optimization for a prosumer in continuous double auction-based peer-to-peer market: A prediction-integration model. *Appl. Energy* **2019**, *242*, 1121–1133. [CrossRef]
37. Kirschen, D.S.; Strbac, G. *Fundamentals of Power System Economics*; John Wiley & Sons: New York, NY, USA, 2004.
38. Munhoz, F.C.; Correia, P.B. Bidding design for price-taker sellers in bilateral electricity contract auctions. *Int. J. Electr. Power Energy Syst.* **2008**, *30*, 491–495. [CrossRef]
39. Rangel L, F. Competition policy and regulation in hydro-dominated electricity markets. *Energy Policy* **2008**, *36*, 1292–1302. [CrossRef]
40. Liu, S.; Yang, Q.; Cai, H.; Yan, M.; Zhang, M.; Wu, D.; Xie, M. Market reform of Yunnan electricity in southwestern China: Practice, challenges and implications. *Renew. Sustain. Energy Rev.* **2019**, *113*, 109265. [CrossRef]
41. Weber, J.D.; Overbye, T.J. An individual welfare maximization algorithm for electricity markets. *IEEE Trans. Power Syst.* **2002**, *17*, 590–596. [CrossRef]
42. Matena, V.; Krishnan, S.; DeMichiel, L.; Stearns, B. *Applying Enterprise JavaBeanse: Component-Based Development for the J2EEe Platform*, 2nd ed.; Addison-Wesley: Boston, MA, USA, 2003.
43. Jatana, N.; Puri, S.; Ahuja, M.; Kathuria, I.; Gosain, D. A survey and comparison of relational and non-relational database. *Int. J. Eng. Res. Technol.* **2012**, *1*, 1–5.
44. Yang, Y.; Bao, M.; Ding, Y.; Song, Y.; Lin, Z.; Shao, C. Review of information disclosure in different electricity markets. *Energies* **2018**, *11*, 3424. [CrossRef]
45. Scharff, R.; Amelin, M. Trading behaviour on the continuous intraday market Elbas. *Energy Policy* **2016**, *88*, 544–557. [CrossRef]

© 2019 by the authors. Licensee MDPI, Basel, Switzerland. This article is an open access article distributed under the terms and conditions of the Creative Commons Attribution (CC BY) license (http://creativecommons.org/licenses/by/4.0/).

energies

MDPI

Article

Husk Energy Supply Systems for Sunflower Oil Mills

Valerii Havrysh [1], Antonina Kalinichenko [2,*], Grzegorz Mentel [3], Urszula Mentel [4] and Dinara G. Vasbieva [5]

1 Department of Tractors and Agricultural Machines, Operating and Maintenance, Mykolayiv National Agrarian University, 54020 Mykolayiv, Ukraine; havryshvi@mnau.edu.ua
2 Institute of Environmental Engineering and Biotechnology, University of Opole, 45-365 Opole, Poland
3 Department of Economics and Finance, University of Information Technology and Management in Rzeszow, 35-225 Rzeszow, Poland; gmentel@wsiz.rzeszow.pl
4 Department of Security Science, Rzeszow University of Technology, 35-959 Rzeszow, Poland; u.mentel@prz.edu.pl
5 Foreign Languages Department, Financial University under the Government of the Russian Federation, Leningradsky prospect 49, Moscow 125993, Russian; dinara-va@list.ru
* Correspondence: akalinichenko@uni.opole.pl; Tel.: +48-787-321-587

Received: 4 December 2019; Accepted: 9 January 2020; Published: 11 January 2020

Abstract: Together with solar, wind, and hydro renewable energy sources (RES), biomass constitutes an integral part of the high-renewables electricity systems. Considerable feedstocks for electricity generation are process-based residues. Ukraine is the world leader in sunflower seed production, therefore, husk (a by-product of oil production) is a promising biofuel for combustion- based power plants. The plants consume primarily electricity and fossil fuels (natural gas or fuel oil) for steam production. Their usage affects the edible oil production cost and impacts on climate change. The above facts force us to look for alternatives. By-product (husk) utilization can reduce exhaustible energy consumption (fossil fuels and grid electricity) and mitigate climate change. The aim of the study is to make an energy and ecological assessment of biomass energy supply systems. Specifically, the electricity and heat consumption of Ukrainian sunflower oil mills is investigated. Different options of cogeneration systems are analyzed. The preferable mode of combustion-based husk combined heat and power plants is to meet their own heat demand and to sell surplus electricity. Relative gross income and carbon dioxide emission reductions are calculated. Our results show that husk utilization can meet electricity and heat requirements of edible plants. The surplus electricity may be sold to the grid. Husk combined heat and power plants may result in reduction of carbon dioxide by 200–300% and an increase of total income by 24.7–65.7% (compared to conventional energy supply systems).

Keywords: renewable energy sources; husk; energy supply; biomass; efficiency; carbon dioxide; emissions

1. Introduction

Energy plays a key role in the development of a modern civilization. Fossil fuel prices, their exhaustibility, and environmental issues have become primary challenges for mankind. This is a reason why renewable energy, including bioenergy has become a priority for most countries, including the European Union (EU) [1–3]. This fact has compelled the EU to put forward the following objectives [4,5]:

- Strengthening of energy security (decreasing imports of primary energy);
- The reduction of greenhouse gas (GHG) emissions.

To reach the above objectives, a set of measures has been identified. They include the development of renewable energy [6,7]. Recently, the targets of the Renewable Energy Directive (2009/28/EC) were updated. The new targets increase the share of renewable energy consumption (up to 27%) and cut in GHG emissions (up to 40% compared to the emissions in 1990) [8,9].

The utilization of field-based and process-based residues can contribute to providing sustainability of the energy sector [10–12]. Therefore, the biomass residue utilization meets the emerging concept of circular economy. It may be a solution related to waste management [13]. The usage of organic waste as biofuel for power generation or cogeneration is vital for development. It fits into the concept of circular economy that is a new alternative model of transforming residues into energy. This concept promotes to sustainable development [14–16].

Edible oil production is an energy intensive industry. A considerable share of energy demand of technology (electricity and heat) is met by fossil fuels and electricity from the grid. Meanwhile, these mills produce by-products. For example, for sunflower oil production it is husk. Husk could be an alternative to conventional fuels. Biofuel has a number of advantages. First of all, it is a renewable energy source. Secondly, a biomass resource management leads to significant greenhouse gas emission savings. The plant captures carbon dioxide out of the air [17]. Thus, there is a closed cycle of carbon dioxide emission and absorption. Moreover, biomass does not contain sulphur. Therefore, biomass power plants do not emit sulphur dioxide into the atmosphere. As can be seen, process-based residues can be an alternative energy source [18].

The sunflower is widely cultivated around the world. Seventy percent of all sunflower seeds are harvested by European countries [19]. Ukraine, the Russian Federation, and Argentina are the top producers of sunflower seeds [20]. Sunflower husk is a by-product of oil production. Its high energy content (around 15.4 MJ/kg) makes it possible to be used as fuel. Due to the fact that husk is carbon neutral, its utilization for energy production could mitigate climate change. However, a large share of sunflower husk is sent to landfills and its energy usage is scarce [21].

This paper examines the energy supply systems of sunflower oil mills. As a basic case, actual technological parameters of Ukrainian sunflower oil mill were considered.

2. Literature Review

Ukraine has a highly developed agricultural industry, especially edible oil production. From 2013 its share of the world in sunflower seed production ranges from 24.8% to 30.16% [22]. Sunflower seed production is stimulated by high demand for vegetable oil and its high price. Over the last five years, Ukrainian farmers have produced more than 10 million tons of sunflower seed (Figure 1). Despite the increase in production, it has been slowing down since 2012, its value is rather high, and Ukraine is ranked first among the world producers.

Both feedstock and energy costs are important factors that affect the production cost of sunflower oil. The sunflower oil industry in Ukraine is energy intensive. Electricity consumption ranges from 96.6 to 198 kWh per ton of oil and heat consumption (steam) ranges from 348 to 1184 kWh per ton of oil [23]. It is higher compared to other edible oil production [24–26]. It impacts on production cost. To raise the competitiveness of the industry, reducing energy costs is paramount.

It can be reached by means of biomass utilisation. Sunflower oil mills produce a by-product (husk). Husk and products (oil and cake) can be used to generate electricity and/or heat. This substitutes the energy resources sold (fossil fuels and electricity) and reduces greenhouse gas emission. Moreover, biomass CHPs could improve the stability and reliability of the electrical grid. In Ukraine there are about 70 husk boilers in operation. However, only three sunflower oil mills use husk-based combined heat and power plants (CHP) [27,28].

Renewable energy is of great importance in terms of preventing climate change and reducing harmful emissions [29]. The European Union (EU) supports the increase of renewable energy production [30–32]. A lot of scientists have analyzed the influence of renewable energy usage on economic indicators and economic growth [33–37].

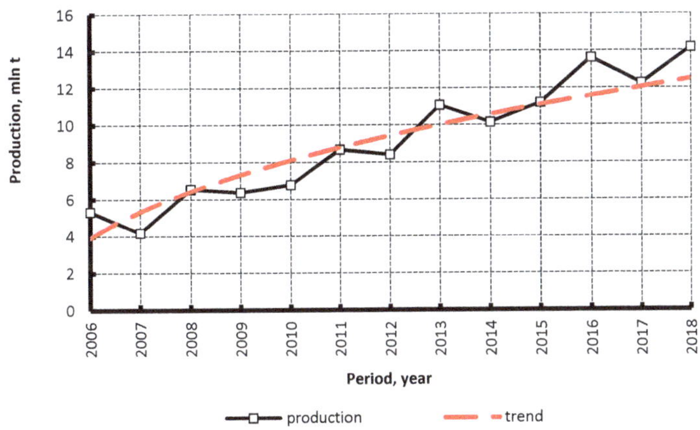

Figure 1. Sunflower seed production history [28].

Previous research has studied the energy supply systems of edible oil plants based on biomass [25,38–40]. They considered a general concept of husk application for energy supply [41,42] and the energy balance of oil mill [40]. However, this problem for sunflower oil mills has been studied insufficiently. Therefore, optimal energy supply schemes are currently being sought. The purpose of the paper is to assess the optimal mode of biomass-based energy supply systems. More specifically, the article aims to reveal specific energy consumption for edible oil production, assess the husk utilization impact on carbon dioxide emission, to analyze different options for biomass utilization.

3. Materials and Methods

The methodology used is as follows. Information and data are collected from the public domain and interactions are made with industrial's officials. Prices of energy, fuels, and products are widespread indicators. Information resources (publications, statistics, and websites) were used. The information was used for further calculations.

All kind of fuels (renewable and fossil) can be compared within the following domains: Energy, environmental, engine efficiency, economics including efficiency of technological equipment.

The energy indicators are as follows: Lower heating value and energy density. The lower heating values were taken from handbooks. The energy density is the amount of energy per unit volume (liter, cubic meter, etc.)

$$DE = LHV \cdot \rho, \text{ MJ/m}^3, \tag{1}$$

where LHV is the lower heating value of the fuel, MJ/kg; ρ is the density of fuel, kg/m^3.

The higher the energy density of fuel, the better the fuel is for consumers.

The ecological indicators may be divided into two groups: Hazardous emissions and carbon dioxide emission. Carbon dioxide emissions are discussed further. This kind of emission has two components: From fuel combustion and in due to electrical consumption. The combustion of hydrocarbon fuels results in the production of carbon dioxide, which is known as a greenhouse gas. This specific value from fossil fuel substitution can be calculated as

$$ERh = HH \cdot \eta_b^{-1} \cdot EF, \text{ t}_{CO2}, \tag{2}$$

where HH is the heat energy of fossil fuel substituted, GJ; EF is the carbon dioxide emission factor for conventional fuel, t_{CO_2}/GJ; η_b is the thermal efficiency of the conventional fuel boiler.

The husk is a carbon dioxide neutral fuel. Its utilization reduces carbon dioxide emission compared to a certain fossil fuel. This specific value on one ton of husk can be calculated as

$$ERs = LHV_h \cdot \eta_h \cdot \eta_b^{-1} \cdot EFm, \mathrm{kg_{CO2}/t}, \quad (3)$$

where LHV_h is the lower heating value of one ton of husk, GJ/t; EFm is the carbon dioxide emission factor for conventional fuel, $\mathrm{kg_{CO2}/GJ}$; η_h is the thermal efficiency of the husk boiler (gasifier).

Emission reduction from onsite electricity production

$$ER_e = EC \cdot EFc,\ \mathrm{t_{CO2}}, \quad (4)$$

where EC is the electricity onsite consumption from grid substituted, MWh; EFc is the emission factor for grid electricity, $\mathrm{t_{CO_2}}$/MWh.

Engine performance indicators can be divided into three groups: Break thermal efficiency or engine efficiency, brake specific fuel consumption, and brake specific energy consumption.

4. Results

4.1. Sunflower Oil Mill Energy Consumption

Top 10 of Ukrainian sunflower oil mills has capacity from 500 to 970 thousand tons of seed per year [43]. Annually they process 7 million tons or 50% of sunflower seed harvest. They use electricity from the grid and thermal energy produced by burning fuel in a boiler (Figure 2). Their specific energy consumption depends on the capacity (Figure 3). Average specific energy consumption is, kWt per ton of oil: Electricity—132.5; heat—779.1.

To reduce energy consumption costs, native mills use husk as a fuel. The output of sunflower seed husk ranges from 15.94% to 18.88% or from 159.4 to 188.8 kg per ton of seeds [44]. To meet their own requirements in heat, powerful Ukrainian vegetable oil mills consume around 46–48% of husk produced (direct burning in boilers). A few mills have their own combined heat and power (CHP) plants. For example, Kirovogradoliya LLC has a husk-based CHP. The CHP was developed to meet all own energy (electricity and heat) requirements. Its electric capacity is 1.7 $\mathrm{MW_e}$ (electric efficiency—around 5%), thermal capacity is 26.7 $\mathrm{MW_t}$. This plant annually consumes up to 42.8 thousand tons of husk (around 57.8% of husk produced). It allows the mill to cover its own energy demands in heat and partially in electricity.

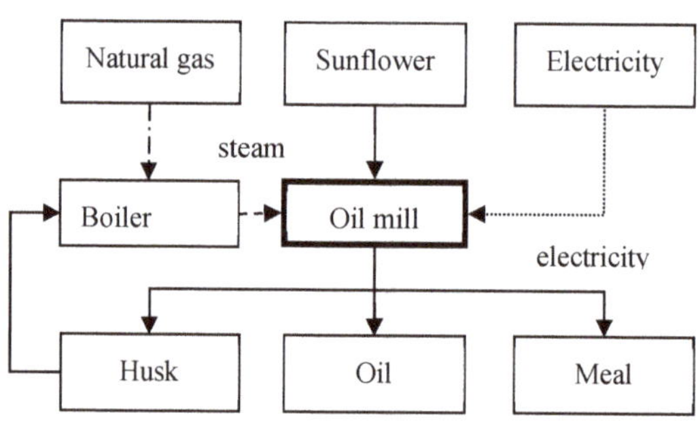

Figure 2. Basic scheme of a Sunflower Seed Oil Mill.

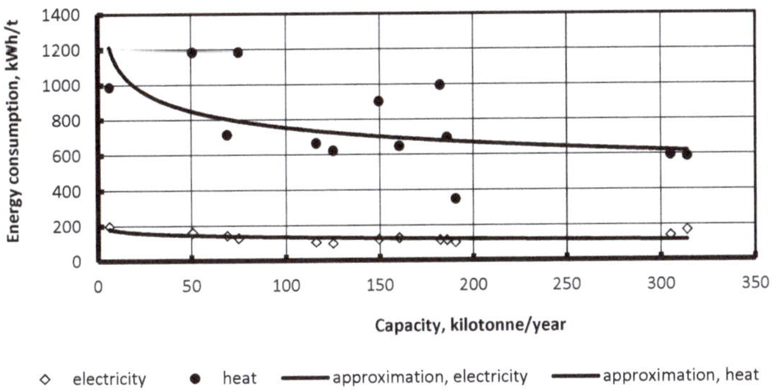

Figure 3. Specific energy consumption (kWh per ton of sunflower oil) vs. annual sunflower oil capacity (thousand tons of seed) [23].

Specific heat to electric consumption ratio of Ukrainian edible oil mills is decreasing with the increase in a plant capacity (Figure 4). The average ratio is 5.97. This means that the thermal efficiency of a CHP unit must be greater than the electrical efficiency by the same value. Surplus electricity can be sold into the grid. It contributes to its stability and to lower carbon dioxide emissions.

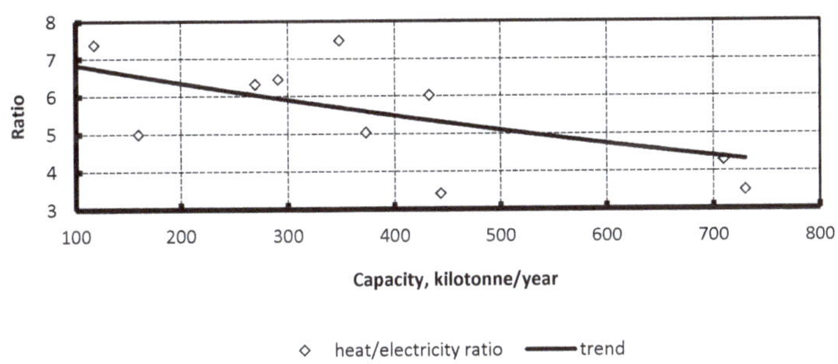

Figure 4. Heat/electricity consumption ratio vs. sunflower oil refinery capacity.

To generate electricity steam turbines and internal combustion engines running on syngas are used. Several oil mills (Galati Sunflower oil factory, Pology oil extraction plant PJSC, Centre Soya Ltd.) have experienced the gasification technology. Last mill has a gasification-based husk CHP plant (rated electric power—700 kW; heat—800 kW) (Flex Technologies Limited, London, UK) [40,45,46].

4.2. Biomass as Fuel

The main results, which are planned to achieve by biomass utilization, can be divided into two groups: Economic and environmental. All products (oil and cake [47]) and by-product (husk) may be used as fuels. Their utilization has distinctive energy and economic efficiency. They can be used in their original form or improved before utilization (gasification, liquefaction, or methanation). Husk can be burnt directly or converted into syngas, biogas, or ethanol [48]. Sunflower meat (cake) can be used as solid fuel or converted into biogas [49]. Table 1 compares parameters of selected fuels.

Table 1. Properties of selected fuels.

Parameter	Unit	Husk	Meal	Oil	Natural Gas	Fuel Oil
LHV	MJ/kg	15.4	28	37	49.75	40.5
Ash content	%	2.1	1,2	-	-	0.05
Carbon content	%	44	64	78	76	88
Bulk density	kg/m^3	95–170	>600	920	0.72	940
Energy density	MJ/m^3	1463–2618	>16,800	34,040	35.82	38,070

Their improvement can get better both environmental and economic indicators of oil production. They can be used in steam boilers to substitute conventional fuel (in Ukraine it is natural gas). If products and by-product substitute natural gas (for steam generation), husk utilization has the best economic result (difference between cost of natural gas substituted and market price of any products and by-product) (Figure 5). The above calculations have been made for one kg. The above differences per one kg of biofuels are, UAH/kg: Husk—4.88; meal—3.32; oil—(−8.22). The sunflower oil is more expensive than natural gas. Therefore, husk as a fuel has an advantage. Husk energy potential is enough to cover energy demand in electricity and heat (Figure 6). It is estimated that the potential electricity production is four-fold higher than the electricity demand. In addition, the heat requirement is a third of the husk potential thermal power production. Excess power and heat can be delivered to external consumers.

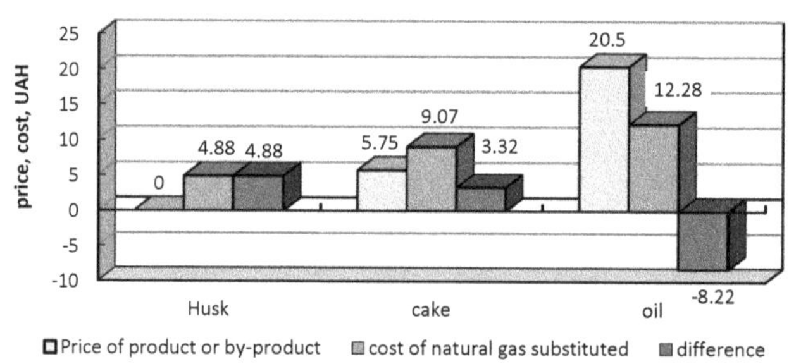

Figure 5. Market prices and cost of natural gas substituted.

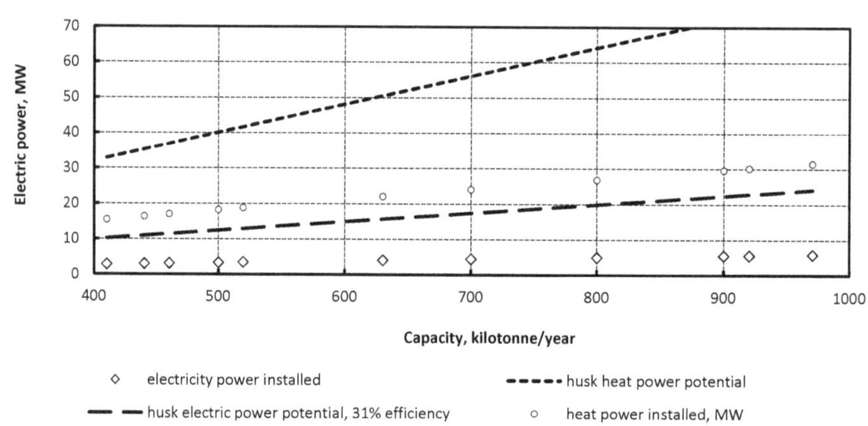

Figure 6. Installed and potential electric power vs. sunflower oil refinery capacity.

4.3. Carbon Dioxide Emission

Ecological indicators may be divided into two groups: Hazardous emissions and carbon dioxide emission. In the study, carbon dioxide emissions were discussed further. This emission consists of two components: Fossil fuel combustion and in due to electricity consumption. For Ukrainian sunflower seed oil mills the first component ranges from 48 to 96 kg per ton of seed processed. The second component ranges from 37 to 77 kg per ton of seed processed.

According to our calculation, one ton of husk (used for steam production) reduces carbon dioxide emission and this range is from 790.10 to 1162.53 kg (Table 2). It corresponds to 51.3–75.5 kg of carbon dioxide per one GJ of thermal energy.

Table 2. Ecological impact of husk utilization.

Conventional Fuel	Volume of Fossil Fuel Substituted by One ton of Husk, t(m^3)	Carbon Dioxide Emission Factor, kg_{CO2}/GJ	Carbon Dioxide Emission Reduced, kg per ton of Husk
Natural gas	428.44	56.10	807.60
Fuel oil	0.38	79.00	1162.53

Electricity consumption for a technological process is the largest sources of emissions for sunflower seed oil mills. Emissions from electricity consumption by any mill are calculated by applying an "emission factor" to the quantity of electricity consumed. Emission factor for grid electricity is taken from an official source. For Ukraine, carbon dioxide emission factor per kWh of electricity consumed is equal to 0.709 [49] or 0.896 $kgCO_2/kWh$ [50].

According to our calculations, power generation from husk (condensate steam turbine, electric efficiency of 31%) reduces carbon dioxide emission by 0.707 kg per kWhe (or 938 kg per ton of husk). Additionally, cogeneration has the potential of 1305 kg per ton of husk. Its carbon dioxide emission reduction potential depends on electrical efficiency (Figure 7). The last factor is prevailing. Gasification-based husk CHP plants allow higher value of carbon dioxide emission reduction compared to combustion-based husk CHP plants.

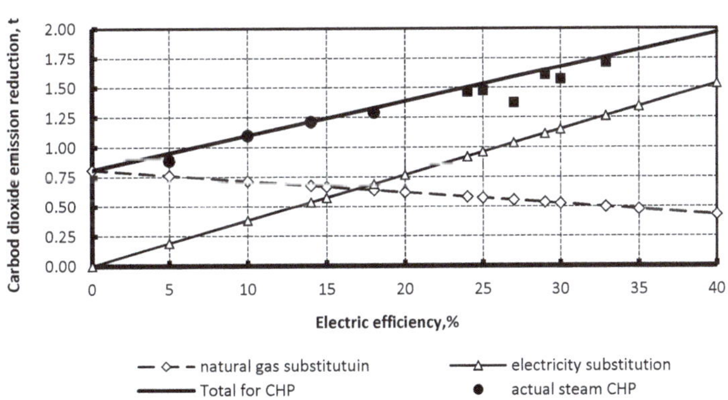

Figure 7. Carbon dioxide emission reduction potential per one ton of husk.

4.4. Energy Supply System Based on Biomass

There are three possible husk utilization pathways: Heat production only, electricity generation only, and combined power and heat generation.

Sunflower husk contains ash (at average 2.1%) [51]. The composition of this by-product of combustion includes calcium, potassium, micro elements, etc. Therefore, the ash can be used as a component to produce fertilizer [52]. Moreover, it is suitable as a filler for the production of ceramics [53]. Its price is more than EUR80/t. Therefore, ash sale can give additional income.

Sunflower husk can be used to generate electricity and heat production. The first pathway substitutes electricity bought, the second—fossil fuels (natural gas, fuel oil, coal, etc.) bought. Their ratio is

$$RCs = \eta_e \cdot \eta_b \cdot Epr \cdot LHV_{ng} \cdot 3,6^{-1} \cdot \eta_h^{-1} \cdot NGpr^{-1},$$

where LHV_g is the lower heating value of natural gas, MJ/m^3; E_{pr} is the price of electricity, UAH/kWh; NG_{pv} is the price of natural gas, UAH/m^3; η_e is the electric efficiency.

Internal combustion engines and gas turbine generators have the highest electric efficiency as compared with steam generators (Figure 8). They can be run on liquid or gaseous fuels. Therefore, husk to be used in the above method must be converted into combustible gas: Syngas or biogas [54,55]. Gasification technology is currently being used and biogas technology is under development.

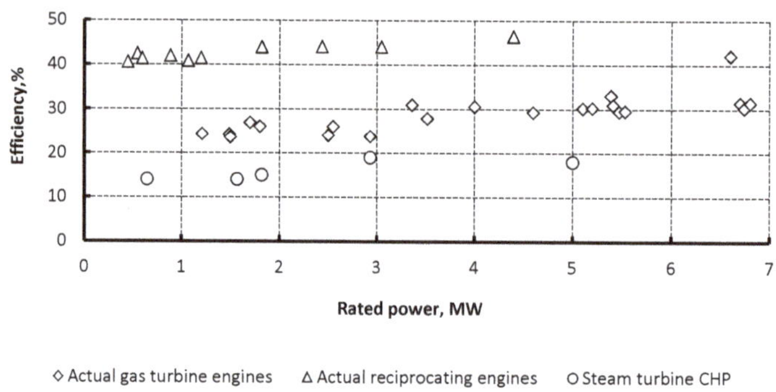

Figure 8. Electric efficiency of gas turbine engines, reciprocating engines, and steam turbine combined heat and power plant CHP (adapted from [56–65]).

Electricity generation can substitute electricity whose cost is lower than solely heat generation (substitution of natural gas). Cogeneration allows mills to reduce more costs of conventional energy bought (Figure 9). As can be seen (Figure 9), the increase in electric efficiency of CHP (and, therefore, the increase of electricity generated) results in the increase of economic benefits.

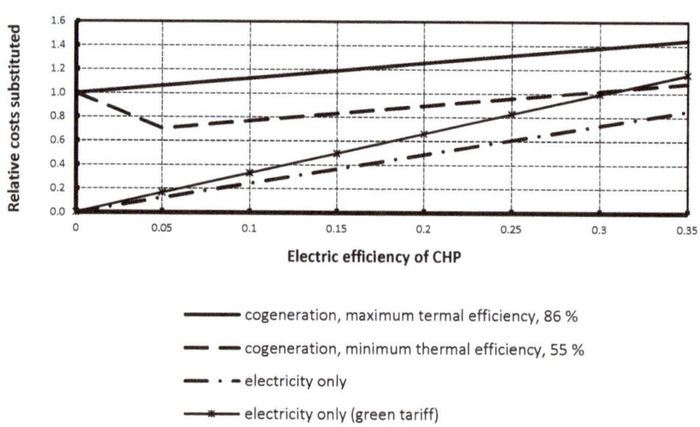

Figure 9. Economic efficiency of CHP (developed by authors).

Husk utilization's economic efficiency depends on the electric efficiency of a CHP plant and can be calculated as

$$RC = \eta_b \cdot LHV_{ng} \cdot \frac{\eta_t \cdot \eta_e \cdot \frac{Epr}{3.6} + NGpr \cdot \frac{(\eta_t - \eta_e)}{\eta_b \cdot LHV_{ng}}}{\eta_h \cdot NGpr},$$

where η_t is the total thermal efficiency of CHP; LHV_h is the lower heating value of husk, MJ/kg.

The more conventional energy resources are substituted, the higher income is made. The possible energy supply technologies are as follows (Figure 10): 1—steam turbine combined heat and power generation plant (CHP) via husk combustion; 2—gas turbine (GT) CHP via syngas combustion; 3—internal combustion engine (ICE) CHP via syngas combustion; 4—gas turbine or internal combustion engine with organic Rankin cycle (ORC).

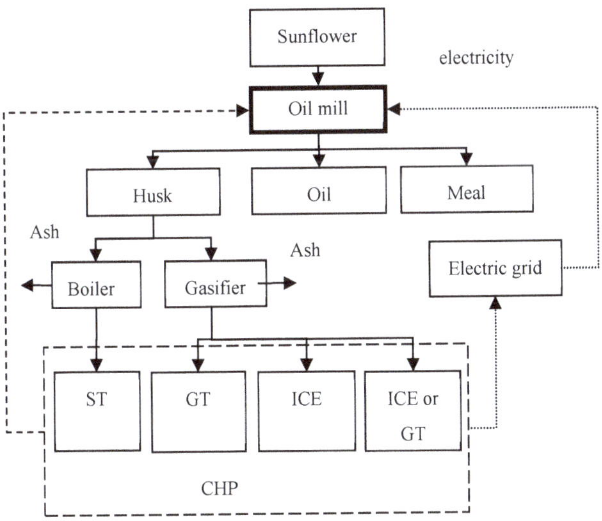

Figure 10. Energy supply technologies of Sunflower Seed Oil Mill (developed by author).

4.5. Principles of Energy Production

Ukrainian sunflower oil refineries use husk for steam production. Only some plants have CHP, but they cover only part of the electricity required.

There are three pathways of energy production by CHP (Table 3). The first pathway is to meet its own electricity requirement. However, in this case CHP cannot cover its own demand in heat. To correct the situation an additional steam boiler must be used. The second pathway is to meet its own heat requirement. In this case a CHP plant generates surplus electricity. It may be sold by green tariffs to electricity grid. For both cases, the remaining husk can be converted into pellets. The third pathway is the following. The husk combustion-based CHP plant can cover requirements in both electricity and heat. This energy supply system could have adapted the heat to electricity production ratio. If there is actual heat to electricity ratio (*HERa*) of a certain sunflower seed oil mill and total efficiency of a CHP plant then necessary electric efficiency (η_t) is

$$\eta_e = \eta_t \cdot (1 + HERa)^{-1}.$$

During operation of a certain mill, the heat to electricity ratio may vary. In this case, surplus electricity can be delivered to the national grid.

Table 3. Relative gross income and carbon dioxide emission reduction.

Pathway of Energy Production	Combustion-Based CHP	Gasification-Based CHP
	Gross income	
The first pathway	1.17	1.16
The second pathway	1.25–1.26	1.64–1.66
The third pathway	1.16	-
Husk for heat only	1	-
	Carbon dioxide emission reduction	
The first pathway	1.60	1.20
The second pathway	2.11–2.16	3.58–3.66
The third pathway	1.85–1.87	-
Husk for heat only	1	-

The gross income comprises three pillars: Cost of natural gas substituted, cost of electricity from grid substituted and sold to the grid by green tariff, and cost of husk pellets produced. Carbon dioxide emission reduction includes two components: Substitution of natural gas (or another conventional fuel) and electricity generation. The relative values of gross income and carbon dioxide reduction (the base is husk utilization to meet heat requirements only) are presented in Table 3. The second pathway of energy production has the best economic and ecological results. In this option, the cogeneration unit uses all available husks. Prerequisites for success are the sale of excess electricity by the green tariffs to the grid and the full use of thermal energy for oil production. Therefore, the prospect energy supply scheme is shown in Figure 11.

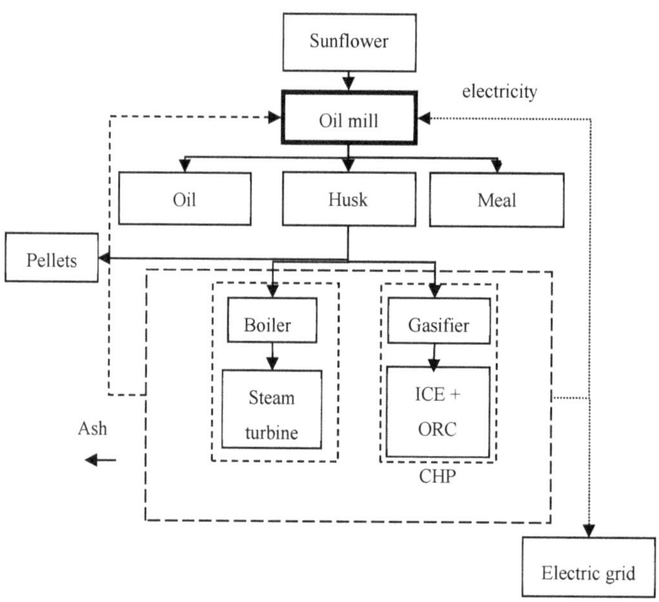

Figure 11. Energy supply scheme of Sunflower Seed Oil Mill (developed by author).

As can be seen, the best strategy for CHP development is to cover heat demand by CHP. As compared with husk utilization for heat production only, it allows mills to increase total income by 21–23% and reduce carbon dioxide emission by 70–73%. Gasification-based technologies may be profitable if their specific investment costs are not more than 40% that of combustion-based technology.

5. Conclusions

The primary goal of biomass utilization is to reduce production costs and carbon dioxide emission. The ecological benefit of husk (as biofuel) is that it is a carbon neutral source of energy. CHP plants ensure the most ecological effect.

The increase of mill capacity results in the decrease of specific energy consumption and heat/electricity consumption ratio. To reduce production costs and carbon dioxide emission, Ukrainian mills utilize sunflower seed husk (primarily to cover requirement in heat). Its consumption ranges from 46% to 57.8% of total husk production.

Heat to electricity consumed ratio of oil mills does not coincide with heat to electricity generated ratio of CHP plants. Therefore, different modes of CHP have been analyzed. The preferable mode is the design of CHP plants to cover heat demand of an oil mill. This option may increase total income by 24.7–65.7% and cut carbon dioxide emission by 201–366%. Either way, excess electricity is sent to the grid. In addition, the remaining husk may be used for pellet or green electricity production.

Energy content of husk exceeds energy demand of any edible oil plant. Husk utilization allows the plants to reduce carbon dioxide emission, kg per ton of oil produced: Heat generation—142; power generation—174; combined power and heat generation—274.

Gasification based husk CHP provides higher electricity efficiency, gross income, and carbon dioxide emission. However, this technology is not mature enough.

At further work, impact of annual operating hours and investment costs on economical indicators are expected to be studied.

Author Contributions: Formal analysis, methodology, V.H. and A.K.; Investigation, G.M., U.M., and D.G.V.; Writing—original draft, V.H. and A.K.; Writing—review and editing, V.H., A.K., G.M., U.M., and D.G.V. All authors have read and agreed to the published version of the manuscript.

Funding: This research received no external funding.

Conflicts of Interest: The authors declare no conflict of interest.

References

1. West, J.; Bailey, I.; Winter, M. Renewable energy policy and public perceptions of renewable energy: A cultural theory approach. *Energy Policy* **2010**, *38*, 5739–5748. [CrossRef]
2. Moula, M.E.; Maula, J.; Hamdy, M.; Fang, T.; Jung, N.; Lahdelma, R. Researching social acceptability of renewable energy technologies in Finland. *Int. J. Sustain. Built Environ.* **2013**, *2*, 89–98. [CrossRef]
3. Röder, M.; Thornley, P. Bioenergy as climate change mitigation option within a 2 °C target—uncertainties and temporal challenges of bioenergy systems. *Energy Sustain. Soc.* **2016**, *6*, 6. [CrossRef]
4. Tol, R.S.J. A cost-benefit analysis of the EU 20/20/2020 package. *Energy Policy* **2012**, *49*, 288–295. [CrossRef]
5. Robbins, A. How to understand the results of the climate change summit: Conference of Parties21 (COP21) Paris 2015. *J. Public Health Policy* **2016**, *37*, 129–132. [CrossRef] [PubMed]
6. Eriksson, O.; Finnveden, G.; Ekvall, T.; Björklund, A. Life cycle assessment of fuels for district heating: A comparison of waste incineration, biomass- and natural gas combustion. *Energy Policy* **2007**, *35*, 1346–1362. [CrossRef]
7. Musall, F.D.; Kuik, O. Local acceptance of renewable energy—A case study from southeast Germany. *Energy Policy* **2011**, *39*, 3252–3260. [CrossRef]
8. Zhou, S.; Matisoff, D.C.; Kingsley, G.A.; Brown, M.A. Understanding renewable energy policy adoption and evolution in Europe: The impact of coercion, normative emulation, competition, and learning. *Energy Res. Soc. Sci.* **2019**, *51*, 1–11. [CrossRef]
9. COM. *A Policy Framework for Climate and Energy in the Period from 2020 to 2030*; European Commission: Brussels, Belgium, 2014; Available online: https://eur-lex.europa.eu/legal-content/EN/TXT/PDF/?uri=CELEX:52014DC0015&from=EN (accessed on 3 December 2019).
10. Nwakaire, J.N.; Obi, F.O.; Ugwuishiwu, B.O. Agricultural waste concept, generation, utilization and management. *Niger. J. Technol.* **2016**, *35*, 957–964. [CrossRef]

11. Chen, Y.; Dai, Y.; Gao, F.; Gao, Y.; Hu, Z.; Li, J.; Liu, M.; Lu, L.; Sun, Q.; Sun, Y.; et al. Utilizations of agricultural waste as adsorbent for the removal of contaminants: A review. *Chemosphere* **2018**, *211*, 235–253. [CrossRef]
12. Harshwardhan, K.; Upadhyay, K. Effective utilization of agricultural waste: A review. *J. Fundam. Renew. Energy Appl.* **2017**, *7*, 237. [CrossRef]
13. Antoniou, N.; Barakat, A.; Ficara, E.; Monlau, F.; Sambusiti, C.; Zabaniotou, A. Contribution to circular economy options of mixed agricultural wastes management: Coupling anaerobic digestion with gasification for enhanced energy and material recovery. *J. Clean. Prod.* **2019**, *209*, 505–514. [CrossRef]
14. Reike, D.; Vermeulen, W.J.V.; Witjes, S. The circular economy: New or refurbished as CE 3.0?—Exploring controversies in the conceptualization of circular economy through a focus on history and resource value retention options. *Resour. Conserv. Recycl.* **2018**, *135*, 246–264. [CrossRef]
15. Honkasalo, A.; Korhonen, J.; Seppala, J. Circular Economy: The concept and its limitations. *Ecol. Econ.* **2017**, *143*, 37–46. [CrossRef]
16. Santos Costa, I.; Massard, G.; Agarwal, A. Waste management policies for industrial symbiosis development: Case studies in European countries. *J. Clean. Prod.* **2010**, *18*, 815–822. [CrossRef]
17. Santoli, L.; Mancini, F.; Nastasi, B.; Piergrossi, V. Building integrated bioenergy production (BIBP): Economic sustainability analysis of Bari airport CHP (combined heat and power) upgrade fueled with bioenergy from short chain. *Renew. Energy* **2015**, *81*, 499–508. [CrossRef]
18. Taylor, M.J.; Alabdrabalameer, H.; Skoulou, V. Choosing Physical, Physicochemical and Chemical Methods of Pre-Treating Lignocellulosic Wastes to Repurpose into Solid Fuels. *Sustainability* **2019**, *11*, 3604. [CrossRef]
19. Food and Agriculture Organization of the United Nations (FAO). Agriculture Data. Available online: http://www.fao.org/faostat/en/#home (accessed on 3 December 2019).
20. Production Volume of Sunflower Seed in Major Producer Countries in 2018/2019. Statista. Available online: https://www.statista.com/statistics/263928/production-of-sunflower-seed-since-2000-by-major-countries/ (accessed on 3 December 2019).
21. Cubitto, M.A.; Gentili, A.R. Bioremediation of crude oil-contaminated soil by immobilized bacteria on an agroindustrial waste—Sunflower seed husks. *Bioremediat. J.* **2015**, *19*, 277–286. [CrossRef]
22. National Sunflower Association. World Supply & Disappearance. 8 February 2019. Available online: http://www.sunflowernsa.com/stats/world-supply/ (accessed on 10 November 2019).
23. Maslikov, M. *Energy Efficiency Benchmarking in the Vegetable Oil Subsector of the Ukraine's Agro-Industrial Sector*; Report; UNIDO: Kyiv, Ukraine, 2012; Available online: https://open.unido.org/api/documents/4677753/download/REPORT%20-%20Energy%20Efficiency%20Benchmarking%20in%20the%20Vegetable%20Oil%20Subsector%20of%20the%20Ukraine%E2%80%99s%20Agro-Industrial%20Sector (accessed on 9 November 2019).
24. Departmental Norms of Technological Design of Small-Capacity Enterprises for the Production of Vegetable Oils from Sunflower Seeds and Rapeseed by Pressing. Moscow. 1993. Available online: http://www.proagro.com.ua/reference/standard/oilprod/20421.html (accessed on 10 November 2019).
25. Sommart, K.; Pipatmanomai, S. Assessment and Improvement of Energy Utilization in Crude Palm Oil Mill. In *2011 International Conference on Chemistry and Chemical Process*; IPCBEE; IACSIT Press: Singapore, 2011; Volume 10, pp. 161–166. Available online: http://www.ipcbee.com/vol10/31-V00054.pdf (accessed on 10 November 2019).
26. Özilgen, M.; Sorgüven, E. Energy and exergy utilization, and carbon dioxide emission in vegetable oil production. *Energy* **2011**, *36*, 5954–5967. [CrossRef]
27. Geletukha, G.; Zheliezna, T.; Kramar, V.; Kucheruk, P. Prospects for the Development of Bioenergy as an Instrument for Natural Gas Replacement in Ukraine—12th Position Paper of UABio. 2015. Available online: http://uabio.org/en/activity/uabio-analytics/2475-uabio-position-paper-12 (accessed on 10 November 2019).
28. Plant Growing. State Statistics Service of Ukraine. Available online: https://ukrstat.org/uk/operativ/operativ2006/sg/sg_rik/sg_u/rosl_u.html (accessed on 10 November 2019).
29. Erbach, G. *Promotion of Renewable Energy Sources in the EU: EU Policies and Member State Approaches*; European Parliament: Brussels, Belgium, 2016. [CrossRef]
30. European Union. *Directive 2009/28/EC of the European Parliament and of the Council of 23 April 2009 on the Promotion of the Use of Energy from Renewable Sources and Amending and Subsequently Repealing Directives 2001/77/EC and 2003/30/EC*; European Union: Brussels, Belgium, 2009; pp. 16–62.

31. European Union. *Directive 2015/1513 of the European Parliament and of the Council of 9 September 2015 Amending Directive 98/70/EC Relating to the Quality of Petrol and Diesel Fuels and Amending Directive 2009/28/EC on the Promotion of the Use of Energy from Renewable Sources*; European Union: Brussels, Belgium, 2015; pp. 1–29.
32. Proposal for a Directive of the European Parliament and of the Council on the Promotion of the Use of Energy from Renewable Sources (Recast). Brussels. 30.11.2016. COM(2016) 767 Final, 2016/0382 (COD). Available online: https://eur-lex.europa.eu/resource.html?uri=cellar:151772eb-b7e9-11e6-9e3c-01aa75ed71a1.0001.02/DOC_1&format=PDF (accessed on 10 November 2019).
33. Armeanu, D.S.; Vintila, G.; Gherghina, S.C. Does Renewable Energy Drive Sustainable Economic Growth? Multivariate Panel Data Evidence for EU-28 Countries. *Energies* **2017**, *10*, 381. [CrossRef]
34. Amri, F. The relationship amongst energy consumption (renewable and non-renewable), and GDP in Algeria. *Renew. Sustain. Energy Rev.* **2017**, *76*, 62–71. [CrossRef]
35. Fotourehchi, Z. Renewable Energy Consumption and Economic Growth: A Case Study for Developing Countries. *Int. J. Energy Econ. Policy* **2017**, *7*, 61–64.
36. Bhattacharya, M.; Paramati, S.; Ozturk, I.; Bhattacharya, S. The effect of renewable energy consumption on economic growth: Evidence from top 38 countries. *Appl. Energy* **2016**, *162*, 733–741. [CrossRef]
37. Cucchiella, F.; D'Adamo, I.; Gastaldi, M. Future Trajectories of Renewable Energy Consumption in the European Union. *Resources* **2018**, *7*, 10. [CrossRef]
38. Booneimsri, P.; Kubaha, K.; Chullabodhi, C. Increasing power generation with enhanced cogeneration using waste energy in palm oil mills. *Energy Sci. Eng.* **2018**, *6*, 154–173. [CrossRef]
39. Azhdari, A.; Ghadamian, H.; Ataei, A.; Yoo, C.K. A new approach for Optimization of Combined Heat and Power Generation in Edible Oil Plants. *J. Appl. Sci.* **2009**, *9*, 3813–3820. [CrossRef]
40. Ion, I.V.; Popescu, F. Improving the Energy Balance in a Sunflower Oil Mill. *Sci. Work. Univ. Food Technol.* **2017**, *64*, 164–171.
41. Clef, E.L.; Kemper, T. Sunflower Seed Preparation and Oil Extraction. In *Sunflower*; Academic Press; AOCS Press: Urbana, IL, USA, 2015; pp. 187–226. [CrossRef]
42. Donaldson, A.; Kadakia, P.; Gupta, M. Production of Energy and Activated Carbon from Agri-Residue: Sunflower Seed Example. *Appl. Biochem. Biotechnol.* **2012**, *168*, 154. [CrossRef]
43. Shamray, M. Who Owns the Oil Extraction Plants? Agrarian Market. 31 January 2018. Available online: https://biz.censor.net.ua/resonance/3047488/kto_vladeet_masloekstraktsionnymi_zavodami (accessed on 10 November 2019).
44. Utilization of Sunflower Seeds Husk for Steam and Power Production at the Oil Extraction Plant OJSC "Kirovogradoliya". Kyiv; 2009. Available online: http://www.seia.gov.ua/seia/doccatalog/document?id=126992 (accessed on 10 November 2019).
45. Kalinichenko, A.; Havrysh, V.; Atamanyuk, I. The Acceptable Alternative Vehicle Fuel Price. *Energies* **2019**, *12*, 3889. [CrossRef]
46. Flex Technology. WBP-750 Sunflower Husk Gasification CHP Plant Rated 700 kWe and 800 kWth. 2019. Available online: https://flextechnologies.co.uk/projects/ (accessed on 10 November 2019)
47. Kachel-Jakubowska, M.; Kraszkiewicz, A.; Krajewska, M. Possibilities of using waste after pressing oil from oilseeds for energy purposes. *Agric. Eng.* **2016**, *20*, 45–54. [CrossRef]
48. Popescu, B.; Şenilă, L.; Vărăticeanu, C.; Şimon, G. Cellulosic bioethanol from sunflower seed hulls—A renewable energy source. *Studia UBB Ambient.* **2013**, *58*, 105–110.
49. Brander, M.; Sood, A.; Wylie, C.; Haughton, A.; Lovell, J. *Electricity-Specific Emission Factors for Grid Electricity*; Econometrica: Edinburgh, UK, 2011; pp. 1–22. Available online: https://ecometrica.com/assets/Electricity-specific-emission-factors-for-grid-electricity.pdf (accessed on 10 November 2019).
50. Gómez, D.R.; Watterson, J.D. Stationary Combustion. In *2006 IPCC Guidelines for National Greenhouse Inventories*; IGES: Japan, 2006; Volume 2, pp. 2.2–2.47. Available online: https://www.ipcc-nggip.iges.or.jp/public/2006gl/pdf/2_Volume2/V2_2_Ch2_Stationary_Combustion.pdf (accessed on 10 November 2019).
51. Maj, G.; Krzaczek, P.; Kuranc, A.; Piekarski, W. Energy properties of sunflower seed husk as industrial extrusion residue. *Agric. Eng.* **2017**, *21*, 77–84. [CrossRef]
52. Paleckienė, R.; Sviklas, F.M.; Šlinkšienė, R.; Štreimikis, V. Complex Fertilizers Produced from the Sunflower Husk Ash. *Pol. J. Environ. Stud.* **2010**, *19*, 973–979.

53. Quaranta, N.E.; Pelozo, G.G.; Cesari, A.; Cristóbal, A.A. Characterization of sunflower husk ashes and feasibility analysis of their incorporation in soil and clay mixtures for ceramics. *WIT Trans. Ecol. Environ.* **2016**, *203*, 13–23. [CrossRef]
54. Kalinichenko, A.; Havrysh, V.; Perebyynis, V. Evaluation of biogas production and usage potential. *Ecol. Chem. Eng. S* **2016**, *23*, 387–400. [CrossRef]
55. Kalinichenko, A.; Havrysh, V. Preliminary assesment of biogas projects. *Arch. Environ. Prot.* **2019**, *45*, 68–83. [CrossRef]
56. We Power the World with Innovative Gas Turbines. Siemens Gas Turbine Portfolio. 2019. Available online: https://assets.new.siemens.com/siemens/assets/public.1551272819.10f4860b140b2456f05d32629d8d758dc00bcc30.gas-turbines-siemens-interactive.pdf (accessed on 10 November 2019).
57. Zorya-Mashproekt. Gas Turbine Engines for Gas Pipelines. 2016. Available online: https://mfa.gov.ua/mediafiles/sites/jordan/files/Commercial_offers/SE_Zorya-Mashproekt_-_Gas_Turbine_Engines_For_Gas_Pipelines.pdf (accessed on 10 November 2019).
58. Veya Investments Ltd. Catalogue. Gas Turbine Units and Industrial Equipment on their Basis. Available online: http://www.veyainvest.com/catalog.files/CatalogGasEn.pdf (accessed on 10 November 2019).
59. Kawasaki Gas Turbine Generator Sets. Available online: https://global.kawasaki.com/en/energy/pdf/Green_Brochure.pdf (accessed on 10 November 2019).
60. PAES-2500, EG-2500T, 2,5 MW Gas-Turbine Power Generating Sets. Available online: http://xn--80aajzhcnfck0a.xn--p1ai/PublicDocuments/0400774.pdf (accessed on 10 November 2019).
61. Saturn 20. Gas Turbine Generator Set. Available online: http://s7d2.scene7.com/is/content/Caterpillar/C10550179 (accessed on 10 November 2019).
62. Centaur 40. Gas Turbine Generator Set. Available online: http://www.kushaindustry.com/pdf/brochure/SOLARTurbines.pdf (accessed on 10 November 2019).
63. Jenbacher Gas Engines. Clarke Energy. Available online: https://www.clarke-energy.com/gas-engines/ (accessed on 10 November 2019).
64. 2G Energy AG. Access Mode. Available online: http://www.2g-energy.com/media/2292/avus-_natural-gas.pdf (accessed on 10 November 2019).
65. Obernberger, I.; Thek, G. Cost assessment of selected decentralized CHP applications based on biomass combustion and biomass gasification. In Proceedings of the 16th European Biomass Conference & Exhibition, Valencia, Spain, 2–6 June 2008.

© 2020 by the authors. Licensee MDPI, Basel, Switzerland. This article is an open access article distributed under the terms and conditions of the Creative Commons Attribution (CC BY) license (http://creativecommons.org/licenses/by/4.0/).

energies

MDPI

Article

Renewable Energy and EU 2020 Target for Energy Efficiency in the Czech Republic and Slovakia

Jacek Brożyna [1,*], Wadim Strielkowski [2,*], Alena Fomina [3] and Natalya Nikitina [4]

1 Department of Quantitative Methods, The Faculty of Management, Rzeszow University of Technology, Aleja Powstańców Warszawy 10/S, 35-959 Rzeszow, Poland
2 Department of Trade and Finance, Faculty of Economics and Management, Czech University of Life Sciences Prague, Kamýcká 129, 165 00 Prague 6, Prague, Czech Republic
3 JSC "Central Research Institute of Economy Management and Information Systems Electronics", Kosmonavta Volkova str. 2, 127299 Moscow, Russian; fomina@ymservices.ru
4 Research Institute of Perspective Directions and Technologies, Russian State Social University, Wilhelm Pieck str. 4/1, 129226 Moscow, Russian; nn.0803@mail.ru
* Correspondence: jacek.brozyna@prz.edu.pl (J.B.); strielkowski@pef.czu.cz (W.S.)

Received: 30 December 2019; Accepted: 19 February 2020; Published: 21 February 2020

Abstract: Our paper focuses on the renewable energy and EU 2020 target for energy efficiency in the Czech Republic and Slovakia. We study the reduction of greenhouse gas (GHG) emissions in these two EU Member States through the prism of the Europe 2020 strategy and the 3 × 20 climate and energy package and economic growth (represented by the Gross Domestic Product (GDP) that allows to measure the national dynamics and provide cross-country comparisons) without attributing specific attention to issues such as the electrification of transport or heating, and thence leaving them outside the scope of this paper. Both Czech Republic and Slovakia are two post-Communist countries that still face the consequences of economic transformation and struggle with the optimal management of natural resources. Both countries encountered profound system transformation after 1989 that are apparent in all three measures of sustainable development used in our study. We show that it is unlikely that the planned increase in renewable energy in the Czech Republic and Slovakia will reach its targets, but they might succeed in reducing their energy consumption and greenhouse gas emissions. Our findings show that the energy intensity of Czech and Slovak economies increased in the early 2000s and then stabilized at a level about twice of the EU average. It appears that this value is likely to remain the same in the forthcoming years. However, implementation of GHG emissions in the Czech Republic and Slovakia may be at risk in case the proper energy policy is not maintained. Moreover, our results show how the increase in the share of renewable energy and improvement in energy efficiency go hand in hand with mining and exploiting the energy sources that is notorious for the transition economies. We also demonstrate that a proper energy policy is required for effectively reducing energy consumption and greenhouse gas emissions. There is a need for commitments made by relevant stakeholders and policymakers targeted at achieving sustainable economic growth and energy efficiency. In addition, we demonstrate that there is a need for maintaining a proper balance between economic development and environmental protection, which is a must for the EU sustainable energy development agenda and all its accompanying targets for all its Member States.

Keywords: renewable energy sources; sustainable development; energy efficiency; economic growth; energy consumption; Czech Republic; Slovakia

1. Introduction

Increasing energy demand stimulates economic growth (represented by the gross domestic product (GDP), but energy consumption also causes greenhouse gas emissions. One can see that GDP allows comparing the dynamics of economic development over time and on a cross-country basis but recently one can hardly assess economic growth without attributing attention to the consumption of natural resources and preserving the environment. In was in the past decades that the increasing attention to global warming and climate change has focused on the relationship between environmental pollutants, energy consumption and economic growth [1–3]. In order to effectively control greenhouse gas (GHG) emissions and ensure the sustainability of economic development, it is important to better understand the relationships between greenhouse gas emissions, energy consumption and economic growth [4–6]. The energy consumption for each mode of transport is calculated as direct energy (consumption of fossil fuels and electricity during transport) and cumulative energy (including the energy consumed during the entire production process (exploration, extraction, transport and production of fuels). The emissions are calculated as carbon dioxide (CO_2) equivalent to taking into account the total GHG potential of emissions from the combustion of fossil fuels. With regard to the above, one can differentiate between the direct CO_2 equivalent, which occurs at the place of energy conversion, and the cumulative CO_2 equivalent, which takes into account the entire production process (exploration, extraction, transport and production of fuel) [7,8].

One of the illustrative examples is the transport sector that, together with heating, constitutes one of the mean areas where profound electrification based on the renewable energy source (RES) became an overall target many governments worldwide committed to. Even though we do not base the results of our paper on this sector and do not analyze it or consider its in-depth implications, a simple reference can be useful here. In general, greenhouse gas emissions from the transport sector rose from around 146 megatons of carbon dioxide equivalents in 2000 to 174 megatons in 2017 [9,10]. Greenhouse gas emissions from passenger cars rose from 81 megatons in 2000 to 94 megatons in 2017. The largest increase from 50 megatons in 2000 to 72 megatons in 2017 is for freight vehicles [11]. It is important that the company monitors both intensity and overall emissions. To keep the indicator set small, only the GHG intensity is included in this toolkit. Carbon offsets or other emissions trading programs are not considered in this toolkit.

With all of the above, it has to be mentioned that the EU is on the forefront of electric transportation with an ambitious plan to operate around 250 million electric vehicles (EVs) by 2025, which represents an effective transition to the climate targets [12]. However, the effective reduction of energy consumption and greenhouse gas emissions that the electric transportation was envisaged to entail, seem to require proper energy policy and careful planning [13]. The plans for transport electrification meet lots of obstacles and introduce several paradoxes. For example, one can observe that very often the electricity for powering the electric vehicles (EVs) is produced at the coal power stations, which creates a negative overall impact for the environment. Thence, it becomes apparent that the transition to clean electric transport should proceed along the lines of green-to-green paradigm and have to be considered from the point-of-view of the sustainability spectrum. It is not easy to provide a justified opinion on how to avoid the negative impact on the environment related to electromobility implementation. One of the possibilities would be the new advances in EV technology using alternative energy sources or improvements in battery storage technology that would allow to transfer large amounts of energy over large spaces.

Overall, one would probably agree with us that economic development and growth in today's globalized and cumbersome world should be based on the optimal management of natural resources that would not induce any harm or burden for the future generations to come and to their natural environment. Thence, the attention should be focused on the resource management that would both ensure the global competitiveness of economies without compromising their economic growth and well-being.

This paper focuses on the renewable energy sources in the EU 2020 target for energy efficiency in the two EU Member States, Czech Republic and Slovakia. We scrutinize the EU national energy efficiency targets for 2020 (which represents an important energy policy task, as Newbery at al. [14] demonstrate) and compare them with those of the two countries in question. Moreover, we employ the Auto Regressive Integrated Moving Average (ARIMA) model to obtain the forecasts for whether the 2020 targets can be achieved.

2. Energy Consumption and Greenhouse Gas Emissions

Household energy consumption is the main reason for the sector's observed greenhouse gas emissions [15,16]. Although the ratio of total energy consumption to GHG emissions is direct, the contribution of electricity consumption to GHG generation compared to other fuels used primarily for thermal purposes is much more significant compared to their share of total energy [17,18]. This can be backed up with the fact that the average carbon intensity (in gCO_2-e/kWh) is internationally used in calculating greenhouse gas (GHG) emissions from the electricity system, and the role of GHG in this system is highlighted in many reports and studies covering a wide spectre of countries, including China, Iran as well as other countries (see, e.g., [19–21]).

One of the main advantages of efficiency improvements is that they slow down the growth in energy consumption and reduce greenhouse gas emissions [22,23]. Energy intensity is the ratio of energy consumption per activity unit (such as floor space and GDP). In a way, energy efficiency is a measure of how effectively energy is used for a specific purpose and an important way of decarbonization [24–26]. The energy evaluation makes a major contribution to ensuring that users are where improvements are needed. Lots can be achieved in an energy assessment, from the disclosure of energy consumption to waste identification and efficient energy use. Efficient use of energy is still an important national and international topic in the discussion of political measures, both in European Union and abroad [27,28]. The assessment of energy efficiency in different countries is important for each country. To improve the efficiency of anyone's home, one should first carefully consider her or his options. An audit assesses electricity bills, insulation, heating and cooling systems, electrical systems as well as devices to determine how much energy your house uses and where energy is wasted. Following the recommendations and specially devised strategies can save 5% to 30% of the electricity bill [29,30].

Speaking about the penetration of the renewable energy sources into the traditional electricity and power systems, one has to look deeper into the specifics. In many countries, hydro sources are often needed to generate energy for almost all fuels and technologies to generate electricity, and energy is needed to treat and transport both water and wastewater [31–33]. A fascinating case study on the subject is the state of California in the United States with its large water supply systems (which require a lot of energy for pumping) that moves water from the relatively humid northern areas of the state to the drier and more populated southern region (including the major metropolitan areas of Los Angeles and San Diego) [34]. Conversely, the majority of the natural gas used in the water system is used for water heating on the consumer side of the water meter. Savings varied significantly across the state's hydrological region, with the largest savings in the populous south coast region (237,200 mg) and the lowest savings in the sparsely populated North Lahontan region (1400 mg) [35]. Since the savings in electricity and greenhouse gas emissions are calculated directly from the water savings, the results of these calculations showed a similar spatial variation.

When it comes to the debate of promoting renewable energy sources (RES) for the future electricity and power systems, one has to consider all possible alternatives [36–38]. Apart from the traditional renewables there are also some carbon-based alternatives to oil (e.g., methane hydrates and the conversion of coal into methane gas, or the use of oil reservoirs and shale oil), but other interesting options present themselves too [39–41]. One of them is the microbial fuel cells (MFCs) that convert biochemical to electrical energy [42–48]. MFCs can be used in biomass-based energy production, even though a plethora of technical challenges has to be solved before they will be practical for renewable energy production [49–51]. Nevertheless, their applications and possible deployment show that there are many less explored possibilities of using renewables in electricity generation, many of those not well-known to the general public or less explored by the researcher who might not be aware of all the possible implications for energy security and energy policy they might present [52–56].

All in all, energy consumption and greenhouse gas emissions constitute an important problem that all the world's largest economies are facing today, However, one can also see that this might be a political, rather than a climate protection, energy efficiency or economic issue. World leaders and important stakeholders are interested in re-election or maintaining their leading positions. Therefore, they want to make sure economic stability and growth are delivered at all costs. However, in the same time they have to face the commitments of tackling the climate changes and global warming, as well as introducing more renewable energy sources into the generation of power and electricity. Somehow, a balance should be reached and maintained to keep both the voters and the international partners satisfied. The European Union (EU) is in a specifically difficult position in this situation due to its complex structure, which lacks the traits of the federal state, and has a complex decision-making process as well as evaluation and acceptance procedures.

3. Europe 2020 Strategy and the 3 × 20 Climate and Energy Package

One would probably agree with is that EU plays a crucial role in the world as a powerful actor and leader in sustainable economic growth. The EU serves as a role model for more governments and actors when it comes to taking real and effective action [57]. The search for ways to increase and improve the use of renewable energies should not stop in 2020. Countries should continue to focus on this area in the coming decades and shape the next steps together. The first deadline for adoption of the package in Parliament was March 2009. However, there have been protests in some countries regarding the modalities to achieve these goals, particularly as a result of the economic and financial crisis that has led to tough negotiations between countries. The European Council of the 11th and 12th of December 2008 finally adopted the package but changed the original measures.

Various EU countries have many issues with meeting their energy efficiency and consumption obligations. For example, the French government admits failing to meet its climate change commitments. In 2017, France achieved 16.3% of its energy consumption from renewable sources, compared to its 23% target for 2020. Wood and hydropower are the main sources of green energy in France, ahead of biofuels [58,59]. The legislative proposals concern energy efficiency, the design of the electricity market and the governance rules for the Energy Union.

The climate package recognizes energy poverty as a major challenge in Europe and, with these proposals, aims to protect vulnerable consumers through targeted socio-political and energy-efficient measures [60,61]. In the package, one can see only minimum requirements for total energy efficiency. They regulate the maximum permissible energy consumption per floor area or room volume in new and existing buildings.

There are also provisions such as energy performance certificates, indicating the energy consumption of an existing or new building or a new building unit, and usually classify it in steps that differ in terms of energy consumption per square meter [62,63]. The certificates are issued by certified energy auditors and must be issued publicly, for example in advertisements for the sale or rental of buildings.

In this regard, the policy stipulates that regular maintenance can lead to significant operational improvements and recommends combining these inspections with certifications. The number of charging stations has been growing faster and faster than the number of EVs that could use them, and their installation is becoming increasingly profitable for electricity suppliers. In order to boost the market for cleaner vehicles, the EU Parliament and the Council agreed in February 2019 to amend the directive on the promotion of clean and energy-efficient vehicles [64]. The directive stipulates that authorities that procure vehicles (e.g., for public transport) must take their CO_2 emissions and the emissions of other pollutants into account. The EU emissions trading system includes emissions from more than 11,000 power plants and industrial plants and, from 2013, emissions from aviation. Around 40% of total EU emissions are covered by the regulation [65]. In addition to the EU27, Croatia, Iceland, Norway and Liechtenstein are also part of the ETS. When the EHS was introduced in 2005, it was the first trading system for greenhouse gases. As already mentioned, the EU should achieve its overall GHG target. The EEA's projections show that it will cut its greenhouse gas emissions by six percentage points above the 2020 target with existing measures and by seven percentage points by adopting additional measures. Since 2016, five EU countries have actually increased their greenhouse gas emissions compared to 1990. Careful monitoring was applied to the primary energy consumption in order to assess progress in energy efficiency in terms of goals and policies for the European Union and its Member States. In 2009, the Europe 2020 Strategy was adapted [66]. It includes very important and timely targets set for the whole European Union are as follows:

- reducing greenhouse gas emissions by at least 20% compared to 1990 levels;
- increasing the share of renewable energy in the final energy consumption to 20%;
- moving towards a 20% increase in energy efficiency (from 2005 levels).

Table 1 shows the national energy efficiency targets for 2020 for the EU28, Czech Republic and Slovakia.

Table 1. National energy efficiency targets for 2020 for the EU28, Czech Republic and Slovakia [43].

EU Member State	Greenhouse Gas Emissions *1 (%)	Share of Renewable Energy *2 (%)	Primary Energy Consumption *3 (Mtoe)	Final Energy Consumption *3 (Mtoe)
Czech Republic	9	13	39.6	25.3
Slovakia	13	14	16.4	9.0
EU28 *4	20	20	1483.0	1086.0

Note: Mtoe—million tonnes of oil equivalent; *1—compared to 2005 levels; *2—share of renewable energy in gross final energy consumption; *3—absolute level of energy consumption in 2020 (Mtoe) as notified from Member States in 2013, in the NEEAP 2014, annual reports or in separate notifications to the European commission in 2015 and 2016 (Mtoe); *4—compared to 1990 levels.

A little explanation should be made here for better clarity of the explanation of our empirical model and its main results and implications that are presented in the next sections. As opposed to final energy consumption, primary energy consumption refers to energy that has not been subject to any conversion or transformation process. Energy intensity represents the amount of primary energy consumption per unit of GDP. The energy intensity indicator depends on the industrial structure of the economy and thus is not an exact proxy for energy efficiency in the EU Member States.

Moreover, several more methodological issues should be explained about the energy intensity (EI), gross inland energy consumption (GIEC), gross domestic product (GDP) and their relationship. All of the above can be expressed in the formula that follows:

$$EI = GIEC/GDP \tag{1}$$

where:

EI—energy intensity;
GIEC—gross inland energy consumption;
GDP—gross domestic product.

4. Methodology

The data used for our empirical models was accessed in December 2019 via Eurostat, a European Statistical Office. Some of the latest data are for 2018 (GDP), but others are for 2017 (e.g., GHG), which is given by the data availability and accessibility.

For forecasting time series, a popular and widely used statistical method called ARIMA [67–69] has been used. ARIMA is an acronym for Auto Regressive Integrated Moving Average. AR is a class of linear model where the variable of interest is regressed on its own lagged values. MA is also class of linear model, where the variable of interest is modeled with its own imperfectly predicted values of current and previous times [70]. The I is an integration—it specifies the number of times the differencing operation is performed on a series to make it stationary.

The Auto Regression (AR) process is written as

$$y_t = \phi_1 y_{t-1} + \phi_2 y_{t-2} + \cdots + \phi_p y_{t-p} + \epsilon_t$$

where:

ϕ_{t-1}—parameters;
y_{t-i}—regressors;
ϵ—error.

Moving Average (MA) can be written in terms of error terms:

$$y_t = \theta_1 \epsilon_{t-1} + \theta_2 \epsilon_{t-2} + \cdots + \theta_q \epsilon_{t-q} + \epsilon_t$$

where:

θ_{t-1}—parameters;
ϵ_{t-i}—regressors—imperfections (errors) in predicting previous terms;
ϵ—error.

The ARMA process has the mathematical form:

$$y_t = \sum_{i=1}^{p} \phi_i y_{t-i} + \sum_{j=1}^{q} \theta_i \epsilon_{t-j} + \epsilon_t$$

As a result, the differencing is the ARIMA process. The "predictors" on the right-hand side include both the lagged values of y_t and the lagged errors. We call this an ARIMA (p, d, q) model, where parameters (p, d, q) describe:

AR: p—periods to lag;
I: d—the degree of differencing;
MA: q—the lag of the error component.

All figures used hereinafter in this paper and employed for comparing the situation in Czech Republic and Slovakia were prepared separately for Czech Republic and Slovakia due to one simple fact that the scale of data was different and it would not look clear and comparable if placed on the same figure.

Moreover, we should also explain that the confidence interval (Lo-Hi) of a forecast (shadow on figures) is the range within which the value we forecast will lie with a certain probability. For example, if, for GHG for Slovakia in 2018, the Lo.95-Hi.95 percent of the forecast confidence interval is between 40.09 and 48.38, then with a probability of 95%, GHG (greenhouse gas emission) will be at least 40.09 Mt and at most 48.38 Mt.

The empirical models used hereinafter is based on our previous similar studies covering other EU countries (e.g., Poland) and focusing on the same issues (see, e.g., [71]).

5. Results and Discussions

Our results are outlined below as follows: First, let us look at the greenhouse gas emission (GHG) in the Czech Republic. The dashed line in Figure 1 represents the GHG emission limit for 2020. For the Czech Republic it is no more than 9% comparing to year 2005 (149.53 Mt). It means that the limit for 2020 equals 162.99 Mt. The emissions are decreasing (even taking into the account the high and low forecast as shown in Figure 1).

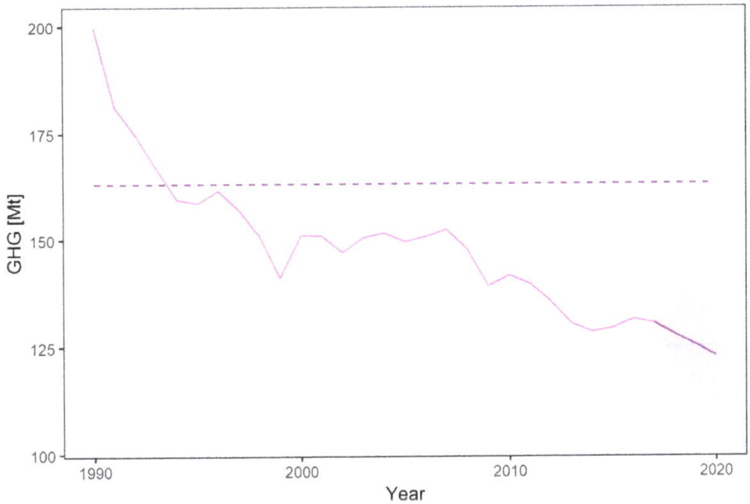

Figure 1. GHG emissions in Czech Republic in 1990–2020 (Source: Own results).

Table 2 depicts the values presented in Figure 1 in more detail, including the forecast, as well as forecast for the values of Hi and Lo at 80% and 95%, respectively, for the Czech Republic.

Table 2. Forecast Auto Regressive Integrated Moving Average (ARIMA) (0,1,0) details for GHG emissions in Czech Republic (Source: Own results).

	Point.Forecast	Lo.80	Hi.80	Lo.95	Hi.95
2018	127.8996	120.8964	134.9028	117.1891	138.6101
2019	125.3328	115.4288	135.2368	110.1859	140.4797
2020	122.766	110.6361	134.8959	104.2149	141.3171

Our key conclusion stemming from the analysis of GHG emissions in Czech Republic is that the country is likely to meet the requirements of Europe 2020 in terms of greenhouse gas emissions (GHG), because from 2007 onwards the trend is towards a continuous reduction in greenhouse gas emissions.

Looking into the case of Slovakia, one can see the following story (see Figure 2 that follows). For Slovakia the GHG limit is no more than 13% comparing to year 2005 (51.28 Mt). It means the limit for 2020 equals 57.95 Mt.

Table 3 depicts the values presented in Figure 1 in more detail, including the forecast, and the Hi and Lo at 80% and 95%, respectively, for Slovakia.

Table 3. Forecast ARIMA (2,1,0) details for GHG emissions in Slovakia (Source: Own results).

	Point.Forecast	Lo.80	Hi.80	Lo.95	Hi.95
2018	44.2312	41.52021	46.9422	40.08509	48.37732
2019	45.10624	40.37804	49.83445	37.87508	52.33741
2020	45.83483	38.46569	53.20397	34.5647	57.10496

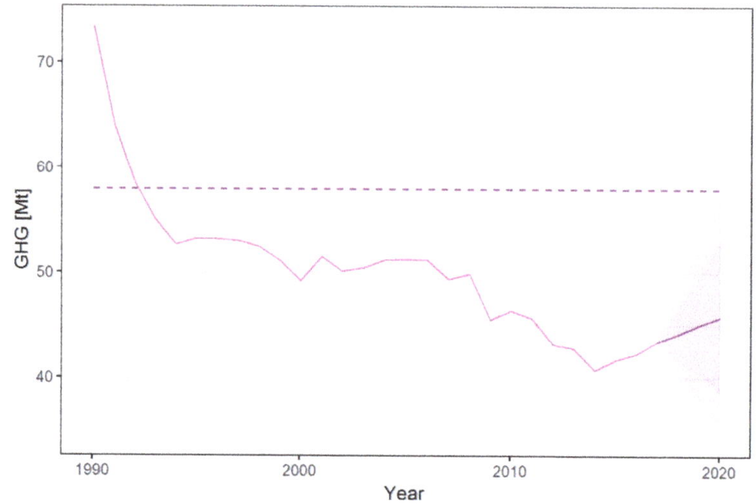

Figure 2. GHG emissions in Slovakia in 1990–2020 (Source: Own results).

The key conclusions for Slovakia that were obtained appear to be similar to in the situation in the Czech Republic. Figure 3 shows the share of renewable energy sources (RES) in the Czech Republic.

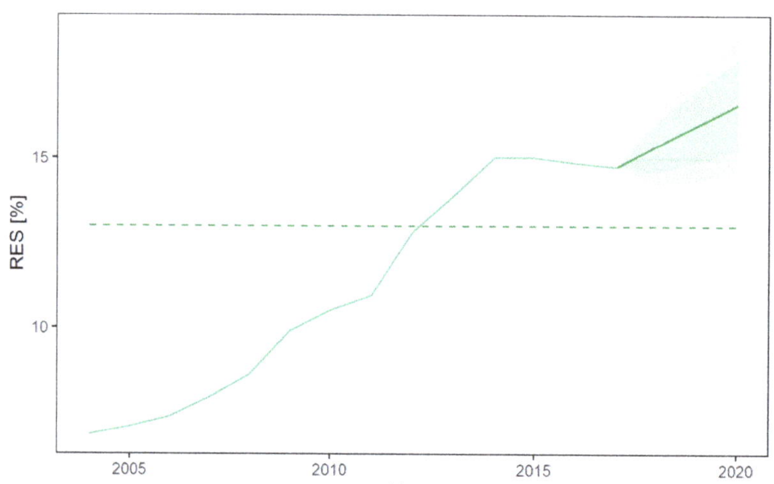

Figure 3. Share of renewable energy sources in gross final energy consumption in the Czech Republic (Source: Own results).

The share of renewable energy in Czech Republic has been growing and since 2005 has always been under Europe 2020 target (see Table 4).

Table 4. Forecast ARIMA (0,1,0) details for the share of renewable energy sources in the Czech Republic (Source: Own results).

	Point.Forecast	Lo.80	Hi.80	Lo.95	Hi.95
2018	15.36808	14.59934	16.13681	14.1924	16.54375
2019	15.97615	14.889	17.06331	14.31349	17.63881
2020	16.58423	15.25274	17.91572	14.5479	18.62056

Even the most pessimistic forecasts (Lo.95) show that the RES will be above the assumed level of 13%. Figure 4 above show the results of the similar simulation for Slovakia.

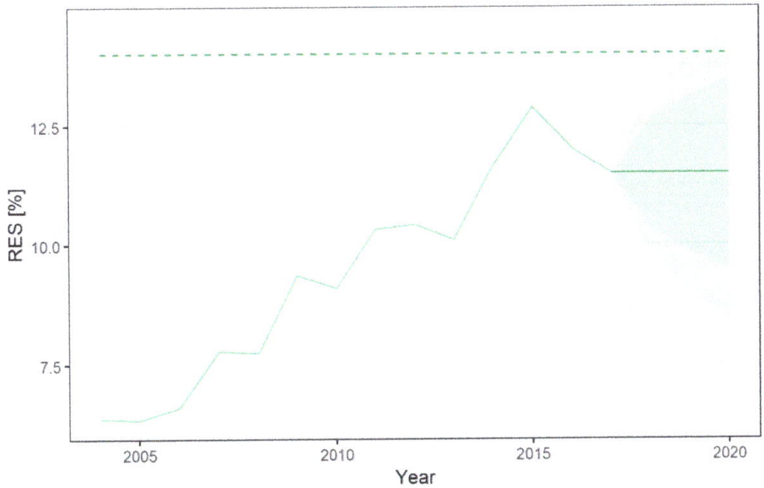

Figure 4. Share of renewable energy sources in gross final energy consumption in Slovakia (Source: Own results).

From Figure 4 and Table 5 one can deduct that the maximum share of renewable energy in Slovakia was in 2015, and since this year has been decreasing. Therefore, it is improbable that Slovakia will achieve Europe 2020 goals in the RES indicator.

Table 5. Forecast ARIMA (0,1,0) details for the share of renewable energy sources in Slovakia (Source: Own results).

	Point.Forecast	Lo.80	Hi.80	Lo.95	Hi.95
2018	11.49	10.33109	12.64891	9.717603	13.2624
2019	11.49	9.851056	13.12894	8.983452	13.99655
2020	11.49	9.482712	13.49729	8.420118	14.55988

Figure 5 and Table 6 shows the primary energy consumption and final energy consumption (PEC, FEC) for the Czech Republic.

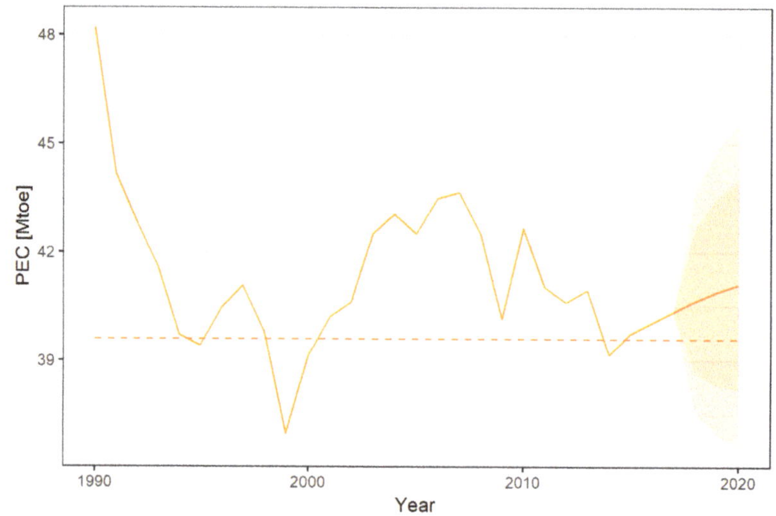

Figure 5. Primary energy consumption in the Czech Republic (Source: Own results).

Table 6. Forecast ARIMA (1,0,0) details for the primary energy consumption in the Czech Republic (Source: Own results).

	Point.Forecast	Lo.80	Hi.80	Lo.95	Hi.95
2018	40.66137	38.64919	42.67354	37.58402	43.73872
2019	40.91168	38.31415	43.50922	36.93909	44.88428
2020	41.11603	38.19276	44.03931	36.64527	45.5868

Overall, it seems that for the Czech Republic primary and final energy consumption have both been fluctuating around their Europe 2020 target (see Figure 6 and Table 7). Based on the forecast, we can assess that the target will be slightly exceeded, but the confidence interval of the forecast gives hope that it could be under the limit. Figure 7 and Table 8 shows the results from a similar simulation for the case of Slovakia.

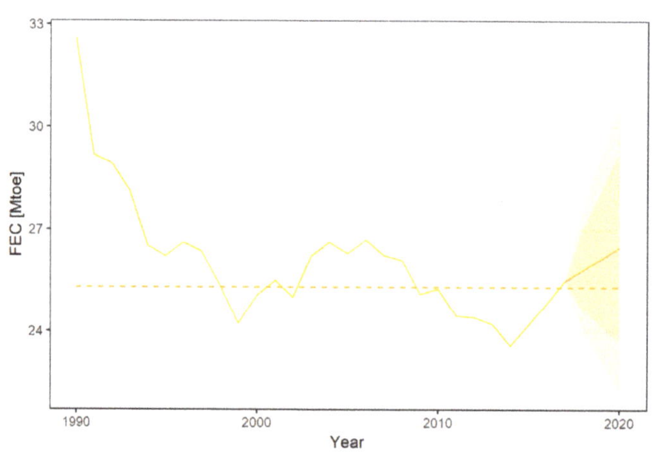

Figure 6. Final energy consumption in the Czech Republic (Source: Own results).

Table 7. Forecast ARIMA (0,2,1) details for the primary energy consumption in the Czech Republic (Source: Own results).

	Point.Forecast	Lo.80	Hi.80	Lo.95	Hi.95
2018	25.81322	24.62623	27.00021	23.99788	27.62857
2019	26.14355	24.17197	28.11513	23.12828	29.15882
2020	26.47387	23.69279	29.25496	22.22057	30.72717

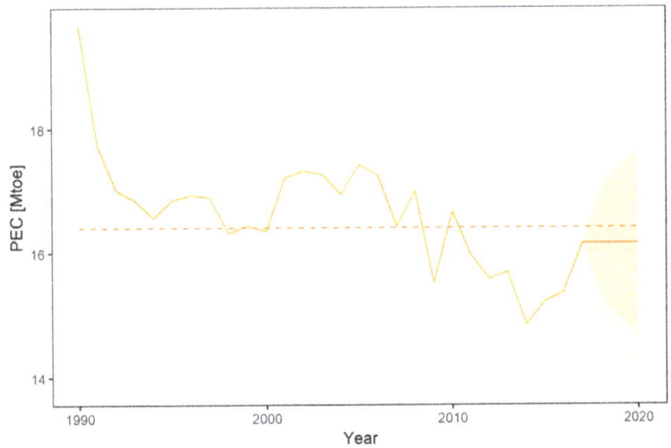

Figure 7. Primary energy consumption in Slovakia (Source: Own results).

Table 8. Forecast ARIMA (0,1,0) details for the primary energy consumption in Slovakia (Source: Own results).

	Point.Forecast	Lo.80	Hi.80	Lo.95	Hi.95
2018	16.14603	15.27609	17.01598	14.81557	17.4765
2019	16.14603	14.91575	17.37632	14.26447	18.0276
2020	16.14603	14.63925	17.65282	13.8416	18.45047

The main conclusions here is that the primary energy consumption in Slovakia has been under the Europe 2020 limit since 2011, but since 2014 we can observe change in the trend—PEC growth. It appears quite difficult to assess what the result in 2020 will be, but our simulations and forecast show it will be very close to the limit.

Final energy consumption for Slovakia was set on an unattainable level for this country. Slovakia has never been close to this level and seems improbable to achieve this level in 2020 (see Figure 8 and Table 9).

Table 9. Forecast ARIMA (0,1,0) details for the final energy consumption in Slovakia (Source: Own results).

	Point.Forecast	Lo.80	Hi.80	Lo.95	Hi.95
2018	11.12881	10.3039	11.95372	9.867218	12.39041
2019	11.12881	9.96221	12.29541	9.344648	12.91298
2020	11.12881	9.700022	12.5576	8.943667	13.31396

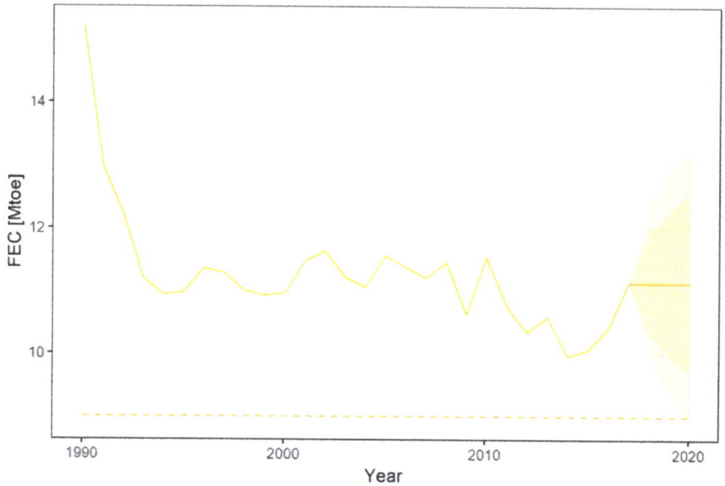

Figure 8. Final energy consumption in Slovakia (Source: Own results).

Figure 9 below shows the levels of the gross domestic product (GDP) in the Czech Republic.

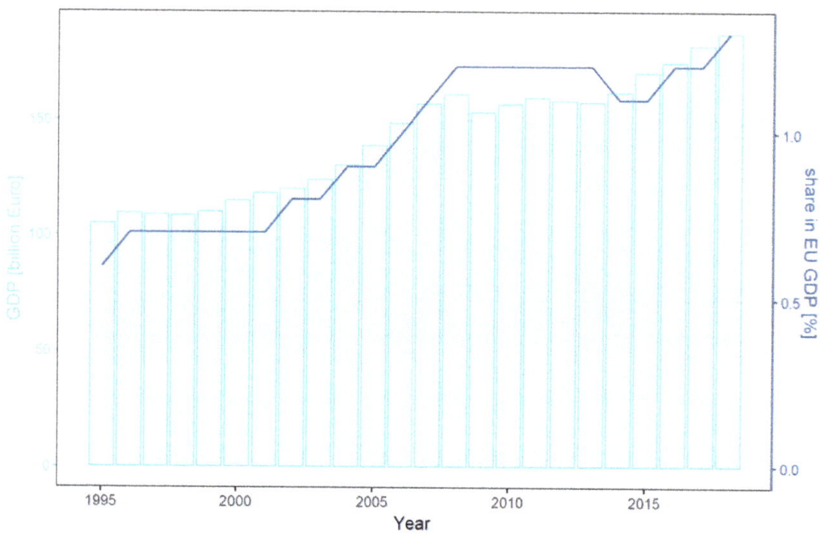

Figure 9. Gross domestic product in current and constant prices in the Czech Republic (Source: Own results).

In general, there is no forecast of GDP, because it does not constitute any importance for Europe's 2020 strategy and its implications. Nevertheless, is seems important to describe how it looked like in the past, because GDP is used for energy intensity calculation and forecast (see the next figures that follow) and is also important for sustainable development. Figure 10 below shows the gross domestic product in current prices in Slovakia.

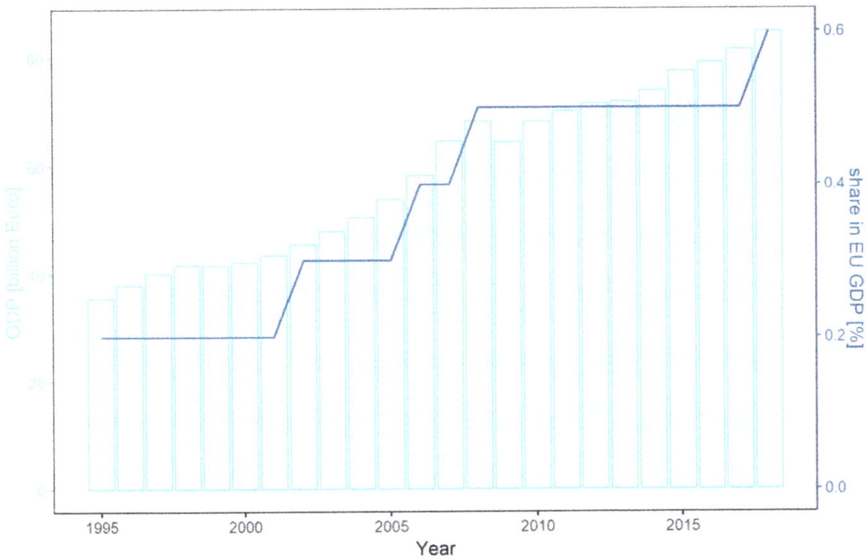

Figure 10. Gross domestic product in current and constant prices in Slovakia (Source: Own results).

The results for Slovakia seem to be very much the same as for Czech Republic. Furthermore, let us look at Figure 11 that shows real GDP growth rate in EU compared to the Czech Republic and Slovakia.

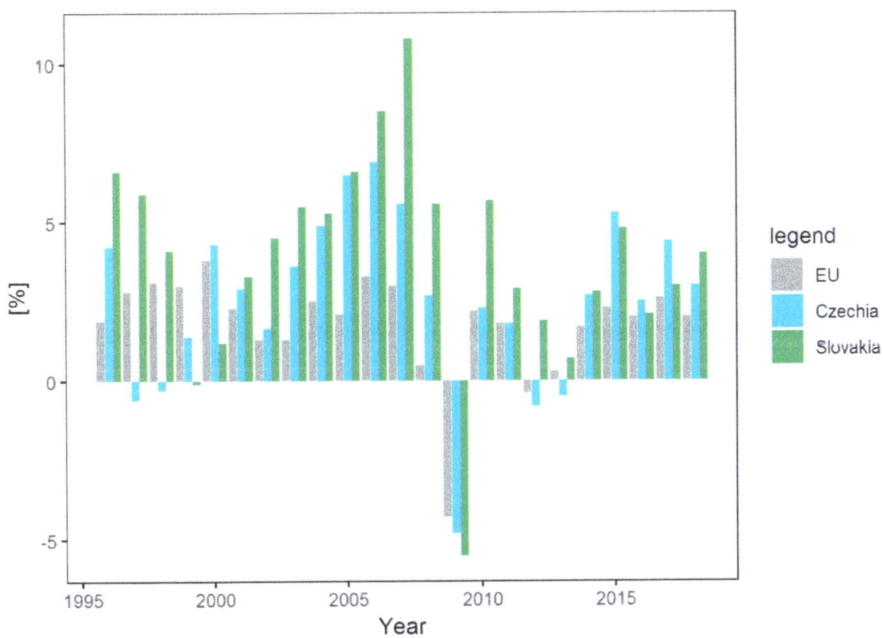

Figure 11. Real GDP growth rate in the EU compared to the Czech Republic and Slovakia (Source: Own results).

The differences between the Czech Republic and Slovakia are quite obvious. Slovakia is growing at a faster pace. This might be attributed to the better and more efficient economic reforms in Slovakia that accepted the Euro as its currency in 2009 while the Czech Republic still keeps its national currency, the Czech koruna.

Figures 12 and 13 that follow shows the energy intensity in the Czech Republic and Slovakia that is calculated as the ratio of gross inland energy consumption (GIEC) to GDP. The two different shapes of energy intensity depict the (i) forecast, (ii) forecast Lo 80% and 95%, as well as (iii) forecast Hi 80% and 95% for each country, respectively (see Tables 10 and 11 for more explanation showing the values for each forecast).

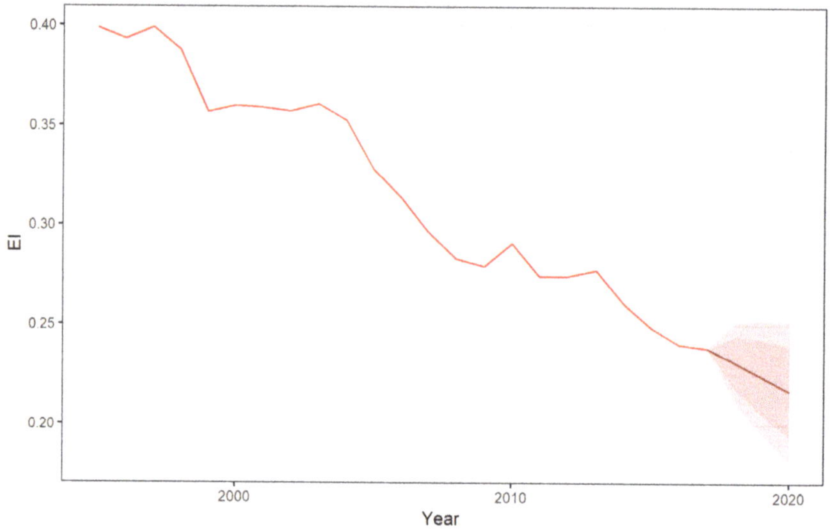

Figure 12. Energy intensity in the Czech Republic expressed as GIEG/GDP with forecast, forecast Lo 80% and 95% and forecast Hi 80% a and 95% (Source: Own results).

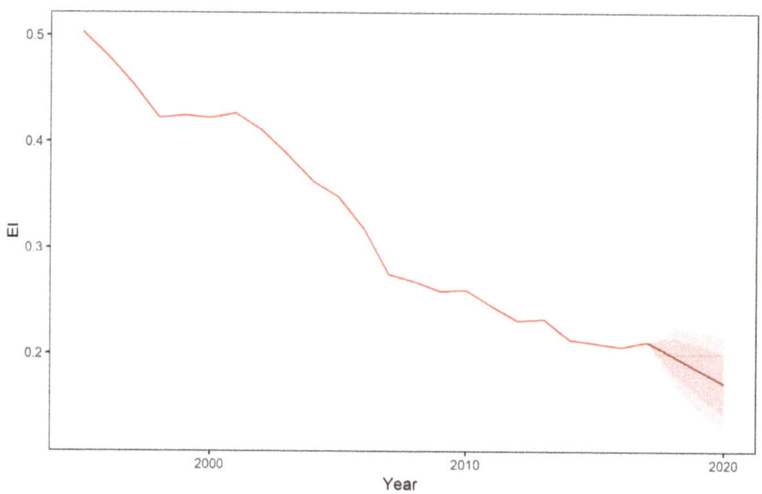

Figure 13. Energy intensity in Slovakia expressed as GIEG/GDP with forecast, forecast Lo 80% and 95% and forecast Hi 80% a and 95% (Source: Own results).

Table 10. Forecast ARIMA (0,1,0) details for the energy intensity in the Czech Republic (Source: Own results).

	Point.Forecast	Lo.80	Hi.80	Lo.95	Hi.95
2018	0.23128835	0.217935688	0.244641013	0.210867215	0.251709485
2019	0.224008806	0.205125289	0.242892322	0.19512896	0.252888652
2020	0.216729261	0.193601771	0.239856751	0.181358817	0.252099704

Table 11. Forecast ARIMA (0,1,0) details for the energy intensity in Slovakia GIEG/GDP with forecast, forecast Lo 95% and forecast Hi 95% (Source: Own results).

	Point.Forecast	Lo.80	Hi.80	Lo.95	Hi.95
2018	0.198569715	0.181460651	0.21567878	0.172403659	0.224735771
2019	0.185333467	0.161137596	0.209529338	0.148329075	0.222337858
2020	0.172097218	0.14246345	0.201730987	0.12677628	0.217418157

The low value of energy intensity speaks of the level of economic development. The average for the EU equals 0.1097. The energy intensity of Czech Republic is twice larger than the average for the EU but it has a decreasing trend. Figure 13 below shows the same situation but using the case of Slovakia. It is apparent that the Slovak energy intensity also exceeds the EU average.

The main conclusions stemming from Figures 12 and 13 and the accompanying tables are that the low value of energy intensity speaks of the modern economy. The EU average equals 0.1097. Energy intensity of Slovakia is twice larger than the average of the EU and decreased fast between 1995 and 2007, but in the last years the decrease is very slow and looks to be stabilizing.

Figure 14 depicts energy consumption and greenhouse gas emissions for the Czech Republic.

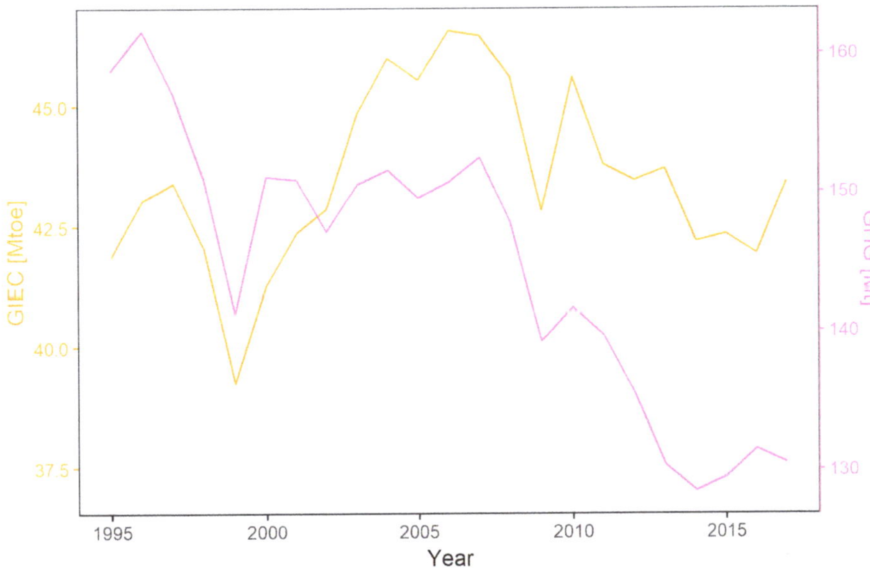

Figure 14. Gross inland energy consumption (GIEC) and GHG emissions in the Czech Republic (Source: Own results).

In the case of the Czech Republic, GIEC and GHG are correlated, but we can see that in the last years GHG emission is decreasing faster than the energy consumption. This is, of course, a positive trend that can be attributed to the improvement in energy policy and strategy.

All in all, also in the case of Slovakia, the GIEC and GHG appear to be correlated. It is apparent from Figure 15 that in the last years, GHG emission is decreasing a little faster than energy consumption, but not as fast as in the case of the Czech Republic that was analyzed above. This was shown on the previous figures describing the renewable energy sharing system.

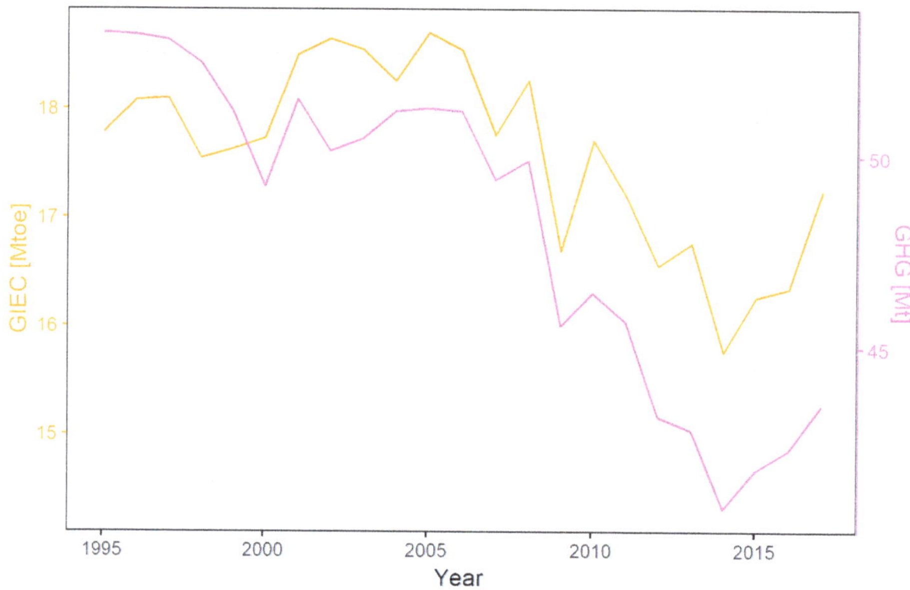

Figure 15. Gross inland energy consumption (GIEC) and GHG emissions in Slovakia (Source: Own results).

6. Conclusions

Recent commitments to sustainable development and mitigating climate changes made by most of the world's governments also found their way into the energy policy of the European Union, becoming the basis of its national energy efficiency targets for 2020 embedded in the Europe 2020 strategy. The Europe 2020 strategy and the 3 × 20 climate and energy package envisage the reduction of greenhouse gas emissions in EU Member States as well as increasing the share of renewable energy for enhancing energy efficiency. Due to the varying level of economic development, different objectives were set for different EU Member States, with the Czech Republic and Slovakia, which constitute the case studies employed in this paper, facing targets that were lower than that in the case of the more economically developed EU countries.

With regard to the above, one has to understand that climate targets were set in order to slow down or even reverse (albeit in the long run) the depletion of natural resources and preventing environmental degradation. In no way were these targets set with a purpose of halting the economic growth (especially when it comes to the economies in transition) but rather to help the countries in question to develop in accordance with the principles of energy efficiency and sustainable economic growth.

Both countries selected for our case study, the Czech Republic and Slovakia, experienced deep system transformation after the fall of Communism in 1989 that are apparent in the GDP, gross inland energy consumption and GHG emissions that constitute the measures of sustainable development used in our research. Our results indicate that it is quite unlikely that the planned increase in renewable energy is going to reach its targets for the Czech Republic and Slovakia (which is similar to the case of other EU Member States that joined after 2004) but it will be possible to reduce energy consumption and greenhouse gas emissions. This is, among other things, due to the fact that gross inland energy consumption and greenhouse gas emissions in the Czech Republic and Slovakia appear to be correlated. Greenhouse gas emissions are going down in both countries in question a little faster than energy consumption, but this pace is more rapid in the Czech Republic than in Slovakia.

In addition, it becomes apparent that implementation of GHG emissions in the Czech Republic and Slovakia may be at risk in case the proper energy policy is not maintained by the stakeholders and governments of the respective countries. Moreover, our findings show that regardless of the mix of fossil and renewable energy, the state of economic development and the geographical location of any EU Member States, a proper energy policy is required for effectively reducing energy consumption and greenhouse gas emissions. The energy intensity of Czech and Slovak economies increased in the early 2000s and then stabilized at a level about twice of the EU average. Our analysis of the energy intensity for the both countries in question shows that in the forthcoming years its value is likely to remain the same.

Overall, our results also demonstrate that maintaining a proper balance between economic development and environmental protection should be kept at all cost regardless of the position of the country. The cases of the Czech Republic and Slovakia scrutinized in this paper confirms that. Both countries have (common) Communist pasts but both underwent a spectacular economic transition and became Member States of the European Union. However, their story might be used by other EU Member States, both constituting the "core" EU and those that joined in 2004 or after. It might be also interesting to study the implications of Brexit and the shift of energy policies during and after the transition period for the United Kingdom. Further progress in maintaining a proper balance between economic development and environmental protection might be ensured by the decisive steps of the Czech and Slovak (as well as other EU) stakeholders and policymakers in terms of investments into renewable energy sources, modernizing the old energy sector and seeking for new solutions for sustainability and energy efficiency.

Author Contributions: Conceptualization, J.B., W.S. and A.F.; methodology, J.B. and N.N.; validation, J.B. and A.F.; formal analysis, J.B. and W.S.; resources, N.N. and A.F.; writing—original draft preparation: J.B., W.S., A.F., and N.N.; project administration, W.S. and J.B. All authors have read and agreed to the published version of the manuscript.

Funding: The paper received no funding.

Conflicts of Interest: The authors declare no conflict of interest.

References

1. Chen, P.Y.; Chen, S.T.; Hsu, C.S.; Chen, C.C. Modeling the global relationships among economic growth, energy consumption and CO_2 emissions. *Renew. Sustain. Energy Rev.* **2016**, *65*, 420–431. [CrossRef]
2. Brożyna, J.; Mentel, G.; Ivanová, E.; Sorokin, G. Classification of Renewable Sources of Electricity in the Context of Sustainable Development of the New EU Member States. *Energies* **2019**, *12*, 2271. [CrossRef]
3. Esso, L.J.; Keho, Y. Energy consumption, economic growth and carbon emissions: Cointegration and causality evidence from selected African countries. *Energy* **2016**, *114*, 492–497. [CrossRef]
4. Brożyna, J.; Mentel, G.; Szetela, B. Renewable Energy and Economic Development in the European Union. *Acta Polytech. Hung.* **2017**, *14*, 11–34. [CrossRef]
5. Magazzino, C. The relationship among economic growth, CO 2 emissions, and energy use in the APEC countries: A panel VAR approach. *Environ. Syst. Decis.* **2017**, *37*, 353–366. [CrossRef]
6. Brożyna, J.; Mentel, G.; Szetela, B. A Mid-Term Forecast of Maximum Demand for Electricity in Poland. *Montenegrin J. Econ.* **2016**, *12*, 73–88. [CrossRef]
7. Millar, R.J.; Nicholls, Z.R.; Friedlingstein, P.; Allen, M.R. A modified impulse-response representation of the global near-surface air temperature and atmospheric concentration response to carbon dioxide emissions. *Atmos. Chem. Phys.* **2017**, *17*, 7213–7228. [CrossRef]
8. Chang, K.; Chang, H. Cutting CO_2 intensity targets of interprovincial emissions trading in China. *Appl. Energy* **2016**, *163*, 211–221. [CrossRef]

9. Elective. Number of Plug-In Cars Climbs to 5.6M Worldwide. 2019. Available online: https://www.electrive.com/2019/02/11/the-number-of-evs-climbs-to-5-6-million-worldwide (accessed on 29 November 2019).
10. Strielkowski, W.; Volkova, E.; Pushkareva, L.; Streimikiene, D. Innovative Policies for Energy Efficiency and the Use of Renewables in Households. *Energies* **2019**, *12*, 1392. [CrossRef]
11. HIS Markit. Number of Fast-Charging Stations for Electric Vehicles Set to Rise to Nearly 200,000 in 2020. 2019. Available online: https://news.ihsmarkit.com/press-release/design-supply-chain-media/number-fast-charging-stations-electric-vehicles-set-rise-nea?page=1 (accessed on 29 November 2019).
12. Strielkowski, W.; Streimikiene, D.; Fomina, A.; Semenova, E. Internet of Energy (IoE) and High-Renewables Electricity System Market Design. *Energies* **2019**, *12*, 4790. [CrossRef]
13. Sotiriou, C.; Michopoulos, A.; Zachariadis, T. On the cost-effectiveness of national economy-wide greenhouse gas emissions abatement measures. *Energy Policy* **2019**, *128*, 519–529. [CrossRef]
14. Newbery, D.; Pollitt, M.G.; Ritz, R.A.; Strielkowski, W. Market design for a high-renewables European electricity system. *Renew. Sustain. Energy Rev.* **2018**, *91*, 695–707. [CrossRef]
15. Nejat, P.; Jomehzadeh, F.; Taheri, M.M.; Gohari, M.; Majid, M.Z.A. A global review of energy consumption, CO_2 emissions and policy in the residential sector (with an overview of the top ten CO_2 emitting countries). *Renew. Sustain. Energy Rev.* **2015**, *43*, 843–862. [CrossRef]
16. Fujimori, S.; Dai, H.; Masui, T.; Matsuoka, Y. Global energy model hindcasting. *Energy* **2016**, *114*, 293–301. [CrossRef]
17. Scarlat, N.; Dallemand, J.F.; Fahl, F. Biogas: Developments and perspectives in Europe. *Renew. Energy* **2018**, *129*, 457–472. [CrossRef]
18. Reuter, M.; Patel, M.K.; Eichhammer, W. Applying ex-post index decomposition analysis to primary energy consumption for evaluating progress towards European energy efficiency targets. *Energy Effic.* **2017**, *10*, 1381–1400. [CrossRef]
19. Khan, I. Temporal carbon intensity analysis: Renewable versus fossil fuel dominated electricity systems. *Energy Sources Part A Recovery Util. Environ. Eff.* **2019**, *41*, 309–323. [CrossRef]
20. Zhao, F.; Liu, F.; Liu, Z.; Hao, H. The correlated impacts of fuel consumption improvements and vehicle electrification on vehicle greenhouse gas emissions in China. *J. Clean. Prod.* **2019**, *207*, 702–716. [CrossRef]
21. Aryanpur, V.; Atabaki, M.S.; Marzband, M.; Siano, P.; Ghayoumi, K. An overview of energy planning in Iran and transition pathways towards sustainable electricity supply sector. *Renew. Sustain. Energy Rev.* **2019**, *112*, 58–74. [CrossRef]
22. Van Vuuren, D.P.; Stehfest, E.; Gernaat, D.E.; Doelman, J.C.; Van den Berg, M.; Harmsen, M.; Girod, B. Energy, land-use and greenhouse gas emissions trajectories under a green growth paradigm. *Glob. Environ. Chang.* **2017**, *42*, 237–250. [CrossRef]
23. Schandl, H.; Hatfield-Dodds, S.; Wiedmann, T.; Geschke, A.; Cai, Y.; West, J.; Owen, A. Decoupling global environmental pressure and economic growth: Scenarios for energy use, materials use and carbon emissions. *J. Clean. Prod.* **2016**, *132*, 45–56. [CrossRef]
24. Bataille, C.; Waisman, H.; Colombier, M.; Segafredo, L.; Williams, J.; Jotzo, F. The need for national deep decarbonization pathways for effective climate policy. *Clim. Policy* **2016**, *16*, S7–S26. [CrossRef]
25. Timmons, D.; Konstantinidis, C.; Shapiro, A.M.; Wilson, A. Decarbonizing residential building energy: A cost-effective approach. *Energy Policy* **2016**, *92*, 382–392. [CrossRef]
26. Wesseling, J.H.; Lechtenböhmer, S.; Åhman, M.; Nilsson, L.J.; Worrell, E.; Coenen, L. The transition of energy intensive processing industries towards deep decarbonization: Characteristics and implications for future research. *Renew. Sustain. Energy Rev.* **2017**, *79*, 1303–1313. [CrossRef]
27. Shove, E. What is wrong with energy efficiency? *Build. Res. Inf.* **2018**, *46*, 779–789. [CrossRef]
28. Renn, O.; Marshall, J.P. Coal, nuclear and renewable energy policies in Germany: From the 1950s to the "Energiewende". *Energy Policy* **2016**, *99*, 224–232. [CrossRef]
29. Barma, M.C.; Saidur, R.; Rahman, S.M.A.; Allouhi, A.; Akash, B.A.; Sait, S.M. A review on boilers energy use, energy savings, and emissions reductions. *Renew. Sustain. Energy Rev.* **2017**, *79*, 970–983. [CrossRef]
30. Cao, X.; Dai, X.; Liu, J. Building energy-consumption status worldwide and the state-of-the-art technologies for zero-energy buildings during the past decade. *Energy Build.* **2016**, *128*, 198–213. [CrossRef]
31. Harish, V.S.K.V.; Kumar, A. A review on modeling and simulation of building energy systems. *Renew. Sustain. Energy Rev.* **2016**, *56*, 1272–1292. [CrossRef]

32. Gu, Y.; Li, Y.; Li, X.; Luo, P.; Wang, H.; Robinson, Z.P.; Li, F. The feasibility and challenges of energy self-sufficient wastewater treatment plants. *Appl. Energy* **2017**, *204*, 1463–1475. [CrossRef]
33. Gude, V.G. Energy and water autarky of wastewater treatment and power generation systems. *Renew. Sustain. Energy Rev.* **2015**, *45*, 52–68. [CrossRef]
34. Nayak, M.A.; Herman, J.D.; Steinschneider, S. Balancing Flood Risk and Water Supply in California: Policy Search Integrating Short-Term Forecast Ensembles with Conjunctive Use. *Water Resour. Res.* **2018**, *54*, 7557–7576. [CrossRef]
35. Spang, E.S.; Holguin, A.J.; Loge, F.J. The estimated impact of California's urban water conservation mandate on electricity consumption and greenhouse gas emissions. *Environ. Res. Lett.* **2018**, *13*, 14016. [CrossRef]
36. Logan, B. *Microbial Fuel Cells*; Wiley: Hoboken, NJ, USA, 2008.
37. Hoogers, G. *Fuel Cell Technology Handbook*; CRC Press: Boca Raton, FL, USA, 2003.
38. U.S. Department of Energy. *Fuel Cell Handbook*; Office of Fossil Energy: Morgantown, WV, USA, 2004.
39. Logan, B.E.; Regan, J.M. Electricity-producing bacterial communities in microbial fuel cells. *Trends Microbiol.* **2006**, *14*, 512–518. [CrossRef] [PubMed]
40. Logan, B.E.; Regan, J.M. Microbial fuel cells—Challenges and applications. *Environ. Sci. Technol.* **2006**, *40*, 5172–5180. [CrossRef] [PubMed]
41. Logan, B.E.; Hamelers, B.; Rozendal, R.; Schroder, U.; Keller, J.; Verstraete, W.; Rabaey, K. Microbial Fuel Cells: Methodology and Technology. *Environ. Sci. Technol.* **2006**, *40*, 5181–5192. [CrossRef]
42. Liu, H.; Ramnarayanan, R.; Logan, B.E. Production of electricity during wastewater treatment using a single chamber microbial fuel cell. *Environ. Sci. Technol.* **2004**, *38*, 2281–2285. [CrossRef]
43. Rabaey, K.; Verstraete, W. Microbial fuel cells: Novel biotechnology for energy generation. *Trends Biotechnol.* **2005**, *23*, 291–298. [CrossRef]
44. Włodarczyk, P.P.; Włodarczyk, B. Microbial fuel cell with Ni-Co cathode powered with yeast wastewater. *Energies* **2018**, *11*, 3194. [CrossRef]
45. Permana, D. Performance of Single Chamber Microbial Fuel Cell (SCMFC) for biological treatment of tofu wastewater. In Proceedings of the IOP Conference Series: Earth and Environmental Science, Saint Petersburg, Russia, 17–18 April 2019; Volume 277, p. 12008. [CrossRef]
46. Chaudhuri, S.K.; Lovley, D.R. Electricity generation by direct oxidation of glucose in mediatorless microbial fuel cells. *Nat. Biotechnol.* **2003**, *21*, 1229–1232. [CrossRef]
47. US EPA. Report; *Clean Watersheds Needs Survey Overview*; United States Environmental Protection Agency: Galloway, NJ, USA, 2008.
48. European Parliament. Directive 2010/75/EU of the European Parliament and of the Council of 24 November 2010 on industrial emissions (integrated pollution prevention and control). *Off. J. Eur. Union* **2010**, *334*, 117–119.
49. Włodarczyk, B.; Włodarczyk, P.P. Analysis of the Potential of an Increase in Yeast Output Resulting from the Application of Additional Process Wastewater in the Evaporator Station. *Appl. Sci.* **2019**, *9*, 2282. [CrossRef]
50. Włodarczyk, P.P.; Włodarczyk, B. Wastewater Treatment and Electricity Production in a Microbial Fuel Cell with Cu–B Alloy as the Cathode Catalyst. *Catalysts* **2019**, *9*, 572. [CrossRef]
51. Gallouj, F.; Weber, K.M.; Stare, M.; Rubalcaba, L. The futures of the service economy in Europe: A foresight analysis. *Tech. Forecast. Soc. Chang.* **2015**, *94*, 80–96. [CrossRef]
52. Włodarczyk, P.P.; Włodarczyk, B.; Włodarczyk, P.P.; Włodarczyk, B. Preparation and Analysis of Ni–Co Catalyst Use for Electricity Production and COD Reduction in Microbial Fuel Cells. *Catalysts* **2019**, *9*, 1042. [CrossRef]
53. Huggins, T.; Fallgren, P.H.; Jin, S.; Ren, Z.J. Energy and performance comparison of microbial fuel cell and conventional aeration treating of wastewater. *J. Microb. Biochem. Technol.* **2013**, *6*, 1–5.
54. Ren, Z.; Yan, H.; Wang, W.; Mench, M.M.; Regan, J.M. Characterization of microbial fuel cells at microbially and electrochemically meaningful time scales. *Environ. Sci. Technol.* **2011**, *45*, 2435–2441. [CrossRef]
55. Ehsani, M.; Gao, Y.; Gay, S.E.; Emadi, A. Fundamentals, Theory and Design. In *Modern Electric, Hybrid Electric and Fuel Cell Vehicles*; CRC Press: Boca Raton, FL, USA, 2005.
56. Hamnett, A. Mechanism and electrocatalysis in the direct methanol fuel cell. *Catal. Today* **1997**, *38*, 445–457. [CrossRef]
57. Brożyna, J.; Mentel, G.; Szetela, B. Influence of double seasonality on economic forecasts on the example of energy demand. *J. Int. Stud.* **2016**, *9*, 9–20. [CrossRef]

58. Schreurs, M.A.; Tiberghien, Y. Multi-level reinforcement: Explaining European Union leadership in climate change mitigation. *Glob. Environ. Politics* **2007**, *7*, 19–46. [CrossRef]
59. Ghosh, D.; Shukla, P.R.; Garg, A.; Ramana, P.V. Renewable energy technologies for the Indian power sector: Mitigation potential and operational strategies. *Renew. Sustain. Energy Rev.* **2002**, *6*, 481–512. [CrossRef]
60. Bouzarovski, S.; Petrova, S.; Sarlamanov, R. Energy poverty policies in the EU: A critical perspective. *Energy Policy* **2012**, *49*, 76–82. [CrossRef]
61. Rosenow, J.; Cowart, R.; Bayer, E.; Fabbri, M. Assessing the European Union's energy efficiency policy: Will the winter package deliver on 'Efficiency First'? *Energy Res. Soc. Sci.* **2017**, *26*, 72–79. [CrossRef]
62. Olaussen, J.O.; Oust, A.; Solstad, J.T. Energy performance certificates-Informing the informed or the indifferent? *Energy Policy* **2017**, *111*, 246–254. [CrossRef]
63. Johansson, T.; Vesterlund, M.; Olofsson, T.; Dahl, J. Energy performance certificates and 3-dimensional city models as a means to reach national targets-A case study of the city of Kiruna. *Energy Convers. Manag.* **2016**, *116*, 42–57. [CrossRef]
64. Tagliapietra, S.; Zachmann, G.; Edenhofer, O.; Glachant, J.M.; Linares, P.; Loeschel, A. The European union energy transition: Key priorities for the next five years. *Energy Policy* **2019**, *132*, 950–954. [CrossRef]
65. Ellerman, A.D.; Marcantonini, C.; Zaklan, A. The European Union emissions trading system: Ten years and counting. *Rev. Environ. Econ. Policy* **2016**, *10*, 89–107. [CrossRef]
66. Europe 2020 Strategy. Available online: https://ec.europa.eu/info/business-economy-euro/economic-and-fiscal-policy-coordination/eu-economic-governance-monitoring-prevention-correction/european-semester/framework/europe-2020-strategy_en (accessed on 12 December 2019).
67. Communication from the Commission to the European Parliament, the Council, the European Economic and Social Committee and the Committee of the Regions. Taking stock of the Europe 2020 Strategy for Smart, Sustainable and Inclusive Growth. Available online: https://eur-lex.europa.eu/legal-content/EN/ALL/?uri=CELEX:52014DC0130 (accessed on 10 December 2019).
68. Lee, C.M.; Ko, C.N. Short–term load forecasting using lifting scheme and ARIMA models. *Expert Syst. Appl.* **2011**, *38*, 5902–5911. [CrossRef]
69. Asteriou, D.I.; Hall, S.G. *ARIMA Models and the Box–Jenkins Methodology. W: Applied Econometrics*, 2nd ed.; Palgrave MacMillan: London, UK, 2011; pp. 265–286.
70. De Andrade, L.C.M.; da Silva, I.N. Very Short-Term Load Forecasting Based on ARIMA Model and Intelligent Systems. In Proceedings of the 2009 15th International Conference on Intelligent System Applications to Power Systems, Curitiba, Brazil, 8–12 November 2009; pp. 1–6.
71. Brożyna, J. Energy consumption and greenhouse gas emissions against the background of Polish economic growth. In *Energy Transformation towards Sustainability*; Elsevier: Amsterdam, The Netherlands, 2019; pp. 51–70. [CrossRef]

© 2020 by the authors. Licensee MDPI, Basel, Switzerland. This article is an open access article distributed under the terms and conditions of the Creative Commons Attribution (CC BY) license (http://creativecommons.org/licenses/by/4.0/).

MDPI
St. Alban-Anlage 66
4052 Basel
Switzerland
Tel. +41 61 683 77 34
Fax +41 61 302 89 18
www.mdpi.com

Energies Editorial Office
E-mail: energies@mdpi.com
www.mdpi.com/journal/energies

www.ingramcontent.com/pod-product-compliance
Lightning Source LLC
Chambersburg PA
CBHW061617220926
51850CB00005B/531
9783039362561